Methods in Enzymology

Volume XLI
CARBOHYDRATE METABOLISM
Part B

METHODS IN ENZYMOLOGY

EDITORS-IN-CHIEF

Sidney P. Colowick   Nathan O. Kaplan

Methods in Enzymology

Volume XLI

Carbohydrate Metabolism

Part B

EDITED BY

W. A. Wood

DEPARTMENT OF BIOCHEMISTRY
MICHIGAN STATE UNIVERSITY
EAST LANSING, MICHIGAN

1975

ACADEMIC PRESS New York San Francisco London
A Subsidiary of Harcourt Brace Jovanovich, Publishers

ACADEMIC PRESS, INC.
111 Fifth Avenue, New York, New York 10003

United Kingdom Edition published by
ACADEMIC PRESS, INC. (LONDON) LTD.
24/28 Oval Road, London NW1

Library of Congress Cataloging in Publication Data
Main entry under title:

Carbohydrate metabolism.

    (Methods in enzymology, v. 9)
    Includes bibliographical references.
    1.   Carbohydrate metabolism.   2.   Enzymes.   I.   Wood,
Willis A., Date   ed.   II.   Series: Methods in
enzymology, v. 9 [etc.]   [DNLM: 1.   Carbohydrates–
Metabolism.   W1ME9615K v. 9]
QP601.C733 vol. 9 574.1′925′08s   [574.1′33]   72-26891
ISBN 0–12–181941–8 (pt. B)

Table of Contents

Section I. Analytical Methods

Section II. Enzyme Assay Procedures

Section III. Preparation of Substrates

Section IV. Oxidation–Reduction Enzymes

Section V. Epimerases and Isomerases

Section VI. Miscellaneous Enzymes

Contributors to Volume XLI

Article numbers are in parentheses following the names of contributors.
Affiliations listed are current.

TESSA ABRAMSKY (50), *Department of Biochemical Genetics, The Rockefeller University, New York, New York*

W. R. ALEXANDER (89), *Institute of Physiological Chemistry, Universität of Göttingen, Göttingen, Germany*

REMI E. AMELUNXEN (58, 59), *Department of Microbiology, University of Kansas, School of Medicine, Kansas City, Kansas*

RICHARD L. ANDERSON (34, 35), *Department of Biochemistry, Michigan State University, East Lansing, Michigan*

GAD AVIGAD (4, 5, 18, 30, 31, 33), *Department of Biochemistry, College of Medicine and Dentistry of New Jersey-Rutgers Medical School, Piscataway, New Jersey*

J. MARTYN BAILEY (100, 101), *Biochemistry Department, George Washington University School of Medicine, Washington, D.C.*

R. L. BENSON (85), *Department of Entomology, Washington State University, Pullman, Washington*

LARRY H. BERNSTEIN (10), *Department of Pathology, University of California at San Diego, School of Medicine, La Jolla, California*

HANS P. BLASCHKOWSKI (105), *Institut für Biologische Chemie, Universität Heidelberg, Heidelberg, Germany*

SHELBLY L. BRADBURY (77), *Laboratory of Biochemistry, National Institute of Dental Research, National Institutes of Health, Bethesda, Maryland*

RAYMOND B. BRIDGES (51), *Departments of Oral Biology and Cell Biology, A. B. Chandler Medical Center, University of Kentucky, Lexington, Kentucky*

HOWARD L. BROCKMAN, JR. (68), *The Hormel Institute, University of Minnesota, Austin, Minnesota*

RONALD W. BROSEMER (53, 60), *Program in Biochemistry and Biophysics and Department of Chemistry, Washington State University, Pullman, Washington*

R. CAMMACK (71), *Department of Plant Sciences, King's College, University of London, London, England*

DANIEL O. CARR (58), *Department of Biochemistry and Molecular Biology, University of Kansas Medical Center, Kansas City, Kansas*

RICHARD CHABY (6), *Institut de Biochimie, Université de Paris-Sud, Orsay, France*

DANIEL CHARON (6, 20), *Institut de Biochemie, Université de Paris-Sud, Orsay, France*

R. CHILLA (46), *Physiologisch-Chemisches Institut der Georg-August-Universität Göttingen, Göttingen, Germany*

T. H. CHIU (19), *Department of Microbiology, University of Pittsburgh, School of Dental Medicine, Pittsburgh, Pennsylvania*

JULIA F. CLARK (77), *Department of Pharmacology, Indiana University School of Medicine, Indianapolis, Indiana*

PHILIP COHEN (47), *Department of Biochemistry, University of Dundee, Dundee, Scotland*

R. A. COOPER (104), *Department of Biochemistry, University of Leicester, Leicester, England*

A. STEPHEN DAHMS (34, 35), *Department of Chemistry, California State University, San Diego, California*

KEITH DALZIEL (48, 57), *Department of Biochemistry, University of Oxford, Oxford, England*

ASIS DATTA (86), *School of Life Sciences, Jawaharlal Nehru University, New Delhi, India*

D. D. DAVIES (61), *School of Biological Sciences, University of East Anglia, Norwich, England*

A. P. DAWSON (56), *School of Biological Science, University of East Anglia, Norwich, England*

EUGENE E. DEKKER (27), *Department of Biological Chemistry, The University of Michigan, Ann Arbor, Michigan*

J. DE LEY (3, 22, 36), *Laboratory of Microbiology and Microbial Genetics, Faculty of Science, State University, Ghent, Belgium*

JEAN DEUPREE (87), *Department of Pharmacology, University of Nebraska Medical Center, Omaha, Nebraska*

K. M. DOERING (90), *Institute of Physiological Chemistry, Universität of Göttingen, Göttingen, Germany*

G. F. DOMAGK (46, 89, 90), *Institute of Physiological Chemistry, Universität of Göttingen, Göttingen, Germany*

CLYDE C. DOUGHTY (78), *Department of Biological Chemistry, University of Illinois at the Medical Center, Chicago, Illinois*

JAMES I. ELLIOTT (81), *Department of Biochemistry, North Carolina State University, Raleigh, North Carolina*

SASHA ENGLARD (18, 30, 31, 33), *Department of Biochemistry, Albert Einstein College of Medicine, Yeshiva University, Bronx, New York*

JOHANNES EVERSE (9, 10), *Department of Chemistry, University of California, San Diego, La Jolla, California*

DER-FONG FAN (19), *William Singer Research Institute, Allegheny General Hospital, Pittsburgh, Pennsylvania*

DAVID SIDNEY FEINGOLD (19), *Department of Microbiology, University of Pittsburgh, School of Medicine, Pittsburgh, Pennsylvania*

RONALD D. FELD (62, 64), *Department of Pathology, University of Iowa, Iowa City, Iowa*

STEVEN C. FINK (53), *Department of Chemistry, and Graduate Program in Biochemistry, Washington State University, Pullman, Washington*

P. H. FISHMAN (100, 101), *Developmental and Metabolic Neurology Branch, National Institute of Neurological Diseases and Stroke, National Institutes of Health, Bethesda, Maryland*

ERNST FREESE (67), *Laboratory of Molecular Biology, National Institute of Neurological Diseases and Stroke, National Institutes of Health, Bethesda, Maryland*

S. FRIEDMAN (85), *Department of Entomology, University of Illinois, Urbana, Illinois*

K. H. GABBAY (37), *Cell Biology Laboratory, Endocrine Division, Children's Hospital Medical Center, Boston, Massachusetts*

WILLIAM O. GIELOW (88), *Department of Biological Science, Biochemistry and Molecular Biology Section, University of California, Santa Barbara, California*

ERWIN GOLDBERG (70), *Department of Biological Sciences, Northwestern University, Evanston, Illinois*

HECTOR GONZALEZ-CEREZO (57), *Instituto de Biologia, Departamento de Bioquimica, Universidad Nacional Antonoma de Mexico, University City, Mexico*

ROBERT W. GRACY (84, 94), *Department of Chemistry, North Texas State University, Denton, Texas*

GEORGE G. GUILBAULT (11), *Chemistry Department, University of New Orleans, New Orleans, Louisiana*

RAGY HANNA (17), *International Regulatory Affairs Division, E. R. Squibb and Sons, Princeton, New Jersey*

FRED C. HARTMAN (95), *Biology Division, Oak Ridge National Laboratory, Oak Ridge, Tennessee*

JAMES A. HAYASHI (78), *Department of Biochemistry, Rush Medical College, Chicago, Illinois*

WOLFGANG HENGSTENBERG (28), *Abt. Molekulare Biologie, Max-Planck Institut für Medizinische Forschung, Heidelberg, Germany*

BENNO HESS (12, 102), *Max-Planck Institut für Ernahrungs-Physiologie, Dortmund, Germany*

SUSUMU HIZUKURI (83), *Department of Agricultural Chemistry, Kagoshima University, Kagoshima-shi, Japan*

R. P. HULLIN (75), *Department of Biochemistry, University of Leeds, Leeds, England*

KEN IZUMORI (98), *Department of Food Science, Faculty of Agriculture, Kagawa University, Miki-Cho, Kagawa-ken, Japan*

WILLIAM B. JAKOBY (77), *Section on Enzymes, National Institute of Arithritis, Metabolic and Digestive Diseases, National Institutes of Health, Bethesda, Maryland*

FRANK W. JANSSEN (39, 79), *Wyeth Laboratories, Inc., Philadelphia, Pennsylvania*

MARILYN SCHUMAN JORNS (74), *Department of Biochemistry, The University of Texas, Southwestern Medical School, Dallas, Texas*

G. R. JULIAN (42), *Chemistry Department, Montana State University, Bozeman, Montana*

MASAYUKI KATAGIRI (72), *Department of Chemistry, Faculty of Science, Kanazawa University, Ishikawa, Japan*

K. KERSTERS (22), *Laboratory of Microbiology and Microbial Genetics, Faculty of Sciences, State University, Ghent, Belgium*

RICHARD M. KERWIN (79), *Wyeth Laboratories, Inc., West Chester, Pennsylvania*

J. H. KINOSHITA (37), *Laboratory of Vision Research, National Eye Institute, National Institutes of Health, and U.S. Department of Health, Education and Welfare, Bethesda, Maryland*

JOACHIM KNAPPE (105), *Institut für Biologische Chemie, Universität Heidelberg, Heidelberg, Germany*

JAMES A. KNOPP (81), *Department of Biochemistry, North Carolina State University, Raleigh, North Carolina*

LENA E. KONDO (2), *Department of Biochemistry, The University of Texas, M.D. Anderson Hospital and Tumor Institute, Houston, Texas*

W. K. G. KRIETSCH (92, 93), *Institut für Physiologische Chemie und Physikalische Biochemie, Universität München, Munich, Germany*

J. W. KUSIAK (100), *Biochemistry Department, The George Washington University Medical School, Washington, D.C.*

EUN WOO LEE (91), *Department of Pharmacology, Thomas Jefferson University, Philadelphia, Pennsylvania*

NANCY LEE (88, 96), *Department of Biological Sciences, University of California, Santa Barbara, California*

H. B. LEJOHN (65), *Department of Microbiology, University of Manitoba, Winnipeg, Canada*

H. RICHARD LEVY (44), *Biological Research Laboratories, Department of Biology, Syracuse University, Syracuse, New York*

GEORGE L. LONG (69), *Department of Chemistry, Pomona College, Claremont, California*

KENNETH W. MCKERNS (43), *Departments of Obstetrics and Gynecology, University of Florida, College of Medicine, The J. Hillis Miller Health Center, Gainesville, Florida*

UMADAS MAITRA (27), *Department of Developmental Biology and Cancer, Division of Biological Sciences, Albert Einstein College of Medicine, Yeshiva University, Bronx, New York*

H. PAUL MELOCHE (25), *The Institute for Cancer Research, Philadelphia, Pennsylvania*

JOSEPH MENDICINO (17), *Department of Biochemistry, University of Georgia, Athens, Georgia*

ROBERT P. METZGER (40), *Department of Physical Science, San Diego State University, San Diego, California*

PATRICK W. MOBLEY (40), *Department of Metabolism and Endocrinology, City of Hope National Medical Center, Duarte, California*

M. L. MORSE (28), *Webb Waring Lung Institute, Department of Biophysics and Genetics, University of Colorado, Medical Center, Denver, Colorado*

R. P. MORTLOCK (24), *Department of Microbiology, University of Massachusetts, Amherst, Massachusetts*

HEDY MULHAUSEN (17), *Chemical Abstracts Service, Columbus, Ohio*

S. MULHERN (100), *Laboratory of Biochemistry, National Heart and Lung Institute, National Institutes of Health, Bethesda, Maryland*

ZIRO NIKUNI (83), *Hayashi-Gakuen Junior Women's College, Ryodakaya, Konan-shi, Aichi-ken, Japan*

I. LUCILE NORTON (95), *Biology Division, Oak Ridge National Laboratories, Oak Ridge, Tennessee*

YOSHIAKI NOSOH (82), *Laboratory of Chemistry of Natural Products, Tokyo Institute of Technology, Meguroku, Tokyo, Japan*

CHARLES OLIVE (44), *Division of Cancer Research, Michael Reese Medical Center, Chicago, Illinois*

CHARLES A. PASTERNAK (103), *Department of Biochemistry, Oxford University, Oxford, England*

JIM PATRICK (96), *Department of Biochemistry, Dartmouth Medical School, Hanover, New Hampshire*

BARBARA M. F. PEARSE (49), *M.R.C. Laboratory of Molecular Biology, Cambridge, England*

EDWARD PENHOET (15), *Department of Biochemistry, University of California, Berkeley, California*

P. G. PENTCHEV (100, 101), *Developmental and Metabolic Neurology Branch, National Institute of Neurological Disease and Stroke, National Institutes of Health, Bethesda, Maryland*

LEWIS I. PIZER (55), *Department of Microbiology, University of Pennsylvania, School of Medicine, Philadelphia, Pennsylvania*

HELEN QUILL (13), *Department of Nutrition and Food Science, Massachusetts Institute of Technology, Cambridge, Massachusetts*

JESSE C. RABINOWITZ (73), *Department of Biochemistry, University of California, Berkeley, California*

W. E. RAZZELL (76), *Vancouver Laboratory, Fisheries and Marine Service, Environment Canada, Vancouver, Canada*

F. J. REITHEL (42), *Chemistry Department, University of Oregon, Eugene, Oregon*

MARIO RIPPA (52), *Istituto di Chimica Biologica, Università degli Studi di Ferrara, Ferrara, Italy*

JOHN M. ROBINSON (2), *W. L. Clayton Research Center, Anderson Clayton Foods, Richardson, Texas*

IRWIN A. ROSE (26), *The Institute for Cancer Research, Fox Chase Cancer Center, Philadelphia, Pennsylvania*

MICHAEL A. ROSEMEYER (47, 49), *Department of Biochemistry, University College, London, England*

I. Y. ROSENBLUM (63), *The Procter & Gamble Co., Miami Valley Laboratories, Cincinnati, Ohio*

HANS W. RUELIUS (39, 79), *Wyeth Laboratories, Inc., Philadelphia, Pennsylvania*

WILLIAM J. RUTTER (15), *Department of Biochemistry and Biophysics, University of California, San Francisco, California*

H. J. SALLACH (62, 63, 64), *Department of Physiological Chemistry, University of Wisconsin, Madison, Wisconsin*

ROBERT S. SARFATI (6), *Institut de Biochimie, Université de Paris-Sud, Orsay, France*

MARLA SARTORIS (19), *Department of Ophthalmology, Eye & Ear Hospital, Pittsburgh, Pennsylvania*

HARRY SCHACHTER (1), *Department of Biochemistry, University of Toronto, Toronto, Canada*

WILLIAM A. SCOTT (41, 50), *Department of Cellular Physiology and Immunology, The Rockefeller University, New York, New York*

MAXWELL G. SHEPHERD (45), *Department of Biochemistry, University of Otago, Dunedin, New Zealand*

GERALD H. SHEYS (78), *Department of Pathology, Cambridge Hospital, Cambridge, Massachusetts*

MARCO SIGNORINE (52), *Istituto di Chimica Biologica, Università degli Studi di Ferrara, Ferrara, Italy*

MICHAEL SILVERBERG (48), *Department of Pathology, Yale University School of Medicine, New Haven, Connecticut*

J. C. SLAUGHTER (61), *Department of Brewing and Biological Sciences, Heriot-Watt University, Edinburgh, Scotland*

ROBERT SNYDER (91), *Department of Pharmacology, Thomas Jefferson University, Philadelphia, Pennsylvania*

DAVID J. SPECTOR (55), *Department of Microbiology, University of Pennsylvania, School of Medicine, Philadelphia, Pennsylvania*

CHARLES R. STEINMAN (77), *Department of Medicine, Mount Sinai School of Medicine, New York, New York*

ROSELYNN M. STEVENSON (65), *Department of Microbiology, University of Manitoba, Winnipeg, Canada*

ALLEN C. STOOLMILLER (23), *Departments of Pediatrics and Biochemistry, University of Chicago, Medical School, Chicago, Illinois*

FREDRIK C. STORMER (106), *National Institute of Public Health, Oslo, Norway*

WALTER A. SUSOR (15), *Department of Biochemistry and Biophysics, University of California, San Francisco, California*

LADISLAS SZABÓ (6, 20, 21), *Institut de Biochemie, Faculté des Sciences, Université de Paris, Orsay, France*

YASUHITO TAKEDA (83), *Department of Agricultural Chemistry, Kagoshima University, Kagoshima-shi, Japan*

SHIGEKI TAKEMORI (72), *Department of Chemistry, Faculty of Science, Kanasawa University, Ishikawa, Japan*

C. J. R. THORNE (56), *Department of Biochemistry, Stanford University Medical Center, Stanford, California*

BILL E. TILLEY (84), *Department of Physiology, University of California, Irvine, California*

FRANÇOIS TRIGALO (6, 21), *Institut de Biochimie, Université de Paris-Sud, Orsay, France*

O. TSOLAS (16), *Roche Institute of Molecular Biology, Nutley, New Jersey*

J. VAN BEEUMEN (3, 36), *Laboratory of Microbiology and Microbial Genetics,*

Faculty of Science, State University, Ghent, Belgium

WEIERT VELLE (48), *Department of Physiology, Veterinary College of Norway, Oslo, Norway*

EARL F. WALBORG, JR. (2), *Department of Biochemistry, The University of Texas, System Cancer Center, M. D. Anderson Hospital and Tumor Institute, Houston, Texas*

MILTON M. WEISER (13), *Massachusetts General Hospital, Boston, Massachusetts*

HAROLD B. WHITE, III (54), *Department of Chemistry, University of Delaware, Newark, Delaware*

RICHARD J. WHITE (103), *Research Laboratories, Lepetit S.P.A., Milan, Italy*

ARNE N. WICK (40), *Department of Chemistry, San Diego State University, San Diego, California*

CHARLES L. WITTENBERGER (51, 66), *Microbial Physiology Section, Laboratory of Microbiology and Immunology, National Institute of Dental Research, National Institutes of Health, Bethesda, Maryland*

FINN WOLD (29), *Department of Biochemistry, University of Minnesota, St. Paul, Minnesota*

TERRY WOOD (7, 8, 14), *Department of Chemistry, Njala University College, Freetown, Sierra Leone, West Africa*

W. A. WOOD (68, 87), *Department of Biochemistry, Michigan State University, East Lansing, Michigan*

CHARLES L. WORONICK (80), *Medical Research Laboratory, Department of Medicine, Hartford Hospital, Hartford, Connecticut*

BERND WURSTER (12, 102), *Department of Biology, Princeton University, Princeton, New Jersey*

KEI YAMANAKA (32, 98, 99), *Department of Food Science, Faculty of Agriculture, Kagawa University Miki-cho, Kagawa-ken, Japan*

AKIRA YOSHIDA (67), *Department of Biochemical Genetics, City of Hope National Medical Center, Duarte, California*

Preface

Volumes XLI and XLII of "Methods in Enzymology" report new procedures appearing in the literature since 1965. As for Volume IX, the procedures included are for dissimilatory reactions between disaccharides and pyruvate. A few important reactions of pyruvate leading to fermentation end products are also included. The originally planned single volume became two volumes in recognition of the greatly increased number of pages needed to adequately cover this expanding field. The distribution of material between the volumes is arbitrary.

I wish to thank all of the authors for their contributions and their cooperation. It is a pleasure to recognize Ms. Patti Prokopp for her expert secretarial assistance.

W. A. WOOD

METHODS IN ENZYMOLOGY

EDITED BY

Sidney P. Colowick and Nathan O. Kaplan

VANDERBILT UNIVERSITY
SCHOOL OF MEDICINE
NASHVILLE, TENNESSEE

DEPARTMENT OF CHEMISTRY
UNIVERSITY OF CALIFORNIA
AT SAN DIEGO
LA JOLLA, CALIFORNIA

METHODS IN ENZYMOLOGY

EDITORS-IN-CHIEF

Sidney P. Colowick   Nathan O. Kaplan

VOLUME VIII. Complex Carbohydrates
Edited by ELIZABETH F. NEUFELD AND VICTOR GINSBURG

VOLUME IX. Carbohydrate Metabolism
Edited by WILLIS A. WOOD

VOLUME X. Oxidation and Phosphorylation
Edited by RONALD W. ESTABROOK AND MAYNARD E. PULLMAN

VOLUME XI. Enzyme Structure
Edited by C. H. W. HIRS

VOLUME XII. Nucleic Acids (Parts A and B)
Edited by LAWRENCE GROSSMAN AND KIVIE MOLDAVE

VOLUME XIII. Citric Acid Cycle
Edited by J. M. LOWENSTEIN

VOLUME XIV. Lipids
Edited by J. M. LOWENSTEIN

VOLUME XV. Steroids and Terpenoids
Edited by RAYMOND B. CLAYTON

VOLUME XVI. Fast Reactions
Edited by KENNETH KUSTIN

VOLUME XVII. Metabolism of Amino Acids and Amines (Parts A and B)
Edited by HERBERT TABOR AND CELIA WHITE TABOR

VOLUME XVIII. Vitamins and Coenzymes (Parts A, B, and C)
Edited by DONALD B. MCCORMICK AND LEMUEL D. WRIGHT

VOLUME XIX. Proteolytic Enzymes
Edited by GERTRUDE E. PERLMANN AND LASZLO LORAND

Methods in Enzymology

Volume XLI
CARBOHYDRATE METABOLISM
Part B

Section I

Analytical Methods

[1] Enzymic Microassays for D-Mannose, D-Glucose, D-Galactose, L-Fucose, and D-Glucosamine

By HARRY SCHACHTER

The methods usually used for the quantitative analysis of mixtures of neutral sugars and hexosamines involve resolution of the mixture into individual sugars by paper chromatography,[1] ion exchange column chromatography,[1-5] automated liquid chromatography with borate buffers,[4,6] or gas–liquid chromatography[4,7,8] followed by analysis of the purified sugars or sugar derivatives. Colorimetric methods of sugar analysis can be applied to mixtures without prior purification, but many of these methods tend to be relatively nonspecific[1-3,9,10]; fairly specific colorimetric reactions are available for fucose[1] and sialic acid.[1,10] Enzymic methods have also been used for sugar analysis; e.g., commercial kits are available for the assay of D-glucose and D-galactose by glucose oxidase[1,11] and galactose oxidase,[12] respectively. The current methods of choice for the quantitative analysis of complex sugar mixtures are automated liquid chromatography using borate buffers[4,6] and gas–liquid chromatography.[4,7,8] However, enzymic methods are useful for the repetitive assay of a limited number of sugars, e.g., when following the action of a glycosidase; further, enzymic methods do not require the prior separation of sugars, and they are stereospecific. This communication presents simple methods for the enzymic microassay of D-mannose, D-glucose, D-galactose, L-fucose, and D-glucosamine in sugar mixtures.[13]

Principles

D-*Mannose*

D-Mannose is phosphorylated to D-mannose 6-phosphate by hexokinase and ATP, and the production of ADP is measured using pyruvate

---

[1] R. G. Spiro, this series, Vol. 8 [1].
[2] E. A. Davidson, Vol. 8 [3].
[3] R. W. Wheat, Vol. 8 [4].
[4] R. G. Spiro, Vol. 28 [1].
[5] T.-Y. Liu, Vol. 28 [4].
[6] Y. C. Lee, Vol. 28 [6].
[7] C. C. Sweeley, W. W. Wells, and R. Bentley, Vol. 8 [7].
[8] R. A. Laine, W. J. Esselman, and C. C. Sweeley, Vol. 28 [10].
[9] G. Ashwell, Vol. 3 [12].
[10] G. Ashwell, Vol. 8 [6].

kinase and lactic dehydrogenase.[14] The sequence of reactions is as follows:

$$\text{D-Mannose} + \text{ATP} \xrightarrow{\text{hexokinase}} \text{D-mannose 6-phosphate} + \text{ADP}$$

$$\text{ADP} + \text{phosphoenolpyruvate} \xrightarrow{\text{pyruvate kinase}} \text{pyruvate} + \text{ATP}$$

$$\text{Pyruvate} + \text{NADH} + \text{H}^+ \xrightarrow{\text{lactate dehydrogenase}} \text{lactate} + \text{NAD}^+$$

The extent of NADH oxidation is measured spectrophotometrically. The method is not specific for D-mannose since D-glucosamine, D-glucose, D-fructose, and 2-deoxy-D-glucose are also phosphorylated by hexokinase.[15-17] This difficulty can be overcome by removing D-glucosamine with a cation exchange resin prior to assay and by measuring D-glucose with glucose-6-phosphate dehydrogenase[18,19]; if glucose is present in the mixture, the appropriate correction must be applied to the mannose assay.

D-*Glucose*

D-Glucose assay involves the following reactions:

$$\text{D-Glucose} + \text{ATP} \xrightarrow{\text{hexokinase}} \text{D-glucose 6-phosphate} + \text{ADP}$$

$$\text{D-Glucose 6-phosphate} + \text{NADP}^+ + \text{H}_2\text{O} \xrightarrow[\text{dehydrogenase}]{\text{glucose-6-phosphate}} \text{6-phospho-D-gluconate} + \text{NADPH} + \text{H}^+$$

The extent of NADP$^+$ reduction is measured spectrophotometrically. This assay is specific for D-glucose.[18,19]

D-*Galactose*

The assay involves the following reaction catalyzed by D-galactose dehydrogenase[20]:

$$\beta\text{-D-Galactose} + \text{NAD}^+ \xrightarrow[\text{dehydrogenase}]{\text{D-galactose}} \text{3-D-galactonolactone} + \text{NADH} + \text{H}^+$$

[11] N. O. Kaplan, Vol 3 [14]; J. H. Pazur, Vol 9 [18].
[12] D. Amaral, F. Kelly-Falcoz, and B. L. Horecker, Vol. 9 [19].
[13] P. R. Finch, R. Yuen, H. Schachter, and M. A. Moscarello, *Anal. Biochem.* **31**, 296 (1969).
[14] T. Bücher and G. Pfleiderer, Vol. 1 [66].
[15] M. R. McDonald, Vol. 1 [32].
[16] R. A. Darrow and S. P. Colowick, Vol. 5 [25].
[17] M. D. Joshi and V. Jagannathan, Vol. 9 [69a]; I. T. Schulze, J. Gazith, and R. H. Gooding, Vol. 9 [69b].
[18] A. Kornberg and B. L. Horecker, Vol. 1 [42].
[19] S. A. Kuby and E. A. Noltmann, Vol. 9 [23].
[20] K. Wallenfels and G. Kurz, Vol. 9 [22].

The assay is run at a relatively high pH such that the galactonolactone is spontaneously hydrolyzed; the reaction thus becomes irreversible and proceeds to completion. The extent of NAD^+ reduction is measured spectrophotometrically.

L-*Fucose*

This sugar can be assayed with L-fucose dehydrogenase,[21,22] as follows:

$$\text{L-Fucose} + \text{NAD}^+ \xrightarrow{\text{L-fucose dehydrogenase}} \text{4-L-fuconolactone} + \text{NADH} + \text{H}^+$$

At the high pH used for the assay, the fuconolactone is spontaneously hydrolyzed and the reaction proceeds to completion. The extent of NAD^+ reduction is measured spectrophotometrically. The assay is not specific for L-fucose since the dehydrogenase also acts on D-arabinose, L-galactose, and other sugars[21]; however, these sugars are not commonly found in biological materials.

D-*Glucosamine*

D-Glucosamine is removed from the sugar mixture by adsorption on a cation exchange resin. The glucosamine can then be eluted off the resin and assayed by the hexokinase-pyruvate kinase-lactic dehydrogenase assay described above for D-mannose. Since D-glucosamine is the only hexosamine acted on by hexokinase, the presence of other hexosamines in the sugar mixture does not interfere with this assay.

Reagents

Standard solutions of chromatographically pure D-mannose, D-glucose, D-galactose, L-fucose, and D-glucosamine hydrochloride, 1 mM
ATP, 50 mM, pH 7.0
Phosphoenolpyruvate, 0.1 M
NAD⁺, 10 mM, pH 7.0
NADH, 10 mM, pH 7.0
NADP⁺, 10 mM, pH 7.0
Tris·HCl buffer, 0.5 M, pH 8.0
Tris·HCl buffer, 1.0 M, pH 8.6

[21] H. Schachter, J. Sarney, E. J. McGuire, and S. Roseman, *J. Biol. Chem.* **244**, 4785 (1969).
[22] P. W. Mobley, R. P. Metzger, and A. N. Wick, this volume [40].

HEPES buffer (N-2-hydroxyethylpiperazine N'-2-ethanesulfonate), 0.2 M, pH 7.2

$MgCl_2$, 0.2 M

Commercial enzyme preparations: these were obtained as crystalline suspensions in ammonium sulfate; they were diluted, if necessary, with ice-cold water to the concentrations indicated shortly before use and were kept at 4°:

Hexokinase, 30 units/ml

Pyruvate kinase, 15 units/ml

Lactate dehydrogenase, 3600 units/ml

D-Glucose-6-phosphate dehydrogenase, 50 units/ml

D-Galactose dehydrogenase, 10 units/ml

L-Fucose dehydrogenase, 0.1 unit/ml. The enzyme was prepared by the method of Schachter et al.[21] from pork liver and was carried to the Sephadex G-100 stage of purification; the fraction eluted from the Sephadex column was concentrated about 10-fold by pressure dialysis.

The enzyme unit used in all the above solutions is defined as the amount which transforms 1 μmole of substrate per minute.

Assay Methods

If the sugar mixture contains D-glucosamine, it must first be passed through a small column of AG 50W-X8, 200–400 mesh, in the hydrogen form; hexosamines adhere to this resin whereas neutral sugars pass through. The size of the column depends on the amount of material being processed, but a column containing about 1 ml of wet resin should be adequate for most analytical purposes. The column is equilibrated with water, and the sugar solution is adjusted to pH 2 and loaded on the column. The column is washed with about 10 volumes of water, the wash is dried by flash-evaporation and reconstituted with water to give sugar concentrations of about 1 mM; the neutral sugar assays are performed on this solution. The AG 50W-X8 column is then washed with 2 volumes of 2 N HCl to elute hexosamines. The wash is dried by repeated flash-evaporation and reconstituted with water to give sugar concentrations of about 1 mM; this solution is used for D-glucosamine assay.

D-Mannose Assay

To a 3-ml test tube is added 20 μl of ATP, 50 μl of Tris·HCl (pH 8.0), 5 μl of $MgCl_2$, 10 μl of hexokinase, 0–0.20 ml of D-mannose, and water to a final volume of 0.30 ml. The solution is incubated at 37° for

15 min, and the reaction is stopped by heating at 100° for 2 min. The following are then added: 25 μl of NADH, 10 μl of phosphoenolpyruvate, 0.50 ml of HEPES, 25 μl of MgCl$_2$, and 70 μl of water. The absorbance at 340 nm is measured, and 10 μl of pyruvate kinase and 10 μl of lactate dehydrogenase are added to start the reaction. The solution is incubated at 37° for 30 min, and the absorbance at 340 nm is again measured; the initial absorbance reading is subtracted to give the change in absorbance.

D-*Glucose Assay*

To a 3-ml test tube is added 20 μl of ATP, 50 μl of Tris·HCl (pH 8.0), 5 μl of MgCl$_2$, 10 μl of hexokinase, 0–0.20 ml of D-glucose, and water to a final volume of 0.30 ml. The solution is incubated at 37° for 15 min, and the reaction is stopped by heating at 100° for 2 min. The following are then added: 50 μl of NADP$^+$, 0.10 ml of MgCl$_2$, 0.20 ml of Tris·HCl (pH 8.0), and 1.35 ml water. The absorbance of the solution at 340 nm is determined and 10 μl of glucose-6-phosphate dehydrogenase is added to start the reaction. The solution is incubated at 37° for 30 min, and the absorbance at 340 nm is again measured.

D-*Galactose Assay*

To a 3-ml test tube is added 50 μl NAD$^+$, 0.50 ml of Tris·HCl (pH 8.6), 0–0.20 ml of D-galactose, and water to a final volume of 0.90 ml. The absorbance at 340 nm is determined, and 25 μl of D-galactose dehydrogenase is added to start the reaction. The solution is incubated at 37° for 30 min, and the absorbance at 340 nm is again determined.

L-*Fucose Assay*

To a 3-ml test tube is added 0.10 ml of NAD$^+$, 0.10 ml of Tris·HCl (pH 8.0), 0–0.20 ml of L-fucose, and water to a final volume of 0.90 ml. The absorbance at 340 nm is measured, and 50 μl of L-fucose dehydrogenase is added to start the reaction. The solution is incubated at 37° for 60 min and the absorbance at 340 nm again determined.

D-*Glucosamine Assay*[23]

To a 3-ml test tube is added 20 μl of ATP, 50 μl of Tris·HCl (pH 8.0), 5 μl of MgCl$_2$, 10 μl of hexokinase, 0–0.20 ml of D-glucosamine hydrochloride, and water to a final volume of 0.30 ml. The solution is

[23] R. Fulford and H. Schachter, unpublished data, 1973.

incubated at 37° for 15 min, and the reaction is stopped by heating at 100° for 2 min. The following is then added: 25 μl of NADH, 10 μl of phosphoenolpyruvate, 0.50 ml of HEPES, 25 μl of $MgCl_2$, and 70 μl of water. The absorbance at 340 nm is measured, and 10 μl of pyruvate kinase and 10 μl of lactate dehydrogenase are added. After 30 min at 37°, the absorbance at 340 nm is again determined.

Fluorometric D-Galactose Assay[24]

Studies on a new testicular galactoglycerolipid[25] prompted the development of a sensitive fluorometric assay for D-galactose based on the above enzymic assay. To a 1-ml cuvette, 1 cm light path, is added 50 μl of NAD^+, 0.50 ml of Tris·HCl (pH 8.6), 0–0.15 ml of D-galactose (0.1 mM), and water to a final volume of 0.98 ml. The cuvette is placed in an Eppendorf fluorometer equipped with a Hg 313 + 366 nm excitation filter and a 400–3000 nm emission filter. The machine is set to zero, and 20 μl of D-galactose dehydrogenase is added to the cuvette with rapid mixing to start the reaction. The reaction is allowed to proceed until complete (about 20 min at 22°), and a final reading is taken.

Comments on the Procedures

All procedures are carried out with 1 cm light path cuvettes. The change in absorbance of fluorometric units obtained from the reagent blank is subtracted from the changes obtained in the presence of sugar. Typical standard curves for the assay of D-mannose, D-glucose, D-galactose, and L-fucose have been published[13]; Figs. 1 and 2 show previously unpublished standard curves for the D-glucosamine assay and the fluorometric D-galactose assay. Assuming a molar extinction coefficient for NADH or NADPH[26] at 340 nm of 6.2×10^3, it is possible to calculate the extent of conversion of substrate to product at completion of the reaction. The percent conversions for the spectrophotometric assays are at least as follows: D-mannose, 96; D-glucose, 94; D-galactose, 84; L-fucose, 95; and D-glucosamine·HCl, 83. Under these conditions, it is possible to measure 0.025 μmole of sugar.

All the assays can in theory be made at least 10 times more sensitive by using fluorometry to measure the interconversions between reduced and oxidized pyridine nucleotides. Fluorometry has been used in this laboratory to measure nanomole amounts of D-galactose (Fig. 2). Above

[24] M. J. Kornblatt, R. K. Murray, and H. Schachter, unpublished data, 1973.
[25] M. J. Kornblatt, H. Schachter, and R. K. Murray, *Biochem. Biophys. Res. Commun.* **48**, 1489 (1972).
[26] B. L. Horecker and A. Kornberg, *J. Biol. Chem.* **175**, 385 (1948).

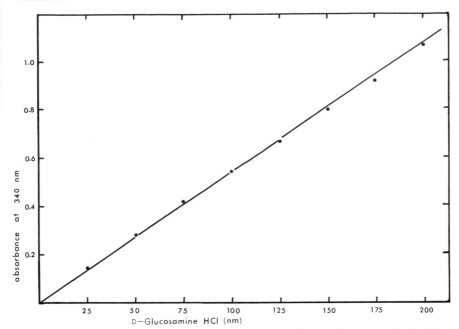

FIG. 1. A plot of absorbance change at 340 nm against D-glucosamine hydrochloride concentration under the assay conditions described in the text; the data are averages of duplicate determinations (R. Fulford and H. Schachter, 1973, unpublished).

2 nmoles of D-galactose, the reproducibility of repeated assays is quite good (less than 6% variation). The assay has been extended down to 0.5 nmole, but repeated assays may differ from each other by as much as 20% at this level of sensitivity.

The above assays have all been tested for interference from other sugars and protein hydrolyzates. The D-mannose assay is not affected by the presence of D-galactose, L-fucose, N-acetyl-D-glucosamine, or a bovine serum albumin hydrolyzate[13]; however, both D-glucose and D-glucosamine are detected by this assay. As mentioned above, this problem is overcome by removing glucosamine with a cation exchange resin and by correcting for D-glucose content on the basis of the D-glucose-6-phosphate dehydrogenase assay. The latter assay is not affected by other sugars or by a protein hydrolyzate.[13] Both the D-galactose and L-fucose assays may be affected by sugars of uncommon occurrence (e.g., D-fucose and L-arabinose are substrates for D-galactose dehydrogenase and L-galactose and D-arabinose are substrates for L-fucose dehydrogenase) ; however, L-fucose does not interfere with assay of D-galactose; D-galactose

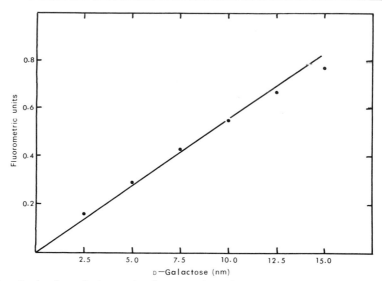

Fig. 2. A plot of change in fluorometric units (arbitrary) against D-galactose concentration under the assay conditions described in the text; the data are averages of duplicate determinations (M. J. Kornblatt, R. K. Murray, and H. Schachter, 1973, unpublished).

does not interfere with assay of L-fucose; and D-glucose, D-mannose, N-acetyl-D-glucosamine, D-glucosamine, and a protein hydrolyzate do not interfere with either the D-galactose or L-fucose assays.[13] The AG 50W-X8 column removes all glucosamine without loss of neutral sugars.[13] The recovery of D-glucosamine hydrochloride following elution from the ion exchange column is over 90%, and D-galactosamine hydrochloride does not interfere with the D-glucosamine assay.[23]

[2] Automated Determination of Saccharides Using Ion-Exchange Chromatography of Their Borate Complexes[1]

By EARL F. WALBORG, JR., LENA E. KONDO, and JOHN M. ROBINSON

Most saccharides form anionic complexes with the borate ion and therefore are amenable to resolution by ion-exchange chromatography.

[1] Supported by grants from the National Institute of Arthritis and Metabolic Diseases (AM 10130), The Robert A. Welch Foundation (G-354), and The Paul and Mary Haas Foundation.

Several automated ion-exchange systems, employing elution with alkaline borate buffers, are capable of resolving and quantitating saccharide mixtures of varying complexity.[2-8] These systems utilize elution conditions (pH 8–10, 50–55°) at which saccharides are subject to alkali-catalyzed rearrangements[9] or other degradative reactions.[10] The system of Hough *et al.*[11] uses borate buffers at pH 7.0, a pH at which borate possesses negligible buffering capacity. Elution is accomplished using a Cl⁻ gradient, a method that produces an increased effluent pH.[8] Such a system also requires extensive regeneration to convert the exchanger to the borate form.[6]

The ion-exchange chromatographic system of Walborg and Kondo[12] employs elution with 2,3-butanediol/borate buffers, which possess greater buffering capacity at neutral pH. This method allows the resolution and quantitation of complex mixtures of neutral mono-, di-, and trisaccharides, under conditions that minimize alkaline rearrangements and other degradative reactions.

Method A

This method[12] was designed for the separation and quantitation of complex mixtures of neutral mono-, di-, and trisaccharides.

Materials

Resin: Aminex A-14, a 4% cross-linked anion-exchange resin of 20 μm particle diameter, was obtained from Bio-Rad Laboratories, Richmond, California.

Sugars: Commercial sugar preparations of the highest quality available were utilized. L-Rhamnose, D-mannose, D-lyxose, L-fucose, D-ribose, D-arabinose, D-galactose, D-xylose, L-sorbose, D-mannohep-

[2] A. Floridi, *J. Chromatogr.* **59**, 61 (1971).
[3] S. Katz, S. R. Dinsmore, and W. W. Pitt, Jr., *Clin. Chem.* **17**, 731 (1971).
[4] Y. C. Lee, see this series, Vol. 28, [6].
[5] C. M. Mundie, M. V. Cheshire, and J. S. D. Bacon, *Biochem. J.* **108**, 51P (1968).
[6] G. N. Catravas, *in* "Automation in Analytical Chemistry: Technicon Symp. 1966," Vol. 1, p. 397. Mediad Inc., White Plains, New York, 1967.
[7] J. I. Ohms, J. Zec, J. V. Benson, Jr., and J. A. Patterson, *Anal. Biochem.* **20**, 51 (1967).
[8] R. B. Kesler, *Anal. Chem.* **39**, 1416 (1967).
[9] J. C. Speck, *Advan. Carbohyd. Chem.* **13**, 63 (1958).
[10] R. L. Whistler and J. N. BeMiller, *Advan. Carbohyd. Chem.* **13**, 289 (1958).
[11] L. Hough, J. V. S. Jones, and P. Wusteman, *Carbohyd. Res.* **21**, 9 (1972).
[12] E. F. Walborg, Jr. and L. E. Kondo, *Anal. Biochem.* **37**, 320 (1970).

tulose, cellobiose, lactose, maltose, turanose, melibiose, melezitose, raffinose, and stachyose were obtained from Schwarz/Mann, Orangeburg, New York. D-Glucose and sucrose were obtained from Merck and Co., Rahway, New Jersey. D-Fructose was obtained from Fisher Scientific Co., Fairlawn, New Jersey. Maltotriose and isomaltose were obtained from Pierce Chem. Co., Rockford, Illinois.

Buffers

Buffer A: 0.15 M boric acid, 0.5 M 2,3-butanediol, 0.1% Brij-35, 0.5 ml toluene/liter, adjusted to pH 7.0 (23°) with NaOH

Buffer B: 0.8 M boric acid, 1.0 M 2,3-butanediol, 0.1% Brij-35, 0.5 ml toluene/liter, adjusted to pH 7.0 (23°) with NaOH. The boric acid and toluene were analytical reagents obtained from Fisher Scientific Co. Brij-35 [polyoxyethylene(23)lauryl ether] was obtained from Pierce Chemical Co. The 2,3-butanediol, obtained from Aldrich Chemical Co., Milwaukee, Wisconisn, was used without further purification.

Aniline/acetic acid/orthophosphoric acid reagent[13,14] was prepared by adding 1600 ml of glacial acetic acid (Fisher Scientific Co.) to 48 ml of redistilled aniline. To this was added 800 ml of 85% orthophosphoric acid (Matheson Scientific). Aniline was prepared by twice distilling analytical grade aniline in the presence of NaOH and zinc dust. The distillation was performed under N_2. The colorless aniline was stored in a glass-stoppered bottle at 4°. Aniline prepared and stored in this manner remained suitable for use at least 2 months. The aniline/acetic acid/orthophosphoric acid reagent is stable for at least 2 months at 23°.

Instrumentation

The Hitachi Perkin-Elmer Model 034-004 Liquid Chromatograph was employed. Jacketed columns, having an internal diameter of 4 mm and a length of 100 cm were fabricated by Glenco Scientific, Inc. (2802 White Oak Drive, Houston, Texas; Cat. No. 3202 — 4 × 100). The pumps were set to deliver buffer at 20 ml/hr and reagent at 60 ml/hr by the use of the appropriate gears. A filter containing Dowex 1-X4 (resin particle diameter, 60–70 μm) was inserted in the effluent line preceding the column. The filter was constructed of an 8-cm length of ¼-inch stainless steel tubing capped with two reducing unions (Swagelok No. 400-6-2-316 with snubber element, 60 μm). This filter removes a discoloration that

[13] E. F. Walborg, Jr. and L. Christensson, *Anal. Biochem.* **13**, 186 (1965).
[14] E. F. Walborg, Jr. and R. S. Lantz, *Anal. Biochem.* **22**, 123 (1968).

would otherwise accumulate at the top of the column resin bed. The resin in the filter was changed monthly. Teflon tubing was used throughout the instrument to carry buffer and/or reagent. The column temperature was controlled with a circulating water bath (Part No. 034-0035), equipped with a temperature change timer (Part No. 034-0109). The mixing manifold was maintained at 60–70° by use of a heating tape. This was necessary to avoid the formation of a precipitate in the mixing manifold. The tubing carrying effluent from the mixing manifold was connected directly to the reaction bath (Part No. 034-0036) without passing through the stainless steel fittings designed for the instrument. The effluent line between the mixing manifold and the reaction bath was insulated to reduce cooling of the column effluent–reagent mixture. The reaction bath contained a Teflon reaction coil of the following specifications: length 13 m; internal diameter, 1.5 mm; and volume, 23 cm^3. The temperature of the reaction coil was maintained at 120° by means of a thermostatted glycerol bath. A back pressure of 4 kg/cm^2 was maintained on the reaction coil to prevent boiling of the reaction mixture. The mixed effluent was submitted to spectrophotometric analysis at 310, 365, and 390 nm, utilizing the UV-VIS Effluent Monitor (Part No. 034-0037), equipped with a 4 mm path length cell. The absorbance at each wavelength was recorded at 48-sec intervals, using the Hitachi Model 183 multipoint recorder (0–5 mV). The recording interval was set by selecting a 4-sec printing speed and employing only one of the four ink pads for each wavelength input. A chart speed of 60 mm/hr was used. Figure 1 shows a functional diagram of the automated system. Although this analytical technique was developed using the Hitachi Perkin-Elmer Model 034, there is no inherent reason why liquid chromatographic instrumentation of other manufacturers could not be adapted to serve as well or better.

Procedure

Preparation of the Resin. The resin (25 g of Aminex A-14) was treated in the following manner:

a. Converted to the OH$^-$ form by washing with 500 ml of 1 N NaOH.

b. Washed with 500 ml water to remove excess OH$^-$.

c. Washed with 500 ml of buffer B (pH unadjusted).

d. Suspended in 250 ml of above solution and the pH adjusted to 7.0 at 23°.

e. Equilibrated with three 100-ml changes of buffer B.

f. The resin slurred in buffer B, was deaerated by boiling gently with stirring.

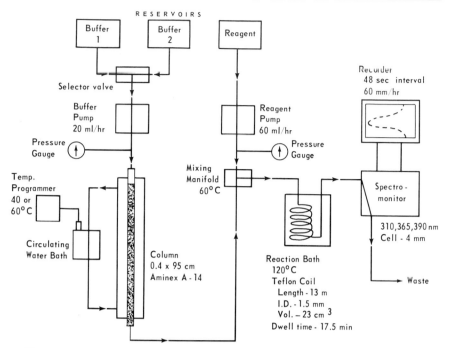

FIG. 1. Functional diagram of automated saccharide analysis using the Hitachi Perkin-Elmer Model 034-004.

g. The resin was allowed to settle, then the total volume was adjusted to give a slurry consisting of one part settled resin and two parts buffer. The slurry was maintained at 70° while the column was being packed.

Packing and Equilibration of Column. The column, maintained at 60°, was packed in 5–7 sections while a buffer flow of 20 ml/hr was maintained. The final height of the resin bed was 90–94 cm. A small disk of glass filter paper (Reeve Angel 934 AH) was placed on top of the resin bed to avoid stirring the resin. The column was equilibrated with 250 ml of eluting buffer prior to starting an analysis.

Operating Conditions. System 1: This chromatographic system utilized elution with buffer B at a column temperature of 60°. System 2: This stepwise, two-buffer system utilized elution with buffer A, followed by buffer B, the buffer change being effected automatically at 4.5 hr. The column temperature was programmed to change from 40° to 60° at 5.5 hr. Column back pressures varied from 15 kg/cm² with buffer B to 28 kg/cm² with buffer A. Repetition of analyses by means of System 2 required that the column temperature be decreased to 40° and the

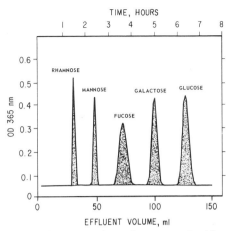

FIG. 2. Chromatographic separation of monosaccharide mixture using Method A (System 1). Monosaccharides were present in the following quantities: 0.8 μmole of rhamnose, 1.0 μmole of mannose, 1.5 μmoles of fucose, 2.0 μmoles of galactose, and 2.5 μmoles of glucose.

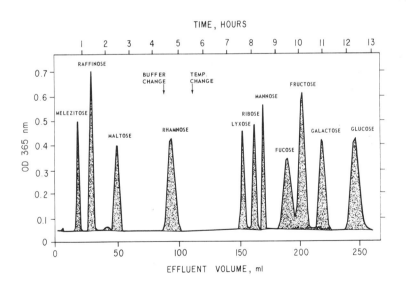

FIG. 3. Chromatographic separation of saccharide mixture using Method A (System 2). Saccharides were present in the following quantities: 0.3 μmole of melezitose, 0.5 μmole of raffinose, 0.8 μmole of maltose, 2.0 μmoles of rhamnose, 1.5 μmoles of lyxose, 1.5 μmoles of ribose, 1.0 μmole of mannose, 1.5 μmoles of fucose, 2.5 μmoles of fructose, 2.0 μmoles of galactose, and 2.5 μmoles of glucose.

column equilibrated with 400 ml of buffer A. Regeneration of the column with NaOH was not required for either System 1 or 2. Samples were applied in 0.5 ml of buffer A using N_2 pressure, 3 kg/cm^2. The sample was followed by two 0.25-ml washes of buffer A.

Resolution

The resolution attained using Systems 1 and 2 is shown in Figs. 2 and 3, respectively. The elution times of all saccharides examined in each of these systems are given in Tables I and II. As an approximation, separation may be achieved for any saccharide pair whose retention times differ by as much as 30 min.

Quantitation

Quantitation was accomplished by integrating the area under the peak using the height–width method. Peak height was measured as absorbance,

TABLE I
CHROMATOGRAPHIC CONSTANTS FOR SYSTEM 1[a]

Saccharide	Height–width constants		Retention times	
	C_{HW}[b]	±SD	Min	±SD
Rhamnose	18.5	0.22	91	1.9
Lyxose	12.8	0.38	134	2.4
Ribose	11.7	0.15	136	0.8
Mannose	14.5	0.33	146	1.6
Isomaltose	18.7	0.18	187	0.8
Fucose	20.0	0.09	223	3.4
Sorbose	13.1	0.27	294	2.7
Xylose	12.6	0.17	295	1.0
Galactose	15.2	0.16	308	3.9
Mannoheptulose	13.3	0.19	347	2.9
Melibiose	29.7	0.24	364	7.2
Glucose	15.0	0.20	392	6.7

[a] The values were calculated on replicate samples containing saccharides present in the following quantities: rhamnose, 0.8 μmole; lyxose, 1.5 μmoles; ribose, 1.5 μmoles; mannose, 1.0 μmole; isomaltose, 0.52 μmole; fucose, 1.5 μmoles; sorbose, 3.0 μmoles; xylose, 2.5 μmoles; galactose, 2.0 μmoles; mannoheptulose, 3.0 μmoles; melibiose, 1.5 μmoles; glucose, 2.5 μmoles.

[b] The peak widths in dots were multiplied by 4 in order to allow direct comparison to Method B. Although not specifically stated, previously reported height–width constants [E. F. Walborg, Jr. and L. E. Kondo, Anal. Biochem. 37, 320 (1970)] were also calculated in this manner.

TABLE II
CHROMATOGRAPHIC CONSTANTS FOR SYSTEM 2[a]

Saccharide	Height–width constants		Retention times	
	C_{HW}[b]	±SD	Min	±SD
Sucrose	34.6	0.91	52	1.3
Melezitose	46.4	0.86	57	0.8
Raffinose	49.3	0.40	90	1.3
Cellobiose	32.3	0.96	96	1.7
Maltotriose	39.9	1.17	127	1.8
Maltose	28.8	0.81	151	3.6
Stachyose	55.0	0.34	243	6.1
Rhamnose	17.4	0.25	287	1.7
Lactose	31.6	0.93	306	5.5
Lyxose	11.8	0.08	457	5.1
Ribose	11.4	0.22	480	5.4
Mannose	14.3	0.28	510	1.0
Fucose	20.6	0.13	569	5.8
Arabinose	12.5	0.24	602	6.0
Fructose	18.1	0.73	604	5.4
Xylose	12.8	0.12	635	5.2
Sorbose	12.7	0.41	650	4.5
Galactose	14.8	0.95	657	4.2
Glucose	15.2	0.16	736	7.6

[a] The values were calculated on replicate samples containing saccharides present in the following quantities: melezitose, 0.3 μmole; raffinose, 0.50 μmole; maltotriose, 0.32 μmole; stachyose, 0.25 μmole; rhamnose, 2.0 μmoles; lyxose, 1.5 μmoles; mannose, 1.0 μmole; fucose, 1.5 μmoles; arabinose, 2.0 μmoles; xylose, 2.5 μmoles; sorbose, 1.88 μmoles; glucose, 2.5 μmoles. The values for sucrose, cellobiose, maltose, lactose, ribose, fructose, and galactose were calculated using four different sample loads, varying in the range between 0.2 and 5.0 μmoles of each saccharide (as monosaccharide).

[b] The peak widths in dots were multiplied by 4 in order to allow direct comparison to Method B. Although not specifically stated, previously reported height–width constants [E. F. Walborg, Jr. and L. E. Kondo, *Anal. Biochem.* **37**, 320 (1970)] were also calculated in this manner.

and peak width in terms of time by counting the number of printed points above the half-height. Height–width constants (C_{HW}) representing area per micromole have been calculated for each saccharide and are shown in Tables I and II. The precision of the method was approximately ±5%. Absolute recoveries, determined by manual colorimetric analysis of column effluents, were at least 85–100%.[15]

[15] E. F. Walborg, Jr., D. B. Ray, and L. E. Öhrberg, *Anal. Biochem.* **29**, 433 (1969).

Qualitative Information Concerning Classes of Sugars

The absorbance is recorded at three different wavelengths: 310, 365, and 390 nm. The ratio of the absorbances, 365/390 nm, has been utilized as a qualitative tool to distinguish between classes of saccharides; e.g., the aldopentoses yield a ratio of 1.78; the aldo- and ketohexoses 1.29; and the 6-deoxyaldohexoses 1.10. The values for the individual saccharides are shown in Table III.

TABLE III
RATIO OF ABSORBANCES AT 365 AND 390 nm

Saccharide	C_{HW} (365 nm)/C_{HW} (390 nm)	
	Ratio	$\pm$SD
Aldopentoses		
Arabinose	1.76	0.03
Lyxose	1.78	0.03
Ribose	1.78	0.02
Xylose	1.79	0.01
Aldohexoses		
Galactose	1.28	0.02
Glucose	1.28	0.01
Mannose	1.29	0.02
Ketohexoses		
Fructose	1.31	0.02
Sorbose	1.32	0.01
Ketoheptose		
Mannoheptulose	1.34	0.01
6-Deoxyaldohexoses		
Fucose	1.10	0.01
Rhamnose	1.10	0.01
Oligosaccharides containing aldo- and/or ketohexoses		
Disaccharides		
Cellobiose	1.28	0.02
Isomaltose	1.27	0.02
Lactose	1.29	0.01
Maltose	1.29	0.02
Melibiose	1.29	0.01
Sucrose	1.29	0.02
Trisaccharides		
Maltotriose	1.30	0.02
Melezitose	1.30	0.03
Raffinose	1.29	0.01
Tetrasaccharide		
Stachyose	1.29	0.01

Method B

This represents a modified version of Method A (System 1) which provides a more rapid and sensitive analysis of the neutral monosaccharides commonly found in glycoproteins: mannose, fucose, galactose, and glucose. Since the materials, instrumentation, and procedures are similar to those of Method A, only the modifications are listed below.

Materials

Resin: Durrum DA-X4, a 4% cross-linked anion-exchange resin with particle size of 20 ± 5 μm diameter, was obtained from Durrum Chem. Corp., Palo Alto, California.

Instrumentation

The Hitachi Perkin-Elmer Model 034 Liquid Chromatograph was modified as follows:

Column: A jacketed column, having an internal diameter of 4 mm and a length of 60 cm was fabricated by Glenco Scientific, Inc. (Cat. No. 3202—4 × 60).

Reaction coil: The reaction bath contained a Teflon reaction coil of the following specifications: length, 57 m; internal diameter, 0.08 mm, and volume, 28 cm³.

Recorder: The absorbance at each wavelength was recorded at 12-sec intervals, using the Hitachi Model 183 multipoint recorder, equipped with a scale expansion device (Part No. 034–0256). Full-scale deflection was 0.2 OD unit.

Operating Conditions

The column, packed with resin to a height of 50 cm, was eluted with buffer B at a flow rate of 60 ml/hr at a column temperature of 60°. The column back pressure was approximately 10–15 kg/cm².

Resolution

The resolution using this system is shown in Fig. 4. The elution times of all the saccharides examined are given in Table IV.

Quantitation

The height–width constants (C_{HW}) are shown in Table IV. The precision of the method was approximately $\pm 10\%$.

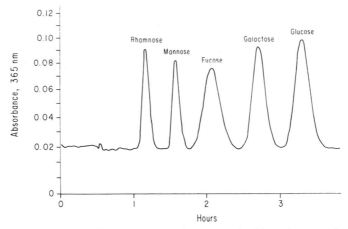

Fɪɢ. 4. Chromatographic separation of monosaccharide mixture using Method B. Monosaccharides were present in the following quantities: 0.15 μmole of rhamnose, 0.20 μmole of mannose, 0.30 μmole of fucose, 0.40 μmole of galactose, and 0.50 μmole of glucose.

TABLE IV
Cʜʀᴏᴍᴀᴛᴏɢʀᴀᴘʜɪᴄ Cᴏɴsᴛᴀɴᴛs ғᴏʀ Mᴇᴛʜᴏᴅ B[a]

Saccharide	Height–width constant, C_{HW}	C_{HW} (365 nm)$/C_{HW}$ (390 nm)	Retention time (min)
Rhamnose	15.2 ± 1.0^b	1.09 ± 0.03^b	70 ± 1^b
Mannose	11.2 ± 0.5	1.29 ± 0.03	96 ± 2
Fucose	16.2 ± 0.7	1.04 ± 0.02	126 ± 1
Galactose	11.5 ± 0.4	1.29 ± 0.03	166 ± 2
Glucose	11.7 ± 0.7	1.30 ± 0.03	205 ± 6

[a] The values were calculated on replicate samples containing saccharides present in the following quantities: 0.16 μmole of rhamnose, 0.2 μmole of mannose, 0.3 μmole of fucose, 0.4 μmole of galactose, and 0.5 μmole of glucose.
[b] Mean ± standard deviation.

Preparation of Acid Hydrolyzates for Analysis

Acid hydrolyzates were prepared for analysis in the following manner:

a. After cooling to 23° an internal sugar standard was added to the hydrolyzate to allow correction for mechanical losses incurred during the neutralization procedure. The choice of an internal standard depends on the nature of the oligo- or polysaccharide being investigated; e.g., rham-

TABLE V

DEGRADATION OF SACCHARIDES ON HEATING FOR 12 HR IN (A) 0.15 M
BORIC ACID/2,3-BUTANEDIOL, pH 7.0, (B) 25 mM BORAX, pH 9.2

Saccharide	Buffer	% Degradation at various temperatures (°C)				
		40°	50°	60°	70°	80°
Mannose	pH 7.0	—	0	—	0	6
	pH 9.2	—	0	7	21	66
Fructose	pH 7.0	—	1	—	7	19
	pH 9.2	—	4	14	39	81
Maltose	pH 7.0	—	0	—	5	27
	pH 9.2	13	20	37	77	—
Turanose	pH 7.0	—	3	—	21	70
	pH 9.2	15	28	71	95	—

nose has been used as an internal standard in hydrolyzates of glycoproteins.[16]

b. The hydrolyzate was neutralized with Dowex 2-X8 in the HCO_3^- form (1 g of moist resin per milliequivalent of acid). In order to avoid the presence of fine resin particles in the neutralizing resin, Dowex 2-X8 (200–400 mesh) was settled in water and only the rapidly settling particles utilized.

c. The Dowex 2-hydrolyzate mixture was slurried and filtered. The hydrolysis tube was rinsed with four 1-ml portions of distilled water, and these washes in turn were used to wash the neutralizing resin.

d. The combined filtrate and washes were frozen and lyophilized in an evacuated desiccator over NaOH.

Note on Alkaline Degradation of Saccharides during Ion-Exchange Chromatography

To test the effectiveness of 2,3-butanediol buffers in minimizing alkaline rearrangements and other degradative reactions, the stability of several saccharides was examined in buffer A and 25 mM borax, pH 9.2. Saccharides were heated at various temperatures for 12 hr, and the degradation of saccharide was quantitated by ion-exchange chromatography using Method B. As shown in Table V the saccharides are more stable in buffer A. Turanose, which is particularly labile to degradation via elimination of a glycoxy anion,[10] is subject to considerable degradation at 50°.

[16] E. F. Walborg, Jr., L. Christensson, and S. Gardell, *Anal. Biochem.* 13, 177 (1965).

[3] Polarographic Determination of 3-Keto Sugars[1]

By J. Van Beeumen and J. De Ley

3-Keto sugars (synonym: 3-uloses) are oxidized sugars formed by chemical synthesis or by enzymic action of hexopyranoside:cytochrome c oxidoreductase from *Agrobacterium*. Whereas the purification and properties of the enzyme are dealt with in another contribution,[2] an experimentally simple assay for qualitative and quantitative analysis of 3-uloses is described here.

Method

Principle. The 3-uloses are determined by classical and derivative polarography, based on the electroreducible character of the free carbonyl group at the dropping mercury electrode (DME). The assay mixture is made alkaline to convert the hydrated electroinactive form of some 3-uloses into the free keto form. The identification of the 3-uloses follows from the position of the half-way potential ($E_{1/2}$) of the polarographic wave. The quantitative determination follows from the diffusion current (i_d) given by the height of the wave. Principles of polarography are described, e.g., in Meites.[3]

Reagents

LiCl, 0.1 M (indifferent electrolyte)
Na_2CO_3, solid
N_2, pure grade
3-Keto sugar, 10 mM to 10 μM in water

Apparatus. We used a PO 4 polarograph (Radiometer, Copenhagen) with a K 501 saturated calomel electrode (SCE) as external reference electrode, but any good polarograph should do. The dropping mercury indicator electrode used by us had an outflow rate, m, of 1.979 mg sec^{-1} and a drop time, t, of 3.383 sec at -1.55 V vs the SCE, at an effective pressure of the mercury of 46.9 cm (values of m and t are the mean of 5 determinations of each 10-drop count). The capillary constant $m^{2/3}$ $t^{1/6}$, a factor determining the diffusion current, was 1.931 mg$^{2/3}$ sec$^{-1/2}$.

[1] J. Van Beeumen and J. De Ley, *Anal. Biochem.* **44**, 254 (1971).
[2] J. Van Beeumen and J. De Ley, this volume [36].
[3] L. Meites, "Polarographic Techniques." Wiley (Interscience), New York, 1965.

Procedure. In a 15-ml vessel maintained at 25° are pipetted 0.1–0.4 ml of a presumed 3-ulose and 4.9–4.6 ml of LiCl. Eighty milligrams of carbonate are then added, giving a pH of 11.3, and the mixture is flushed with nitrogen during 5 min. We define this state as "standard conditions." The current-potential polarogram is now recorded at a scanning rate of 0.4 V/min, a damping setting of 4, and a residual current compensation of 0.25–0.35 μA/V depending on the sample composition (see section on application of the method). A derivative polarogram is recorded immediately thereafter, using an oscillation damping of 8 or 9. It may be necessary to record the polarogram every 5 min for half an hour (see below).

The diffusion current is calculated as the difference in wave height with the base line (the "residual current"), taking the currents as the midpoint of the oscillations (shown as an example in Fig. 1C). The apparent value of $E_{1/2}$ is at the peak in the derivative polarogram (see Fig. 1B). The real value of $E_{1/2}$ is 40 or 50 mV less negative when the damping setting is either 8 or 9.

Interpretation of the Polarograms

Qualitative Identification of 3-Uloses. There are two main diagnostic criteria to identify 3-uloses (see Fig. 1). The first one is that $E_{1/2}$ falls within the range between -1.50 and -1.57 V. The second one is that i_d decreases in function of time. The half-life is about 1 hr for 3-keto-hexoses and 5 hr for 3-keto disaccharides. Spectrophotometric examination has shown that the decrease is due to decomposition of the 3-ulose, via an intermediate reductone, under the alkaline assay conditions.[4]

Three additional polarographic characteristics relate to the extent of hydration of the 3-ketocarbonyl group at neutral pH. The first one is that the maximal value of i_d is reached only a certain time after the addition of Na_2CO_3. The more hydrate there is in the equilibrium with the free ketone, the longer will it take to obtain a maximal diffusion current. 3-Ketolactose, e.g., is hydrated over 98%[4] and reaches a maximal i_d 20 min after the removal of oxygen (Fig. 1, A,B). The second characteristic is seen in 0.1 M LiCl without carbonate. The "limiting current" i_l is much smaller than i_d (at pH 11) when the 3-ulose is highly hydrated (Fig. 1, D). Using some basic equations,[4] it is possible to calculate the amount of free ketone in the hydrate/ketone equilibrium. The third characteristic is that i_d at pH 11.3 is missing when an excess of boric acid or tetraborate has been added to the 3-ulose, prior to the Na_2CO_3. This phenomenon is very likely due to complex formation of the hydrate with

[4] J. Van Beeumen and J. De Ley, *Bull. Soc. Chim. Belg.* **80,** 683 (1971).

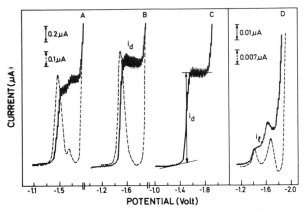

FIG. 1. Polarograms of 0.255 μmole of 3-ketolactose per milliliter of assay mixture, measured under "standard conditions" (A–C) and in 0.1 M LiCl without Na₂CO₃ (D). (A), (B), and (C) were recorded, respectively, 7, 25, and 80 min after the addition of Na₂CO₃. The direct analyses (——) were carried out with an oscillation damping $D = 4$ and a compensation of the residual current $C = 0.25$ μA/V. In the derivative polarograms (- - - -), only the midpoint of each oscillation $(D = 8)$ is shown. (C) The way the diffusion current can be calculated. Current i_l stands for the limiting current at pH 7. The second wave in (D) disappears in alkali and is of no interest here. Modified from J. Van Beeumen and J. De Ley, *Anal. Biochem.* **44**, 254 (1971).

boric acid.[4] The excess needed is inversely proportional to the extent of hydration. For example, it is 200-fold for 3-ketolactose and 600-fold for 3-ketosucrose. There is no borate effect on a free ketone as in methyl β-3-keto-D-glucopyranoside.

Quantitative Determination of 3-Uloses. The relationship current/concentration is given by the Ilkovič equation (see, e.g., in Meites[3])

$$i_d = 607 \, n \, D^{1/2} \, m^{2/3} t^{1/6} \, C \qquad (1)$$

where the capillary constant $m^{2/3}t^{1/6}$ (see above) depends on the equipment and $n \, D^{1/2}$ on the compound used; n equals 2 for 3-ketolactose and probably for other 3-uloses as well,[4] D is the diffusion coefficient, and C is the concentration of 3-ulose expressed as micromoles per milliliter of electrolyte solution, i_d is in μA. The decomposition of 3-ulose in the alkaline standard conditions is about 10%,[1,4] so that Eq. (1) becomes

$$i_d = 1091 \, D^{1/2} \, m^{2/3} t^{1/6} \, C \qquad (2)$$

Experimentally, we found for any capillary and using crystalline 3-ketolactose as reference product

$$i_d = 2.63 \, m^{2/3} t^{1/6} \, C \qquad (3)$$

and for crystalline methyl β-3-keto glucoside

$$i_d = 2.72 \, m^{2/3} t^{1/6} C \tag{4}$$

Since D, by virtue of the Stokes-Einstein diffusion equation,[5] is proportional to molecular weight, Eq. (3) is most likely valid for any 3-keto disaccharide and Eq. (4) for any 3-keto methylhexoside. The constants for any 3-ulose in Eqs. (3) and (4) can be calculated from

$$1091 \times D^{1/2} \tag{5}$$

They are, respectively, 2.84 for 3-ketohexoses and 3-ketohexosamines, 2.77 for 3-keto methylthiohexosides, 2.65 for 3-keto-p-arbutine, 2.63 for 3-ketosalicine, 2.53 for 3-ketobionic acids, and 2.39 for 3-ketotrihexoses.

Equation (3) is valid within the range of 1 mM to 3 μM 3-ketolactose. The smallest measurable concentration is 1 μg/ml. Below concentrations of 10 μM, $E_{1/2}$ shifts from -1.50 to -1.56 V.

Interfering Effects. Proteins, even at low concentrations after removal with chloroform, change $E_{1/2}$ by -0.03 V and decrease i_d for at least 15%. They also increase the slopes of both the plateau and the residual current, but these phenomena can partially be obviated by a compensation setting of 0.35 μA/V. Traces of 5-methyl phenazinium methyl sulfate (PMS) are likewise deleterious, but the effect on i_d depends on the amount of 3-ulose under assay: for 50 μM 3-ketolactose, 34 μg of phenazine decrease i_d for some 50%. The degree of eventual interference can be assessed by remeasuring the polarogram after a known amount of a crystalline reference 3-ulose (e.g., 3-ketolactose) has been added.

Application of the Method

We have used the method to study the substrate specificity of the 3-keto sugar forming enzyme hexopyranoside:cytochrome c oxidoreductase from *Agrobacterium*.[2] Most of the reaction products can be identified qualitatively by typical color formation on thin layer plates of cellulose.[2] The polarographic method allows identification and quantification of the 3-uloses.

Preparation of the Samples. The 3-uloses are prepared in a Warburg respirometer using the preparation from at least step 3 of the purification procedure of the oxidoreductase.[2] Each Warburg vessel contains 1.3 ml of 50 mM Tris·HCl buffer, pH 7.9, 0.65 mg of PMS in 0.1 ml of water, 0.3 ml of enzyme (step 3) in 10 mM phosphate buffer, pH 6.9, and 0.2

---

[5] S. M. Cantor and Q. P. Peniston, *J. Amer. Chem. Soc.* **62**, 2113 (1940).

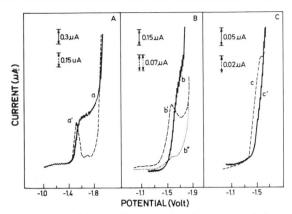

Fig. 2. Direct (———) and derivative (- - - -) polarograms of the oxidation products of melibiose (A), glucosamine (B), and 2-deoxyglucose (C) formed enzymically by hexopyranoside:cytochrome c oxidoreductase from *Agrobacterium*. (A) Curves a, and a', 0.2-ml aliquot of a Warburg reaction mixture, analyzed under standard conditions. (B) Aliquot, 0.2 ml, measured under standard conditions (curves b and b'); curve b'' is obtained after a 130-fold excess of boric acid is added prior to the carbonate. It indicates that the keto group of the putative 3-ulose is hydrated at neutral pH. (C) Aliquot, 1 ml, deproteinized 8 times with chloroform; curves c and c' are recorded 5 min as well as 6 hr after the addition of carbonate. Modified from J. Van Beeumen and J. De Ley, *Anal. Biochem.* **44**, 254 (1971).

ml of 0.1 M substrate. The central well contains 0.1 ml of 20% KOH. The oxidation is allowed to proceed until 0.5 μmole of oxygen per micromole of substrate is consumed. A 0.1–0.3 ml aliquot of the reaction mixture is diluted to 0.6 ml with a stock solution of LiCl (0.1 M LiCl final). This solution is deproteinized by at least three treatments with 0.6 ml of chloroform. The precipitate is removed by centrifugation for 5 min at 3000 g. The decomposition products of phenazine are removed in the chloroform layer.

Polarography of the Enzymic Reaction Products. The polarograms of the oxidized substrates can be classified in three types. The first type belongs to the 3-uloses, produced in high yield, from lactulose (85%), lactose (80%), lactobionic acid (75%), methyl-β-D-thiogalactose (72%), and melibiose (65%). Their half-wave potential is at -1.50 V, and the plateau of the diffusion current is not steep (Fig. 2A). The second type is shown by the putative 3-uloses, produced in much lower yield, from glucosamine (31%), methyl-α-D-glucose (22%), methyl-β-D-glucose (19%), maltose (17%), trehalose (17%), cellobiose (16%), glucose (14%), maltobionic acid (14%), sucrose (12%), cellobionic acid (12%), galactose (12%), leucrose (6%), glucose 1-phosphate (4%), mannose

(4%), and p-arbutine (3%). Their $E_{1/2}$ value is at -1.57 V and the plateau of the polarographic wave is very steep. Figure 2B illustrates the case of melibiose. The third type is shown by oxidation products formed in extremely low yields from 2-deoxyglucose (3%) and 2-deoxy-galactose (1%). The action of the enzyme on these substrates is so weak that even the derivative recordings show only a faint shoulder at -1.57 V (Fig. 2C). Since i_d does not change upon prolonged incubation under the assay conditions, the definite conclusion that these oxidation products are 3-keto sugars awaits chemical proof.

Since the polarographic method does not differentiate between individual 3-uloses, other methods should be used to identify them, e.g., thin-layer chromatogrpahy.[2]

[4] Colorimetric Ultramicro Assay for Reducing Sugars[1]

By GAD AVIGAD

Assay Method

Principle. Ferrocyanide produced as the result of the reduction of ferricyanide by a reducing sugar is assayed indirectly by the formation of ferrous ions. These subsequently interact with an iron chelate to yield a product with high color intensity. The reagent for iron employed in the following procedure is 2,4,6-tripyridyl-s-triazine.[2]

Reagents

Potassium ferricyanide (A). An 0.7 mM solution is made with 10 mM sodium hydroxide that contains 2% of sodium carbonate; kept in a brown bottle, it is stable for at least 2 weeks at room temperature.

Ferric chloride (B). A 2 mM reagent is made with a solution containing 0.2 M citric acid and 2 M sodium acetate in 5 M acetic acid. It is advisable to dissolve the ferric salt in a small volume of the acid citrate–acetate solution, so as to provide amounts of reagents sufficient for only a day or two of use. Blank values tend to increase somewhat with older solutions.

[1] G. Avigad, *Carbohyd. Res.* **7**, 94 (1968).
[2] Abbreviation used: TPTZ, 2,4,6,-tripyridyl-s-triazine or tris-(2-pyridyl)-1,3,5-tri-azine.

TPTZ³ (C). A 2.5 mM solution (0.78 mg/ml) in 3 M acetic acid; it is stable for at least a week at room temperature.

Procedure. Reactions should be shielded from direct light. Aliquots (1.0 ml) of solutions containing 2–100 nmoles of reducing sugar are placed in test tubes (1.5 × 12.5 cm). Reagent A (1.0 ml) is added, and the tubes are kept for 10 min in a boiling-water bath. Reagent B (1.0 ml) and reagent C (2.0 ml) are then added immediately, without cooling. The solutions are stirred vigorously with a Vibromixer; after 5 min in the dark, the absorbance at 595 nm of the violet Fe (TPTZ)$_2^{2+}$ is read against a reagent blank.

Comments

The color produced is stable for at least 6 hr when the mixture is kept in the dark, and also when it is diluted with several volumes of water. The molar extinction coefficient for D-glucose obtained in the present procedure was 75,000 ± 3%, corresponding to 3.3 mole equivalents of Fe(TPTZ)$_2^{2+}$ (cf. Collins *et al.*[4]). In comparison with D-glucose (100%), the relative molar yields of color with representative sugars were: D-fructose, 104; 2-deoxy-D-*arabino*-hexose, 80; D-xylose, 95; D-glucuronic acid, 100; 2-amino-2-deoxy-D-glucose, 135; maltose, 140; melibiose, 110; D-glucitol, <0.1; methyl α-D-glucoside, <0.2; and sucrose, <0.2%.

Samples taken from biological solutions after deproteinization with trichloroacetic acid or perchloric acid can be assayed without difficulty. Mercaptans, Cu^{2+}, Ag^+, CN^-, NO^-, or high concentrations of EDTA may interfere with formation of color, but other ions commonly found in biological solutions do not affect the reaction. The violet Fe(TPTZ)$_2^{2+}$ complex can be extracted into organic solvents, a property that can be advantageous in the determination of reducing sugars and of end groups of polysaccharides in turbid solutions, in suspensions, or in solutions in nonpolar solvents. In general, the reaction, similar to most other methods for reducing sugars, lacks specificity and is susceptible to interference if other reducing components are present in the solution. The present procedure, however, affords a stable, soluble color which has obvious advantages over measurements based on the unstable, Prussian Blue

---

[3] The reagents alone should have an A_{595} of <0.25. Tris value is probably attributable to the presence of ferrous ions and other reducing contaminants in the reagents. If desired, most of these contaminations can be removed by prior reaction with TPTZ and extraction of the complex with nitrobenzene.[4] This treatment is obligatory if the method is to be adapted for scaling down of volumes to increase its sensitivity.

[4] P. F. Collins, H. Diehl, and G. F. Smith, *Anal. Chem.* **31,** 1862 (1959).

colloid.[5] The method also provides one of the most sensitive assays available for the determination of reducing sugars.

We have found that 3-(2-pyridyl)-5,6-diphenyl-1,2,4-triazine[6,7] can substitute for TPTZ as the reagent for ferrous ions used for the determination of reducing sugars. It should also be noted that the violet $Fe(TPTZ)_2^{2+}$ is readily oxidized by periodate, thus providing a sensitive method for the quantitative assay of this widely used oxidant.[8]

A procedure for the determination of ferrocyanide by direct UV-spectrophotometry in an assay for reducing sugars has been described.[9] This method may be employed only when no components other than ferrocyanide that adsorb significantly in the far ultraviolet range are present in the solution analyzed.

[5] J. T. Park and M. J. Johnson, *J. Biol. Chem.*, **181**, 149 (1949). See also this series, Vols. 3 [58], 8 [1], and 28 [139].
[6] A. A. Schilt and P. J. Taylor, *Anal. Chem.* **42**, 220 (1970).
[7] P. Carter, *Anal. Biochem.* **40**, 450 (1971).
[8] G. Avigad, *Carbohyd. Res.* **11**, 119 (1969).
[9] D. K. Kidby and D. J. Davidson, *Anal. Biochem.* **55**, 321 (1973).

[5] Colorimetric Assays for Hexuronic Acids and Some Keto Sugars

By GAD AVIGAD

Reaction of the Reducing Sugars with an Acid-Copper Reagent[1]

Principle. Hexuronic acids and some keto sugars rapidly reduce dilute Cu^{2+} solutions (15.6 mM) at acid pH in presence of high salt concentrations to suppress Cu^+ reoxidation. The Cu^+ produced is measured colorimetrically with the arsenomolybdate reagent. The method is useful for the determination of hexuronic acids and certain ketoses in the presence of other sugars, aldoses in particular. It may also be employed for a specific measurement of hexuronic acids at the reducing ends of oligosaccharides, such as those appearing in enzymic digests of polysaccharides.

Reagents

Copper solution (A). Dissolve by boiling and vigorous stirring 280 g of anhydrous sodium sulfate and 40 g of sodium chloride in 720

[1] Y. Milner and G. Avigad, *Carbohyd. Res.* **4**, 359 (1967).

ml of water. Complete dissolution will occur when 100 ml of 2 M acetate buffer (pH 5.1) and 65 ml of 0.32 M cupric sulfate are added to the salt solution at about 60°. The pH of the reagent obtained should be 4.8; if necessary, the pH is carefully adjusted to this value by dropwise addition of 10 N hydrochloric acid or sodium hydroxide. The volume is then completed with water to 1 liter. The reagent, stored at 37° in a well-stoppered brown bottle, is stable for at least 2 months.

Arsenomolybdate reagent (B), prepared according to Nelson's method.[2]

Standard sugar solution (C). A 10 mM stock solution of hexuronic acid (or other sugar) is prepared in 0.05% benzoic acid and kept at 4°.

Procedure. To a 0.5-ml sample containing 20–500 nmoles of hexuronic acid in standard test tubes, add 1.5 ml of reagent A. Mix, cover with glass marbles, and immerse in a boiling water bath for exactly 10 min. Cool immediately in an ice bath without shaking. Add 1.0 ml of reagent B. Shake gently and add 2.0 ml of water. If the color intensity is too high, water can be added to 10 ml without impairing the proportionality of the absorption. Read the absorbance at a wavelength between 600 and 700 nm against a reagent blank.

Comments. Relative rates of reaction for different sugars were found to be: D-glucuronic acid, 100; D-galacturonic acid, 75; D-mannuronic acid, 97; D-aldohexoses, <1; D-aldopentoses, <3; 2-amino-2-deoxy-D-glucose, 8; 2-acetamido-2-deoxy-D-glucose, 2.0; D-fructose, 13; L-sorbose, 22; D-*threo*-2,5-hexodiulose, (5-*keto*-D-fructose) 355; D-*arabino*-hexulosonic acid (2-ketogluconic acid), 57; D-*xylo*-5-hexulosonic acid (5-ketogluconic acid), 670; DL-glyceraldehyde, 91; pyruvic acid, 13; gluconic acid, 0; maltose, cellobiose, lactose, melibiose, <1.

The best sensitivity and selectivity of the reaction is maintained at pH 4.6–5.0. When the sample of sugar to be analyzed contains a strong buffer of pH > 5.0, it should be adjusted to pH 4.8 with acetic acid before addition of reagent A. If the solution examined contains a component, such as citrate, which chelates copper, it should be titrated with an equi-

[2] The reagent is prepared as follows: To 25 g of $(NH_4)_6Mo_7O_{24} \cdot 4H_2O$ dissolved in 450 ml of water, add with mixing 21 ml of analytical grade concentrated H_2SO_4, followed by 3 g of $Na_2HASO_4 \cdot 7H_2O$ dissolved in 25 ml of water. The reagent is kept in a glass-stoppered brown bottle and is ready for use after an incubation of 24 hr at 37° or of about 72 hr at room temperature. It is stable for at least one year, and before use may be diluted with one volume of 1.5 N H_2SO_4 without impairing its reactivity [N. Nelson, *J. Biol. Chem.* 153, 375 (1944)]. See also Vol. 1 [29], Vol. 3 [12], and Vol. 8 [1].

molar amount of cupric sulfate prior to the application of copper reagent A.

Reaction with *m*-Hydroxydiphenyl[3]

Principle. Hexuronic acids, free or as glycosides, produce a distinctive chromogen when treated with tetraborate in concentrated sulfuric acid and subsequently allowed to react with *m*-hydroxydiphenyl.

Reagents

> H₂SO₄-tetraborate (A): 12.5 mM sodium tetraborate in analytical grade, concentrated sulfuric acid.
> *m*-Hydroxydiphenyl (B): A 0.15% solution in 0.5% NaOH. Reagent is stable for about 1 month when kept refrigerated and protected from light.

Procedure. To 0.2 ml sample containing 10–100 nmoles hexuronic acid, add 1.2 ml of reagent A. The tubes are refrigerated in crushed ice. The mixture is stirred well, then immersed for 5 min in a boiling water bath. After cooling in ice, 20 μl of reagent B is added, and after 5–10 min, absorbance is read at 520 nm against reagent blanks.

Comments. The color produced is stable for several hours. Absorbance for D-glucuronic, D-galacturonic, and L-iduronic acids is about 1.5 times higher than that obtained for equimolar amounts of D-mannuronic acid. The reaction is much faster to perform and is also 2–3 times more sensitive to hexuronic acid than the widely used orcinol or carbazole procedures.[4] In addition, interference by aldohexoses and aldopentoses, which often creates a serious problem with the other colorimetric procedures, is significantly reduced in the reaction with *m*-hydroxydiphenyl.

The method is adaptable for the assay of uronic acid containing biopolymers in small samples of biological material.[3]

---

[3] N. Blumenkrantz and G. Asboe-Hansen, *Anal. Biochem.* **54**, 484 (1973).
[4] See Vol. 3 [12] and Vol. 8 [3].

[6] Estimation of 3-Deoxy-2-ketoaldonic Acids

By Richard Chaby, Daniel Charon, Robert S. Sarfati,
Ladislas Szabó, and François Trigalo

Estimation with Thiobarbituric Acid[1]

Principle. Treatment of 3-deoxy-2-ketoaldonic acids with periodate yields 2,4-dioxobutyrate which condenses with 2-thiobarbituric acid to give a red dye[2] having an absorption maximum at 549 nm.

Reagent

> Thiobarbiturate reagent[3]: 300 mg of reagent grade thiobarbituric acid are suspended in 50 ml of water and dissolved by addition of 30 ml of 0.1 N NaOH. When all solids are dissolved, 2.5 ml of 1 N HCl are added, the pH is adjusted to 2 by addition of 0.1 N HCl or NaOH, as required, and the volume is completed to 100 ml by addition of water. It is kept in bottles wrapped in aluminum foil and is stable for months at room temperature.

Procedure.[4] An ice-cold sample (15–35 μg) of 3-deoxy-2-ketoaldonic acid in water (1 ml) is mixed with an ice-cold solution of sodium metaperiodate (10–13 mM, 1 ml), 0.2 N with respect to sulfuric acid; the mixture is kept in an ice bath (0°) in the cold room. For 3-deoxy-D-*manno*-oct-2-ulosonic acid and for 3-deoxy-D-*erythro*-hex-2-ulosonic acid, the periodate treatment is allowed to proceed overnight; for 3-deoxy-D-*arabino*-hept-2-ulosonate 24 ± 2 hr and for 3-deoxy-D-*threo*-hex-2-ulosonate 48 ± 3 hr are required. For aldulosonic acids of unknown structure, the time required for the quantitative formation of the dioxobutyrate should be established by either taking aliquots from the oxidation mixture and determining the time when the thiobarbiturate reaction is at its maximum or by following the reduction of periodate by Avigad's colorimetric assay.[5] The first method is direct, but requires much more material than the second one. After the time indicated or determined, sodium arsenite solution (2% w/v, 0.25 ml), 0.5 N with respect to HCl,

[1] A. Weissbach and J. Hurwitz, *J. Biol. Chem.* **234,** 705 (1959).

[2] R. Kuhn and P. Lutz, *Biochem. Z.* **338,** 554 (1963).

[3] P. R. Srinivasan and D. B. Sprinson, *J. Biol. Chem.* **234,** 716 (1959).

[4] D. Charon, R. S. Sarfati, D. R. Strobach, and L. Szabó, *Eur. J. Biochem.* **11,** 364 (1969).

[5] G. Avigad, *Carbohyd. Res.* **11,** 119 (1969).

is added to a sample (0.25 ml) of the reaction mixture and, after the disappearance of the yellow color (iodine), thiobarbituric acid reagent (1 ml) is added. After thorough mixing, the tubes are heated in a boiling water bath for 10 min and cooled in tap water; then 0.3 N HCl (1 ml) is added, and the absorbancy is measured at 549 nm against a reagent blank. In these conditions all the 3-deoxy-2-ketoaldonic acids had a molar absorption coefficient of $92 \pm 5 \times 10^3$.

Comments. Only free, i.e., unsubstituted, 3-deoxy-2-ketoaldonic acids can be estimated reliably with periodate/thiobarbiturate. Indeed, to obtain quantitative results, one molar equivalent of 2,4-dioxobutyrate ("β-formyl pyruvate") must be released during the oxidation reaction, and this reaction must take place at a rate considerably greater than the destruction of the dioxobutyrate itself by periodate.

In the case of phosphorylated aldulosonic acids with 5–8 carbon atoms, dephosphorylation is carried out with acid phosphatase prior to the periodate cleavage: the incubation with the enzyme is carried out in a volume of 0.1–0.3 ml, which is then diluted to the appropriate volume with the cold $NaIO_4/H_2SO_4$ solution.

In the case of 5-O-substituted (e.g., 5-O-methyl,[6] 5-O-glycosyl[7]) 3-deoxy-2-ketooctonic acids, the conditions of Weissbach and Hurwitz[1] were used: the molar absorption coefficient was found to be about 13.000 for both compounds, and it is probable that this value will be valid for all 5-O-substituted 3-deoxyoctulosonates. The molar absorption coefficient in the conditions of the cold acid method has not been determined.

In only one case was the molar absorption coefficient of a 4-O-substituted 3-deoxy-2-ketoaldonic acid, namely that of 4-O-methyl-D-*arabino*-heptulosonic acid[8] determined. In Weissbach and Hurwitz's conditions, this was found to be about 5000.

It can be concluded that in the absence of appropriate standards the estimation of *substituted* 3-deoxy-2-ketoaldonic acids by the periodate/thiobarbiturate method is likely to be very misleading.

Estimation with Diphenylamine[9]

Principle. 3-Deoxy-D-*manno*-octulosonic acid is heated with diphenylamine in acid solution: a condensation product of unknown composition is formed whose absorption is used for the estimation of the aldulosonate.

[6] D. Charon and L. Szabó, *Eur. J. Biochem.* **29**, 184 (1972).
[7] R. S. Sarfati and L. Szabó, unpublished results, 1972.
[8] D. Charon and L. Szabó, *Carbohyd. Res.,* **34**, 271 (1974).
[9] R. Chaby, R. S. Sarfati, and L. Szabó, *Anal. Biochem.* **58**, 123 (1974).

Reagent

Diphenylamine reagent: 1 g of reagent grade diphenylamine is dissolved in a mixture of 10 ml of distilled 96% EtOH, 90 ml of glacial acetic acid, and 100 ml of conc. HCl solution (*d*, 1.19). The reagent should be stored in a brown bottle; it is stable for at least one month at room temperature.

Procedure. A sample (1 ml) of 3-deoxy D-*manno*-octulosonate (15–250 μg) is mixed with the reagent (2 ml), and the mixture is heated for 30 min in closed tubes in a boiling water bath. The tubes are cooled in tap water (30 min) and the absorption is measured in 10-mm cells against a reagent blank. If other sugars are absent, the reading is done at 425 nm; 1 μmole/ml of sample of 3-deoxyoctulosonic acid has an absorbancy of 1.11. In the presence of other sugars (pentoses, hexoses, heptoses, 2-deoxyribose, rhamnose), the absorbancy (*A*) is measured at 420, 470, 630, and 650 nm. The octulosonate content is given by the equation:

$$k[\text{octulosonate}] = A_{420} - A_{470} + A_{630} - A_{650}$$

the value of the equation being 0.67 for 1 μmole of 3-deoxy-D-*manno*-octulosonic acid per milliliter of sample.

Comment. While pentoses, hexoses, and heptoses react with diphenylamine in this test, their absorption spectra all have a minimum in the 400–480 nm range, and their interference is largely abolished by the use of the equation given. Hexosamines do not react in the test, but neuraminic acids interfere. Fairly accurate estimation (±5%) is achieved in the presence of up to a 10-fold excess of pentoses and hexoses; the error is ±10% in the presence of a 10-fold excess of D-*glycero*-L-*manno*-heptose. The method can be applied successfully for the accurate estimation of 5-*O*-glycosylated 3-deoxyoctulosonic acids, and, very probably, to that of free and *O*-substituted 3-deoxy-2-ketoaldonic acids in general, provided that the *O*-substituent is acid labile. The 3-deoxyoctulosonate content of bacterial endotoxins cannot be estimated by this method because of the very large excess of sugars present.

[7] Determination of Sedoheptulose 7-Phosphate

By TERRY WOOD

Assay Method

Principle. D-Sedoheptulose 7-phosphate is phosphorylated in the presence of fructose-6-phosphate kinase and ATP to give D-sedoheptulose

1,7-diphosphate. The disphosphate is cleaved by aldolase giving D-erythrose 4-phosphate and dihydroxyacetone phosphate. The latter reacts with NADH in the presence of α-glycerophosphate dehydrogenase to give glycerol 3-phosphate and NAD causing a fall in absorbance at 340 nm.[1]

$$\text{Sedoheptulose 7-phosphate} + \text{ATP} \rightarrow \text{sedoheptulose 1,7-diphosphate} + \text{ADP} \quad (1)$$
$$\text{Sedoheptulose 1,7-diphosphate} \rightarrow \text{erythrose 4-phosphate}$$
$$+ \text{ dihydroxyacetone phosphate} \quad (2)$$
$$\text{Dihydroxyacetone phosphate} + \text{NADH} + \text{H}^+ \rightarrow \text{glycerol 3-phosphate}$$
$$+ \text{ NAD}^+ \quad (3)$$

Reagents

ATP, disodium salt, 40 mM, pH adjusted to between 7.0 and 7.4. Store at $-20°$.

NADH, 25 mM

Buffer: 200 mM triethanolamine-HCl containing 20 mM EDTA and 24 mM magnesium chloride, pH 7.5

A mixture (2 mg/ml) of α-glycerophosphate dehydrogenase (140 units/ml) and triosephosphate isomerase (550 units/ml) from rabbit muscle diluted 10 times with a solution of 1 mg/ml bovine serum albumin containing 1 mM EDTA

Aldolase from rabbit muscle, crystalline suspension in 3.2 M ammonium sulfate (100 units/ml, 10 mg/ml)

Fructose-6-phosphate kinase from rabbit muscle, crystalline suspension in 1.3 M ammonium sulfate, pH 7.2, containing 50 mM glycerol 3-phosphate, 4 mM AMP, and 1 mM dithiothreitol (1000 units/ml, 5 mg/ml)

Procedure. The sample containing up to 0.2 μmole of sedoheptulose 7-phosphate, 0.01 ml of ATP (0.4 μmole), 0.01 ml of NADH (0.25 μmole), 0.01 ml of the diluted α-glycerophosphate dehydrogenase–triosephosphate isomerase mixture (0.14 unit dehydrogenase, 0.55 unit isomerase), 0.01 ml of fructose-6-phosphate kinase (10 units), and 1.00 ml of buffer are placed in a cuvette; the volume is completed to 1.99 ml with water. The absorbance at 340 nm is recorded over at least 5 min to allow the phosphorylation to occur. When this time has elapsed and the absorbance is steady, 0.01 ml of aldolase (0.9 unit) is added, and the rapid decrease in absorbance that results is followed until the value is once more steady. The assay is repeated in the absence of fructose-6-phosphate kinase to measure the sum of fructose and sedoheptulose diphosphates in the sample. Fructose 6-phosphate, if present, is determined with glucose-6-phosphate isomerase and glucose-6-phosphate dehydrogenase.[2]

[1] T. Wood and W. M. Poon, *Arch. Biochem. Biophys.* **141**, 440 (1970).
[2] H.-J. Hohorst *in* "Methods of Enzymatic Analysis" (H.-U. Bergmeyer, ed.), p. 134. Academic Press, New York, 1965.

The fall in absorbance at 340 nm in the absence of fructose-6-phosphate kinase is subtracted from the fall in its presence, and the resulting value is used to calculate the sedoheptulose 7-phosphate content of the sample. A correction is made for fructose 6-phosphate present, remembering that fructose 6-phosphate yields two equivalents of triose phosphate in the assay whereas sedoheptulose 7-phosphate gives only one.

Notes

1. Suspensions of fructose-6-phosphate kinase in 1.3 M $(NH_4)_2SO_4$ lose their activity slowly over several weeks when stored. All difficulties experienced with this assay have been traced to the use of old, partially inactivated kinase preparations.

2. Sedoheptulose 7-phosphate is phosphorylated by fructose-6-phosphate kinase at about one-tenth the rate of fructose 6-phosphate at 0.1 mM concentrations, and the rates at saturating concentrations (2 mM) are approximately equal.[3]

3. Sedoheptulose 1,7-diphosphate is cleaved by aldolase at about half the rate at which it acts on fructose 1,6-diphosphate.[4]

4. Using fresh, fully active enzyme preparations, the assay is complete at 5–15 min after the addition of aldolase. This is considerably shorter than the times of up to 100 min required for the assay of sedoheptulose 7-phosphate with transaldolase.[1,5]

5. In the absence of substrate, but in the presence of all other constituents, the absorbance of NADH in the assay mixture decreased by less than 0.008 over 30 min.

6. The assay has been applied to the measurement of sedoheptulose 7-phosphate in purified preparations,[1] and in a mixture of products obtained by incubating tissue extracts with ribose 5-phosphate.[3] In the latter case the method gave more consistent values than the colorimetric cysteine-sulfuric acid procedure of Dische.[6]

[3] T. Wood, unpublished results, 1972.
[4] B. L. Horecker, P. Z. Smyrniotis, H. H. Hiatt, and P. A. Marks, *J. Biol. Chem.* **212**, 827 (1955).
[5] J. Cooper, P. A. Srere, M. Tabachnik, and E. Racker, *Arch. Biochem. Biophys.* **74**, 306 (1958).
[6] Z. Dische, *J. Biol. Chem.* **204**, 983 (1953)

[8] Preparation and Analysis of Mixtures of D-Ribose 5-Phosphate, D-Ribulose 5-Phosphate, and D-Xylulose 5-Phosphate

By Terry Wood

For the coupled assay at 340 nm of transketolase with α-glycerophosphate dehydrogenase and triosephosphate isomerase,[1,2] or with glyceraldehyde phosphate dehydrogenase,[3] as auxiliary enzymes, a mixture of D-ribose 5-phosphate and D-xylulose 5-phosphate is needed. Similarly, a mixture of D-ribose 5-phosphate and D-ribulose 5-phosphate is employed for the assay of D-ribulose-5-phosphate 3-epimerase by coupling it to the transketolase reaction.[4,5] A rapid and simple procedure is described for preparing mixtures of the pentose phosphates for the assay of the above two enzymes.

$$\text{Ribose 5-phosphate} \rightleftarrows \text{ribulose 5-phosphate} \rightleftarrows \text{xylulose 5-phosphate} \qquad (1)$$

Preparation of Assay Mixture

Reagents

D-Ribose 5-phosphate disodium salt, 100 mM. Store at 2°.
D-Ribose 5-phosphate ketol isomerase, from spinach, 200 units/ml, 2.7 mg/ml. Store at −20°.
D-Ribulose-5-phosphate 3-epimerase from Baker's yeast, sulfate free, 50 units/ml, 1 mg/ml. Store at −20°.
Triethanolamine-HCl buffer, 20 mM, pH 7.4

Procedure. Ribose 5-phosphate solution (0.30 ml, 30 μmoles) is mixed with 2.2 ml buffer, and the pH is adjusted to 7.0–7.4 using a pH meter. The volume is then made up to 2.97 ml with buffer, and the solution is placed in a cuvette in the cell compartment at 37° of a Beckman DB double-beam spectrophotometer connected to a Varicord 43 linear-log recorder (Photovolt Corporation, 115 Broadway, New York 10010). When thermal equilibrium has been reached, 0.03 ml of ribose-5-phos-

[1] B. L. Horecker, P. Z. Smyrniotis, and H. Klenow, *J. Biol. Chem.* **205**, 661 (1953).
[2] M. E. Kiely, E. L. Tan, and T. Wood, *Can. J. Biochem.* **47**, 455 (1969).
[3] G. A. Kochetov, L. I. Nikitushkina, and N. N. Chernov, *Biochem. Biophys. Res. Commun.* **40**, 873 (1970).
[4] B. L. Horecker, P. Z. Smyrniotis, and J. Hurwitz, *J. Biol. Chem.* **223**, 1009 (1956).
[5] W. Yaphe, D. D. Christensen, J. E. Biaglow, M. A. Jackson, and H. Z. Sable, *Can. J. Biochem.* **44**, 91 (1966).

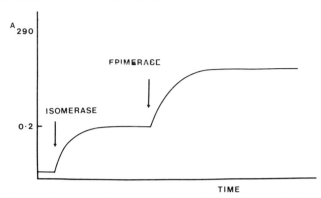

Fig. 1. Preparation of a mixture of D-ribose 5-phosphate, D-ribulose 5-phosphate, and D-xylulose 5-phosphate.

phate ketol isomerase is added, and the absorbance is recorded until there is no further increase and the value, read against buffer, remains steady, or very nearly so, at around 0.200 (Fig. 1). At this point the mixture is either ultrafiltered or xylulose 5-phosphate is generated by the addition of 0.03 ml of ribulose-5-phosphate 3-epimerase. After the addition of the epimerase, a further increase in absorbance occurs and the incubation is continued until 10–15 min after the point where the absorbance remains constant (Fig. 1).

After the completion of either the isomerase or the epimerase reaction, the mixture is ultrafiltered through a PM-30 membrane in a 10-ml Diaflo ultrafiltration cell (Amicon Corp., 280 Binney St., Cambridge, Massachusetts 02142) placed either in a warm room at 37°, or in a blast of warm air from a hair-drier. Ultrafiltration should be complete after 5 min. The ultrafiltrate is returned to the cell and passed through a second time to remove any traces of enzyme. The second ultrafiltrate is cooled in ice, 2 N acetic acid is added to bring the pH to 5.9–6.4, and the solution is stored frozen at −20°. The absence of epimerase and isomerase may be verified by adding a suitable volume to the reaction mixture for the assay of these enzymes,[6] whereupon there should be no change in absorbance at 290 nm.

Notes

1. Although the commercial purified epimerase is specified, it need not be free of the isomerase, and a crude concentrate of the two enzymes

[6] This volume [14].

from spleen, muscle, or liver, may be used provided that transketolase is absent.

2. The ultrafiltration is carried out at, or near, 37°, to prevent the composition of the equilibrium mixture changing after removal from the spectrophotometer.

3. The procedure may also be used with a 20 mM solution of ribose 5-phosphate. At this higher concentration, however, further spectral changes of the isomerase product will occur[7] and may take place before the epimerase is added, obscuring the progress of the enzymic reaction.

Analyses

Most authors[8-11] are agreed that the equilibrium of the isomerase reaction at 37° leads to a ribose 5-phosphate/ribulose 5-phosphate ratio of approximately 3. There is less agreement concerning the equilibrium of the epimerase reaction, but two laboratories[9,12] have reported a value as high as 3 for the xylulose 5-phosphate/ribulose 5-phosphate ratio, and Kauffman et al.[13] reported, more recently, a value of 2.4. Assuming limiting ratios of 3:1:3 for ribose 5-phosphate:ribulose 5-phosphate:xylulose 5-phosphate, then after treatment of 10 mM ribose 5-phosphate with isomerase we would expect 7.5 mM ribose 5-phosphate and 2.5 mM ribulose 5-phosphate, assuming no dilution during the manipulations. Similarly, after both isomerase and epimerase we may expect 4.3 mM ribose 5-phosphate, 1.4 mM ribulose 5-phosphate, and 4.3 mM xylulose 5-phosphate. In practice, the concentration of xylulose 5-phosphate is usually found to be smaller, and that of ribulose 5-phosphate greater, than the above values.

Ribose 5-phosphate is measured, without interference from the ketopentose phosphates, by the colorimetric phloroglucinol method.[14]

Ribulose 5-phosphate is measured in ribose 5-phosphate-ribulose 5-phosphate mixtures with cysteine-carbazole[8]; a heating time of 2 hr at 37° is used to develop the color. Under these condtions 0.1 μmole of ribulose 5-phosphate gives an absorbance at 540 nm of 0.32.[15]

[7] F. C. Knowles, Ph.D. Dissertation, University of California, Riverside, 1968.
[8] B. Axelrod and R. Jang, *J. Biol. Chem.* **209**, 847 (1954).
[9] M. Tabachnik, P. A. Srere, J. Cooper, and E. Racker, *Arch. Biochem. Biophys.* **74**, 315 (1958).
[10] M. Urivetzky and K. K. Tsuboi, *Arch. Biochem. Biophys.* **103**, 1 (1963).
[11] F. H. Bruns, E. Noltmann, and E. Valhaus, *Biochem. Z.* **330**, 483 (1958).
[12] Z. Dische and H. Shigeura, *Biochim. Biophys. Acta* **24**, 87 (1956).
[13] F. C. Kauffman, J. G. Brown, J. V. Passonneau, and O. H. Lowry, *J. Biol. Chem.* **244**, 3647 (1969).
[14] Z. Dische and E. Borenfreund, *Biochim. Biophys. Acta* **23**, 639 (1957).
[15] T. Wood, unpublished results, 1972.

Xylulose 5-phosphate is measured, either alone, or in conjunction with ribulose 5-phosphate by the spectrophotometric method described below.

Spectrophotometric Determination of D-Xylulose 5-Phosphate and D-Ribulose 5-Phosphate

The method is similar to that described by Cooper *et al.*,[16] but uses α-glycerophosphate dehydrogenase and triose phosphate isomerase to measure the glyceraldehyde 3-phosphate formed. One molecule of keto-pentose phosphate causes the oxidation of 1 molecule of NADH.

Reagents

Dithiothreitol, 50 mM. Store at −20°.
Thiamine pyrophosphate, 10 mM. Store at −20°.
D-Ribose 5-phosphate disodium salt, 100 mM. Store at 2°.
NADH, 25 mM
Glycylglycine-Na buffer, 100 mM containing 5 mM magnesium chloride, pH 7.4
A mixture (2 mg/ml) of α-glycerophosphate dehydrogenase (140 units/ml) and triosephosphate isomerase (550 units/ml) from rabbit muscle diluted 10 times with 1 mg of bovine serum albumin per milliliter containing 1 mM EDTA.
Transketolase from Baker's yeast, sulfate free, 25 units/ml, 1.2 mg/ml
D-Ribulose-5-phosphate 3-epimerase from Baker's yeast, sulfate free, 50 units/ml, 1 mg/ml

Procedure. The reaction mixture contains 1.00 ml of buffer, 0.02 ml ribose 5-phosphate (2 μmoles), 0.02 ml thiamine pyrophosphate (0.2 μmole), 0.04 ml of dithiothreitol (2 μmoles), 0.01 ml NADH (0.25 μmole), 0.01 ml of diluted α-glycerophosphate dehydrogenase (0.14 unit)–triose phosphate isomerase (0.55 unit) mixture, not more than 0.2 μmole of keto-pentose phosphate, and water to a final volume of 1.99 ml. The cuvette is placed in the cell compartment at 37° of a Beckman DB double-beam spectrophotometer connected to a recorder. The recorder is started, and after thermal equilibrium is reached and the absorbance has become constant, 0.01 ml of transketolase (0.25 unit) is added. The absorbance falls as the xylulose 5-phosphate reacts and after 5–20 min is once again constant. Ribulose-5-phosphate 3-epimerase (5 μl, 0.25 unit) is added, and the absorbance is recorded until it is again steady after a further 15–40 min.

---

[16] J. Cooper, P. A. Srere, M. Tabachnik, and E. Racker, *Arch. Biochem. Biophys.* **74**, 306 (1958).

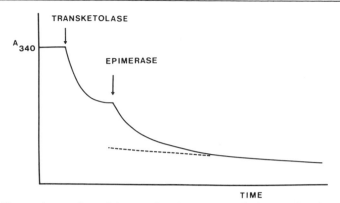

FIG. 2. Assay of D-xylulose 5-phosphate and D-ribulose 5-phosphate.

If the absorbance does not reach a final steady value (owing to traces of isomerase in the enzymes used) the recording is continued long enough to allow the final linear rate to be extrapolated back to the moment the epimerase was added (Fig. 2). The values obtained for the ribulose 5-phosphate and xylulose 5-phosphate in the sample are corrected for any absorbance changes obtained in a control run with the sample absent.

Notes

1. Values obtained are reproducible to within ±5% of the mean.

2. The determination of ribulose 5-phosphate by this procedure can be time-consuming but xylulose 5-phosphate reacts readily. If its reaction with transketolase appears sluggish, one may try the effect of reversing the order of addition of the sample and transketolase.

3. The total pentose phosphate in a sample may be determined by omitting ribose 5-phosphate from the reaction mixture and adding ribose-5-phosphate ketol isomerase, epimerase, and transketolase together. Under these conditions 2 molecules of pentose phosphate gives rise to the oxidation of 1 molecule of NADH.

[9] Enzymic Determination of Lactic Acid

By JOHANNES EVERSE

Several methods have been described for the chemical determination of lactic acid concentrations in tissue extracts, sera, and other biological systems.[1,2] These methods are based on the chemical conversion of lactic acid to acetaldehyde, which is subsequently coupled to an aromatic compound to produce a colored product. The intensity of the color is determined spectrophotometrically

Although these methods yield satisfactory results in most cases, the reactions are not specific for lactate. Other hydroxyacids and aldehydes may give a similar reaction with the reagents, and erroneously high values may be obtained. Furthermore, it is generally necessary to deproteinize the solutions before a lactate determination can be carried out, and long incubation times may be required for full color development.

The advantages of an enzymic determination are, first, that the enzyme is generally quite specific for its substrate, and few erroneous values are obtained. Second, enzymic reactions that can be followed spectrophotometrically may be carried out rapidly and inexpensively without deproteinization of the solutions.

Enzymic Method

Principle. Lactate dehydrogenase catalyzes the reaction

$$\text{Lactate} + \text{DPN}^+ \rightleftharpoons \text{pyruvate} + \text{DPNH} + \text{H}^+$$

reversibly, and the equilibrium constant at pH 7.0 is 4×10^5 in favor of lactate formation. Because of this unfavorable equilibrium, the oxidation of lactate catalyzed by lactate dehydrogenase is usually carried out at relatively high pH values. Even under these conditions, however, one does not achieve a complete oxidation of all the lactic acid; in fact only about 50% of the lactate is oxidized at equilibrium, when a concentration as high as 2 mg/ml of DPN$^+$ is used.

A much more complete oxidation of lactic acid may be achieved with the use of analogs of DPN$^+$ that have a more positive redox potential. In the method that is described here the 3-acetylpyridine analog of DPN$^+$ is used as the coenzyme. This analog has a redox potential of -0.240, as compared to -0.320 for DPN$^+$. (The redox potential of pyruvate/lactate is -0.190.) Hence, in the presence of 3-acetylpyridine-DPN$^+$ at a concentration of 2 mg/ml the oxidation of lactate is 98.5% complete at pH 10, if the lactate concentration is 5 μg/ml or less. This appears to be sufficiently accurate for most routine lactate determinations.

[1] S. B. Baker, this series, Vol. 3 [42].
[2] A. H. Free and H. M. Free, *in* "Clinical Pathology of the Serum Electrolytes" (F. William Sunderman and F. William Sunderman, Jr., eds.), p. 107. Thomas, Springfield, Illinois, 1966.

Reagents

Borate buffer, 0.01 M, pH 9.2, which is prepared by dissolving 0.38
g of sodium borate ($Na_2B_4O_7 \cdot 10H_2O$) in 100 ml of water
3-Acetylpyridine-DPN$^+$, 20 mg/ml, dissolved in water
Lactate dehydrogenase, beef heart, 5 mg/ml

The solution containing the unknown amount of lactic acid is diluted
with water until the estimated concentration of lactic acid is between
20 and 50 μg/ml.

Procedure. Transfer into a 1-ml cuvette: 0.8 ml of 0.01 M borate
buffer, 0.1 ml of 3-acetylpyridine-DPN$^+$ solution, 0.01 ml of lactate dehydrogenase solution, and 0.1 ml of the diluted lactic acid solution.

After the reaction has reached equilibrium, the optical density of the
solution is determined at 363 nm. The reference cuvette contains the same
ingredients as the assay cuvette, except that water has been added instead
of the enzyme.

The lactate concentration (μg/ml) of the original solution may be
calculated from the following equation, assuming that the original solution was a-fold diluted to bring the lactate concentration within the measurable range.

$$X = [(\Delta \text{ in optical density}) (10a) (90)]/9.1$$

in which 90 represents the molecular weight of lactic acid, and 9.1 is
the millimolar extinction coefficient of 3-acetylpyridine-DPNH at 363 nm.

General Remarks

The enzymic method for the determination of lactic acid has been
used in our laboratory for several years; it yields highly satisfactory
results if the manipulations are properly performed. Two aspects of the
method need special attention in order to prevent the occurrence of erroneous values.

The absorbance at 363 nm that is observed during a lactic acid determination is due to the formation of 3-acetylpyridine-DPNH. The oxidized
coenzyme does not absorb light at 363 nm. However, commercial preparations of 3-acetylpyridine-DPN$^+$ sometimes produce a slight absorbance
at 363 nm at alkaline pH values. It is therefore necessary to use the
coenzyme analog in the reference cuvette, as described in the procedure,
in order to compensate for any absorbance that is of a nonenzymic origin.

One should also be aware of the fact that a certain amount of lactate
is usually present on the skin. Erroneously high values for lactate may
be found if one mixes the contents of the cuvette by hand without cover-

ing the cuvette first with a clean piece of plastic film. The presence of lactate on the skin requires that solutions do not come in contact with the skin, or with any material that previously has been touched by uncovered hands.

The reaction in the cuvette should reach equilibrium within 2 or 3 min, and the optical density at 363 nm may be determined at that time. The reduced coenzyme is fairly stable under the assay conditions, but it is advisable to take the measurements within 5 min after the equilibrium is attained, in order to minimize any spurious nonspecific reactions. It is advisable to keep the conditions as constant as possible for maximum reproducibility.

Section II

Enzyme Assay Procedures

[10] Determination of the Isoenzyme Levels of Lactate Dehydrogenase

By LARRY H. BERNSTEIN and JOHANNES EVERSE

The determination of the isoenzyme content of lactate dehydrogenase (LDH) preparations is conventionally done using electrophoresis. Electrophoretic separation of the isoenzymes has the advantage that an evaluation may be made of the amount in which each of the five characteristic enzyme bands are present. (The five bands represent the two homologous isoenzyme tetramers H_4 and M_4 LDH and the three hybrid forms H_3M, H_2M_2, and HM_3 LDH.) The technique has the disadvantage that it is time consuming and it is difficult to obtain quantitative results.

Various other procedures for the determination of the isoenzyme content of LDH, i.e., the percentage H-type and M-type LDH, have recently been described. These procedures are based on the fact that the two isoenzymes possess different physical and kinetic properties. Differences in the physical properties of the H- and M-type LDH include differences in their thermal stability,[1] solubility in organic solvents,[2] and stability against urea denaturation.[3] Other procedures employ the effect of inhibitors like oxamate or oxalate,[4,5] differences in Michaelis constant,[6] and differences in the immunological properties[7] of the two types of LDH.

The use of kinetic methods for determining the levels of the two isoenzymes has certain advantages. Kinetic determinations of enzyme levels are usually quite accurate; they can be carried out rapidly and inexpensively, and they are readily adaptable to automated procedures.

In this chapter we will describe two methods for a rapid determination of the isoenzyme content of LDH. Both methods are based on the fact that the H-type LDH is significantly inhibited by high concentrations of pyruvate in the presence of NAD^+, whereas the activity of the M-type enzyme is not significantly affected under these conditions. The first method involves carrying out two LDH assays under different conditions, which may be done in any conventional spectrophotometer in a relatively

[1] A. L. Latner and A. W. Skillen, *Proc. Ass. Clin. Biochem.* **2**, 100 (1963).
[2] A. L. Latner and D. M. Turner, *Lancet* **1**, 1293 (1963).
[3] A. E. H. Emery, *Biochem. J.* **105**, 599 (1967).
[4] D. T. Plummer and J. H. Wilkinson, *Biochem. J.* **87**, 423 (1963).
[5] P. M. Emerson and J. H. Wilkinson, *J. Clin. Pathol.* **18**, 803 (1965).
[6] E. S. Vesell and A. G. Bearn, *J. Clin. Invest.* **40**, 586 (1961).
[7] J. S. Nisselbaum, M. Schlamowitz, and O. Bodansky, *Ann. N.Y. Acad. Sci.* **94**, 970 (1961).

short time. The second procedure requires the availability of a stopped-flow apparatus. The latter procedure has the advantage that only one assay is required for each unknown preparation, no incubation is necessary, and the results are obtained within seconds.

Principle of the Methods

Lactate dehydrogenases catalyze the reversible reduction of pyruvate using NADH as the coenzyme

$$\text{Pyruvate} + \text{NADH} + \text{H}^+ \rightleftarrows \text{lactate} + \text{NAD}^+$$

The substrate for both lactate dehydrogenase isoenzymes is the keto form of pyruvate. The equilibrium constant of the LDH reaction is 4×10^5 at pH 7.0 in favor of lactate production.

The mechanism of the lactate dehydrogenases is ordered, and the reaction proceeds as follows:

$$
\begin{array}{ccccc}
\text{E} & \xrightarrow{} & \text{E} \cdot \text{NADH} & \xrightarrow{} & \text{E} \cdot \text{NADH} \cdot \text{pyr} \rightarrow \\
\uparrow & & & \uparrow & \\
\text{NADH} & & & \text{pyruvate} &
\end{array}
$$

$$
\begin{array}{ccccc}
\text{E} \cdot \text{NADH} \cdot \text{lact} & \xrightarrow{} & \text{E} \cdot \text{NADH} & \xrightarrow{} & \text{E} \\
\downarrow & & \downarrow & & \\
\text{lactate} & & \text{NAD}^+ &
\end{array}
$$

The transition state is thus a ternary complex of enzyme, coenzyme, and substrate. In addition to the transitory complexes, lactate dehydrogenases can form abortive ternary complexes. These abortive complexes consist of enzyme, reduced coenzyme, and lactate, or of enzyme, oxidized coenzyme, and pyruvate. The complexes are abortive in that they are not intermediates to the catalytic reaction. They, in fact inhibit the catalytic process, since the enzyme that is bound in the form of an abortive complex is not available for catalysis.

Studies in a number of laboratories have shown that the H-type LDH forms such abortive complexes much more readily than the M type enzyme.[8] Hence, when a mixture of H- and M-type LDH is incubated in the presence of NAD$^+$ and pyruvate, a large portion of the H-type enzyme is converted to the abortive complex, whereas the M-type LDH remains largely unaffected. This distinction in the properties of the two isoenzymes therefore allows for a resolution of the isoenzyme content of an LDH preparation by kinetic methods.

[8] J. Everse and N. O. Kaplan, *Advan. Enzymol. Relat. Areas Mol. Biol.* **37**, 61 (1973).

Determination of Lactate Dehydrogenase Isoenzymes Using Standard Spectrophotometry

Reagents

NADH, 13 mM: Dissolve 10 mg of reduced nicotinamide adenine dinucleotide, disodium salt, in 1.0 ml of 0.1 M potassium phosphate buffer, pH 7.5.

NAD$^+$, 1.3 mM: Dissolve 2 mg of oxidized nicotinamide adenine dinucleotide, disodium salt, in 2.0 ml of distilled water.

Pyruvate, 0.1 M: Dissolve 11 mg of sodium pyruvate in 1.0 ml of 0.1 M potassium phosphate buffer, pH 7.5.

Procedure

ASSAY FOR THE TOTAL AMOUNT OF LDH (FIRST ASSAY). The following amounts of the reagents are placed in a 3-ml quartz cuvette: 2.90 ml of 0.1 M potassium phosphate buffer, pH 7.5; 0.03 ml of NADH, 13 mM; 0.02 ml of pyruvate, 0.1 M. The reaction is initiated by the addition of an appropriate amount of enzyme, and the change in absorption at 340 nm is recorded as a function of time. The amount of enzyme used should be adjusted so as to give a change in absorption of 0.150–0.300 per minute.

ASSAY FOR ISOENZYME CONTENT (SECOND ASSAY). The following amounts of the reagents are placed in a 3-ml cuvette: 2.60 ml of 0.1 M potassium phosphate buffer, pH 7.5; 0.03 ml of NAD$^+$, 1.3 mM; 0.30 ml of pyruvate, 0.1 M. Add the same amount of enzyme that was used in the first assay. The reagents are thoroughly mixed, and allowed to incubate at room temperature for 15 min. After this incubation period, the reaction is initiated by the addition of 0.03 ml of NADH, 13 mM, and the change in absorbance at 340 nm is recorded as a function of time.

Calculations. The total enzyme activity is calculated according to the formula

$$\text{IU/ml} = \frac{(\Delta \text{ in absorbance per min})(\text{total volume in cuvette})}{6.22 \times (\text{volume of enzyme solution used for assay})}$$

using the change in absorbance obtained during the first assay.

In order to calculate the isoenzyme content of an unknown solution, the inhibition of the unknown sample is compared with that of the pure H$_4$ and M$_4$ isoenzymes under identical conditions. Suppose that the rates obtained with the pure H$_4$ isoenzyme are a in the presence of low

[9] N. O. Kaplan, J. Everse, and J. Admiraal, *Ann. N.Y. Acad. Sci.* **151**, 400 (1968).

pyruvate (first assay) and b in the presence of high pyruvate levels (second assay; a and b are expressed as ΔOD per minute). Similarly, the rates obtained with an equimolar amount of the M_4 isoenzymes are c and d, respectively. The analysis of a sample of unknown composition yields ΔOD_L as its rate at the low pyruvate concentration, and ΔOD_H in the presence of high pyruvate. The activity of the unknown sample, ΔOD_L as well as ΔOD_H, represent the sum of the rates of the two isoenzymes that are present in the sample. Also, the activities of the H isoenzyme with respect to that of the M-type LDH are expressed in the rates obtained with the pure isoenzymes. The following two equations therefore apply:

$$\Delta OD_L = a[H] + c[M]$$
$$\Delta OD_H = b[H] + d[M]$$

in which [H] and [M] represent the concentrations of the H and M-type LDH, respectively, in the unknown solution, expressed in some arbitrary units. The equations are then solved for [H] and [M], and the percentage H-type is determined by the equation:

$$\%\text{H-type LDH} = \frac{[H]}{[H] + [M]} \times 100\%$$

Note. Some general remarks are presented at the end of this chapter, which are applicable to both methods discussed here.

Determination of LDH Isoenzymes Using Rapid Kinetics

It has been well established that the H-type lactate dehydrogenase is inhibited in the presence of high concentrations of pyruvate, even when the enzyme is not preincubated in the presence of NAD^+.[8,9] Under these conditions LDH forms an abortive complex with pyruvate and the NAD^+ that is produced during the reaction. Since no NAD^+ is present at the onset of the reaction, no inhibition of the enzymic activity is found at that time. After some NAD^+ has been produced, one observes a decrease in the catalytic rate until a steady state is reached. This phenomenon is illustrated in Fig. 1, which demonstrates the effect on the human LDH isoenzymes. The assay of the H-type LDH shows no inhibition during the first second of the reaction. This is followed by a decrease in the activity, until a steady state is reached when 5 or 6 sec have elapsed. No significant inhibition is observed with the M-type enzyme under these conditions, whereas mixtures of the isoenzymes yield the expected intermediate values.[10]

[10] M. J. Bishop, J. Everse, and N. O. Kaplan, *Proc. Nat. Acad. Sci. U.S.* **69**, 1761 (1972).

Human M₄ LDH   Human H₄ LDH

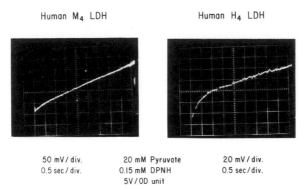

50 mV/div.	20 mM Pyruvate	20 mV/div.
0.5 sec/div.	0.15 mM DPNH	0.5 sec/div.
	5V/OD unit	

Fig. 1. Pre-steady state kinetics of the human M_4 and the human H_4 LDH. An Aminco Morrow stopped-flow instrument was used for the determination.

Given a kinetic curve which is intermediary to those shown in Fig. 1, it is clear that from the initial part of the curve one may calculate the total amount of enzyme that is present in the sample. Furthermore, after the reaction has reached the steady state, the slope of this portion of the curve is proportional to the ratio in which the isoenzymes are present in the preparation. Hence, if the reaction is performed in a stopped-flow apparatus, one may calculate the total activity as well as the isoenzyme content from the data obtained from a single assay. For standardization purposes, it is necessary to perform assays on the pure isoenzymes under identical conditions.

Stock Solutions

NADH, 13 mM: Dissolve 10 mg of reduced nicotinamide adenine dinucleotide, disodium salt, in 1.0 ml of 0.1 M potassium phosphate buffer, pH 7.5.

Pyruvate, 1 M: Dissolve 110 mg of sodium pyruvate in 1.0 ml of 0.1 M potassium phosphate buffer, pH 7.5.

Procedure. Mix 0.4 ml of the pyruvate solution with 0.2 ml of the NADH solution and add 9.4 ml of 0.1 M phosphate buffer, pH 7.5, to adjust the total volume to 10 ml. This solution is placed in reservoir 1 of a stopped-flow apparatus. Reservoir 2 contains the undiluted serum, or an appropriate dilution of the tissue extract or enzyme preparation that is to be tested.

The change in optical density at 340 nm with time may be either displayed on an oscilloscope, or (preferably) obtained in a digitized form. From the observed data one retrieves the activity of the enzyme at 0.3 sec after the onset of the reaction, and again after 10 sec. Similar sets of observations are obtained with the pure H_4 and M_4 isoenzymes.

Calculations. The concentration of pyruvate used in this procedure (20 mM) is significantly above the K_m for pyruvate of both isoenzymes. Under these conditions the two isoenzymes of chicken as well as those of human appear to have the same specific activity (SA). Hence, once the initial activity of a known concentration of enzyme has been established under the applied conditions, the concentration of any unknown solution of LDH may be calculated directly from its initial activity.

$$\text{IU/ml} = \frac{\Delta\text{OD sample}}{\Delta\text{OD standard}} \times \text{SA of standard in IU/ml}$$

The calculation of the isoenzyme content is analogous to that described in the first part of this chapter. Using fast kinetics, a, c, and ΔOD_L represent the activities during the initial part of the reactions, whereas b, d, and ΔOD_H represent the activities obtained after 10 sec have elapsed.

General Remarks

In this chapter we have previously pointed out that the inhibition of LDH at high pyruvate concentrations is caused by the formation of an abortive complex, consisting of enzyme, NAD+, and pyruvate. The pyruvate involved in the formation of the abortive complex is the enol form of pyruvate, whereas the keto form is the actual substrate for the enzymic reaction. It is thus necessary that these two forms of pyruvate be at equilibrium with each other in order to obtain reproducible results. The rate of conversion between the two forms of pyruvate, however, is very slow. As a result, the enol-pyruvate content of various commercial pyruvate preparations varies significantly.

It is therefore necessary to prepare the pyruvate solution at least 24 hr prior to use. The solution should be kept at room temperature during this time in order to allow the tautomers to reach equilibrium. Phosphate ions, among others, catalyze the enolization of pyruvate, and for this reason the pyruvate is dissolved in phosphate buffer.

The differences in catalytic activity that are observed with the H- and M-type LDH in the presence of high pyruvate concentrations vary significantly among different species. Large differences between the two types are observed with the LDH's from cow and rabbit, whereas the difference between the two human enzymes is less pronounced. In quantitating the amount of the isoenzymes present in an unknown preparation, it is therefore necessary that the pure isoenzymes from the same source as the unknown preparation are used as the standards.

[11] Fluorometric Determination of Dehydrogenase Activity Using Resorufin

By GEORGE G. GUILBAULT

Assay Method

Principle. A single, rapid fluorometric method is described for measuring the activity of dehydrogenases. The method is based on the conversion of the nonfluorescent material, resazurin ($3H$-phenoxazin-3-one, 7-hydroxy-, 10-oxide), to the highly fluorescent compound, resorufin, in conjunction with the nicotinamide adenine dinucleotide (NAD+–NADH) system.[1,2] The rate of change

$$\text{Acid} \;+\; \text{NAD}^+ \xrightarrow{\text{dehydrogenase}} \alpha\text{-Keto acid} \;+\; \text{NADH} \;+\; \text{H}^+$$

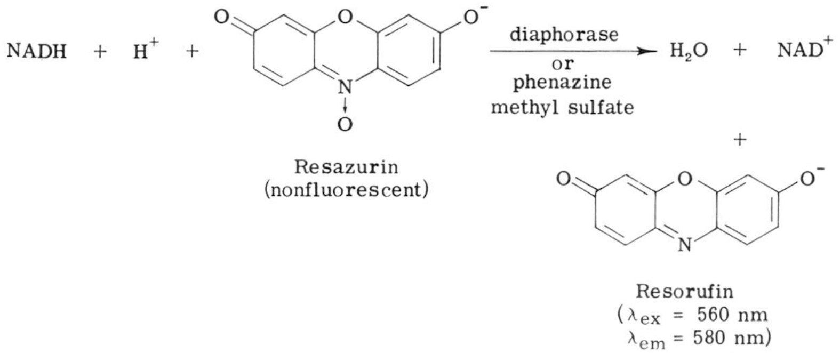

Resazurin
(nonfluorescent)

Resorufin
(λ_{ex} = 560 nm
λ_{em} = 580 nm)

in the fluorescence of the solution with time ($\Delta f/\min$), is proportional to the amount of dehydrogenase present.

Reagents

Enzyme. Diaphorase, type II (Sigma Chemical Co., St. Louis, Missouri) activity 16 units/mg. One unit equals a decrease in absorbancy of 1.00 per minute of 2,6-dichloroindophenol at 25°.

Buffer, glycine-hydrazine: 1 M glycine–0.4 M hydrazine, pH 9.5. This solution is stable for 2 weeks.

Substrates. Resazurin (Eastman Organics Co., Rochester, New York). A stock 0.2 mM solution was prepared in methyl Cellosolve. This solution should be nonfluorescent, with a bluish-red

---

[1] G. G. Guilbault and D. N. Kramer, *Anal. Chem.* **36**, 2497 (1964).
[2] G. G. Guilbault and D. N. Kramer, *Anal. Chem.* **37**, 1219 (1965).

hue. A sample of resazurin that is contaminated with resorufin may be purified by acetylation of the compound with acetic anhydride and pyridine by conventional procedures,[3,4] followed by isolation of the ester in water and recrystallization. Addition of base to the ester will then give pure resazurin.

Sodium maleate, 0.1 M, and sodium lactate, 10 mM, prepared by dissolving the neutralized acid in glycine-hydrazine buffer

Ethanol solutions, 50 mM, prepared in glycine-hydrazine buffer

Sodium glutamate, 10 mM. A stock solution was prepared by dissolving the purified compound in glycine-hydrazine buffer.

Glucose 6-phosphate. A 1 mM solution was prepared in 0.1 M Tris buffer, pH 8.5, with the purified material (Calbiochem, 80.6% pure).

Glycerol and L-(—)-α-glycerol phosphate (Calbiochem), 1 mM solutions prepared in 0.1 M Tris buffer, pH 9.0

Nicotinamide adenine dinucleotide (NAD$^+$) and its reduced form (NADH), obtained from Calbiochem, were found to be 86.3 and 80.3% pure. Stock 10 mM solutions were prepared in triply distilled water.

Nicotinamide adenine dinucleotide phosphate (NADP$^+$) and its reduced form (NADPH). Stock 1 mM solutions were prepared by dissolving the compound (Calbiochem, 70.1 and 68.7% pure) in triply distilled water.

Phenazine methyl sulfate. An aqueous 1 mM solution of the purified substance (Sigma Chemical Co., St. Louis, Missouri) is stable for 4 days if kept refrigerated in a dark bottle.

Apparatus. All fluorescence measurements were made with an Aminco-Bowman spectrophotofluorometer (SPF), equipped with a thermoelectric cooler to maintain a constant temperature at 25°. The resorufin produced was measured at excitation and emission wavelengths of 560 and 580 nm, respectively. Although the excitation and emission wavelengths are close together, scattered radiation was not found to be a problem, as was pointed out in previous papers.[4,5] No blank correction was needed.

Procedures. Two milliliters of the appropriate substrate (sodium maleate for MADH, sodium glutamate for GADH, etc.) of the concentration and pH specified above, 0.1 ml of 0.2 mM NAD$^+$ (NADP$^+$ in the determination of G-6-PDH), 0.1 ml of 0.2 mM resazurin, and 1 ml of diaphorase (0.08 unit) or 1 ml of phenazine methyl sulfate, are placed

[3] G. G. Guilbault and D. N. Kramer, *Anal. Chem.* **36**, 409 (1964).
[4] G. G. Guilbault and D. N. Kramer, *Anal. Chem.* **37**, 120 (1965).
[5] D. N. Kramer and G. G. Guilbault, *Anal. Chem.* **36**, 1662 (1964).

in the fluorescence cell in the SPF, and the instrument is adjusted to read zero. At zero time, 0.1 ml of the solution of the dehydrogenase to be analyzed (containing approximately 0.0001 to 0.50 unit/ml) is then added, and the change in fluorescence with time, $\Delta f/\Delta t$, is automatically recorded at the excitation and emission wavelengths given. From calibration plots of Δf/minute vs concentration, the quantity of the dehydrogenase present in solution is calculated.

There should be no blank in this procedure—i.e., no increase in fluorescence with time without the dehydrogenase added. If such an increase is observed, fresh solutions of the reagents, especially the NAD⁺ and diaphorase, should be prepared.

Results

The results of the determination of lactic, alcohol, malic, glutamic, glucose-6-phosphate, α-glycerol phosphate, and glycerol dehydrogenases, based upon five or more determinations, were as follows: LDH, 0.000300 to 0.100 unit/ml; ADH, 0.000303 to 0.151 unit/ml; MADH, 0.00105 to 0.510 unit/ml GADH, 0.000103 to 0.0330 unit/ml; milliliter G-6-PDH, 0.00202 to 0.340 unit/ml; GPDH, 0.0105 to 1.10 unit/ml; and GDH, 0.00500 to 0.105 unit/ml; these values may be determined with standard deviations of ± 1.1, 0.8, 1.4, 0.9, 1.1, 1.1, and 0.9%, respectively.

Discussion

General. The recent research in enzyme assay has been directed toward two immediate ends, to replace the long and tedious procedures required in most previous assays with simple, rapid measurements, and to use more sensitive procedures in all analyses. These have been our goals in the development of the present procedure for dehydrogenases. The previous indirect, lengthy methods[6,7] have been replaced with the rapid measurement of the initial rate of production of resorufin, $\Delta f/\Delta t$, which is a measure of the concentration of the enzymes present. Since resorufin is a highly fluorescent compound (fluorescence coefficient = 1.56×10^7, compared to 1.40×10^6 for quinine sulfate in 0.1 N H₂SO₄), the method is extremely sensitive, and small concentrations of enzymes may be determined (10^{-4} unit). The fluorescence coefficient is defined as the fluorescence in units observed on the SPF divided by the concentration in moles per liter.

Stability of Reagents. The stock resazurin solution 0.2 mM in methyl Cellosolve, is stable for at least a year, and the NAD⁺ and NADH solu-

[6] O. H. Lowry, N. R. Roberts, and M. Chang, *J. Biol. Chem.* **222**, 97 (1956).
[7] O. H. Lowry, N. R. Roberts, and I. Kapphahn, *J. Biol. Chem.* **224**, 1047 (1957).

tions are stable for a week when stored at 5°. The substrate solutions are stable for months at room temperature. The only unstable reagent is diaphorase, and solutions must be prepared fresh each day. The calibration plots, however, need not be repeated daily, provided they are initially determined with an enzyme solution that was freshly prepared.

Other Determinations

Guilbault *et al.*[8] have shown that the procedure can be applied to the assay of mixtures of organic acids. Rapid fluorometric methods were described using six enzyme systems for the determination of mixtures of 21 organic acids. Lactate (types II and IV), malate, glutamate, isocitrate, and β-hydroxybutyrate dehydrogenases were used, coupled with NAD, phenazine methyl sulfate, and resazurin, in fluorometric procedures for the determination of acetic, adipic, benzilic, butyric, D-α- and D-β-hydroxybutyric, chloroacetic, DL-citric, formic, L-glutamic, glutaric, glycolic, *threo*-D-isocitric, L-lactic, L-malic, malonic, oxalic, phthalic, DL-succinic, and L-tartaric acids in the concentration range 0.1–500 μg with an accuracy and precision of about 2%.

Nakhodkina and Pechenkina[9] suggested resazurin as an indicator of microbiological contamination. Otaka *et al.*[10] used it as a test of the microbiological quality of meat, and Komkov *et al.*[11] assayed for bacteria in milk, using this dye. In all cases, the production of the rose-red color of resorufin served as an indication of living microorganisms, either reductases or dehydrogenases. Similar studies were done by Otskuka and Nakae[12] for determining the quality of raw milk, by Abo-Elnaga[13] in testing for *Pseudomonas* bacteria in milk, and by Sato *et al.*,[14] who determined the activity of *L. fermenti* growth zones on an assay plate.

Green[15] patented a diagnostic test for dental caries using resazurin, which correlated well with a direct determination using oxidation-reduction potentials. When saliva is mixed with resazurin a change from blue to red indicates dental caries.

It is believed that this new procedure will prove to be generally useful for assay of all dehydrogenase and reductase enzyme systems.

[8] G. G. Guilbault, S. H. Sadar, and R. McQueen, *Anal. Chim. Acta* **45**, 1 (1969).

[9] V. Z. Nakhodkina and M. Pechenkina, *Sakh. Prom.* **45**, 30 (1971).

[10] F. Otaka, M. Maruyama, and M. Fujita, *Nippon Shokuhin Kogyo, Gakkai-Shi* **11**, 499 (1964).

[11] I. P. Komov, S. V. Sokolovskaya, G. Kazakova, I. Arkhangel'skii, and M. Yatsuk, *Tr. Vses. Nauch. Issled. Inst. Vet. Sanit.* **24**, 12 (1964).

[12] G. Otskuka and T. Nakae, *J. Dairy Sci.* **52**, 2041 (1969).

[13] I. Abo-Elnaga, *Arch. Lebensmittelhyg.* **18**, 77 (1967).

[14] A. Sato, T. Itagaki, N. Inakoshi, and T. Tsukahana, *Bitamin* **34**, 19 (1966).

[15] G. H. Green, U.S. Patent 3,332,743 (1965).

[12] Quantitative Determination of the Anomerase Activity of Glucosephosphate Isomerase from Baker's Yeast

By Bernd Wurster and Benno Hess

Glucosephosphate isomerase from yeast not only catalyzes the isomerization of D-glucose 6-phosphate to D-fructose 6-phosphate, but also the anomerization of D-glucose 6-phosphate,[1-3] D-glucose 6-sulfate,[4] D-fructose 6-phosphate,[3] and D-mannose 6-phosphate.[5] The anomerase activity toward D-glucose 6-phosphate could also be shown for glucosephosphate isomerase from *Escherichia coli*,[6] *Rhodotorula gracilis*,[6] potato tubers,[6] rat liver,[6] rat kidney,[6] rat muscle,[6] and rabbit muscle.[7]

Assay Method

Principle. Because of the specificity of glucose-6-phosphate dehydrogenase for β-D-glucopyranose 6-phosphate,[1,8-11] the spontaneous and enzyme-catalyzed anomerization of α- to β-D-glucopyranose 6-phosphate can be determined quantitatively, using the following test system[1,2]:

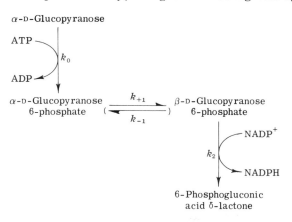

[1] M. Salas, E. Viñuela, and A. Sols, *J. Biol. Chem.* **240**, 561 (1965).
[2] B. Wurster and B. Hess, *Hoppe-Seyler's Z. Physiol. Chem.* **354**, 407 (1973).
[3] K. J. Schray, S. J. Benkovic, P. A. Benkovic, and I. A. Rose, *J. Biol. Chem.* **248**, 2219 (1973).
[4] C. W. Carlson, S. L. Lowe, and F. J. Reithel, *Enzymologia* 33, 192 (1967).
[5] I. A. Rose, E. L. O'Connell, and K. J. Schray, *J. Biol. Chem.* **248**, 2232 (1973).
[6] B. Wurster and B. Hess, *FEBS Lett.* 38, 33 (1973).
[7] O. H. Lowry and J. V. Passonneau, *J. Biol. Chem.* **244**, 910 (1969).
[8] J. M. Bailey, P. H. Fishman, and P. G. Pentchev, *J. Biol. Chem.* 43, 4827 (1968).
[9] J. M. Bailey, P. H. Fishman, and P. G. Pentchev, *Biochemistry* 9, 1189 (1970).

The reactions are initiated by the addition of freshly dissolved
α-D-glucopyranose. In the hexokinase reaction, α-D-glucopyranose 6-phos-
phate is produced with constant reaction velocity (v_{system}); in the following anomerization reaction—spontaneous and catalyzed by glucosephosphate isomerase—α-D-glucopyranose 6-phosphate is converted to
β-D-glucopyranose 6-phosphate, and finally β-D-glucopyranose 6-phosphate
is oxidized in the glucose-6-phosphate dehydrogenase reaction. If the
maximal velocity of the glucose-6-phosphate dehydrogenase reaction
($V_{glucose-6-phosphate\ dehydrogenase}$) is much higher than the velocity of the hexokinase reaction ($v_{system} = k_0$) and, in addition, the reaction velocity constant of the glucose-6-phosphate dehydrogenase reaction

$$k_2 = (V_{glucose-6-phosphate\ dehydrogenase})/(K_{m(\beta-D-glucopyranose\ 6-phosphate)})$$

is much larger than the reaction velocity constant of the anomerization
reaction, k_{+1}, the steady state concentration of β-D-glucopyranose
6-phosphate approaches zero, and therefore the backward reaction
$k_{-1} \times$ [β-D-glucopyranose 6-phosphate] can be neglected. At concentrations of α-D-glucose, ATP, and NADP$^+$ saturating hexokinase and glu-
cose-6-phosphate dehydrogenase, respectively, the hexokinase reaction is
zero order, and the anomerization reaction and the glucose-6-phosphate
dehydrogenase reaction are first order. Under steady-state conditions the
velocity of the irreversible reaction sequence is given by:

$$v_{system} = k_0 = k_{+1} \times [\alpha\text{-D-glucopyranose 6-phosphate}]$$
$$= k_2 \times [\beta\text{-D-glucopyranose 6-phosphate}] = \frac{d[\text{NADPH}]}{dt} = \text{constant}$$

From the steady-state reaction velocity of the generation of NADPH,
v_{system} can be obtained, and quenching the reactions after reaching steady
state, the steady-state concentrations of α-D-glucopyranose 6-phosphate
can be determined; then, k_{+1} can be computed from v_{system} and [α-D-gluco-
pyranose 6-phosphate].[2] In the presence of glucosephosphate isomerase,
the isomerization of D-glucose 6-phosphate to D-fructose 6-phosphate does
not influence the anomerization of α- to β-D-glucopyranose 6-phosphate.
In the steady state, the concentration of D-fructose 6-phosphate is
constant.[2]

   *Calculation and Definition of the Anomerase Activity Constant of Glucose-
phosphate Isomerase.* The velocity constant of the anomerization reaction
k_{+1} is the sum of the velocity constant for the spontaneous anomerization reaction $k_{+1(s)}$ and the velocity constant (activity constant) for the

[10] S. P. Colowick and E. B. Goldberg, *Bull. Res. Counc. Isr. Sect. A* **11**, 373 (1963).
[11] J. E. Smith and E. Beutler, *Proc. Soc. Exp. Biol. Med.* **122**, 671 (1966).

glucosephosphate isomerase-catalyzed anomerization reaction $k_{+1(e)}$, $k_{+1} = k_{+1(s)} + k_{+1(e)}$; $k_{+1(e)}$ is obtained by subtraction of $k_{+1(s)}$ from k_{+1}. Under the experimental conditions described below a value of $k_{+1(s)} = 3.8 \pm 0.2$ min^{-1} is obtained.[2] With increasing isomerase activity (concentration) of glucosephosphate isomerase, a good proportionality between the isomerase activity of glucosephosphate isomerase and the anomerase activity constant of glucosephosphate isomerase $k_{+1(e)}$ is found.[2]

Under the experimental conditions described below for the enzyme from baker's yeast an isomerase activity of 1 unit/ml corresponds to an anomerase activity constant of glucosephosphate isomerase, expressed as pseudo-first order reaction velocity constant, of 1.1 ± 0.1 min^{-1}.[2]

Reagents

Buffer: 50 mM imidazole-HCl, 50 mM KCl, 8 mM MgSO$_4$, pH 7.6
The following reagents are dissolved in this buffer:
α-D-Glucose, 75 mM
ATP, 100 mM
NADP$^+$, 100 mM
Hexokinase from yeast, 1.5–5 units/ml
Glucose-6-phosphate dehydrogenase from yeast, 1000 units/ml
Glucosephosphate isomerase from yeast, 125–500 units/ml; enzymes
  were freed from ammonium sulfate by dialysis.
HClO$_4$, 4 N
KOH, 4 N

Procedure. The reaction mixture (final volume 1 ml), containing 0.88 ml of buffer, 0.02 ml of ATP solution, 0.02 ml of NADP$^+$ solution, 0.02 ml of hexokinase, 0.02 ml of glucosephosphate isomerase, and 0.02 ml of glucose-6-phosphate dehydrogenase, is incubated for 10 min in a cuvette holder (connected to a thermostat) at 25° ($\pm$0.2°). The reaction is initiated by addition of 0.02 ml of α-D-glucose (freshly dissolved in ice-cold buffer). In the steady state, after a reaction time of 1 min,[2] the reaction mixture of 1 ml volume is acid quenched by the addition of 0.2 ml of 4 N HClO$_4$. After neutralization with 4 N KOH, the steady-state concentration of α-D-glucopyranose 6-phosphate is determined. In analogous experiments without glucosephosphate isomerase the velocity constant for the spontaneous anomerization reaction $k_{+1(s)}$ can be determined.

The isomerase activity of glucosephosphate isomerase is determined under identical conditions of buffer and temperature at 2 mM D-fructose 6-phosphate, 2 mM NADP$^+$, and 10 units of glucose-6-phosphate dehydrogenase per milliliter.[2]

Applicability of the Test System

Using the test system described above, the glucosephosphate isomerase-catalyzed anomerization of α- to β-D-glucopyranose 6-phosphate can be determined only quantitatively in preparations of glucosephosphate isomerase which contain very little activity of the enzymes (1) hexokinase, which is used to produce a constant reaction velocity of the test system; (2) ATPase, which causes hydrolysis of the substrate ATP during the incubation period; (3) aldose 1-epimerase, which gives rise to anomerization of α- to β-D-glucopyranose; (4) 6-phosphogluconolactonase, which accelerates the hydrolysis of 6-phosphogluconic acid δ-lactone to 6-phosphogluconate, which inhibits glucosephosphate isomerase[12]; (5) 6-phosphogluconate dehydrogenase, which causes a consecutive reduction of $NADP^+$; (6) fructose-6-phosphate kinase, which traps D-fructose 6-phosphate; and (7) glucose-6-phosphate 1-epimerase.[6,13,14]

The experiments are carried out at pH 7.6 and 25°. Under these conditions the apparent equilibrium constant of the glucose-6-phosphate dehydrogenase reaction is $K_{app} = 20 \pm 4$, with respect to β-D-glucopyranose 6-phosphate $K'_{app} \approx 32$, and the reaction velocity constant for the hydrolysis of 6-phosphogluconic acid δ-lactone is $k_h = 0.07$ min^{-1}.[2] The quasi-irreversible treatment of the glucose-6-phosphate dehydrogenase reaction is justified considering the following points: High $NADP^+$ concentration drives the reaction in the direction of the products. Furthermore, the experiments have to be carried out at low reaction velocity of the overall reaction system (v_{system}) but still provide for measurable concentrations of α-D-glucopyranose 6-phosphate in the steady state. With smaller v_{system}, the steady state is attained after a smaller turnover of D-glucose 6-phosphate and $NADP^+$; therefore, only a slight reverse reaction of 6-phosphogluconic acid δ-lactone and NADPH should occur. After a reaction time of $3 \times (1/k_{+1})$, the steady state is reached within approximately 5%. Thus, at $k_{+1(s)} = 3.8$ min^{-1}, the transient would be over after about 0.8 min. For these reasons, experimental reaction times longer than 1 min were avoided.

Within the reaction time of 1 min, the spontaneous hydrolysis of 6-phosphogluconic acid δ-lactone to 6-phosphogluconate and a subsequent inhibition of glucose-phosphate isomerase[12] can be neglected. The spontaneous anomerization of α-D-glucopyranose to β-D-glucopyranose and subsequent phosphorylation of β-D-glucopyranose is also negligibly small;

---

[12] C. W. Parr, *Nature (London)* **178**, 1401 (1956).

[13] B. Wurster and B. Hess, *FEBS Lett.* **23**, 341 (1972).

[14] B. Wurster and B. Hess, this volume [102].

glucosephosphate isomerase does not catalyze the anomerization of α- to β-D-glucopyranose.[2]

[13] Estimation of Fructokinase (Ketohexokinase) in Crude Tissue Preparations

By MILTON M. WEISER and HELEN QUILL

Fructose + ATP → fructose 1-P + ADP

Assay Method

Principle. Fructokinase activity is measured by estimating the appearance of the specific product, fructose-1-P. This is difficult to do in crude tissue preparations if the contribution of the hexokinases to the phosphorylation of fructose (fructose-6-P) is not nullified. The assay utilizes two properties of hexokinases which differentiate them from fructokinase: (1) hexokinase activity is more labile to [H+] than is fructokinase, and (2) hexokinase activity is markedly inhibited by N-acetyl-D-glucosamine while fructokinase is relatively unaffected.[1]

Reagents

HCl, 5 N
KOH, 1 N
Tris·HCl, 1.0 M, pH 7.4 (50 mM)[2]
MgCl$_2$, 0.1 M (10 mM)
ATP, 0.1 M (10 mM)
D-[UL-^{14}C]Fructose, 0.1 M (usually 200–300 cpm/nmole) (10 mM)
N-Acetyl-D-glucosamine, 1.0 M (50 mM)

The crude tissue extract may be prepared in a number of different buffer solutions. Since half the assay volume is the tissue extract as enzyme source, the tissue extract buffer will contribute significantly to the final concentrations of ions and buffers used. We have used the following procedure for initial crude tissue preparation. Intestinal mucosal scrapings or minced liver fragments are suspended in 2.5–10 volumes (v:w) of 25 mM potassium phosphate buffer, pH 7.4, containing 5 mM EDTA

---

[1] M. M. Weiser and H. Quill, *Anal. Biochem.* **43**, 275 (1971).
[2] Final concentrations to be achieved in the assay mixture are enclosed in parentheses.

and 0.5 mM dithiothreitol. After homogenization with a Potter-Elvehjem motor-driven homogenizer, the material is centrifuged for 1 hr at 105,000 g. An aliquot of the 105,000 g supernatant is then brought to pH 6.1 with 5 N HCl. Care must be taken not to go below pH 6.0 since this will result in a loss of fructokinase activity. This acid treatment destroys 50–60% of hexokinase activity with no loss of fructokinase activity.[1] After acidification the resultant precipitate is removed by centrifugation (20,000 g, 20 min), and the supernatant is readjusted to pH 7.4 with 1 N KOH.

Procedure. The reaction mixture consists of this acid-treated supernatant as enzyme source. A 0.1 ml of enzyme preparation is added to a small test tube (75 mm $\times$ 12 mm OD) placed in ice. The total reaction volume of 0.2 ml is completed by adding 0.02 ml of MgCl$_2$, 0.01 ml of Tris·HCl, 0.02 ml of ATP, 0.01 ml of N-acetyl-D-glucosamine, 0.02 ml of D-[UL-^{14}C]fructose, and an appropriate volume of water. The tubes are then placed in a 37° water bath and incubated for 30 min. The reaction is stopped by heating at 100° for 2 min with tubes covered by a marble or glass teardrop. The precipitate is then removed by centrifugation and 30 μl of the supernatant applied to Whatman 3 MM filter paper (57 cm long). High voltage paper electrophoresis is performed at 3000 to 4500 V for 20–40 min using a pyridine/acetate buffer, pH 6.5.[3] Migration is checked with appropriate standards (fructose and fructose-1-P or any phosphorylated monosaccharide) using a AgNO$_3$[4] or periodate-benzidine method.[5] Migration can also be checked by radiochromatogram scanning. Fructose stays at the origin and the phosphorylated derivatives move 15–25 cm. Appropriate areas are then cut from the paper strip; this should include areas of substrate as well as product. The cut strips are then counted in a liquid scintillation counter using a toluene-based scintillation system (e.g., Liquifluor, New England Nuclear Corp. or equivalent toluene-PPO-POPOP system).

Alternate Method of Measuring Product. Aliquots (20–30 μl) of the supernatant from the heat-treated reaction mixture are spotted on DEAE-impregnated paper disks (Whatman DE-81) and treated according to the method of Breitman.[6] The paper disks are washed twice in 1.0 mM ammonium formate, once in distilled water and once in absolute ethanol. After drying, the disks are placed in a liquid scintillation vial, covered with 10–15 ml of the toluene-based scintillation system and counted in a liquid scintillation counter.

Units. The detected radioactivity of the product, [UL-^{14}C]fructose

[3] M. M. Weiser, and K. J. Isselbacher, *Biochim. Biophys. Acta.* **208**, 349 (1970).
[4] See R. W. Wheat, this series, Vol. 8 [4].
[5] H. T. Gordon, W. Thornburg, and L. N. Werum, *Anal. Chem.* **28**, 849 (1956).
[6] T. R. Breitman, *Biochim. Biophys. Acta* **67**, 153 (1963).

1-phosphate is converted to nanomoles on the basis of the specific activity of the [UL-^{14}C]fructose used. Appropriate expression of activity would then be nanomoles per milliliter per minute or nanomoles per milligram of protein per minute.

Comment. The usual fructokinase assays do not directly measure the specific product fructose 1-phosphate,[7,8] and appear to be accurate for fructokinase activity only when the enzyme is in a semipurified state.[9] In a crude homogenate or in a crude 105,000 *g* supernatant, any fructokinase assay is inaccurate owing to the activity of nonspecific hexokinases which may contribute 30% of the total fructose phosphorylating activity depending on physiological factors such as diet.[10–12] The assay for fructokinase described here appears to overcome this problem.

[7] See H. G. Hers, this series, Vol. 1 [34].
[8] R. C. Adelman, P. D. Spolter, and S. J. Weinhouse, *J. Biol. Chem.* **241**, 5467 (1966).
[9] R. C. Adelman, F. J. Ballard, and S. J. Weinhouse, *J. Biol. Chem.* **242**, 3360 (1967).
[10] M. M. Weiser, H. Quill, and Isselbacher, K. J. *J. Biol. Chem.* **246**, 2331 (1971).
[11] M. M. Weiser, H. Quill, and Isselbacher, K. J. *Amer. J. Physiol.* **221**, 844 (1971).
[12] R. J. Grand and S. Jaksina, *Gastroenterology* **64**, 429 (1973).

[14] Assay for D-Ribose-5-phosphate Ketol Isomerase and D-Ribulose-5-phosphate 3-Epimerase

By TERRY WOOD

$$\text{D-Ribose 5-phosphate} \xrightleftharpoons[\text{isomerase}]{} \text{D-ribulose 5-phosphate}$$

$$\xrightleftharpoons[\text{epimerase}]{} \text{D-xylulose 5-phosphate} \quad (1)$$

The assay for the isomerase is based upon the increase in absorbance observed at 290 nm when ribose-5-phosphate ketol isomerase acts upon ribose 5-phosphate to form ribulose 5-phosphate.[1,2] When the reaction has reached equilibrium, the addition of epimerase gives rise to a further increase in absorbance as xylulose 5-phosphate is formed.[2] In the presence of an excess of the isomerase, the second rate of absorbance increase is proportional to the activity of the epimerase.

Reagents

Triethanolamine·HCl buffer, 100 m*M*, pH 7.4
D-Ribose 5-phosphate, disodium salt, 100 m*M*. Store at 2°
D-Ribose-5-phosphate ketol isomerase, 200 units/ml, 2.7 mg/ml

[1] F. C. Knowles and N. G. Pon, *J. Amer. Chem. Soc.* **90**, 6536 (1968).
[2] T. Wood, *Anal. Biochem.* **33**, 297 (1970).

Assay of Ribose-5-phosphate Ketol Isomerase

The reaction mixture contains 1.78 ml 50 mM triethanolamine·HCl buffer and 0.20 ml 100 mM ribose 5-phosphate (20 μmoles). It is placed in the cell compartment at 37° of a double-beam spectrophotometer connected to a recorder and read against a control cuvette containing an equal volume of buffer. When temperature equilibrium is reached, 0.02 ml of a suitable dilution of isomerase is added to both cuvettes, and the increase in absorbance at 290 nm is recorded. The initial velocity is linear over at least 5 min for up to 0.2 unit of isomerase.

Assay of Ribulose-5-phosphate 3-Epimerase

The assay mixture is set up as before, and 0.01 ml (2 units) of the isomerase is added. The absorbance increases rapidly at first and then levels off after 12–15 min. If small amounts of epimerase are present in the isomerase the recorder trace will show a slight upward slope. A suitable dilution of epimerase in 0.01 ml is added, and the further increase in absorbance at 290 nm is recorded (Fig. 1). The initial velocity is linear over 5 min for up to 0.2 unit of epimerase.

Calculation of Isomerase and Epimerase Activity

Careful determination of the absorption coefficient for the conversion of ribose 5-phosphate to ribulose 5-phosphate gave a value of 72.[2] Ribulose 5-phosphate and xylulose 5-phosphate differ only in their configuration about carbon-3, and addition of epimerase does not change the absorbance of pure ribulose 5-phosphate. Consequently, the same absorption coefficient is used in calculating the activity of both enzymes.

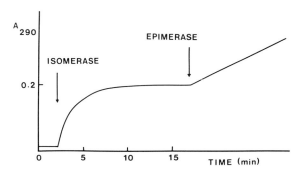

FIG. 1. Assay of D-ribulose-5-phosphate 3-epimerase.

If y = initial velocity in absorbance units per minute, then for a 2.00-ml volume:

$$\text{Rate in micromoles/min} = y \times 2.00/0.072$$

Notes

1. The commercially available sodium salt can give an absorbance less than 0.04 at 290 nm. Ketopentose phosphates in the ribose 5-phosphate may be destroyed by treatment with alkali if desired.[3]

2. The commercial spinach enzyme (75 units/mg) is very suitable. Once prepared, the solution may be stored at $-20°$ and thawed when required. Repeated freezing and thawing does not appear to affect the enzyme and may be beneficial by inactivating traces of contaminating ribulose-5-phosphate 3-epimerase. The yeast enzyme (230 units/mg) contains less epimerase impurity but is inactivated by repeated freezing and thawing.

3. The chromophore has its absorption peak at 278–280 nm. A wavelength of 290 nm is used to reduce interference from proteins and nucleic acids. The exact structure of the absorbing chromophore is not certain, but the absorption coefficient is in the range of values for simple straight-chain ketose phosphates.[4] After approximately 65 min at 37° in 50 mM triethanolamine·HCl buffer, pH 7.4, the absorbance of the isomerase product begins to increase spontaneously and nearly doubles over the next 4 hr. This increase occurs even after ultrafiltration and is independent of the presence of enzyme.[5] For this reason the addition of epimerase after equilibrium has been reached should not be long delayed. In 50 mM sodium phosphate buffer the absorbance began to increase spontaneously after only 20 min, and it is recommended that the use of this buffer be avoided, although it is employed in the isomerase assay at 280 nm described by Knowles *et al.*[6]

4. The validity of the isomerase assay has been checked against the phloroglucinol[2] and cysteine-carbazole[6] colorimetric methods and that of the assay for epimerase against the spectrophotometric method at 340 nm.[2]

5. A number of substances such as EDTA, EGTA, 8-hydroxyquinoline, dithiothreitol, and mercaptoethanol, interfere by reacting with the product of the isomerase reaction to give chromophores of greater absorb-

[3] F. Dickens and D. H. Williamson, *Biochem. J.* **64**, 567 (1956).
[4] G. R. Gray and R. Barker, *Biochemistry* **9**, 2454 (1970).
[5] T. Wood, unpublished results, 1972.
[6] F. C. Knowles, M. K. Pon, and N. G. Pon, *Anal. Biochem.* **29**, 40 (1969).

ance at 290 nm. Thus, 10 μl of 1 mM EDTA can mimic the reaction of 0.15 unit of epimerase. Consequently, the absence of artifacts should be checked by testing, in the assay, the buffers used to dissolve the enzymes.

6. The method for isomerase gave values close to those obtained by the 340 nm assay with undialyzed extracts of rat muscle, heart, kidney, intestine, brain, lung, spleen, and uterus, but spuriously high values were obtained with extracts of rat liver and rat blood hemolysates.[5]

Applications

The isomerase assay has been used to follow the purification of the enzyme from spinach[7] and other sources,[8] in the determination of Michaelis constants,[7,8] and for the assay of isomerase activity in animal tissues.[8,9] The epimerase assay has been used to follow the purification of the enzyme and determine its Michaelis constant.[8]

[7] F. C. Knowles and N. G. Pon, *Anal. Biochem.* **24**, 305 (1968).
[8] M. E. Kiely, A. L. Stuart, and T. Wood, *Biochim. Biophys. Acta* **293**, 534 (1973).
[9] F. C. Kauffman, *J. Neurochem.* **19**, 1 (1972).

[15] Fructose-diphosphate Aldolase, Pyruvate Kinase, and Pyridine Nucleotide-Linked Activities after Electrophoresis

By WALTER A. SUSOR, EDWARD PENHOET, and WILLIAM J. RUTTER

A number of methods have been used to localize specific enzyme activities after electrophoresis. The widely used method of choice of most enzymes that can be coupled to the reduction of NAD or NADP is the coupled reduction of a tetrazolium dye to form an insoluble pigment. This method as applied to the detection of fructose-diphosphate aldolase[1] is described in detail below.

There are in addition several enzymes which are more readily coupled to the oxidation of NADH or NADPH. These can be assayed by the disappearance of fluorescence or UV absorbance of the reduced nucleotides. The second method presented here was designed for the detection of pyruvate kinase,[2] an enzyme that can be coupled to the oxidation of

[1] E. Penhoet, T. Rajkumar, and W. J. Rutter, *Proc. Nat. Acad. Sci. U.S.* **56**, 1275 (1966).
[2] W. A. Susor and W. J. Rutter, *Anal. Biochem.* **43**, 147 (1971).

NADH. The method can also be applied to the assay of aldolase and other enzyme systems which result in a net oxidation of NADH or NADPH to NAD or NADP.

Detection of Aldolase by Tetrazolium Dye Reduction

Proteins in the sample with differing electrophoretic mobilities are resolved by electrophoresis on cellulose polyacetate strips. Bands of aldolase activities may then be detected by placing the strip in contact with an agar film containing the assay reagents. The aldolase on the strip comes in contact with its substrate fructose 1,6-diphosphate and catalyzes its cleavage to 3-phosphoglyceraldehyde and dihydroxyacetone phosphate. In the presence of arsenate, 3-phosphoglyceraldehyde is then oxidized to 1-arseno-3-phosphoglycerate by glyceraldehyde-3-phosphate dehydrogenase with the concomitant reduction of NAD to NADH. The NADH produced then reduces oxidized phenazine methosulfate; this in turn reduces oxidized nitroblue tetrazolium to the reduced, highly colored blue state. Thus, wherever aldolase is present on the strip, a blue band appears in the adjacent agar.

Electrophoresis of Aldolase. The aldolase sample (tissue extract or purified enzyme) is prepared as described by Penhoet and Rutter.[3] Electrophoresis is carried out on 1 × 6 inch cellulose polyacetate strips (Gelman Sepraphore III or equivalent) in a chamber having about 15 cm between the anode and cathode buffers (Gelman No. 51101). The electrophoresis buffer consists of 60 mM barbital-sodium, 10 mM 2-mercaptoethanol, pH 8.6; it is prepared by dissolving 10.8 g of sodium barbital and 1.5 g of barbituric acid to 1000 ml with distilled water, cooling to 4°, and adding 0.7 ml of neat 2-mercaptoethanol immediately prior to use.

The cellulose polyacetate strips are "activated" prior to use by floating the dry strips on the surface of the buffer until they are thoroughly wetted and sink into the buffer. After about 20 min, the strip is placed on a piece of white photo blotter, and excess buffer is allowed to drain into the blotter. About 3 μl of sample is loaded in a band midway between the ends of the paper with a sample applicator similar to Gelman No. 51220. The sample application site can be conveniently marked by gently touching a freshly sharpened No. 1 lead pencil to the edge of the strip. The strip is placed in the chamber with the sample band midway between the electrodes and electrophoresis carried out for 90 min at 250 V, constant voltage, at 4°.

---

[3] This series, Vol. 42 [38].

Enzyme Detection. Bands of aldolase activity are detected by placing the strips containing the electrophoresed sample on the surface of an agar plate. The assay plates are prepared by dissolving 125 mg of Noble agar in 20 ml of 50 mM tris(hydroxymethyl)aminomethane, 10 mM sodium arsenate, pH 8.0, in a boiling water bath. The solution is immediately cooled to 42–45° in a 42° water bath, and the following ingredients are rapidly mixed in: 2 ml of 100 mM sodium fructose 1,6-diphosphate, 1 ml of 10-mg/ml nitroblue tetrazolium chloride, 2 ml of 15-mg/ml NAD, 0.6 ml of 1-mg/ml phenazine methosulfate, and 0.1 ml of 10-mg/ml glyceraldehyde-3-phosphate dehydrogenase. The phenazine methosulfate is light sensitive; it should be made up fresh and kept in the dark. After adding the last ingredient, pour the mixture into five or six 100-mm plastic petri plates and place in a refrigerator as they solidify. The plates may be stored in the dark for up to 24 hr before use. After electrophoresis, the strips are removed from the chamber with a forceps and placed on the agar surface. The plates are then incubated at 37° for 10–15 min for color development. Record the results by photographing the inverted plates.

The principal artifacts observed with this method come from particle-bound activities, which do not move from the origin, and alcohol dehydrogenase, which uses the alcohol of crystallization of the nitroblue tetrazolium as substrate. Particulate activities can be removed by adequate centrifugation. Alcohol dehydrogenase bands can be detected on a control plate prepared without fructose diphosphate.

Detection by UV Absorbance

Pyruvate kinase activity is detected by NADH oxidation produced by coupling the enzyme with lactate dehydrogenase. Aldolase may be detected either by NADH oxidation or NAD reduction by coupling with triosephosphate isomerase and α-glycerophosphate dehydrogenase or glyceraldehyde-3-phosphate dehydrogenase, respectively. The enzymes are first resolved by electrophoresis on cellulose polyacetate strips, then the strips are applied to a thin agar film containing the assay reagents. The change in the oxidation state of the pyridine nucleotide substrate is followed visually by observing changes in fluorescence excited by 340 nm light, and the result is recorded by contact printing on photographic paper. Reduced pyridine nucleotides, which absorb at 340 nm and fluoresce yellow, appear as light areas on the photographic paper, while oxidized pyridine nucleotides transmit the light, exposing the photographic paper to produce darkened areas.

Electrophoresis of Pyruvate Kinase. The pyruvate kinase sample is prepared from rat tissue extract in 0.25 M sucrose, 25 mM tris(hydroxymethyl)aminomethane (Tris·Cl), pH 7.5, 2.5 mM EDTA, and 0.1 M dithiothreitol and assayed spectrophotometrically for enzyme activity[4] in the presence of 1 mM 2-phosphoenolpyruvic acid, tricyclohexylammonium salt (PEP), 1 mM NADP, and 0.2 mM fructose 1,6-disphosphate (FDP). The sample should be diluted to 0.5–5 enzyme units/ml with 0.5 M sucrose, 20 mM Tris·Cl (pH 7.5), 0.1 M 2-mercaptoethanol, and 1 mg of fraction V bovine serum albumin per milliliter. An enzyme unit is defined as that amount which will oxidize 1 μmole of NADH per minute at 25° in the spectrophotometric assay.

Cellulose polyacetate strips (Sephraphore III, Gelman Instrument Co., work well) are "activated" prior to use by soaking for about 30 min in electrophoresis buffer containing 1 mg of fraction V bovine serum albumin per milliliter. A strip is then placed on a piece of blotter paper (white photo blotter serves well), and 3 μl of the sample is applied in a band midway between the ends of the strip with a sample applicator similar to Gelman No. 51220. Electrophoresis is carried out at 4° for 3 hr at 250 V (17 V/cm) in a buffer containing 0.5 M sucrose, 20 mM Tris·HCl (pH 7.5), 0.1 mM FDP and 0.1 M 2-mercaptoethanol. FDP is added to effect separation of pyruvate kinase A and C, which have nearly identical mobilities in its absence.[5] In the presence of FDP, pyruvate kinase C migrates farther toward the cathode while the mobility of pyruvate kinase A is essentially unchanged; 0.1 mM FDP is sufficient to give good separation with rat tissue, although higher concentrations may occasionally be indicated (1 mM FDP gives better resolution with rabbit extracts). Sucrose and 2-mercaptoethanol are required to stabilize the enzyme during electrophoresis. Serum albumin is added to the activation buffer to reduce tailing and to permit electrophoresis of purified pyruvate kinase since pyruvate kinase adheres tightly to the cellulose polyacetate strip in the absence of other proteins.

Preparation of Assay Plates. The position of the enzyme activity after electrophoresis is determined by placing the strip in contact with an agar film containing the components of the spectrophotometric assay medium. Glass lantern slide covers (3.25 × 4 inches) are used as (1) support and (2) as cover glass to press the agar into a thin film. The support glasses for the agar film are prepared by immersing the slide covers for 2 min in a fresh silicone (Siliclad, Clay-Adams, Inc.) solution diluted 1 part to 50 parts distilled water. They are then rinsed with distilled water, and dried at 100°. These Siliclad-treated glasses are wrapped with 2 lb test nylon monofilament fishing line, dividing the plates into three sections. The nylon line serves as a spacer for the agar film and allows the assay of three 1-inch cellulose acetate strips per plate. The cover-glass

[4] T. Bücher and G. Pfleiderer, this series, Vol. 1, p. 435.
[5] W. A. Susor, *Fed. Proc., Fed. Amer. Soc. Exp. Biol.* 29, 729 (1970).

plates are left untreated. The treated and wrapped support glasses and untreated cover glasses are warmed to 45° a short time prior to use.

The assay medium for pyruvate kinase consists of a solution of 9 mg of Noble agar (Difco) per milliliter, 50 mM Tris·Cl (pH 7.5), 10 mM MgCl$_2$, 60 mM KCl, 5 mM EDTA, 2 mM PEP, 2 mM ADP, 1 mM NADH, 0.2 μM FDP and 30 units of pig heart lactate dehydrogenase per milliliter (Sigma, type VI). The agar is dissolved in distilled water to one-half of the final volume, in a boiling water bath, then cooled to 45°. The remaining regents, at 40°, are then added to the agar solution, mixed, and immediately applied to the support glass. About 2 ml of this assay mixture is pipetted onto the surface of a wrapped silicone-treated glass, and an untreated glass cover is carefully placed over the still-fluid agar solution, to form a silicone-treated glass–agar–glass sandwich. The solidified agar solution will adhere more firmly to the silicone-treated glass than to the untreated cover, allowing the latter to be removed, exposing a thin, uniform agar surface. Completed plates can be kept for several days in a humid chamber at 4°. The upper cover glass is removed immediately prior to use.

Enzyme Detection. After electrophoresis, the cellulose acetate strips are placed on the surface of the agar-reagent assay plate and covered with a sheet of plastic film, such as Handi-Wrap (Dow Chemical Co., Inc.). (When Handi-Wrap is used it should be freshly removed from the spool since the film slowly oxidizes on exposure to air to form an UV-absorbing material which interferes with the assay.)

The course of the reaction is followed visually by observing the decrease (or increase) in fluorescence of the pyridine nucleotide on illumination with 340-nm light. After distinct bands have appeared (1–5 min), they are recorded by contact-printing on photographic enlarging paper (Kodabromide F-5, Eastman Kodak Company), with the Handi-Wrap side of the assay plate in contact with the enlarging paper. Exposure is made with a long-wave UV lamp of the hand-held type (Blak-Ray UVL-22) fitted with a 340-nm broad band-pass filter (Corning 7-60). The lamp should be secured about 30 cm above the sensitized paper on a mount having a 5 mm diameter hole between the bulb and the sensitized paper. Under these conditions, about 15-sec exposure is required. Figure 1 is a sketch showing the physical relationship of the components of the assay system.

Aldolase and Other Enzymes

Aldolase may be detected by the above method, both by coupling to NAD reduction through glyceraldehyde-3-phosphate dehydrogenase

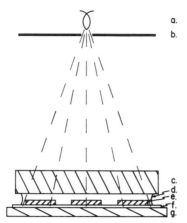

Fig. 1. Cross sections showing arrangement of assay components: (a) 340 nm light source, (b) dark slide with 5 mm diameter opening, (c) silicone-treated lantern slide glass, (d) reagents in agar film, (e) electrophoresis strip, (f) plastic film, (g) enlarging paper. The relative dimensions of the drawing are exaggerated.

(NAD method) and by coupling to NADH oxidation through triose-phosphate isomerase and α-glycerophosphate dehydrogenase (NADH method). The detection method for NAD and NADH consists of an agar film prepared as described above for pyruvate kinase. The detection medium for the NAD method contains 9 mg of Noble agar per milliliter 50 mM Tris·Cl (pH 7.8), 10 mM NaAsO$_4$ (pH 7.8), 5 mM NaFDP, 1 mM NAD, and 0.1 mg of glyceraldehyde-3-phosphate dehydrogenase per milliliter. The detection medium for the NADH method contains 9 mg of Noble agar per milliliter, 50 mM Tris·Cl (pH 7.5), 2.5 mM NaFDP, 1 mM NADH, and 0.1 mg of α-glycerophosphate dehydroge-nase–triosephosphate dehydrogenase per milliliter (Calbiochem).

The resolution and detection of aldolase A-C hybrid set of enzymes by the three methods is shown in Fig. 2. The resolution and sensitivity of these methods are similar. The NBT method retains the advantage, however, that the colored formazan product is a precipitate and does not diffuse; thus the plates can be read several days later. The plates produced by the NADH and NAD methods, on the other hand, must be read and recorded within a few minutes since the products remain in solution and diffuse rather rapidly.

Resolution of closely spaced bands can be further enhanced, or the identification of members of a hybrid set in the absence of a complete set simplified, by the simultaneous application of the test sample to half of the electrophoresis strip and a control sample to the other half. The

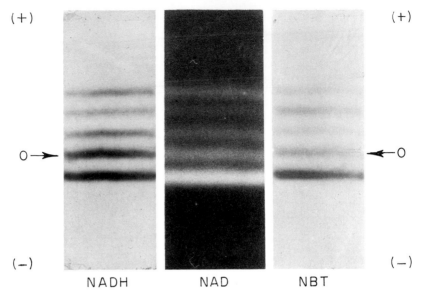

FIG. 2. Electropherograms of rat brain aldolase. NADH: Aldolase detected by coupling with triosephosphate isomerase and α-glycerophosphate dehydrogenase to NADH oxidation. NAD: Aldolase detected by coupling with glyceraldehyde-3-phosphate dehydrogenase to NAD reduction. NBT: Aldolase detected by coupling with glyceraldehyde-3-phosphate dehydrogenase to NAD reduction, which is in turn coupled with phenazine methosulfate to nitroblue tetrazolium chloride (NBT) reduction. Adult rat brain tissue was homogenized in 2 vol of 0.25 M sucrose, 25 mM Tris·Cl, and 2.5 mM EDTA, pH 7.5, and centrifuged at 100,000 g for 30 min. NADH, NAD, and NBT samples consisted of 1.2×10^{-3}, 2.5×10^{-3} and 2.5×10^{-3} unit of aldolase activity, respectively. Note that one equivalent of reduced nucleotide is produced by the NAD and NBT assays, and two equivalents of oxidized nucleotide are produced by the NADH method.

two samples are placed on opposite halves of the sample applicator and are allowed to mix in the center of the applicator. The partial mixing provides a control for interfering components, for example, high salt or protein interactions in the homogenate which may affect the mobilities. A sample of this technique is shown in Fig. 3.

The method for detection of pyruvate kinase described here is adapted from the spectrophotometric assay of Bücher and Pfleiderer.[4] A method which is similar in principle was also developed by von Fellenberg, Richterich, and Aebi[6] for the detection of pyruvate kinase after agar-gel electrophoresis. The latter involves flooding the surface of the electrophoresis gel with the assay reagents dissolved in agar. Bockelmann *et*

[6] R. von Fellenberg, R. Richterich, and H. Aebi, *Enzymol. Biol. Clin.* 3, 240 (1963).

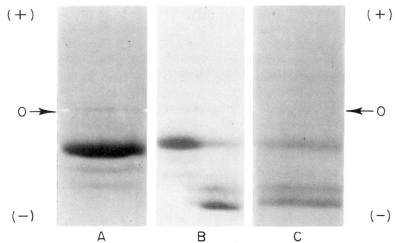

Fig. 3. Electropherograms of 13- and 21-day fetal rat liver aldolase: (A) 13-day fetal rat liver, (B) 13- and 21-day samples simultaneously applied to opposite halves of the strip, (C) 21-day fetal rat liver. Samples A and C contained 4.7×10^{-3} and 6.4×10^{-3} unit of aldolase activity, respectively. Although three bands are seen, in both strips A and C the nonidentity of the more cathode bands is clearly demonstrated by strip B.

al.[7] were able to detect pyruvate kinase activity by coupling through hexokinase and glucose-6-phosphate dehydrogenase to NAD reduction, which was in turn coupled to the formation of an insoluble blue formazan from tetrazolium bromide. Adenylate kinase, however, interferes, requiring additional controls. Alcohol dehydrogenase also interferes with the tetrazolium method, since the reagent contains alcohol of crystallization.[8]

The many enzymes whose reactions involve a pyridine nucleotide cofactor or whose products can be coupled to a pyridine nucleotide oxidation or reduction can also be detected by the incorporation of the specific assay mixture into the agar film, adjusting the pyridine nucleotide to 1 mM and adding sufficient substrate to allow complete oxidation or reduction of the pyridine nucleotide. The customary controls should be performed, such as testing the reactions on an assay plate prepared without substrate.

[7] W. Bockelmann, U. Wolf, and H. Ritter, *Humangenetik* **6**, 78 (1968).
[8] C. R. Shaw and A. C. Koen, *Science* **156**, 1516 (1967).

Section III

Preparation of Substrates

[16] Sedoheptulose 1,7-Biphosphate

By O. TSOLAS

Various methods for the preparation of sedoheptulose 1,7-biphosphate (SBP) have been published. They depend on the enzymic conversion of precursor sugar phosphates, catalyzed by aldolase and transketolase or transaldolase. One such method involves the preparation of SBP from fructose biphosphate by the following series of reactions[1]:

$$\text{Fructose 1,6-biphosphate} \xrightleftharpoons{\text{aldolase}} \begin{array}{c} \text{glyceraldehyde 3-phosphate} \\ + \\ \text{dihydroxyacetone phosphate} \end{array} \qquad (1)$$

$$\begin{array}{c} \text{Sedoheptulose 7-phosphate} \\ + \\ \text{glyceraldehyde 3-phosphate} \end{array} \xrightleftharpoons{\text{transaldolase}} \begin{array}{c} \text{erythrose 4-phosphate} \\ + \\ \text{fructose 6-phosphate} \end{array} \qquad (2)$$

$$\begin{array}{c} \text{Erythrose 4-phosphate} \\ + \\ \text{dihydroxyacetone phosphate} \end{array} \xrightleftharpoons{\text{aldolase}} \text{sedoheptulose 1,7-biphosphate} \qquad (3)$$

An alternative method, which relied entirely on commercially available substrates and was used for preparative purposes, was based on the following reactions[2]:

$$\text{Fructose 1,6-biphosphate} \xrightleftharpoons{\text{aldolase}} \begin{array}{c} \text{dihydroxyacetone phosphate} \\ + \\ \text{glyceraldehyde 3-phosphate} \end{array} \qquad (1)$$

$$\text{Fructose 6-phosphate} \xrightarrow{\text{transketolase}} \begin{array}{c} \text{"active glycolaldehyde"} \\ + \\ \text{erythrose 4-phosphate} \end{array} \qquad (4)$$

$$\begin{array}{c} \text{Dihydroxyacetone phosphate} \\ + \\ \text{erythrose 4-phosphate} \end{array} \xrightleftharpoons{\text{aldolase}} \text{sedoheptulose 1,7-biphosphate} \qquad (3)$$

The overall yield in the above method was 30%. An alternative method,[3] based on the availability of erythrose 4-phosphate and transaldolase and giving yields close to theoretical, will be described below.

Principle. Sedoheptulose 1,7-biphosphate is prepared enzymically by three reactions performed in one step. Fructose 1,6-biphosphate is converted to dihydroxyacetone phosphate with aldolase and triosephosphate isomerase. In the presence of a slight excess of erythrose

---

[1] B. L. Horecker, P. Z. Smyrniotis, H. H. Hiatt, and P. A. Marks, *J. Biol. Chem.* **212**, 827 (1955).

[2] P. Z. Smyrniotis and B. L. Horecker, *J. Biol. Chem.* **218**, 745 (1956).

[3] J. Kowal, T. Cremona, and B. L. Horecker, *Arch. Biochem. Biophys.* **114**, 13 (1966).

4-phosphate, all the dihydroxyacetone phosphate is converted to sedoheptulose 1,7-biphosphate.[3] The sequence of reactions is shown in the following equations:

$$\text{Fructose 1,6-biphosphate} \xrightarrow[]{\text{aldolase}} \begin{array}{l} \text{dihydroxyacetone phosphate} \\ + \\ \text{glyceraldehyde 3-phosphate} \end{array} \qquad (1)$$

$$\text{Glyceraldehyde 3-phosphate} \xrightarrow[]{\substack{\text{triosephosphate} \\ \text{isomerase}}} \text{dihydroxyacetone phosphate} \qquad (5)$$

$$\begin{array}{l} \text{2 Dihydroxyacetone phosphate} \\ + \\ \text{2 erythrose 4-phosphate} \end{array} \xrightarrow[]{\text{aldolase}} \text{2 sedoheptulose 1,7-biphosphate} \qquad (3)$$

$$\begin{array}{l} \text{Fructose 1,6-biphosphate} \\ + \\ \text{2 erythrose 4-phosphate} \end{array} \rightleftarrows \text{2 sedoheptulose 1,7-biphosphate} \qquad (6)$$

Reagents

Fructose 1,6-biphosphate (158 mM). Monocalcium salt. Dissolve 418 mg in 5.0 ml of distilled water. Assay with aldolase, triosephosphate isomerase, α-glycerophosphate dehydrogenase, and DPNH.[4]

Erythrose 4-phosphate (25 mM). Prepare by oxidation of glucose 6-phosphate with lead tetraacetate.[5] Also available commercially.

Aldolase–triosephosphate isomerase mixture: Mix aldolase from rabbit muscle, crystallized according to Taylor,[6] also available commercially, 9.5 mg, 2.3 ml, and triosephosphate isomerase, obtained from Boehringer-Mannheim, New York, New York, 0.8 mg, 0.4 ml; dialyze against 1 liter of H_2O in the cold, over a period of 6 hr, change dialysis water once. Centrifuge to remove the slight precipitate, and collect the supernatant solution (2.8 ml).

Procedure. To 50 ml of erythrose 4-phosphate, 1250 μmoles, pH 2, add 3.48 ml, 550 μmoles, of fructose 1,6-biphosphate. Adjust the pH to 7.2 with 10 N NaOH and 1 N NaOH (total 3 ml). Add 2.8 ml of the aldolase–triosephosphate isomerase mixture and incubate at room temperature for 1 hr 15 min. To stop the reaction and precipitate the protein, add 3 ml of chloroform and 0.3 ml of n-butanol. Stir vigorously for 2 min. Separate in a separatory funnel and collect the aqueous layer.

Adjust the pH of the aqueous layer to 5.0 with 2 N acetic acid. Add 1 ml of 1 M barium acetate, centrifuge off the slight precipitate, and

[4] H. U. Bergmeyer (ed.), "Methods of Enzymatic Analysis." Academic Press, New York, 1965.
[5] See this series, Vol. 9 [6].
[6] See this series, Vol. 1 [39].

collect the clear supernatant. To this solution add 23 ml of saturated barium hydroxide to bring the pH to 6.3 (volume 90 ml). Add 90 ml of absolute ethanol, and transfer the cloudy solution to a glass centrifuge bottle. Leave in the cold for 2 hr and then centrifuge. Decant, drain carefully, and dry the precipitate in a desiccator attached to a pump. When the material is dry (overnight), transfer to a weighed bottle and weigh: 1298 mg. Expected: SBP-2 Ba (MW 644.9), 1.1 mmole, 710 mg. For assay, dissolve 66 mg in 0.5 ml of distilled water and add 2 drops of 2 N HCl. Add 0.5 ml of 0.5 M Na_2SO_4, centrifuge off the white precipitate, wash it with distilled water, and combine the supernatant solutions. Adjust to pH 6 with 1 N NaOH and make up the volume to 1.5 ml. Assay enzymically with aldolase and α-glycerophosphate dehydrogenase. Expected: 1.10 mmole SBP. Found: 1.10 mmole SBP. Contaminations (expressed as mole percent of SBP): glucose 6-phosphate, 3.9%; erythrose 4-phosphate, 8.2% triose phosphate, 0.5%.

Comments. The contaminating sugar phosphates can be eliminated by chromatography on Dowex 1-formate, according to Smyrniotis and Horecker.[2] Glucose 6-phosphate is a contaminant introduced with the preparation of erythrose 4-phosphate. The chemically synthesized compound[7] would be free of hexose phosphate and can be used in the above preparation.

[7] See series, Vol. 6 [70].

[17] Aldose 1,6-Diphosphates

By JOSEPH MENDICINO, HEDY MULHAUSEN,
and RAGY HANNA

D-Glucose 6-P (D-ribose 5-P, D-glucosamine 6-P, D-Galactose 6-P, D-Mannose 6-P)
+ acetic anhydride + sodium acetate or pyridine →
$$\beta\text{-D-glucose 6-P tetraacetate} \quad (1)$$
β-D-Glucose 6-P tetraacetate + crystalline phosphoric acid →
$$\alpha\text{-D-Glucose 1,6-}P_2 \text{ triacetate} \quad (2)$$
$$\alpha\text{-D-Glucose 1,6-}P_2 \text{ triacetate} + OH^- \rightarrow \text{acetate} + \alpha\text{-D-glucose 1,6-}P_2 \quad (3)$$

A simple general procedure for the synthesis of terminally phosphorylated α-anomeric diphosphate derivatives of aldoses was developed.[1] Fully acetylated hexose 6-phosphates, or pentose 5-phosphates prepared as shown in reactions (1) and (2) were phosphorylated at the anomeric carbon with crystalline phosphoric acid by a modification of the procedure of MacDonald.[2]

[1] R. Hanna and J. Mendicino, *J. Biol. Chem.* **245**, 4031 (1970).

Assay Methods

A microanalytical procedure was developed to measure the effect of time of incubation and temperature on the rate and extent of the phosphorylation reaction.[3] In this assay the sugar diphosphate product was partially purified and the rate of formation of acid-labile phosphate was measured by the method of Fiske and SubbaRow[4] as modified by King.[5] Total phosphate was determined by the procedure of Ames and Dubin.[6] Reducing sugar was analyzed by the method of Somogyi-Nelson.[7]

Reagents

Disodium D-glucose 6-P, 5 g (15.4 mmoles)

Anhydrous pyridine dried by distillation over calcium hydride

Acetic anhydride

Dry tetrahydrofuran stored over type 4A molecular sieve (Linde Corporation, Division of Union Carbide, Birmingham, Alabama)

LiOH, 1.0 N

Magnesium acetate, 50 mM

Ammonium hydroxide, 0.1 M

Dowex 1-HCO$_3^-$, prepared by treating Dowex 1-acetate with 1 M NaHCO$_3$ and washing the resin with 10 volumes of distilled water

Dowex 50-NH$_4^+$

Dowex 50-H$^+$

Crystalline phosphoric acid[8] was dried and stored over magnesium perchlorate

Analytical Assay Procedure. About 100 mg of dry crystalline phosphoric acid was dissolved in 10 ml of dry tetrahydrofuran, and 10 μmoles of the D-sugar 6-P tetraacetate was dissolved in 10 ml of chloroform or tetrahydrofuran. Aliquots of the two solutions were mixed, and the solvent was removed under reduced pressure. This procedure ensured complete mixing of very small amounts of the two reactants in the resulting melt. There is no reaction until the solvent is removed. The mixture was incubated under vacuum for various times and at different temperatures.[3]

[2] D. L. MacDonald, *J. Org. Chem.* **27**, 1107 (1962).

[3] J. Mendicino and R. Hanna, *J. Biol. Chem.* **245**, 6113 (1970).

[4] C. H. Fiske and Y. SubbaRow, *J. Biol. Chem.* **66**, 375 (1925).

[5] E. J. King, *Biochem. J.* **26**, 292 (1932).

[6] B. N. Ames and D. T. Dubin, *J. Biol. Chem.* **235**, 769 (1960).

[7] M. Somogyi, *J. Biol. Chem.* **160**, 61 (1945).

[8] Crystalline phosphoric acid was obtained from Fluka, AG, Buchs, S. G., Switzerland, Crystallization of 85% phosphoric acid can be carried out by adding some of the crystals to this supersaturated solution at 3°.

To stop the reaction the syrup was dissolved in 10 ml of dry tetrahydrofuran and the solution was poured into 10 ml of 0.5 N LiOH at 3°. Insoluble lithium phosphate was removed by centrifugation and residual inorganic phosphate was removed as magnesium ammonium phosphate by the addition of 1 ml of magnesium acetate and 1 ml of ammonium hydroxide. The suspension was centrifuged and the solution was passed through a Dowex 50-NH$_4^+$ column (3 × 2 cm). The amount of sugar diphosphate formed was estimated by determination of acid-labile phosphate. The acid-labile anomeric phosphate group was hydrolyzed by heating the sample in 0.1 N HCl at 100° for 15 min.

Preparation of α-D-Glucose 1,6-Diphosphate

The fully acetylated derivative of D-glucose-6-P was prepared by a modification of the procedure used for the synthesis of completely acetylated derivatives of free hexoses with acetic anhydride in pyridine[9] at 3°. Five grams (15.4 mmoles) of the disodium salt of D-glucose-6-P[10] were dissolved in 50 ml of water and the solution was passed through a Dowex 50-H$^+$ column (2 × 20 cm) to remove sodium ions. The column was washed with distilled water until all the sugar monophosphate was removed, and the eluate was neutralized by the addition of 20 ml of pyridine. The solution was concentrated under vacuum at room temperature. The syrup was dissolved in 20 ml of anhydrous pyridine, and it was concentrated under vacuum. This procedure was repeated again with 50 ml of dry pyridine to remove traces of water from the syrup.

The dried pyridinium salt of D-glucose-6-P was then dissolved in 30 ml of pyridine and 10 ml of acetic anhydride. The mixture was stirred at room temperature and the sugar phosphate gradually went into solution. The reaction flask was stored for 3 days at 2°. The solution was then concentrated under vacuum to remove acetic anhydride and pyridine, and the resulting syrup was dissolved in 20 ml of dry tetrahydrofuran. It was not necessary to isolate the tetraacetyl-D-glucose-6-P at this stage, since phosphorylation could be carried out directly on this sample. The yield of the acetylated product (15.2 mmoles) was nearly quantitative as determined by phosphate analysis. The compound gave a positive reaction in the Somogyi-Nelson test, since rapid deacetylation occurred under the conditions of this assay. The ratio of total phosphate

[9] M. L. Wolfrom and A. Thompson, *in* "Methods in Carbohydrate Chemistry" (R. L. Whistler and M. L. Wolfrom, eds.), Vol. 2, p. 211. Academic Press, New York, 1963.
[10] Calbiochem, Grade A, chromatographically homogeneous.

to reducing sugar to acetyl groups was 1.00:0.97:3.92 in the isolated sample.

The phosphorylation of the acetate derivative was carried out according to the procedure of MacDonald.[2] Ten grams of crystalline phosphoric acid, which was dried under vacuum over magnesium perchlorate, were dissolved in a solution containing 25 ml of dry tetrahydrofuran and 15 mmoles of D-glucose tetraacetate. The tetrahydrofuran was removed under vacuum and the resulting melt was evacuated and placed in a water bath at 50° for 2 hr. After this time the thick brown syrup was dissolved in 30 ml of dry tetrahydrofuran, and it was poured into ice-cold 1 N LiOH. The pH was adjusted to 10 with 1 N LiOH, and the suspension was left at room temperature overnight to saponify the acetyl groups. The precipitated lithium phosphate was removed by filtration, and the solution was diluted to 1 liter and passed into a Dowex 1-HCO$_3$ column (5 × 10 cm). The column was washed with 200 ml of water, and monophosphates were eluted with 0.01 N HCl. D-Glucose-6-P and small amounts of P$_i$ were eluted together after the effluent solution reached pH 2. The column was then washed with 500 ml of water and α-D-glucose-1,6-P$_2$ was eluted with a linear gradient of from 50 mM to 0.5 M LiCl, starting with 1 liter of each solution. The first peak, appearing at approximately 0.13 M LiCl, contained the sugar diphosphate ester. The fractions were pooled and adjusted to pH 7.5 with 1 N LiOH, and the sample was evaporated to dryness under reduced pressure. Lithium chloride was removed by extracting the dry solid repeatedly with a solution containing 10 ml of ethanol and 40 ml of acetone until the supernatant obtained by centrifugation was free of chloride ions. The final precipitate was washed with ether and dried over P$_2$O$_5$ under vacuum. The yield of the tetralithium salt was 5 mmoles, or about 33%, based on the amount of acetylated sugar monophosphate used.

Characterization of α-D-Glucose 1,6-Diphosphate

The product was chromatographically and electrophoretically homogeneous, and its characteristics were identical with those of samples of α-D-glucose-1,6-P$_2$ prepared by other methods[11] or isolated from α-D-glucose-1-P.[12] The ratio of acid-labile phosphate to acid-labile reducing sugar to total phosphate was 1.0:0.95:1.97. The preparation contained no free P$_i$ or reducing sugar.

The cyclohexylammonium and barium salts of α-D-glucose-1,6-P$_2$

---

[11] T. Posternak, *J. Biol. Chem.* **180**, 1269 (1949).

[12] K. Hanabusa, H. W. Dougherty, C. DelRio, T. Hasimoto, and P. Handler, *J. Biol. Chem.* **241**, 3930 (1966).

were prepared from the tetralithium salt.[11] The concentrations for optical rotation measurements were determined by phosphate analysis, and the calculations were based on the anhydrous molecular weight of the free acid in each case. The sodium salts of several samples of α-D-glucose-1,6-P_2 prepared by this procedure had $[\alpha]_D^{25} + 82.3°$ to $83.9°$ (c, 0.600, water), which compare well with that reported for α-D-glucose-1,6-P_2 $[\alpha]_D + 83°$ (c, 0.229, water).[11] The oxidation of a sample with sodium periodate at pH 6.5 resulted in the utilization of 2 equivalents of periodate and the formation of 1 equivalent of formic acid. No P_i was released during the oxidation.

The first-order rate constant for the hydrolysis of the compound in 1 N H_2SO_4 at 30° was found to be 7.7×10^{-4} min^{-1}. The reported first-order rate constant for α-D-glucose-1,6-P_2 is 7.8×10^{-4} min^{-1}, whereas that for β-D-glucose-1,6-P_2 is 3.15×10^{-3} min^{-1}, under these conditions.[11] Evidence to show that the β anomer was completely absent was obtained by rechromatography of a sample on a column of Dowex 1-formate (1 $\times$ 14 cm). The sugar diphosphate was eluted with 2 liters of a linear gradient of from 0 to 2 N ammonium formate, pH 3.0. The use of this gradient has been shown to result in a partial separation of the α and β anomers of D-glucose-1,6-P_2.[13] The compound was eluted from this column in a nearly symmetrical peak, and all the fractions had the same hydrolysis rate constant.[1]

Preparation of Other Aldose Diphosphates

The above procedure with slight modifications has been used to prepare α-D-ribose 1,5-diphosphate, α-D-galactose 1,6-diphosphate, α-D-mannose 1,6-diphosphate, and N-acetyl-α-D-glucosamine 1,6-diphosphate.[1] All the sugar diphosphates prepared by this procedure were active when they were assayed for enzymic activity with rabbit muscle phosphoglucomutase.[14] A modified microprocedure has also been used to prepare α-D-[1-^{32}P]glucose 1,6-diphosphate and α-D[6-^{32}P]glucose-1,6-P_2.[1,3]

Some precautions must be taken when preparing sugar diphosphates other than α-D-glucose-1,6-P_2. Several sugar monophosphate samples contain varing amounts of D-glucose-6-P. To remove this impurity, 5-g samples of D-galactose-6-P and D-mannose-6-P were treated with D-glucose-6-P dehydrogenase and NADP$^+$ to convert traces of D-glucose 6-P to 6-P-gluconic acid. The mixture was passed into a Dowex 1-Cl column (4.5 $\times$ 5 cm), and the resin was washed with water. The D-galactose 6-P

[13] J. W. Ray, Jr., and G. A. Roscelli, *J. Biol. Chem.* **239**, 1228 (1964).
[14] H. Mulhausen and J. Mendicino, *J. Biol. Chem.* **245**, 4038 (1970).

or mannose 6-P was then eluted with 0.01 N HCl, and the sample was neutralized by the addition of excess pyridine. The 6-P-gluconic acid remained in the column under these conditions. The tetraacetate and diphosphate derivatives of these sugar monophosphates were then prepared by essentially the same procedures used to acetylate and phosphorylate D-glucose-6-P.

One other precaution which must be taken in the preparation of α-D-mannose-1,6-P$_2$ and α-D-ribose-1,5-P$_2$ is to reduce the concentration of HCl from 10 to 5 mM during the removal of the residual sugar monophosphate from the Dowex 1-HCO$_3^-$ column (5 × 10 cm). These sugar diphosphates are considerably more labile to acid hydrolysis than α-D-glucose-1,6-P$_2$.[1]

[18] 5-Keto-D-fructose and Its Phosphate Esters[1]

By GAD AVIGAD and SASHA ENGLARD

Many strains of acetic acid bacteria, mainly species systematically classified within the genus *Gluconobacter*,[2] produce 5-ketofructose (5KF) when grown on D-fructose.[3-6] The same product also accumulates when L-sorbose, D-sorbitol, or D-mannitol are the sole carbon sources in the growth medium. Reports that 5KF is produced by the basidiomycetes *Trametes sanguinea* and *Polyporus obtusus* have appeared.[7,8] 5KF is produced by oxidation of D-fructose through all phases of bacterial growth.[9] However, it is consumed by growing cells until the pH of the medium drops to 3.5–3.8. When growth ceases, D-fructose is converted stoichiometrically to 5KF, which then accumulates in the medium.[6,9-11] A product

[1] Abbreviations used: 5KF, 5-*keto*-D-fructose, or D-*threo*-2,5-hexodiulose.

[2] Y. Yamada, K. Aida, and T. Uemura, *J. Gen. Appl. Microbiol.* **15**, 181 (1969).

[3] O. Terada, K. Tomizawa, and S. Kinoshita, *Nippon Nogei Kagaku Kaishi* **35**, 127 (1961).

[4] M. Kulhánek and Z. Ševičková, *Folia Biol. (Prague)* **7**, 288 (1962); **10**, 362 (1965).

[5] T. G. Carr, R. A. Coggins, and E. C. Whiting, *Chem. Ind. (London)* p. 1279 (1963).

[6] K. Sato, Y. Yamada, K. Aida, and T. Uemura, *Agr. Biol. Chem.* **33**, 1606, 1612 (1969).

[7] Y. Yamada, K. Iizuka, K. Aida, and T. Uemura, *Agr. Biol. Chem.* **30**, 97 (1966).

[8] F. W. Janssen and H. W. Ruelius, *Biochim. Biophys. Acta* **167**, 501 (1968).

[9] S. L. Mowshowitz, G. Avigad, and S. Englard, *Fed. Proc., Fed. Amer. Soc. Exp. Biol.* **29**, 675 (Abstract) (1970); *J. Bacteriol.* **118**, 1051 (1974).

[10] O. Terada, K. Tomizawa, S. Susuki, and S. Kinoshita, *Nippon Nogei Kagaku Kaishi* **35**, 131 (1961).

[11] S. Englard and G. Avigad, *J. Biol. Chem.* **240**, 2297 (1965).

isolated from *Acetobacter suboxydans* cultures grown on D-fructose, and whose structure was initially proposed to be 6-aldo-D-fructose,[12] was later identified to be 5KF.[13,14] Production of 5KF from L-sorbose by *Pseudomonas convexa* has also been reported.[14a]

5-Keto-D-fructose

Growth of Bacteria. Gluconobacter cerinus subsp. *ammoniacus* Asai, IFO 3267,[15] is grown on a medium containing 0.5% Difco yeast extract and 5–10% fructose. Cultures of 600 ml each in 2-liter Erlenmeyer flasks are incubated at 30° on a rotary shaker at 160 rpm. Cells from one 24-hr agar slant are used to inoculate each flask. Cultures starting with 5% fructose are incubated for 6–7 days, and those starting with 10% fructose are incubated for 12–14 days. Under these conditions, 60–70% of the fructose is converted ultimately to the new sugar, and only a trace or no fructose remains as shown by chromatographic examination and colorimetric analysis of the medium (see below).

Isolation of 5-Ketofructose.[6,10,13,14] One liter of culture medium (pH about 3.8) is processed as follows. The bacteria are removed by centrifugation, 25 g of Norit A and 25 g of Celite 545 are added to the centrifugate, the solution is filtered on a Büchner funnel and then filtered through an Amberlite MB-3 resin (H+, OH-form) column (3.2 × 50 cm). The colorless effluent at about pH 6.5 is filtered on a Büchner funnel with Super-Cel filter aid, flushed with nitrogen, and then reduced in volume (1 mm at 30°) to a light syrup. Absolute ethanol is added until a slight turbidity appears, which is removed by filtration. The clear filtrate is reduced in volume to a syrup, fresh amounts of ethanol are added, and the solution is again evaporated. This procedure is repeated several times until a semicrystalline material appears. This material is triturated with cold ethanol or 1-propanol in a Waring Blendor and then manually in a mortar until it appears as a fine semicrystalline, white material. This substance is collected rapidly by filtration or centrifugation, washed with cold absolute ethanol followed by 1-propanol, and dried in a vacuum desiccator over $CaSO_4$ (Drierite). Any unnecessary delays in the procedure will then cause rapid browning of the material.

5KF is recrystallized from hot ethanol–water (about 80%) and

[12] R. Weidenhagen and G. Bernsee, *Chem. Ber.* **93**, 2924 (1960).

[13] G. C. Whiting and R. A. Coggins, *Chem. Ind. (London)* p. 1925 (1963).

[14] G. Avigad and S. Englard, *J. Biol. Chem.* **240**, 2290 (1965).

[14a] R. P. Longley and D. Perlman, *Biotechnol. Bioeng.* **14**, 843 (1972).

[15] Obtained from the collection of the Institute of Applied Microbiology, The University of Tokyo, Bunkyo-Ku, Tokyo, and also from the American Type Culture Collection, Rockville, Maryland 20852, as ATCC 19441.

allowed to stand overnight at 0°. The colorless needles that are obtained have a melting point of 160–162°; $[\alpha]_D^{18} - 86°$ (in H_2O). After several recrystallizations, a preparation is obtained with a m.p. of 172–175° (O. Terada, personal communication). The overall yield of 5KF is 60–70% of the initial fructose used. Addition of acetic acid to ethanolic solutions of 5KF will facilitate the recovery and crystallization of the sugar,[13] and crystals obtained from acetic acid water solutions have a m.p. of 175–176° and an $[\alpha]_D^{18} -81° \rightarrow -89°$ (in H_2O). However, no detectable mutarotation has been observed for preparations obtained by other investigators.[10–12]

The bisphenylhydrazone derivative[2,8,10–13] of 5KF has a m.p. of $138 \pm 5°$ (as reported for different preparations), $[\alpha]_D^{16} - 164°$ (in pyridine). The bis-p-nitrophenylhydrazone derivative of 5KF[12,13] has a m.p. of 176–177°.

Partial reduction[13,14,16] of 5KF, such as by boron hydrides, yields a mixture of L-sorbose and D-fructose; on complete reduction, a mixture of D-mannitol, D-sorbitol, and L-iditol is obtained.

Aqueous neutral solutions of 5KF tend to develop browning on storage. This process is significantly enhanced in the presence of alkali. It is, therefore, advisable to prepare fresh solutions of 5KF immediately before use or to keep stock solutions at a slightly acidic pH (3.0–5.0) and well refrigerated.

Methods of Analysis

Chromatography.[3,12–14] On paper or thin-layer chromatography, 5KF moves slower than D-fructose and L-sorbose and also yields different colored products with the various spray reagents used to detect reducing hexoses. Appropriate chromatographic solvents should be selected according to convenience and the type of absorbent employed.[17] Among the many spray reagents available, urea-H_3PO_4 for ketoses[18] and p-anisi-

---

[16] O. Terada, S. Suzuki, and S. Kinoshita, *Nippon Nogei Kagaku Kaishi* **35**, 178 (1961).

[17] L. Hough and J. K. N. Jones, *Methods Carbohyd. Chem.* **1**, 21 (1962); R. E. Wing and J. N. BeMiller, *Methods Carbohyd. Chem.* **6**, 42 (1972); B. A. Lewis and F. Smith, *in* "Thin Layer Chromatography" (E. Stahl, ed.), 2nd ed., p. 807, Springer-Verlag, Berlin and New York, 1969; J. Sherma and G. Zweig, "Paper Chromatography and Electrophoresis" Vol. 2, p. 152, Academic Press, New York, 1971. See also this series, Vol. 3 [11].

[18] Reagent is composed of: 3% urea in n-butanol : n-propanol : water : 85% phosphoric acid, 70:20:5:5 (v/v). This solution is stable for several months at room temperature. To obtain colored sugar spots, chromatograms are first sprayed and then kept for 5–10 min at 110° [according to C. S. Wise, R. T. Dimler, H. A. Davis, and C. E. Rist, *Anal. Chem.* **27**, 33 (1955)].

dine[19] and aniline-diphenylamine[20] for reducing sugars are especially suitable since they yield distinctive colors with different sugars, thus facilitating identification.

Colorimetric Analysis. 5KF does not react with Dische's cysteine-carbazole reagent[21] for ketohexoses, but it can react specifically with *o*-aminobiphenyl.[22] Consequently, these two reagents can be used to measure quantitatively both the disappearance of L-sorbose or D-fructose and the appearance of 5KF in bacterial cultures. Roe's resorcinol reagent[21] yields with D-fructose a colored product with an absorption maximum at 520 nm, with 5KF the peak is shifted to 425 nm, and at 520 nm the equivalent molar intensity for 5KF is 12% that of D-fructose.[14] A method using *o*-phenylenediamine for the assay of 5KF[23] is less specific or sensitive.

Enzymic Assays. 5KF can be determined spectrophotometrically using specific NADPH-dependent 5KF reductases.[24,25] Other common ketohexoses do not react and the kinetic and thermodynamic properties of these enzymes favor a complete and stoichiometric reduction of 5KF at very low substrate concentrations. Two highly purified NADPH-reductases of this type are available; the *G. cerinus*[24] and yeast[25] enzymes reduce 5KF to D-fructose or L-sorbose, respectively.

5KF is also a substrate for the NAD^+-linked sorbitol dehydrogenase from liver,[26] the NAD^+- and $NADP^+$-dependent polyol dehydrogenases

---

[19] Reagent is composed of: 2% *p*-anisidine-hydrochloride in *n*-butanol:ethanol:2 *N* HCl, 8:1:1 (v/v). This solution with 0.05% of added sodium hydrosulfite is stable for several weeks when refrigerated. To detect colored sugar spots, chromatograms are sprayed and then kept for 5–10 min at 110° [according to L. Hough, J. K. N. Jones, and W. H. Wadman, *J. Chem. Soc.*, p. 1702 (1950)].

[20] Reagent is composed of: 1% diphenylamine and 1% aniline in ethyl acetate to which phosphoric acid is added to a final concentration of 8%. This solution is stable for several days when kept refrigerated. To develop colored sugar spots, chromatograms are sprayed and then kept at 80° for about 10 min [according to T. Kočoureck, M. Tichá, and T. Koštir, *J. Chromat.* 24, 117 (1966)].

[21] See this series, Vol. 3 [12].

[22] The reaction is performed as follows[14]: 5 ml of reagent (0.4% *o*-aminobiphenyl and 2% thiourea in glacial acetic acid, freshly prepared) is added to 1.0 ml of solutions containing from 0.05 to 1.0 µmole 5KF. Tubes are immersed in a boiling water bath for 25 min and then cooled. Absorption of the stable, colored product is measured at 415 nm or at 440 nm if aldoses are also present. Sensitivity of assay for 5KF is more than 20-fold higher than that of D-fructose, L-sorbose, or D-mannose.

[23] K. Imada, K. Sato, S. Oga, and K. Asano, *Agr. Biol. Chem.* 30, 1173 (1966).

[24] G. Avigad, S. Englard, and S. Pifko, *J. Biol. Chem.* 241, 373 (1966). See also this volume [30].

[25] S. Englard, G. Kaysen, and G. Avigad, *J. Biol. Chem.* 245, 1311 (1970). See also this volume [31].

[26] S. Englard, G. Avigad, and L. Prosky, *J. Biol. Chem.* 240, 2302 (1965).

from *Acetobacter melanogenum*[27] and *G. cerinus*,[11] and the D-mannitol dehydrogenase from *Lactobacillus brevis*.[28] However, the kinetic parameters of these enzymes and their relative lack of specificity limit their usefulness as analytical tools for the quantitative assay of 5KF in the presence of other ketoses and hexitols.

5-Keto-D-fructose 1-Phosphate

Principle. In the presence of ATP, 5KF is effectively phosphorylated by yeast hexokinase[29] or by liver fructokinase[30] to yield 5KF-1-P. For preparative purposes, the most convenient procedure for the preparation of 5KF-1-P is to couple the 5KF-hexokinase system with catalytic amounts of ATP and a phosphoenolpyruvate-pyruvate kinase coupling system. The reaction is allowed to proceed to completion as determined by the appearance of pyruvate. 5KF-1-P is then recovered as the only phosphate ester in solution either by chromatography, or if desired, as the barium salt.

Procedure.[29,31] A reaction mixture (77.8 ml) contains (in millimoles): neutralized phosphoenolpyruvate, 3.98; Tris·HCl buffer (pH 7.4), 5.0; $MgCl_2$, 0.2; KCl, 4.0; ATP, 0.15; 5KF, 4.82; muscle pyruvate kinase, 120 units; and yeast hexokinase, 150 units. Incubation is carried out at 30°, and at various time intervals small aliquots are withdrawn for the estimation of pyruvic acid by the lactic dehydrogenase assay system.[32] After 4 hours, 4.0 mmoles of pyruvate are produced and phosphoenolpyruvate can no longer be detected, indicating that the reaction has reached completion. The solution is passed through a column (20 × 2 cm) of Dowex 50-(H^+)-X8 followed by washing with 150 ml of water. The eluate is reduced in volume by lyophilization. The yellowish concentrate (20 ml) is repeatedly extracted with ether (five 100-ml portions) and then neutralized with 0.01 *N* NaOH to bring the pH to 7.0. Five milliliters of an aqueous solution containing 4.15 mmoles of barium acetate is added, followed by 2 volumes of ethanol. The flocculent precipitate is collected by filtration, washed with 66% ethanol and then with abso-

---

[27] K. Sasajima and M. Isono, *Agr. Biol. Chem.* **32**, 161 (1968).

[28] See Vol. 9 [28].

[29] G. Avigad and S. England, *J. Biol. Chem.* **243**, 1511 (1968).

[30] S. Englard, I. Berkower, and G. Avigad, *Biochim. Biophys. Acta* **279**, 229 (1972).

[31] Purified enzymes and substrates used in our studies were obtained from Boehringer Mannheim, New York, New York or the Sigma Chemical Co., St. Louis, Missouri.

[32] The reaction mixtures (1.0 ml) in cuvettes with 1-cm light path contain HEPES or phosphate buffer, pH 7.4, 50 m*M*; NADH, 0.15 m*M*; and muscle lactate dehydrogenase, 2 units. The decrease in absorption at 340 nm is a measure of the amount of pyruvate present. See also Vol. 13 [68].

lute ethanol, and finally with ether and placed in a vacuum desiccator to dry. About 1.5 g of a white material analyzing as $C_6H_9O_9P$ $Ba \cdot H_2O$ (MW 411.50) is obtained.

Characterization of 5KF-1-P. The barium salt is dissolved in HCl at pH 2.0, and the barium is removed by the addition of a slight excess of Na_2SO_4 or by passage through a Dowex 50-Na^+ column in order to obtain the sodium salt of 5KF-1-P. Purity of the phosphate ester can be checked by chromatography and chemical analyses.[29,30]

5KF-1-P as Substrate for Various Enzymes.[29] 5KF-1-P inactive either as a substrate or inhibitor when present at concentrations of up to 2 mM in the standard spectrophotometric assay systems for yeast phosphoglucose isomerase, transaldolase, glucose-6-P dehydrogenase, muscle aldolase, and phosphoglucomutase.

5KF-1-P is a relatively poor substrate for the specific NADPH:5KF reductases.[24,25] Although 5KF-1-P does not undergo aldolytic cleavage, it inhibits liver aldolase[30] in a competitive manner with respect to Fru-1-P (K_i 0.46 mM). 5KF-1-P is a substrate for phosphofructokinase[33] ($K_m = 0.22$ mM); its maximal rate of phosphorylation is about 25% of that observed with F-6-P.

5-Keto-D-fructose 1,6-Biphosphate

Principle. In the presence of ATP, yeast, and rabbit-muscle, phosphofructokinase can effectively phosphorylate 5KF-1-P to 5KF-1,6-P_2.[30,33] Since phosphofructokinase is inhibited by high ATP concentrations, it is important to maintain a low ATP level during the synthesis of the biphosphate ester on a preparative scale. This is achieved by phosphorylating 5KF in a system containing yeast hexokinase and muscle phosphofructokinase which is coupled with phosphenolpyruvate and pyruvate kinase. The reaction is allowed to proceed to completion as determined by the appearance of two pyruvate equivalents for each mole of 5KF initially present.

5KF-1,6-P_2 has been reported as product of the aldol condensation of hydroxypyruvaldehyde-phosphate and dihydroxyacetone-phosphate in the presence of high concentrations of muscle aldolase.[34]

Procedure.[33] A reaction mixture (40 ml) contains (in millimoles): HEPES, pH 7.6, 3.1; KCl, 2.0; $MgCl_2$, 0.2; ATP, 0.1; neutralized dicyclohexylammonium phosphoenolpyruvate, 2.5; 5KF, 1.0; yeast hexokinase, 140 units; muscle phosphofructokinase, 200 units and muscle pyruvate kinase, 150 units. The reaction mixture is incubated at room tem-

[33] G. Avigad and S. Englard, *Biochim. Biophys. Acta* **343**, 330 (1974).
[34] M. T. Healy and I. Christen, *J. Amer. Chem. Soc.* **94**, 7911 (1972).

perature, and at various time intervals small aliquots are withdrawn for the spectrophotometric estimation of pyruvate.[32] After approximately 6 hr, 2.0 mmoles of pyruvate are produced. To avoid the presence of large amounts of 5KF-1-P, it is important to let the reaction proceed to completion. The solution is passed through a column (1.6 × 20 cm) of Dowex 50W-(H+)-X8 followed by washing with 100 ml of water. The acidic effluent is reduced in volume by lyophilization, and the yellowish concentrate (about 10 ml) is extracted repeatedly with diethylether (five 50-ml portions) and then carefully neutralized to pH 6.0 with 10 mM NaOH. This solution of 5KF-1,6-P$_2$ is stored at −20° and provides a stock solution of relatively high purity. 5KF-1,6-P$_2$ can be further purified, especially from contamination by 5KF-1-P, by chromatography on Dowex 1 (Cl−) column.[33] However, this procedure results in significant losses of material by degradation and the presence of high salt concentrations in the sugar biphosphate solution. The dibarium salt of 5KF-1,6-P$_2$ is sparingly soluble even at pH 2.0 and its preparation[33] is not recommended as a purification procedure. The more soluble tricyclohexylammonium salt is prepared as follows: A sample (3 ml) of the ether-extracted Dowex 50W (H+) effluent containing 0.2 mmole of 5KF-1,6-P$_2$ is added to 1.0 ml of a 12% aqueous solution of cyclohexylamine. The solution is reduced in volume (to approximately 0.5 ml) by lyophilization. Ethanol (2 ml) is added followed by (about 5 ml) of a mixture of ethanol: ethyl acetate (1:1 v/v) until the appearance of a slight turbidity. After 24 hr at 4°, the precipitate which forms is collected by centrifugation, washed with ether, and dried at room temperature. The yield is about 80 mg, and the product analyzes as $C_6H_{12}O_{12}P_2(C_6H_{13}N)_3 \cdot H_2O$ (MW 653.6).

5KF-1,6-P$_2$ as Substrate for Various Enzymes.[33] The biphosphate ester inhibits competitively the action of muscle (K_i 0.27 mM) and liver aldolases on Fru-1,6-P$_2$. It is a relatively poor substrate for liver fructose 1,6-phosphatase. Like Fru-1,6-P$_2$, it can serve as an allosteric effector in the activation of liver pyruvate kinase and the inhibition of liver and yeast 6-phosphogluconic dehydrogenase.

[19] L-Rhamnulose 1-Phosphate[1]

By Der-Fong Fan, Marla Sartoris, T. H. Chiu,
and David Sidney Feingold

L-Rhamnulose 1-phosphate may be prepared as described previously[2] or by the improved method given here, which uses two mutant strains of *Escherichia coli*. In either case, final purification and crystallization of the compound should be performed as described in this chapter.

Deficient mutants are obtained as follows: *E. coli* K12 HfrP72 is grown overnight with shaking at 37° in Difco nutrient broth to the stationary phase. The mutagen N-methyl-N-nitroso-N'-nitroguanidine is added to a final concentration of 80 μg/ml, and the culture is held without shaking at 37° for 1 hr. It is then diluted with 100 volumes of fresh nutrient broth, grown to stationary phase, and plated (after appropriate dilution with 0.85% NaCl) on Endo's agar (Difco Laboratories, Detroit, Michigan) plates containing 1% L-rhamnose. After 2 days at 37°, L-rhamnose-utilizing colonies turn red; the white, nonutilizing colonies are picked and purified on nutrient agar. The parent strain is methionine-negative (met⁻), that is, it requires methionine for growth. Mutants obtained by the above procedure are converted to a strain with no methionine requirement (met⁺) by transduction with a phage P1kc which has been grown on met⁺ *E. coli* K12.[3] met⁺ Bacteria are selected by plating on mineral glucose [in grams per liter; D-glucose, 5; NH_4Cl, 1; $MgSO_4 \cdot 7$ H_2O, 0.2; $FeSO_4 \cdot 7$ H_2O, 0.01; $CaCl_2$, 0.01; trace elements (Mn, Cu, Co, Zn as inorganic salts, 0.02–0.05 mg of each)] agar and are purified by twice restreaking on the same medium. Before use, they are rechecked for the L-rhamnose-negative character by observing growth failure on mineral rhamose (prepared as above with L-rhamnose in place of glucose) agar. Each of the mutant strains is grown with shaking at 37° in a medium composed of (grams per liter): K_2HPO_4, 7.0; KH_2PO_4, 3.0; $(NH_4)_2SO_4$, 1.0; $MgSO_4 \cdot 7$ H_2O, 0.1; casamino acids, 10.0; and L-rhamnose, 4.0. The sugar is sterilized by filtration and added aseptically to the sterile salts solution. The cells are harvested by centrifugation, and disrupted by sonication in 20 mM potassium phosphate buffer (pH 7); debris is removed by centrifugation in the cold. The supernatant fluid is examined for the presence of L-rhamnose isomerase (EC 5.3.1.14) and

[1] T. H. Chiu, R. Otto, J. Power, and D. S. Feingold, *Biochim. Biophys. Acta* **127**, 249 (1966).

[2] This series, Vol. 9 [84].

[3] S. E. Luria, J. N. Adams, and R. C. Ting, *Virology* **12**, 348 (1960).

L-rhamnulokinase (EC 2.7.1.5) by methods previously described.[2,4] L-Rhamnulose 1-phosphate aldolase is detected, after removal of nucleoproteins with $MnCl_2$, with a coupled assay system.[5] Two such mutants isolated were used in the preparation described here: Rha^{-58} (kinaseless) and Rha^{-54} (aldolaseless). They are maintained on nutrient agar slants.

Growth of Cells. The medium described above is used. Nine liters of medium contained in a New Brunswick 10-liter microferm fermenter (New Brunswick Scientific Co., New Brunswick, New Jersey) is inoculated with 1 liter of starter culture. The latter is either strain Rha^{-54} or Rha^{-58} in late log phase. Growth is at 37° with 1.5 ft^3/min aeration and 200 rpm agitation. After 12–16 hr, the culture is in late log phase.

Preparation of L-Rhamnulose. Cells of Rha^{-58} from 10 liters of culture medium are harvested by centrifugation, washed with 0.85% NaCl, and suspended in 140 ml of sodium borate, pH 8.0. To this cell suspension, 14 ml of 20% L-rhamnose and 3 ml of 0.5 M $MnCl_2$ are added. After incubation at 37° for 30 hr with shaking, cells are removed by centrifugation and again suspended in the borate buffer and incubated with L-rhamnose as described. The supernatant fluids obtained from both incubations are combined and concentrated at 37° to about 15 ml. This concentrated syrup is then chromatographed on a Sephadex G-10 column (2.4 cm $\times$ 80 cm) prepared and eluted with distilled water. L-Rhamnulose is located in the eluate by paper chromatography in n-propanol:ethylacetate:water (7:1:2 v/v/v) followed by alkaline silver nitrate spray.[2] The L-rhamnulose-containing fractions are pooled, concentrated, diluted with 50 ml of methanol and taken to near dryness at 37°. This procedure is repeated five times. The final concentrated syrup contains about 80% L-rhamnulose and 20% L-rhamnose.

Preparation of L-Rhamnulose 1-Phosphate. Cells of the *E. coli* strain Rha^{-54} from 10 liters of culture medium are suspended in 75 ml of 20 mM potassium-phosphate buffer, pH 7.0, and disrupted in the French pressure cell.[6] This and all subsequent operations are performed at 0–4°. The supernatant fluid obtained by centrifugation of the disrupted cell suspension at 30,000 g is diluted with the same buffer to 150 ml. After addition of 0.05 volume of 0.5 M $MnCl_2$ with stirring, the precipitate is removed by centrifugation. The supernatant fluid is then fractionated with solid ammonium sulfate. The protein which precipitates between 30 (167 g/l) and 65% (an additional 233 g/l) $(NH_4)_2SO_4$ saturation is dissolved in 30 ml of 0.4 M Tris·HCl, pH 8.0. This preparation usually contains about 10 international units of rhamnulokinase assayed by the

[4] This series, Vol. 9 [101].

[5] This series, Vol. 42 [41].

[6] This series, Vol. 1, [9].

method described previously. To this enzyme preparation, a mixture containing 3 mmoles of L-rhamnulose, 5 mmoles of ATP, and 2.5 mmoles of KF is added. The reaction mixture is then diluted to 60 ml by adding 0.4 M Tris·HCl, pH 8.0. After 3 hr at 30°, the mixture is held at 100° for 2 min and precipitated protein is removed by centrifugation.

Purification of L-Rhamnulose 1-Phosphate. The supernatant fluid is concentrated to about 12 ml at 37° and chromatographed on a Sephadex G-10 column (2.4 cm × 80 cm) prepared and eluted with distilled water. Fractions containing L-rhamnulose 1-phosphate, detected by paper chromatography in butanone:acetic acid:water (15:5:2 v/v/v) followed by molybdic acid spray,[7] are pooled and concentrated at 37°. This fraction is then chromatographed on sheets of Whatman 3 MM paper, using the solvent described above. L-Rhamnulose 1-phosphate, located by spraying a guide strip, is eluted from the paper with water, concentrated, rechromatographed in the same way, eluted with water, and concentrated at 37° to 10 ml. The yellowish solution is shaken vigorously with 500 mg of acid-washed Norit A. The charcoal is then removed by filtration and washed repeatedly with a total volume of 100 ml of distilled water. The colorless filtrate is concentrated to 2 ml at 37° and neutralized with saturated Ba(OH)$_2$. Four volumes of absolute ethanol is added and the solution is held at −20° for 12 hr, yielding a white amorphous precipitate.

Preparation of Crystalline Dicyclohexylammonium Salt of L-Rhamnulose 1-Phosphate. The amorphous barium salt of L-rhamnulose 1-phosphate obtained is compacted by centrifugation and washed twice with a small volume of ice-cold 80% ethanol. Upon drying overnight in a vacuum desiccator, it crystallizes. The crystalline material is redissolved in a small volume of water, 4.0 volumes of absolute ethanol is added, and recrystallization is carried out as described above.

A 0.15-g sample of the barium L-rhamnulose 1-phosphate so obtained is dissolved in 1.5 ml of water and passed through a 1.2 cm × 24 cm column of Amberlite IR-120 (H+ form). The column is eluted with water; 30 ml of acidic eluate is collected and concentrated under vacuum to about 2 ml. Sufficient 0.5 M alcoholic cyclohexylamine (about 2 ml) is added to bring the pH to 7, and the solution is evaporated to dryness under vacuum. The crystalline salt which forms is washed with a small volume of ice-cold absolute ethanol, dissolved in 0.4 ml of hot 80% ethanol, allowed to cool to room temperature, and then stored overnight at −10°. The resulting crystalline dicyclohexylammonium salt of L-rhamnulose 1-phosphate is filtered off and recrystallized by repetition of the same procedure. The long colorless needles obtained are filtered off and

---

[7] This series, Vol. 3, p. 113.

dried under vacuum at room temperature; m.p. 171–173° (decomp. uncorrected). The material so prepared is homogeneous when examined by paper chromatography in the system described above or by paper electrophoresis at pH 3.6 and 5.8.

[20] 3-Deoxy-2-ketoaldonic Acids

By DANIEL CHARON and LADISLAS SZABÓ

Principle. The α-hydroxy group of a 3-deoxyaldonic acid is selectively oxidized by chlorate in the presence of vanadium oxide and phosphoric acid,[1] and the keto acid is isolated by chromatography on cellulose powder.

Method[2]

Ammonium 3-Deoxy D-arabino-Hept-2-ulosonate. Methyl 3-deoxyheptonate is prepared from commercial 2-deoxy-D-*arabino*-hexose[3] (2-deoxyglucose) according to Sprinson *et al.*[4] It should be noted that the melting point of the crystalline material varies from one preparation to another owing to the presence of varying amounts of epimers (D-*gluco*- and D-*manno*-) in the sample. The ratio of the epimers does not affect the yield or the quality of the final product.

The mixed methyl esters (4.5 g) are treated with somewhat less than the calculated amount of 2 N NaOH solution; when the pH of the reaction mixture has dropped to 7 small amounts of the base are added until a stable value of pH 8 is attained. This solution is added to a mixture of sodium chlorate (740 mg), vanadium oxide (commercial V_2O_5, 120 mg) and phosphoric acid (85%, d 1.71; 0.1 ml). The mixture contained in a stoppered flask is magnetically stirred for 5 days during which time the originally yellow color turns blue-green. The pH is now brought to 9 by addition of saturated aqueous $Ba(OH)_2$, and the precipitate formed is removed by centrifugation. The supernatant, which should give no precipitate upon addition of one drop of $Ba(OH)_2$, is decationized by passage through a column of Amberlite IR-120 or Dowex 50 (H^+) resin (3.5 × 35 cm). The effluent and washings are combined, neutralized (pH 7) with dilute NH_4OH solution, and concentrated *in vacuo*

[1] P. P. Regna and P. Caldwell, *J. Amer. Chem. Soc.* **66**, 243 (1944).
[2] D. Charon and L. Szabó, *J. Chem. Soc. Perkin* I., 1175 (1973).
[3] Sigma Chemical Company, P.O. Box 14508, St. Louis, Missouri 63178.
[4] D. B. Sprinson, J. Rothschild, and M. Sprecher, *J. Biol. Chem.* **238**, 3170 (1963).

(bath temperature 35°) to a thick, yellow syrup. This is placed on the top of a cellulose powder (Whatman CF 11 fine: CF 1 coarse, 1:1, v/v) column (4 × 40 cm) equilibrated with acetone/water (85:15, v/v) and eluted with the same solvent. Fractions of 10 ml are collected at a rate of 40–50 ml/hr. Small aliquots (a few μl) of the fractions are spotted sequentially on filter paper strips and tested with Warren's periodate/ thiobarbiturate spray.[5] Fractions containing the deoxy aldulosonic acid are pooled and concentrated to a syrup which is triturated with an equal volume of methanol in which it dissolves. The mixture is left in the cold room for a week, during which time a crystalline mass of ammonium 3-deoxy D-*arabino*-heptulosonate monohydrate is formed. The compound crystallizes readily but very slowly. Yield: 1 g. It can be recrystallized from methanol and then has an m.p. of 96° and $[\alpha]_D^{22} = +42.4°$ ($c = 2.2$, water) at 4 and at 30 min. In the periodate/thiobarbiturate test (this volume [6]) its molar absorption coefficient is 95.000; in the semicar-bazide test[6] the molar absorption coefficient is 10.000, but this value is attained only after 60 min of incubation at 40°.

Ammonium 3-Deoxy D-*threo-Hex-2-ulosonate*[2]

GALACTOMETASACCHARINIC ACID LACTONE.[7] This lactone is a mixture of 3-deoxy D-*xylo*- and D-*lyxo*-hexonic acid lactones. D-Galactose (100 g) is dissolved in 1 liter of deionized water, calcium hydroxide (50 g) is added, the flask is stoppered, the contents thoroughly mixed and kept for 4 weeks at room temperature. The mixture is shaken frequently during the first week. After the time indicated, the mixture, containing a volu-minous precipitate, is filtered and the filtrate is gently boiled for 3 hr, its initial volume being maintained by occasional addition of water. The precipitate formed is filtered off and the filtrate saturated with carbon dioxide. The mixture is again heated, and, after addition of some char-coal, filtered; the filtrate is concentrated to a small volume (about 100 ml, syrupy consistency). Ethanol is added until no more oily precipitate is formed. After standing overnight, the supernatant is decanted from the viscous oil, which is triturated with acetone (100–200 ml) until the oil has completely solidified. Several changes of solvent and additional standing under acetone may be necessary to obtain a completely solid product. The yield is about 20 g. This material is dissolved in water (200 ml; a brown solution is obtained), the pH is adjusted to 8 (dilute NaOH), and the solution is passed through a column (4 × 20 cm) of

[5] L. Warren, *Nature (London)* **186,** 237 (1960).
[6] J. MacGee and M. Doudoroff, *J. Biol. Chem.* **210,** 617 (1954).
[7] *Cf.* R. L. Whistler and J. N. BeMiller, *Methods Carbohyd. Chem.* **2,** 483 (1963).

AG 1 × 8 100/200-mesh ion-exchange resin[8] in the formate form. Elution is carried out with 0.25 N formic acid at a rate of 80 ml/hr, and 50 ml fractions are collected. The fractions containing the 3-deoxy galactonates (detected by the periodate/benzidine spray[9]) are pooled and brought to dryness *in vacuo* (bath temperature, 50°); the white residue crystallizes spontaneously. It is dissolved in the minimal amount of warm methanol, the solution is cooled, and ether is added until incipient turbidity. The sample is seeded with the crystalline material and allowed to stand overnight in the cold; an 8–10 g yield of the mixed lactones, m.p. 138–141°, is obtained.

AMMONIUM 3-DEOXY D-*threo*-HEX-2-ULOSONATE. The above lactone (3.2 g) is first hydrolyzed with sodium hydroxide and then treated with sodium chlorate (740 mg), vanadium oxide (120 mg), and phosphoric acid (0.1 ml) as described for the 7-carbon homolog. By the same isolation procedure, a 1.3 g yield of the crystalline ammonium salt (anhydrous) is obtained which, after recrystallization from methanol, has a melting point of 145° and $[\alpha]_D^{20} = +16°$ at 5 min and $+14°$ at equilibrium ($c = 1.8$, water). In the periodate/thiobarbiturate test (this volume [6]), it has a molar absorption coefficient of 95.000 and in the semicarbazide test[6] 10.000 (15 min incubation at 40°).

Comments. The preparation of galactometasaccharinic acid lactones is lengthy but easy and not really time-consuming. The above preparation has been run on a 500-g scale; proportionally the same yield was obtained. 3-Deoxy D-*erythro*-hex-2-ulosonate could not be reproducibly prepared from "glucometasaccharinic acid" in the above conditions.

Detection of 3-Deoxyaldulosonic Acids on Chromatograms[5]

Reagents

NaIO$_4$, 20 mM

Mixture of ethylene glycol:acetone:conc. H$_2$SO$_4$, 50:50:0.3, by volume 6%

Aqueous sodium 2-thiobarbiturate solution; prepared by mixing 1.2 g of 2-thiobarbituric acid with 8.2 ml of 1 N NaOH solution and sufficient water to make a total volume of 20 ml. The mixture is heated on the water bath until a clear solution is obtained; after cooling, its pH should be 2–3.

Procedure. The paper is sprayed with periodate, 15 min later with glycol, and again 10 min later with thiobarbiturate. Upon 5 min heating at 100°, red spots indicate the location of 3-deoxyaldulosonic acids.

[8] Bio-Rad Laboratories, 32 and Griffin Avenue, Richmond, California 94804.
[9] J. A. Cifonelli and F. Smith, *Anal. Chem.* 26, 1132 (1954).

Detection of Polyhydroxy Compounds on Paper Chromatograms[9]

Reagents

Saturated aqueous KIO_4 solution

Benzidine reagent: 1.84 g of benzidine are dissolved in 100 ml of 50% aqueous EtOH and the solution is mixed with 20 ml of acetone and 10 ml of 0.2 N HCl. The solution will keep for about 1 month.

Procedure. The papers are first *lightly* sprayed with periodate and, after 3–5 min, with the benzidine reagent: white spots on a light blue background locate compounds containing vicinal diols.

Comment. Most, but not all, compounds bearing vicinal diols will react in the test; those reducing periodate slowly may require repeated *light* spraying with periodate *before* the benzidine spray is applied. Benzidine is carcinogenic and should be handled accordingly.

[21] Phosphorylated 3-Deoxy-2-ketaldonic Acids

By FRANÇOIS TRIGALO and LADISLAS SZABÓ

Principle. The α-hydroxy group of a phosphorylated 3-deoxyaldonic acid is selectively oxidized by chlorate in the presence of vanadium oxide and phosphoric acid[1] and the keto acid is isolated by ion exchange chromatography.

3-Deoxy-D-erythro-hex-2-ulosonic Acid 6-Phosphate.[2] To a mixture of glucometasaccharinic acid 6-phosphate,[3] lithium salt (930 mg), commercial vanadium oxide (V_2O_5, 9 mg), and potassium chlorate (129 mg), 3 ml of water containing 0.105 ml of phosphoric acid (85%, *d*, 1.71) is added. The mixture is well shaken, and its pH is brought to 4.6–4.8 by addition of pyridine or 8.5% phosphoric acid solution, as required. The contents of the stoppered tube are magnetically stirred for 5 days; a color change from yellow to blue-green is observed. The mixture is passed through a column (100 ml) of Amberlite IR-120 (or Dowex 50) ion exchange resin in the H^+ form, and the pH of the acidic effluent is brought to 7.5 by addition of 1 N ammonium hydroxide solution. According to the semicarbazide test[4] the solution contains about 1 mmole of α-keto

[1] P. P. Regna and P. Caldwell, *J. Amer. Chem. Soc.* **66**, 243 (1944).
[2] F. Trigalo and L. Szabó, unpublished results, 1972.
[3] S. Lewak and L. Szabó, *J. Chem. Soc.* p. 3975 (1963).
[4] J. MacGee and M. Doudoroff, *J. Biol. Chem.* **210**, 617 (1954).

acid. It is passed through a column (10 ml) of Dowex 1-X8 ion exchange resin (100–200 mesh, chloride form), the column washed with water (100 ml) and then eluted with 0.01 N HCl at a rate of 80 ml/hr. The fractions (12.5 ml) are tested for total phosphorus content,[5] and when no more phosphate is eluted (about 1200 ml of eluent), the elution is continued with 0.02 N HCl. The fractions are analyzed for both phosphorus and α-keto acid content, and those containing the phosphorylated aldulosonic acid pooled, their pH brought to 6.9 by addition of 1 N LiOH and the solution is concentrated in vacuo to about 5 ml (bath temperature, 35°). The pH of the concentrate is carefully adjusted to 7.6 with 1 N LiOH and 100 ml of EtOH are added. After some standing, the precipitate is collected by centrifugation, washed with EtOH until the supernatant is free of chloride ions (3–4 washings are required) and then once with acetone. After drying in vacuo at room temperature over phosphoric oxide, 300 mg of the Li salt (dihydrate) having $[\alpha]_D^{22}$ of $+6.5°$ ($c = 1$, water) are obtained. After enzymic dephosphorylation, the compound has a molar absorption coefficient of 93.000 in the periodate/thiobarbiturate test (this volume [6]). In the semicarbazide test[4] the phosphorylated acid's molar absorption coefficient is 10.000.

3-Deoxy-D-arabino-hept-2-ulosonic Acid 7-Phosphate

3-DEOXYHEPTONIC ACID 7-PHOSPHATE. 2-Deoxy D-arabino-hexose 6-phosphate[6] ("2-deoxy D-glucose 6-phosphate"), 1.2 g, Li salt, are dissolved in water (5 ml) and KCN (780 mg) is added. The mixture is kept in a closed vessel in the cold room for 48 hr and then decationized by passing through a column (100 ml) of Amberlite IR-120 ion exchange resin in the H+ form. The pH of the acid effluent is brought to 8.5 by addition of Ba(OH)₂, and the reaction mixture is heated on a boiling water bath for 15 min. The turbid solution is treated with charcoal, filtered, and finally concentrated in vacuo to about 10 ml. EtOH (150 ml) is added to precipitate the Ba salt of the 3-deoxyheptonic acid 7-phosphate. After some standing, the precipitate is collected by centrifugation, washed with EtOH, and dried over phosphoric oxide in vacuo at room temperature. Yield: 1.5 g, $[\alpha]_D^{22} = -3.6°$ ($c = 1$, HCl 0.1 N).

3-DEOXY D-arabino-HEPT-2-ULOSONATE 7-PHOSPHATE. The above Ba salt (530 mg) is suspended in water (10 ml), and sufficient Amberlite IR-120 H+ resin is added to bind all the Ba ions. The mixture is well stirred, and the resin is filtered off and thoroughly washed with water.

[5] M. Macheboeuf and J. Delsal, Bull. Soc. Chim. Biol. **25**, 116 (1934).

[6] Cf. (a) I. W. Hughes, W. G. Overend, and M. Stacey, J. Chem. Soc., 2846 (1949). (b) A. L. Remizov, J. Gen. Chem. USSR, **31**, 354 (1961); Chem. Abstr. **57**, 9952 (1962). (c) M. L. Wolfrom and N. E. Franks, J. Org. Chem. **29**, 3645 (1964). Commercially available from Sigma Chemical Company, Inc., P.O. Box 14508, St. Louis, Missouri 63178.

The pH of the combined filtrate and washings is brought to 9 with 1 N LiOH and the mixture is kept on the boiling water bath. More base is added until a stable value of pH 9 is obtained. The solvent is then entirely removed *in vacuo*. To the dry residue (about 350 mg of the Li salt) are added potassium chlorate (43 mg), commercial vanadium oxide (V_2O_5, 3 mg) and water (1 ml) containing phosphoric acid (85%, d, 1.71; 0.035 ml) and the mixture is then treated exactly as described for the 6-carbon analog except that, after the chromatographic separation, the fractions which contain the phosphorylated keto acid are treated with lime-water instead of LiOH, because the Li salt of the 7-carbon acid is often difficult to obtain in the solid state; no such difficulty is encountered with the Ca salt; 150 mg of the anhydrous Ca salt, having an $[\alpha]_D^{22}$ of $+18°$ ($c = 0.5$, water) are obtained. In the semicarbazide test[4] this has a molar absorption coefficient of 10.000, which is attained after 60 min of incubation at 40° (instead of the usual 15 min). In the thiobarbiturate test (this volume [6]) the dephosphorylated acid has a molar absorption coefficient of 90.000.

Comments. Although 2-deoxyglucose 6-phosphate is commercially available, it can also be easily obtained from the relatively inexpensive 2-deoxyglucose by preparing its α-methyl pyranoside,[6a] which is transformed into the 6-phosphate.[6b,c] After hydrolysis, the Li salt of the phosphorylated free sugar can be isolated by adjusting the acid solution's pH to 6.95 with LiOH, concentrating it to a small volume and adding a large amount of EtOH. The preparation contains a very small amount of inorganic phosphate which does not interfere with the oxidation reaction.

[22] 2-Keto-3-deoxy-D-gluconate

By K. KERSTERS and J. DE LEY

Principle. 2-Keto-3-deoxy-D-gluconate (KDG) is prepared by enzymic dehydration of D-gluconate.[1] This reaction can be carried out with a crude enzyme preparation. The compound is purified on a Dowex 1 formate column and crystallized as K salt.[2]

---

[1] J. De Ley, K. Kersters, J. Khan-Matsubara, and J. M. Shewan, *Antonie van Leeuwenhoek, J. Microbiol. Serol.* **36**, 193 (1970). See also this series, Vol. 42 [48].
[2] The chemical synthesis of KDG is described by D. Portsmouth [*Carbohyd. Res.* **8**, 193 (1968)]. KDG can also be prepared enzymically by dehydration of D-glucosaminic acid (J. M. Merrick and S. Roseman, this series, Vol. 9 [117]).

Reagents

Sodium D-gluconate, 1 M
Tris·HCl buffer, 0.2 M, pH 8.0
Crude D-gluconate dehydratase from *Alcaligenes* species M250,[3] in phosphate buffer, 10 mM, pH 7.0, containing 1 mM MgCl$_2$
Dowex 1 formate, X8, 200–400 mesh[4]

Procedure. The reaction mixture (350 ml) contains: 25 ml of sodium D-gluconate, 100 ml Tris·HCl buffer, 185 ml of water and 40 ml particle-free extract of *Alcaligenes* species M250[3] (step 2[3]; 11 mg of protein per milliliter). Incubation is carried out at 30° with occasional shaking. The progress of the dehydration is followed by analyzing small samples for KDG formation by the periodate-thiobarbituric acid[5] or the semicarbazide[6] assay methods. After 15–20 hr of incubation, more than 90% of the D-gluconate was converted to 2-keto-3-deoxy acid. The reaction mixture is then deproteinized by addition of 25 ml of 70% perchloric acid. After centrifugation, the supernatant is neutralized at 0° with 10 N KOH. The resulting precipitate of K perchlorate is removed by filtration. The end product is adsorbed on a Dowex 1 formate X8, 200–400 mesh column (3 cm in diameter $\times$ 50 cm), and eluted with a linear gradient of increasing formic acid concentration. The reservoir contains 600 ml of 0.4 M formic acid, and the mixing chamber contains 600 ml of water. A small peak of residual gluconate is eluted at about 0.12 M formic acid. The content of the tubes with KDG (eluting at about 0.3 M formic acid) are pooled and formic acid is removed by repeated evaporations under reduced pressure. The solution is neutralized with 5% KOH, concentrated to a syrup *in vacuo* and solubilized in warm methanol, containing 1% of water. Crystallization of the K salt occurs overnight at 4°. The crystals are collected by filtration in the cold, washed with cold ether, recrystallized twice under similar conditions, and dried *in vacuo* (over CaCl$_2$) for 1 day at 4° and several days at room temperature. The first crystallization is sometimes difficult. The overall yield is approximately 70%. Alternatively, KDG can be crystallized as Ca salt.[2,7]

Our compound has been identified as KDG by a number of chemical analyses.[1] KDG can easily be detected on cellulose thin-layer chromato-

---

[3] For full details: see this series, Vol. 42 [48]. The particle-free extract (step 2) of the gluconate-grown bacteria can be used. Synonymous strain numbers are ATCC 9220, and AB 61 from Dr. H. Lautrop, Copenhagen, Denmark. See also this series, Vol. 42 [48], footnote 4.

[4] See this series, Vol. 9 [6].

[5] A. Weissbach and J. Hurwitz, *J. Biol. Chem.* **234**, 705 (1959).

[6] J. MacGee and M. Doudoroff, *J. Biol. Chem.* **210**, 617 (1954).

[7] R. Bender, J. R. Andreesen, and G. Gottschalk, *J. Bacteriol.* **107**, 570 (1971).

grams by spraying with either periodate-thiobarbituric acid[8] or o-phenyl-enediamine.[9] The K salt is probably a monohydrate[1] and is stable for several years when stored dry at −12°.

[8] L. Warren, *Nature* (*London*) **186**, 237 (1960).
[9] T. Wieland and E. Fischer, *Naturwissenschaften* **36**, 219 (1949).

[23] DL- and L-2-Keto-3-deoxyarabonate[1,2]

By ALLEN C. STOOLMILLER

DL-2-Keto-3-deoxyarabonate

Principle. DL-2-Keto-3-deoxyarabonate is prepared by the condensation of oxaloacetate and glycolaldehyde.[3]

Procedure. To 40 ml of 50 mM K_2HPO_4 is added 12 mmoles of oxaloacetic acid. The solution is neutralized with KOH (pH 7), and 8 mmoles of glycolaldehyde are added. After 15 hr, the solution is adjusted to pH 3 by addition of Dowex 50 (20–50 mesh, hydrogen form) which is then removed by filtration. The filtrate is "degassed" for 5 min under reduced pressure and adjusted to pH 5 with KOH.

The solution containing DL-2-keto-3-deoxyarabonate is applied to a column (4.0 × 40 cm) of Dowex 1-X8 (200–400 mesh, chloride form) which has been washed with water. The column is eluted with 150 ml of 1 mM HCl and then 800 ml of 50 mM HCl. The fractions containing DL-2-keto-3-deoxyarabonate are pooled[4] and adjusted to pH 5 with KOH. This solution is divided into two equal portions, and each is chromatographed on a column (2.2 × 30 cm) of Dowex 1-X8 (200–400 mesh, formate form). The DL-2-keto-3-deoxyarabonate is eluted with a convex gradient of pyridinium-formate buffer, pH 3. A 2-liter mixing chamber containing 0.1 *M* formic acid is adjusted to pH 3.0 with pyridine, and a 4-liter reservoir contains 0.5 *M* formic acid adjusted to pH 3.0 with pyridine. The DL-2-keto-3-deoxyarabonate is eluted with approximately 1.5 liters of the gradient buffer.[5] The fractions containing the product

[1] A. C. Stoolmiller and R. H. Abeles, *J. Biol. Chem.* **241**, 5764 (1966).
[2] According to standard chemical nomenclature, DL-2-keto-3-deoxyarabonate would be designed as 3-deoxy-2-oxo-DL-arabonate; however, the present nomenclature is used for consistency with earlier publications.
[3] This reaction is analogous to the condensation of oxaloacetate and methylglyoxal described by M. Henze, *Hoppe-Seyler's Z. Physiol. Chem.* **189**, 121 (1930).
[4] When 10-ml fractions are collected, the DL-2-keto-3-deoxyarabonate peak is between fractions 60 and 90.

are pooled, and the solution is passed through a column (1.4 × 30 cm) of Dowex 50 (20–50 mesh, hydrogen form) at a flow rate of 3 ml/min to remove pyridine. The column effluent is concentrated to a syrup under reduced pressure to remove most of the formic acid. Final traces of acid are removed by repeated evaporation (4 times) of 10-ml quantities of water. The approximate yield of DL-2-keto-3-deoxyarabonate is 7 mmoles (85–90% based on glycolaldehyde).

L-2-Keto-3-deoxyarabonate

Principle. L-2-Keto-3-deoxyarabonate is isolated from incubations of cell-free extracts of *Pseudomonas saccharophila* with L-arabonate.[6] These compounds are intermediates in the metabolism of L-arabinose by *P. saccharophila.*[7]

Reagents

Tris-chloride buffer, 0.1 M, pH 8.0
Potassium L-arabonate[8]
MgSO$_4$, 50 mM
EDTA, 0.1 M, adjusted to pH 8
Enzyme: A streptomycin sulfate-treated sonic extract of *P. saccharophila* (26 mg of protein per milliliter). See this series, Vol. 42 [50].

Procedure. The reagents are added in the following order and brought to a final volume of 12 ml with water: 2.4 ml of Tris-chloride buffer, 2.4 mmoles of L-arabonate, 2.4 ml of MgSO$_4$, and 1 ml of EDTA. The reaction is initiated by addition of 3 ml of enzyme and incubated at 30° for 3 hr. The reaction is terminated by acidification to pH 3 with Dowex 50 (20–50 mesh, H$^+$ form). The resin and denatured protein are removed by filtration, and the filtrate is concentrated to 5 ml under reduced pressure. An equal volume of absolute ethanol is added, the additional precipitate is removed by filtration, and the filtrate is further concentrated to approximately 1 ml.

The syrup is applied to 8 sheets (23 × 57 cm) of Whatman No. 3 MM chromatography paper and chromatographed for 18 hr in descending fashion with 1-butanol–pyridine–H$_2$O (6:4:3) as solvent. The L-2-keto-

---

[5] When 15-ml fractions are collected, the DL-2-keto-3-deoxyarabonate peak is between fractions 105 and 135.

[6] R. Weinberg, *J. Biol. Chem.* **234**, 727 (1959).

[7] R. Weinberg and M. Doudoroff, *J. Biol. Chem.* **217**, 607 (1955).

[8] L-Arabonate (K salt) was prepared by the hypoiodite oxidation of L-arabinose in aqueous methanol according to S. Moore and K. P. Link, *J. Biol. Chem.* **133**, 293 (1940).

3-deoxyarabonate is located on a test-strip cut from the center of each chromatogram with a periodate-benzidine spray.[9] Areas of the chromatograms corresponding to the location of L-2-keto-3-deoxyarabonate, R_f's 0.24–0.32, are cut out and eluted with water, and the eluent is concentrated under reduced pressure. L-2-Keto-3-deoxyarabonate is obtained in approximately 15% yield.

Determinations

DL- and L-2-keto-3-deoxyarabonate are assayed either chemically or enzymically. The carbonyl group is assayed with 2,4-dinitrophenylhydrazine[10]; the α-keto acid function with semicarbazide[11]; formaldehyde[12] and β-formylpyruvate[13] are determined colorimetrically following periodate oxidation[14]; and the lactone, which forms upon heating at 100° in 0.1 M HCl for 15 min, is measured with hydroxylamine-ferric chloride.[15]

The enzymic assay is based on the absorbance increase which occurs during the NADP-dependent conversion of L-2-keto-3-deoxyarabonate to α-ketoglutarate.[16] Reagents are added to a 1-ml quartz cell with a light path of 1 cm in the following order and brought to a final volume of 0.9 ml with water: 0.4 ml of K_2HPO_4–KH_2PO_4, pH 7.4, containing 20 mM 2-mercaptoethanol, 0.1 ml of 5 mM $NADP^+$, 0.03–0.04 unit of aldehyde dehydrogenase, and 0.01 unit of L-2-keto-3-deoxyarabonate dehydratase. After reading the initial optical density at 340 nm against a water blank, the reaction is initiated by adding 0.1 ml of a solution containing 0.01–0.2 μmole of L-2-keto-3-deoxyarabonate, and the absorbance increase at 340 nm due to NADPH formation is measured spectrophotometrically. The reaction is complete within 5–10 min.

[9] J. A. Cifonelli and F. Smith, *Anal. Chem.* **26**, 1132 (1954).
[10] H. Bohme and O. Winkler, *Z. Anal. Chem.* **412**, 1 (1954).
[11] J. MacGee and M. Doudoroff, *J. Biol. Chem.* **210**, 617 (1954).
[12] D. A. McFadyen, *J. Biol. Chem.* **158**, 107 (1945).
[13] A. Weissbach and J. Hurwitz, *J. Biol. Chem.* **234**, 705 (1959).
[14] W. R. Frisell, L. A. Meech, and C. G. Mackenzie, *J. Biol. Chem.* **207**, 709 (1954).
[15] S. Hestrin, *J. Biol. Chem.* **180**, 249 (1949).
[16] See this series, Vol. 42 [50].

[24] D-Ribulose

By R. P. Mortlock

D-Ribulose may be prepared by chemical or enzymic methods. Normally enzymic means of preparation yield D-ribulose possessing higher biological activity as determined by measuring the percentage of sugar utilized by a specific enzyme. Methods for the enzymic or chemical preparation of D-ribulose have been described previously.[1] The chemical method involves refluxing D-ribose with dry pyridine and then separation of the small yield of D-ribulose from excess D-ribose and contaminating sugars by means of a Dowex borate column or by preparation of the o-nitrophenyl hydrazone. The enzymic method which has been described requires the purification of ribitol dehydrogenase to catalyze the oxidation of ribitol to D-ribulose. Recently a procedure has been described utilizing whole cells of a strain of *Klebsiella aerogenes* to oxidize ribitol (adonitol) to D-ribulose. By this method the D-ribulose is collected in the cell-free supernatant after removal of the cells by centrifugation. The advantages of the latter method are the high yield of D-ribulose, approaching 100% of the ribitol added to the cell suspension, and the ease of recovery of the D-ribulose from the cell-free supernatant. A mutant strain of *Aerobacter aerogenes* which can be employed for this purpose is commercially available.[2]

Principle. *Klebsiella* (*Aerobacter*) *aerogenes* possesses an inducible enzyme pathway for the degradation of ribitol. Ribitol dehydrogenase catalyzes the oxidation of ribitol to D-ribulose, which is then phosphorylated by D-ribulose kinase with the formation of D-ribulose 5-phosphate. Mutants constitutive for the enzymes of this pathway can be isolated by utilizing an uncommon pentitol, xylitol, as the sole carbon and energy source for growth. A further isolation of a D-ribulokinase-negative mutant results in a strain which is constitutive for ribitol dehydrogenase but is unable to grow utilizing ribitol as the carbon and energy source. Such a mutant will oxidize ribitol to D-ribulose with the accumulation of D-ribulose in the medium.

Method. The mutant strain of *Aerobacter aerogenes* PRLR3 utilized for the production of D-ribulose is a uracil-requiring auxotroph, constitutive for ribitol dehydrogenase but negative for D-ribulose kinase activity.[3] The organism is grown in 200 ml of medium in a 1-liter flask on a New Brunswick rotary shaker at 30°. The salts medium consists of 1.5 g of KH_2PO_4, 7.2 g of Na_2HPO_4, 3.0 g of $(NH_4)_2SO_4$, 0.20 g of $MgSO_4$, 0.005 g of $FeSO_4$, in 1 liter of distilled water. Uracil is added to a concentration of 0.05 g per liter, and casein hydrolyzate to a concentration of 10 g per liter. Concentrated solutions of magnesium sulfate, uracil, and casein

[1] This series, Vol. 9 [39].

[2] Sigma Chemical Co., St. Louis, Missouri.

[3] E. J. Oliver, T. M. Bisson, D. J. LeBlanc, and R. P. Mortlock, *Anal. Biochem.* **27,** 300 (1969).

hydrolyzate are autoclaved separately and added to the basic salts medium after cooling. After inoculation of the medium, growth is at 30° under aerobic conditions. When growth is complete the cells are harvested by centrifugation, resuspended in distilled water to one-tenth of their original volume, collected by centrifugation a second time, and then resuspended in one-half of the original volume of sterile 5 mM phosphate buffer at pH 7.5. Ribitol is added to a concentration of 0.5%, and the cell suspension is incubated under the same aerobic conditions at 30°.

Samples are removed at time intervals to measure D-ribulose formation by means of the cysteine-carbazole test of Dische and Borenfreund.[4] Using D-ribulose-o-nitrophenylhydrazone as a standard,[2] when the quantity of D-ribulose present is equal to the amount of ribitol initially added, or if the amount of D-ribulose present reaches a stationary value, the cells are removed by centrifugation and the cell-free supernatant is saved. If centrifugation is carried out using aseptic technique, the cell pellet may be resuspended in fresh phosphate buffer and additional ribitol added for the production of more D-ribulose. The absence of uracil during the oxidation of ribitol to D-ribulose ensures that D-ribulose kinase-positive revertants will not be selected during this procedure.

Occasionally with this strain of *Aerobacter aerogenes*, the percentage of ribitol converted to D-ribulose did not reach 100%, and after reaching a stationary value the amount of D-ribulose present in the medium began to decrease. It was observed that D-ribulose could be utilized by the L-fucose catabolic pathway and the enzymes of this inducible pathway were normally present at high basal levels. An additional mutation was added to the strain to make it unable to utilize L-fucose as a growth substrate. With this latter strain, yields of D-ribulose approaching 100% have been consistently obtained. In two separate experiments using 0.2% and 0.5% ribitol, respectively, over 95% of the ribitol added was converted to D-ribulose within 4 hr incubation.

When the oxidation of the ribitol to D-ribulose is completed, the cells are removed by centrifugation and the cell-free supernatant concentrated to about one-fifth of its original volume by evaporation with the temperature maintained under 40°. The solution is deionized by passage through Dowex 50 (H⁺) and Dowex 3 (CO_3^{2-}). After further concentration by evaporation under vacuum, the preparation obtained is sufficiently pure to be utilized in most experiments requiring D-ribulose as a substrate. Traces of ribitol and D-arabinose increase with storage of

---

[4] This series, Vol. 3 [12].

the frozen solution. If additional purity is required, the o-nitrophenyl-hydrazone derivative can be prepared and crystallized as described by Cohen.[5]

[5] S. S. Cohen, *J. Biol. Chem.* **201**, 71 (1953).

[25] An Enzymic Synthesis Yielding Crystalline Sodium Pyruvate Labeled with Isotopic Hydrogen

By H. Paul Meloche

Principle. Selected lyases catalyze reactions proceeding through enzyme-bound pyruvyl enolates. When such reactions are carried out in water labeled with isotopic hydrogen, an exchange reaction between the methyl protons of pyruvate and hydrogen isotope occurs. Since the three methyl hydrogens of pyruvate are symmetrical, the reaction equilibrates at the exchange of three equivalents of label into pyruvate. This article details the use of 2-keto-3-deoxy-6-P-gluconate (KDPG) aldolase[1] for such a reaction since this enzyme is fully active in a saturated pyruvate solution.[2] A mathematical treatment of initial data is presented, which allows one to predict the incubation time required to approach equilibration between the three methyl hydrogens of pyruvate and hydrogen isotope of solvent. In addition, an experiment is described showing how one converts the ratio of deuteration to tritiation into tritium and deuterium isotope effect values.

General Method

Reagents

Crystalline sodium pyruvate
Purified pyruvate lyase (2-keto-3-deoxy-6-P-gluconate aldolase)[1]
TOH or D_2O
Ethanol (95%)

[1] H. P. Meloche, J. M. Ingram, and W. A. Wood, this series, Vol. 9, p. 520.
[2] H. P. Meloche and Lillian Lin, *Abstr. 73rd Meeting, Amer. Soc. Microbiol.* **176**, P 215 (1973).

Procedure. A saturated (4.55 M) solution is prepared by dissolving 500 mg of crystalline sodium pyruvate in 1.0 ml of L_2O (where L is H, D, or T), with warming. To this is added concentrated enzyme solution in H_2O (20–50 μl). The above mixture is then transferred to a vial suitable for evacuation and sealing under vacuum. The sample is frozen in a dry-ice bath, removed from the bath, and evacuated to less than 100 microns pressure. During this process the solution will thaw, allowing efficient degassing to occur. Care should be exercised to minimize "bumping." While under vacuum, the vial is sealed using a gas-oxygen torch. An alternative approach is to evacuate the air and replace it with pre-purified nitrogen. After repeating this process three times, the vial is sealed under reduced pressure. It is most important to exclude air from the vial during incubation; otherwise the bulk of pyruvate decomposes and residual product cannot be easily recovered as a crystalline salt. This will be discussed further below. Incubation, if carried out at an elevated temperature, should be carried out in a constant temperature oven, in preference to a water bath, to avoid "distillation" of solvent within the sealed vial. At the end of the incubation period, the vial is opened and its contents are transferred to a distillation tube. The solvent is recovered by lyophilization and collected, while the residue is dissolved in 1 ml of water. To this solution is added 4 ml of 95% ethanol to affect crystallization, which initiates immediately. The ethanol also serves to inactivate KDPG aldolase. The preparation is routinely stored overnight in a refrigerator, then the salt is recovered by low speed centrifugation, dried, and the crystallization step is repeated. The second crystals, recovered by centrifugation, are dried *in vacuo*. In our hands, this procedure has been carried out a number of times with an average sodium pyruvate yield of 70%.

Determination of Incubation Time

In our original article[3] we described an experiment for determining incubation time. In this study, 31.45 IU of aldolase (20 μl) were introduced to 1 ml of saturated sodium pyruvate in TOH having a specific activity of 3.78×10^3 cpm/μg atom H. Incubation was carried out for 16 hr at 25°. The recovered crystalline sodium pyruvate was assayed for radioactivity by liquid scintillation counting and α-keto acid using semicarbazide.[4] The specific activity of the [3-^{3}H]pyruvate was found to be 7.5×10^3 cpm/μmole, corresponding to the exchange of 1.985 equivalents of tritium into pyruvate.

[3] H. P. Meloche, *Anal. Biochem.* **38**, 389 (1970).
[4] J. MacGee and M. Doudoroff, *J. Biol. Chem.* **210**, 617 (1954).

From these data one can calculate a first-order rate constant for exchange using the relationship

$$k = 1/t \ln [a/(a - x)] \tag{1}$$

where a is the concentration of protiopyruvate $(4.55\ M)$ and x is the concentration of tritiopyruvate $(4.55 \times 1.985/3 = 3.01\ M)$. Equation (1) becomes

$$k = 1/16 \ln [4.55/(4.55 - 3.01)]$$

and calculates to a k value of 0.068 hr^{-1}. Substituting this value into Eq. (1), one solves for 90% equilibration in 34 hr, and 99% in 68 hr at an incubation temperature of 25°, using 31.45 units of enzyme.

A kinetic primary isotope effect is anticipated in hydrogen isotope exchange reactions, where proton activation is rate limiting, owing to the differences in zero-point energies of C—H vs C—D vs C—T bonds.[5] For enzyme-catalyzed reactions of the type discussed in this article, the isotope effect will be seen if the enolate protonation step is low relative to dissociation of the enzyme–pyruvate complex. Under conditions wherein a primary kinetic isotope effect occurs, deuterium exchanges appreciably faster than tritium. If the value for the tritium isotope effect were known, one could then solve for the magnitude of the deuterium isotope effect. The ratio of isotope effects would then be related to the magnitude of k in Eq. (1), in solving for time required for deuteration of pyruvate. The relationship between the isotope effects is[6]

$$k_H/k_T = (k_H/k_D)^{1.44} \tag{2}$$

where k_H/k_T is the tritium isotope effect, and k_H/k_D is the deuterium isotope effect. The reader should note that from the simple exchange studies described above, one cannot determine whether an isotope effect occurs, or its magnitude, since there is no way of measuring k_H.

Thus, as an addendum, an appropriate means of determining the occurrence and value of tritium and deuterium isotope effects for exchange is presented as an illustration, as catalyzed by KDPG aldolase. The reaction was carried out in tritiated D_2O, where the final concentration of D_2O was 91.07% and the specific activity of tritium was 6200 cpm/μg atom D(H). Incubation was carried out at 25° in the presence of 31.45 IU of KDPG aldolase for 7 hr. By scintillation counting and enzymatic assay the isolated crystalline sodium pyruvate was found to have a spe-

[5] W. P. Jencks, "Catalysis in Chemistry and Enzymology," Chapter 4. McGraw-Hill, New York, 1969.
[6] C. G. Swain, E. C. Stivers, J. F. Reuwer, Jr., and C. J. Schaad, *J. Amer. Chem. Soc.* **80**, 5885 (1958).

cific activity of 6270 cpm/μmole, equivalent to the incorporation of 1.01 tritons. By nuclear magnetic resonance spectroscopy,[7] the methyl group of the pyruvate was found to have 1.26 protons per mole, indicating the incorporation was $(3.0 - 1.26)/0.9107 = 1.91$ deuterons. The ratio of deuterium exchange to tritium exchange then is 1.91/1.01 or 1.89, showing that hydrogen-isotope discrimination occurs. From this ratio and Eq. (2) one can solve for k_H/k_D and k_H/k_T since

$$(k_H/k_D)^{\frac{1}{0.44}} = D \text{ incorporated}/T \text{ incorporated} \tag{3}$$

and

$$k_H/k_T = (k_H/k_D) (D \text{ incorporated}/T \text{ incorporated}) \tag{4}$$

Thus, k_H/k_D is $(1.89)^{\frac{1}{0.44}}$ or 4.25, and k_H/k_T is (4.25×1.89) or 8.03.

Comments

The lability of sodium pyruvate incubated in the presence of air is seen below. A solution containing 4.5 mmoles of sodium pyruvate was incubated 72 hr at 35° in a sealed vial. After incubation, only 37.2% of the pyruvate was recovered as shown by enzymic assay. An attempt to crystallize this material resulted in the formation of an oil. After storage overnight at room temperature, solvents were removed in vacuo. Enzymic assay of the residue showed that only 20.9% of the original pyruvate remained. Consequently, decomposition of pyruvate occurs during incubation in air. The further loss of pyruvate appears to be accelerated during attempts at crystallization.

LABILITY OF PYRUVATE

Step	Pyruvate (μmoles)	Recovery (%)
Original	4500	100
72 Hr incubation	1675	37.2
After attempted crystallization	942	20.9

In contrast, saturated pyruvate solutions incubated either in vacuo or under N_2 for at least 4 days at 35° will yield crystalline sodium pyruvate in 70% yield.

[7] The author is indebted to Dr. A. S. Mildvan for his assistance in performing this analysis.

If appreciable amounts of pyruvate are lost during incubation, the remaining pyruvate can be recovered by chromatography of Dowex 1 (chloride form) eluted with a 0–0.1 N HCl linear gradient. Pyruvate appears at about 30% of the elution profile. Tubes containing pyruvate are pooled, adjusted to pH 4–4.5 with NaOH, and lyophilized. The residue is dissolved in water, and the crystalline salt is recovered by the addition of ethanol. Yields are poor.

The purity of tritiated pyruvate was confirmed by Sephadex G-10 chromatography.[3] Virtually all the radioactivity was associated with the α-keto acid. In general, about 0.1% of the tritium is found as TOH in freshly prepared water solutions of the reagent. It should be noted, however, that slow exchange, presumably via enolization, occurs on prolonged storage of water solutions kept under refrigeration. Tritiated sodium pyruvate prepared by exchange in TOH at a specific activity of 1 Ci/ml is stable for at least a year upon storage as the crystalline salt at room temperature as revealed by enzyme assay or chromatography on Sephadex G-10. Whether radiolytic decomposition of higher specific activity preparations occurs upon storage is not known as of this writing. Presumably, deuterated and lower specific activity tritiated preparations would remain stable indefinitely if stored cold and dry.

Finally, the methodology described in this article can be adapted to general α-keto acid labeling using hydrogen isotopic water and appropriate enzymes, which can turn over high concentrations of substrate at high rates affecting efficient synthesis and good yields of labeled product.

[26] Preparation of Phosphoenolpyruvate and Pyruvate Specifically Labeled with Deuterium and Tritium

By IRWIN A. ROSE

Specific labeling of the hydrogens of phosphoenolpyruvate (PEP) and pyruvate can be readily accomplished with commercially available reagents and enzymes,[1] using all the reactions of glycolysis. The pertinent reactions with their stereospecificities[2-6] are shown in Scheme 1.

[1] The enzymes were obtained from Boehringer Mannheim Co. and Sigma Chemical Co.
[2] I. A. Rose and E. L. O'Connell, *Biochim. Biophys. Acta* 42, 159 (1960).
[3] Y. J. Topper, *J. Biol. Chem.* 225, 419 (1957).
[4] C. K. Johnson, E. J. Gabe, M. R. Taylor, and I. A. Rose, *J. Amer. Chem. Soc.* 87, 1802 (1965).

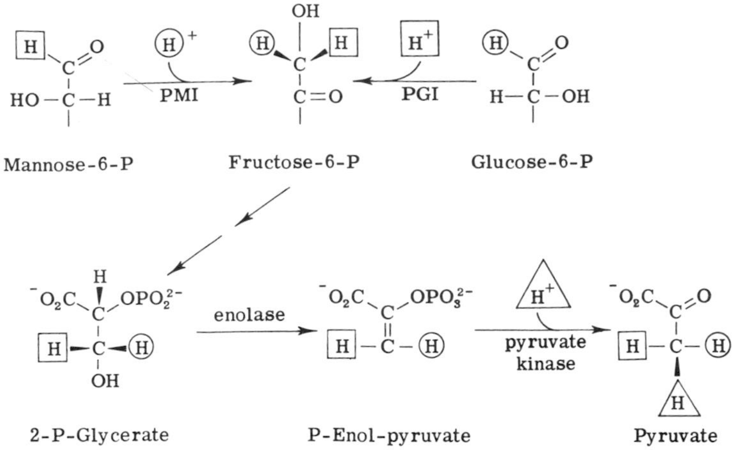

SCHEME 1

Isotope (tritium and deuterium) may be introduced into fructose-6-P (isolated as fructose diphosphate) by the phosphomannose isomerase (PMI) reaction with [1-³H]mannose-6-P and ²H₂O or by the phosphoglucose isomerase (PGI) reaction with [1-³H]glucose-6-P and ²H₂O. By performing the pyruvate kinase step in H₂O, these routes yield R-pyruvate and S-pyruvate, respectively. By performing the isomerase steps in H₂O and the pyruvate kinase step in ²H₂O, these routes yield S- and R-pyruvate, respectively. Since for many stereochemical experiments it is desirable to have both the R- and S-compounds for comparisons, two of the four alternate routes will be needed.

Preparation of Z[3-³H]PEP from [1-³H]Glucose[7]

The [1-³H]glucose is converted to fructose diphosphate (FDP) in H₂O in a 10-min incubation containing [1-³H]glucose (2.4 μmoles, 10⁷ cpm), Tris chloride (50 μmoles, pH 7.5), ATP, and MgCl₂ (5 μmoles each), and about 1 unit each of hexokinase, phosphoglucose isomerase, and P-fructokinase in 1 ml. The 1S-[1-³H]FDP produced is recovered on a Dowex 1 (chloride form) column (1 × 6 cm) by elution with 0.1 N HCl after removal of any monophosphate esters with 0.04 N HCl (yield,

[5] M. Cohn, J. E. Pearson, E. L. O'Connell, and I. A. Rose, *J. Amer. Chem. Soc.* **92**, 4095 (1970).

[6] I. A. Rose, *J. Biol. Chem.* **245**, 6052 (1970).

[7] I. A. Rose, E. L. O'Connell, P. Noce, M. F. Utter, H. G. Wood, J. M. Willard, T. G. Cooper, and M. Benziman, *J. Biol. Chem.* **244**, 6130 (1969).

75%). The FDP is neutralized with NaOH, concentrated, and converted to glycerate-3-P as follows: 2 ml contain the FDP (1.8 μmoles, 9×10^6 cpm), Tris chloride (50 mM, pH 7.5), EDTA (5 mM), Na_2HAsO_4 (5 mM), DPN (2.5 mM) glyceraldehyde-P dehydrogenase (3.6 units), triose-P isomerase (240 units), aldolase (1 unit), and lactate dehydrogenase (1 unit). When the increase in absorbance at 340 nm begins to slow, pyruvate (1 μmole) is added stepwise in order to remove the DPNH. After about 1 hr of incubation the glycerate-3-P (3-PGA) can be isolated on Dowex 1 (Cl⁻) (1 $\times$ 6 cm) by elution with 0.03 N HCl followed by evaporation *in vacuo*. The equilibrium between 3-PGA, 2-PGA, and PEP is established at 55° to increase the yield of PEP in the next step as follows: 3-PGA (7×10^6 cpm), Tris chloride (0.1 M, pH 7.5), $MgCl_2$ (5 mM), EDTA (2 mM), glycerate-2,3-di-P (20 μM), and 2 units each of enolase and phosphoglycerate mutase are incubated in 1 ml. After 1 hr the reaction is terminated with 200 μmoles of HCl. The neutralized and diluted solution is placed on Dowex 1 (Cl⁻) for separation of glycerate phosphates (5.6×10^6 cpm) and PEP (8.9×10^5 cpm) according to Bartlett[8] using 20 mM HCl and 40 mM HCl, respectively. PEP may be recovered without loss by direct lyophilization of the aqueous acid. Since the larger substituent atoms at C_2 and C_3 are in a *cis* relation, this is designated $Z[3-^3H]$PEP.[9]

Preparations of E[3-³H]PEP and E[3-²H]PEP

The tritiated species can be made most conveniently by a similar series of steps except that [1-³H] mannose and phosphomannose isomerase replace the glucose and phosphoglucose isomerase. The following procedure is satisfactory for preparation of $1R$-[1-³H]FDP: [1-³H]mannose (1 mM), triethanolamine-Cl (100 mM, pH 8), ATP (0.4 mM), $MgCl_2$ (5 mM), KCl (20 mM), PEP (3 mM), phosphofructokinase (6 units), and 1 unit each of yeast hexokinase, phosphomannose isomerase, and pyruvate kinase are incubated per milliliter for 1 hr or until the expected yield of FDP is found by assay of an acidified sample. Since the enzymes are free of phosphoglucose isomerase, no exchange of tritium into water should be observed. Tritiated FDP is recovered in good yield by ion-exchange chromatography as before, and the conversion of FDP and PEP follows the previous procedures. The steps from FDP to PEP were as before. An alternate route for E[3-³H]PEP is from glucose-6-P and ³H_2O in the reaction with phosphoglucose isomerase.[7]

[8] G. R. Bartlett, *J. Biol. Chem.* **234**, 459 (1959).

[9] J. E. Blackwood, C. L. Gladys, K. L. Loening, A. E. Petrarca, and J. E. Rush, *J. Amer. Chem. Soc.* **90**, 509 (1968).

In making the deuterated PEP, two sources of dilution by protium have to be avoided. The first problem results from the fact that about half of the F6P formed in the PGI reaction in ^{2}H$_2$O will have received a proton directly from C-2 of the G6P and the remainder from the medium.[10] Thus both the G6P and F6P must be equilibrated by the isomerase reaction in ^{2}H$_2$O before the trapping system for conversion to FDP is added. The degree of isotopic equilibration may be assessed by adding ^{3}H$_2$O to a part of the isomerase incubation and determining the radioactivity fixed into nonvolatile form with time or preferably by adding [2-^{3}H]glucose-6-P, if available, to an aliquot of the ^{2}H$_2$O incubation and following the release of tritium. Complete equilibration with deuterium in this step is important if chiral pyruvate is to be prepared for stereochemical studies, since each tritium-containing methyl group must contain a deuterium and protium in order to make use of the full isotope effect of a test reaction. The second source of dilution occurs owing to the formation of PGA from both halves of FDP in glycolysis. This dilution does not alter the deuterium content of the tritiated species and so can be tolerated in preparations of chiral pyruvate for stereochemical studies. In cases where a high deuterium content is required in PEP, as for certain nuclear magnetic resonance and isotope effect studies, the dihydroxyacetone-P of the aldolase reaction is separated after oxidation of glyceraldehyde-3-P and is then converted to PEP.

The following procedure has been used[5] to prepare E[2-^{2}H]PEP: G6P (2000 μmoles of the K$_2$ salt), 50 μmoles of MgCl$_2$, 10 μmoles of adenosine diphosphate, 200 μmoles of Tris chloride (pH 8.0) in 10 ml of ^{2}H$_2$O (99%) and ^{3}H$_2$O (1540 cpm/μatom of H) are incubated with 1 mg of P-glucose isomerase (390 units) for 3 hr at 25° and 10 hr at 3°. Complete equilibration is shown by the specific activity of G6P + F6P (1800 μmoles), 1540 cpm/μmole. To this solution are added P-glucose isomerase (1 mg, 390 units), pyruvate kinase (2 mg, 250 units), and P-fructokinase (2 mg, 200 units). K-PEP 2400 μmoles in ^{2}H$_2$O is added stepwise to the solution at 25° over a period of about 2 hr. By using the pyruvate kinase reaction to generate ATP, it is possible to maintain ATP, a strong inhibitor of P-fructokinase, at low concentration. FDP formation (1600 μmoles) is followed on acidified samples. The ^{2}H$_2$O may be recovered by freeze-drying if desired and the residue treated with 50 mM HgCl$_2$ (10 ml) to destroy excess PEP and inactivate the enzymes. The FDP is purified by ion exchange on Dowex 1 (Cl$^-$) (2 $\times$ 11 cm) by elution with 0.1 N HCl and precipitated as the Ba salt, 1430 μmoles, 1300 cpm/μmole. FDP is converted to the triose phosphate mixture in the presence of hydrazine

---

[10] I. A. Rose and E. L. O'Connell, *J. Biol. Chem.* **236**, 3086 (1961).

to displace the reaction toward products. FDP (1400 μmoles), bovine serum albumin (30 mg), EDTA (300 μmoles), hydrazine sulfate (20 nmoles), NaHCO$_3$ buffer (2.5 mmoles), and 30 mg of muscle aldolase (420 units) purified to be free of triose-P isomerase[11] are incubated at pH 8.5, 25°, in 25 ml for 60 min, at which time the increase in alkali labile phosphate (10 min at 25° in 1 N NaOH) due to triose phosphates is complete. Protein is precipitated with HClO$_4$ (20 ml of 70%) and the solution is neutralized in the cold with KOH and treated with five 20-ml portions of benzaldehyde to remove the hydrazine and then with ether to remove benzaldehyde. The neutral mixture of triose phosphates is diluted to 1.1 and poured on a Dowex 1 (Cl⁻) column (2 × 15 cm); the mixture of dihydroxyacetone-P (800 μmoles) and D-glyceraldehyde-3-P (670 μmoles) is eluted with 0.025 N HCl in 400 ml. This solution is treated with excess Br$_2$ at pH 5 (25 mM sodium acetate buffer) for 12 hr in the cold in order to oxidize the glyceraldehyde-P. The Br$_2$ is extracted with ether; the neutralized solution is poured on Dowex 1 (Cl⁻) column (2 × 15 cm), and dihdroxyacetone-P (570 μmoles) eluted with 0.02 N HCl. It precedes the appearance of 3-PGA by 100 ml. To the concentrated solution of dihydroxyacetone-P in 20 ml containing 2 M Tris chloride pH 8.5, is added 2 mg of triose-P isomerase and after 30 min at 25° the solution, now largely the Tris adduct of glyceraldehyde-P, is put through Dowex 50 (H⁺) to acidify the product and remove the Tris. The combined column effluent and wash are treated with Br$_2$ as above and the 3-PGA (380 μmoles, 1490 cpm/μmole), recovered from a Dowex 1 (Cl⁻) column with 0.02 N HCl, and taken to dryness *in vacuo*. PEP is formed at 55° in 15 ml of solution adjusted to pH 8 and containing: MgCl$_2$, 150 μmoles; EDTA, 30 μmoles; 2,3-diphosphoglycerate, 0.3 μmole; 15 mg of bovine serum albumin; enolase, 1 mg, 27 units; and phosphoglycerate mutase (1 mg, 18 units). After 1 hr the reaction is terminated with acid and the 3-PGA and PEP are separated on Dowex 1 (Cl⁻). The combined PEP (180 μmoles, 1250 cpm/μmole), which is neutralized immediately after elution, is adsorbed on Dowex 1 (Cl⁻) and eluted as before, and the HCl is removed by lyophilization.

Conversions to Pyruvate

The reactions of phosphoglycerate mutase, enolase, and pyruvate kinase are coupled together for the conversion of PGA to pyruvate. The production of pyruvate must be done with careful attention to detail owing to the possibility of excessive exchange with medium protons in

[11] O. C. Richards and W. J. Rutter, *J. Biol. Chem.* **236**, 3185 (1961)

the pyruvate kinase step depending on pH, nature and concentration of mono- and divalent cations, and accumulation of pyruvate.[12] If the reaction is to be done in ^2H_2O, all reagents must be dissolved and enzymes diluted in ^2H_2O. A typical reaction mixture using phosphoglycerate as the starting material for the conversion at 25° is the following: imidazole buffer (50 mM, pH 6.5), KCl (50 mM), MgCl$_2$ (5 mM), ADP (0.2 mM), 2,3-diPGA (5μM) EDTA (1 mM), glucose (5 mM), and 0.2 unit/ml of each of phosphoglycerate mutase, enolase, hexokinase, and pyruvate kinase, and the [^2H, ^3H]PGA samples (0.5 mM to 1.5 mM). The glucose and hexokinase are present to regenerate ADP consumed in the pyruvate kinase reaction. Pyruvate formation is monitored on small samples with lactate dehydrogenase. The reaction should be completed in about 10 min. The use of higher pH or much higher concentrations of K$^+$ will lead to some scrambling of the hydrogens in the pyruvate due to enolization of the pyruvate prior to its dissociation from pyruvate kinase[12] and should be avoided. High concentrations of pyruvate kinase and prolonged contact with the pyruvate will lead to enolization of free pyruvate. This is minimized by removal of ATP by hexokinase. The pyruvate formed is isolated on a Dowex 1 (Cl$^-$) column by elution with 10 mM HCl.

[12] J. L. Robinson and I. A. Rose, *J. Biol. Chem.* **247**, 1096 (1972).

[27] DL-2-Keto-4-hydroxyglutarate[1]

By EUGENE E. DEKKER and UMADAS MAITRA

2-Keto-4-hydroxyglutarate may be prepared by any of the following procedures: (a) nonenzymic transamination of 4-hydroxyglutamate with pyridoxal[2]; (b) enzymic transamination of 4-hydroxyglutamate, catalyzed by glutamate–aspartate transaminase[3,4]; (c) chemical condensation of glyoxylate with oxaloacetate followed by acidic conditions for decarboxylating β-keto acids[5]; and (d) enzymic condensation of glyoxylate with pyruvate, catalyzed by KHG-aldolase.[6,7] Methods (a) and (b), which start with 4-hydroxyglutamate, can be used to obtain DL-, L-, or D-KHG depending on the specific isomer(s) of 4-hydroxyglutamate util-

[1] Abbreviation used: KHG, 2-keto-4-hydroxyglutarate.
[2] U. Maitra and E. E. Dekker, *J. Biol. Chem.* **238**, 3660 (1963).
[3] U. Maitra and E. E. Dekker, *Biochim. Biophys. Acta* **81**, 517 (1964).
[4] A. Goldstone and E. Adams, *J. Biol. Chem.* **237**, 3476 (1962).
[5] This series, Vol. 17B [175B].
[6] This series, Vol. 42 [45] and [46].
[7] This series, Vol. 17B [175D].

ized.[8] Glutamate–aspartate transaminase from either pig heart[9] or rat liver[3] catalyzes the reaction in procedure (b); only the L-*erythro* and L-*threo* isomers of 4-hydroxyglutamate are effective substrates, yielding either D-KHG or L-KHG, respectively. The only reasonably simple method for obtaining *erythro*-4-hydroxy-L-glutamate (the L-*threo* isomer can be purchased[10]) is by reductive amination of DL-KHG with NADH, NH_4^+, and glutamate dehydrogenase, followed by separation of the two L-epimers (*erythro* and *threo*) of 4-hydroxyglutamate by ion exchange column chromatography.[11] The chemical condensation of glyoxylate with oxaloacetate (method c) yields DL-KHG. So also, however, does the aldo-lase-catalyzed condensation of glyoxylate with pyruvate (method d) since the enzyme from rat[12,13] or bovine liver catalyzes the cleavage or formation of D- and L-KHG at nearly equal rates. KHG-aldolase from *Escherichia coli* is not absolutely stereospecific either, although it strongly favors (by a factor of roughly 9:1) utilization or formation of L-KHG as opposed to D-KHG. (See purification of KHG-aldolase from bovine liver and *E. coli* extracts.[6])

The preparation of DL-KHG by condensation of glyoxylate with oxaloacetate is described in Vol. 17B [175B]. This is the best procedure for making KHG labeled with carbon-14 in positions 4 or 5 using [2-^{14}C]glyoxylate or [1-^{14}C]glyoxylate, respectively. Radioactive KHG with carbon-14 in atoms 1, 2, or 3 can be specifically prepared by aldo-lase-catalyzed condensation of glyoxylate with appropriately labeled pyruvate. The nonenzymic transamination of 4-hydroxyglutamate with pyridoxal, as described here, represents a simple and routine way for synthesizing KHG. *threo*-4-Hydroxy-L-glutamate is commercially avail-able[10]; a preparative scheme for isolating this compound from leaves of *Phlox decussata*, the source in which 4-hydroxyglutamate was first de-tected,[14] has been outlined.[15] By nonenzymic transamination, this isomer

---

[8] See these references for an explanation of the nomenclature and the configurational relationships between the isomers of 4-hydroxyglutamate and the KHG derived therefrom: E. E. Dekker and U. Maitra, *J. Biol. Chem.* **237**, 2218 (1962); Goldstone and Adams[4]; U. Maitra and E. E. Dekker, *J. Biol. Chem.* **239**, 1485 (1964). In brief, *threo*-4-hydroxy-L-glutamate → L-KHG; *erythro*-4-hydroxy-L-glutamate → D-KHG; by chemical procedures, either the *threo*- or the *erythro*-racemate of 4-hydroxyglutamate → DL-KHG.

[9] See this series Vol. 5 [94a] for glutamate-aspartate transaminase from pig heart.

[10] Calbiochem, La Jolla, California.

[11] This series, Vol. 17B [175B].

[12] U. Maitra and E. E. Dekker, *J. Biol. Chem.* **239**, 1485 (1964).

[13] This series, Vol. 17B [175D].

[14] A. I. Virtanen and P. K. Hietala, *Acta Chem. Scand.* **9**, 175 (1955).

[15] E. E. Dekker, *Biochem. Prep.* **9**, 69 (1962).

of 4-hydroxyglutamate yields L-KHG, which is readily utilized by liver KHG-aldolase and is the preferred substrate for the same aldolase from *E. coli*. The details of a straightforward chemical synthesis, which provides gram quantities of 4-hydroxyglutamate and has been checked several times in the author's laboratory, have already been presented.[16] [2-^{14}C]Diethylacetamidomalonate, for the synthesis of [2-^{14}C]4-hydroxyglutamate and thence [2-^{14}C]KHG, is commercially available.[17]

Synthesis

Reagents

Synthetic *threo*- or *erythro*-4-hydroxy-DL-glutamic acid
Pyridoxal hydrochloride, commercial
Cupric chloride, dihydrate
Potassium hydroxide
Dowex 50-X8, hydrogen phase, cation exchange resin

Procedure. 4-Hydroxyglutamic acid (4 mmoles), cupric chloride (2 mmoles), and pyridoxal hydrochloride (4 mmoles) are dissolved in water and the pH of the solution adjusted to 5.0 with 2.5 N potassium hydroxide. Acetate buffer (16 mmoles), pH 5.0, is added, the volume is made to 100 ml, and the solution is heated in a boiling-water bath for $1\frac{1}{4}$ hr. The pH of the cooled solution is then adjusted to 6.5 with potassium hydroxide, and the mixture is filtered. The filtrate is applied to a column (2.4 × 58 cm) of Dowex 50-X8 (hydrogen phase) ion exchange resin. Fractions (10 ml) of effluent fluid are collected as the column is washed with water. The presence of 2-keto-4-hydroxyglutarate in effluent fractions (approximately tubes 15–30) is first detected by acidity to Congo red test paper and then confirmed by reaction with 2,4-dinitrophenylhydrazine, as outlined under Determinations. Positive effluent fractions are promptly pooled and concentrated to minimum volume three times *in vacuo* (water pump) with a rotary evaporator (bath temperature, 40°). The pH of the final solution is adjusted to about 6.8 with potassium hydroxide and stored in the cold. For prolonged storage, the sample should be kept at −15°. The yield is of the order of 35–45%. Solutions of this compound, neutralized to pH 6.5–7.0, are stable at −15° for months. KHG decomposes in acidic or basic solutions. Individual isomers (D- and L-) of KHG prepared in this manner also retain their configuration for months when stored in neutral solution at −15°. If a crystalline product

[16] L. P. Bouthillier and L. Benoiton, *Biochem. Prep.* **9**, 74 (1962).
[17] Amersham/Searle Corporation; Arlington Heights, Illinois.

is desired, the final concentrated preparation is diluted with a small volume of water and titrated to pH 7.4 with 20 mM calcium hydroxide solution. After this solution is again concentrated under reduced pressure, acetone is added to incipient cloudiness and the mixture cooled. The precipitate of the calcium salt can be recrystallized from a water–acetone mixture.

Determinations

KHG can be determined chemically or enzymically. For qualitative purposes, the compound is detected either by (1) paper chromatographic methods as the open-chain acid, the lactone, and the 2,4-dinitrophenyl-hydrazone derivative,[2,4] (2) hydrogenation of the 2,4-dinitrophenylhydrazone yielding 4-hydroxyglutamate, or (3) oxidative decarboxylation with hydrogen peroxide forming malate. In procedures (2) and (3), the products are identified paper chromatographically.[2,4] The conversion of KHG to malate by hydrogen peroxide has also been used to establish the stereoconfiguration of KHG isomers.[4,11] Furthermore, KHG is cleaved by the aldolase and the glyoxylate or pyruvate liberated can be measured.[6]

Quantitatively, two methods are best. Determination of the color intensity of strong alkaline solutions of the 2,4-dinitrophenylhydrazone derivative is the first.[18] A suitable small aliquot of KHG solution, made to 1 ml with water, is mixed with 1 ml of 2,4-dinitrophenylhydrazine solution (0.1% in 2 N hydrochloric acid) and allowed to stand at room temperature for 20 min. The intensity of the colored solution obtained by adding 5 ml of 2.5 N sodium hydroxide is measured with a Klett-Summerson photoelectric colorimeter, using a No. 54 filter. A solution of dried, reagent grade 2-ketoglutarate serves as a standard; 1 μmole has an absorbance of 1.0 (500 Klett units) in this procedure. Alternatively, the concentration of KHG solutions can be determined enzymically by reductive amination with NADH, NH_4^+, and glutamate dehydrogenase. This procedure has been described.[5] Both isomers of KHG serve as substrate. The reaction is essentially quantitative with the equilibrium far in the direction of amino acid formation.[2,19] The concentration of KHG is readily calculated if the amount of NADH oxidized and its molar extinction coefficient are known.

[18] T. E. Friedemann and G. E. Haugen, *J. Biol. Chem.* **147**, 415 (1943).
[19] R. G. Rosso and E. Adams, *J. Biol. Chem.* **242**, 5524 (1967).

[28] o-Nitrophenyl-β-D-galactopyranoside 6-Phosphate

By Wolfgang Hengstenberg and M. L. Morse

Principle. The procedure for phosphorylating primary hydroxyl groups in unprotected nucleosides[1] is applied to o-nitrophenyl-β-D-galactoside (ONPG) to synthesize the 6-phosphate. Phosphorus oxychloride in trimethyl phosphate containing a small amount of water yielded ONPG-6-P in 50% yield. The method should be applicable to the synthesis of the 6-phosphates of other ONP glycosides.

Reagents

o-Nitrophenyl-β-D-galactopyranoside (Sigma)
Trimethyl phosphate (Aldrich)
Phosphorus oxychloride
Ammonia
1:1 Charcoal–Celite mixture
Pyridine
Dowex 50
Cyclohexylamine

Procedure. The procedure given here is much simpler than that employed initially[2] to synthesize ONPG 6-P: 30 mmoles of o-nitrophenyl β-D-galactoside was dissolved in 75 ml of trimethyl phosphate containing 30 mmoles of water and 90 mmoles of phosphorus oxychloride. The mixture was stirred and kept for 3 hr in an ice bath. Crushed ice was then added, Phosphoric and hydrochloric acids were neutralized with concentrated ammonia, and the yellow solution was evaporated under diminished pressure. o-Nitrophenol was removed by evaporating water several times from the solution. The colorless solution contained 14.65 mmoles of ONPG 6-P as determined enzymically. Alkaline hydrolysis of the reaction mixture gave 18.2 mmoles of product, indicating no substantial formation of by-products, since the sum = 18.35 mmoles. The mixture was absorbed on an acid-washed, 1:1 charcoal-Celite mixture. The supernatant of the charcoal absorption was checked enzymically for residual phosphate for the absorption; ∼1000 ml of wet charcoal–Celite mixture were needed. After absorption of the glycosides on the charcoal, the slurry was poured into a column of 10 cm diameter. The inorganic salts were eluted with 4–5 liters of water until no chloride ion was detectable.

---

[1] M. Yoshikawa, T. Kato, and T. Takenishi, *Tetrahedron Lett.* **50**, 5065 (1967).

[2] W. Hengstenberg and M. L. Morse, *Carbohyd. Res.* **7**, 180 (1968).

With about 3 liters of 1:2 (v/v) pyridine–water, the 6-phosphate containing traces of ONPG was eluted with small loss. After evaporation the mixture was passed through a column of Dowex 50 (H⁺ form). The eluate was brought to pH 9 with cyclohexylamine, evaporated to a syrup, and dissolved in a small volume of ethanol. Addition of ether until the solution became cloudy produced crystals of the dicyclohexylammonium salt; yield 13 mmoles. The product decomposed at ∼180°, and had $[\alpha]_D^{20}$ −40° (c 2, water); calc. for the free acid, $[\alpha]_D^{20}$ −60.6°.

Analysis. Calculations for $C_{24}H_{43}N_3O_{11}P$ (dicyclohexylammonium salt): C, 49.65; H, 7.47; N, 7.24; P, 5.33. Found: C, 49.31; H, 7.42; N, 7.25; P, 5.63. Enzymic analysis with 6-phospho-β-D-galactosidase, found: 4.06 μM. Calc: 4.18 μM.

Analytical Procedures. Most of the procedures used were described previously.[2] Enzymic determination of the ONPG 6-phosphate was performed with 10 μg of a highly purified, electrophoretically homogeneous, enzyme preparation in 2.5 ml of the buffer described earlier.[2] For high-voltage paper electrophoresis, 3 mg of ONPG-6P as the dicyclohexylammonium salt, was hydrolyzed in the standard buffer with 10 μg of 6-phosphogalactosidase.[3] The solution was applied to a Whatman 3 MM paper strip 120 cm in length and subjected to electrophoresis for 1 hr at 33 V/cm in an 50 mM sodium borate buffer, pH 9.5, at 30°. The apparatus used was a Gilson High-Voltage Electrophorator, Model DW. The spots on the paper were made visible by spraying with alkaline silver nitrate solution.

[3] See this series, Vol. 42 [73].

[29] Active Site-Specific Reagents and Transition-State Analogs for Enolase

By FINN WOLD

Some work has been done on the preparation of substrate analogs which may serve to explore the enolase active site with regard both to structure and mechanism of action. From studies on a large number of competitive inhibitors, it appears that a carboxylate group and phosphate ester group are required for specific binding at the active site of the enzyme[1]; consequently, the first approach for most workers has been to prepare active site-specific reagents in which these groups are retained.

[1] For a review, see F. Wold, in "The Enzymes" (P. Boyer, ed.), 3rd ed., Vol. 5, p. 499. Academic Press, New York, 1971.

Chlorolactic acid phosphate has been prepared in good yield by phosphorylating the benzyl ester of chlorolactic acid with dibenzylphosphate and subsequently removing the blocking groups by catalytic hydrogenation; iodolactic acid phosphate was also prepared by the same sequence of steps, albeit in very poor yield.[2] Neither of these compounds formed covalent derivatives with enolase. In fact there was evidence that the enzyme catalyzed a very slow dehydrohalogenation of the two halolactic acid phosphates.

To date there is only one active site-specific reagent known to form a covalent derivative with the enzyme, namely, glycidol phosphate (1,2-epoxipropanol 3-phosphate). In addition there are a pair of analogs, D-tartronate semialdehyde 2-phosphate and 3-amino enolpyruvate phosphate, which form very tight noncovalent complexes with the active site of enolases. Because of the strong UV absorbance of the complexes, these compounds have been useful in establishing the number of active sites in the enzyme, and, as likely transition state analogs, their interaction with the enzyme has shed some light on the mechanism of catalysis.

Glycidol phosphate is readily prepared by phosphorylation of commercially available glycidol with $POCl_3$,[3] and it satisfies the major requirements for an active site-specific reagent for rabbit muscle enolase in spite of the fact that it does not contain one of the specificity groups (the carboxylate) required for strong interaction with the enzyme: It inactivates rabbit muscle enolase with a stoichiometry of 1 mole of reagent incorporated per active site; the inactivation shows saturation kinetics, is prevented by the presence of substrate in a competitive manner, and the rate of inactivation is enhanced by the presence of Mg^{2+}. Based on the observation that alkali or, even better, alkaline hydroxylamine solutions readily liberate the reagent (^{32}P-labeled reagent was used) from the enzyme, it is proposed that a carboxyl group in the active site is esterified in the reaction with the epoxide.[3] The proposed reaction of enolase with glycidol phosphate can thus be written as:

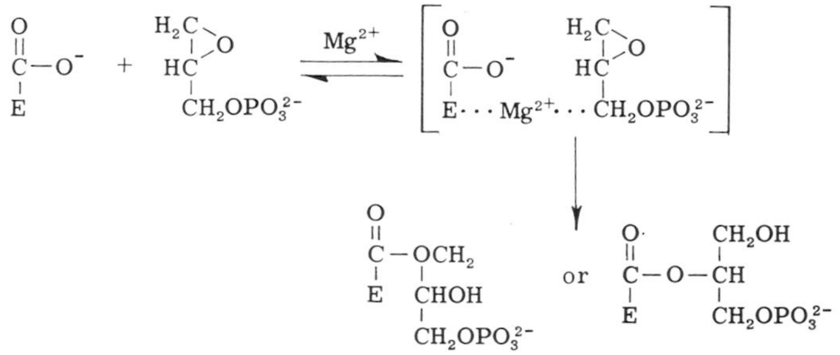

[2] R. P. Cory and F. Wold, unpublished results.
[3] I. A. Rose and E. L. O'Connell, *J. Biol. Chem.* **244**, 6548 (1969).

Tartronate Semialdehyde Phosphate (Ttal 2P) and 3-Amino Enolpyruvate Phosphate (Am-ePrvP).[4] The synthesis of Ttal 2P from commercially available D-gluconolactone (through tri-O-isopropylidene gluconic acid to methyl 3:4, 5:6-di-O-isopropylidene gluconate, phosphorylation at the 2-position with diphenylchlorophosphate, removal of the blocking group by catalytic hydrogenation, saponification and acid hydrolysis of the acetals to give gluconate 2-phosphate, which through periodate oxidation is converted to the desired product in good yield) has been described.[5] Ttal 2P is quite labile and should be freshly prepared from the stable gluconate 2-phosphate. However the aldehyde is sufficiently stable to be purified by ion exchange chromatography.[6] Am-ePrvP is prepared from Ttal 2P by treatment with excess ammonia at pH 9–10.[6] The enamine product can be stored in the presence of ammonia and at low temperature, but upon removal of the excess ammonia or lowering of the pH below neutrality, the derivative decomposes rapidly.

Both these compounds react with the active site of enolases in the presence of Mg^{2+} to give very stable complexes with characteristic UV spectra (see Fig. 1). One use of these compounds has consequently been

DISSOCIATION CONSTANTS AND STOICHIOMETRY OF
ENOLASE-INHIBITOR COMPLEXES[a,b]

| Enzyme source | K_m | Ttal 2P | | Am-ePrvP | | |
		K_I	K_d	K_I	K_d	n
Rabbit muscle	45	4.1	2.5	0.02	~0.1	1.95
Coho salmon muscle	40	—	1.6	—	~0.04	1.94
Yeast	100	14 (68)	15	—	~0.5	2.00
Escherichia coli	100	19	15	0.09	~0.5	2.11

[a] Data from F. Wold, *in* "The Enzymes" (P. Boyer, ed.), 3rd ed., Vol. 5, p. 499, Academic Press, New York, 1971; and T. G. Spring and F. Wold, *Biochemistry* **10**, 4649 (1971).

[b] Ttal 2P, tartronate semialdehyde phosphate; Am-ePrvP, 3-amino enolpyruvate phosphate. All K values are given in units of micromolar concentrations. The K_m values were determined as part of this experiment to allow direct internal comparison of all results. The K_I values were obtained by kinetic measurements, and K_d values from direct spectrophotometric titrations. Because of the tight binding of Am-ePrvP, the K_d values could not be determined precisely, but the end point n (moles of inhibitor bound per mole of enzyme) could be evaluated most precisely with Am-ePrvP.

[4] Abbreviations suggested by Dr. W. E. Cohen, Director, Office of Biochemical Nomenclature.
[5] F. C. Hartman and F. Wold, *Biochim. Biophys. Acta* **141**, 445 (1967).
[6] T. G. Spring and F. Wold, *Biochemistry* **10**, 4649 (1971).

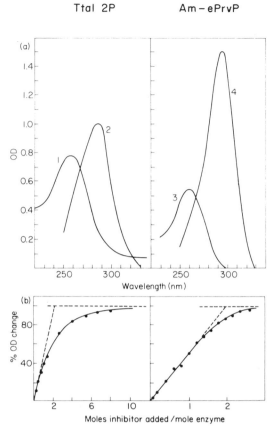

Fig. 1. Spectral properties of the enolase inhibitors tartronate semialdehyde phosphate (Ttal 2P) and 3-amino enolpyruvate (Am-*e*PrvP). (a) Curve 1: The ultraviolet spectrum of the enolate ion–Mg^{2+} complex of Ttal 2P produced in 10 mM NaOH in the presence of 1 mM Mg^{2+} (the free aldehyde at neutral pH has no absorption above 230 nm); $\lambda_{max} = 258$ nm, $\epsilon_{max} = 12,000$ M^{-1} cm^{-1}. Curve 2: The difference spectrum for the rabbit muscle enolase-Ttal 2P complex (enzyme contribution subtracted); $\lambda_{max} = 285$ nm, $\epsilon_{max} = 15,000$ M^{-1} cm^{-1}. Curve 3: The spectrum of Am-*e*PrvP; $\lambda_{max} = 260$ nm, $\epsilon_{max} = 8400$ M^{-1} cm^{-1}. Curve 4: The difference spectrum for the rabbit muscle enolase-Am-*e*PrvP complex; $\lambda_{max} = 295$ nm, $\epsilon_{max} = 23,200$ M^{-1} cm^{-1}. The four spectra have been normalized to equimolar concentrations. (b) The spectrophotometric titration of rabbit muscle enolase with Ttal 2P following the OD change at 285 nm (left) and with Am-*e*PrvP, following OD change at 295 nm (right), showing saturation with 2 moles of inhibitor bound per mole of enzyme. The titration curves and the enzyme-inhibitor spectra were obtained in pH 7.0 buffer containing 1 mM Mg^{2+}. The data are taken from F. Wold [*in* "The Enzymes" (P. Boyer, ed.), 3rd ed., Vol. 5, 499, Academic Press, New York, 1971] and T. G. Spring and F. Wold *Biochemistry* **10**, 4649 (1971).

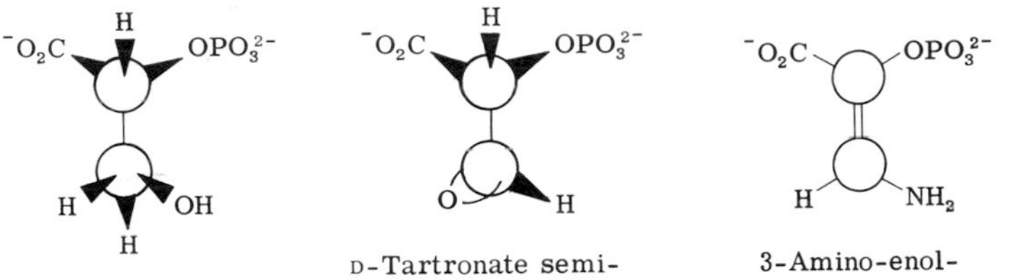

D-Glycerate-2-P

D-Tartronate semi-
aldehyde 2-phosphate
(Ttal 2p)

3-Amino-enol-
pyruvate phosphate
(Am-ePrvP)

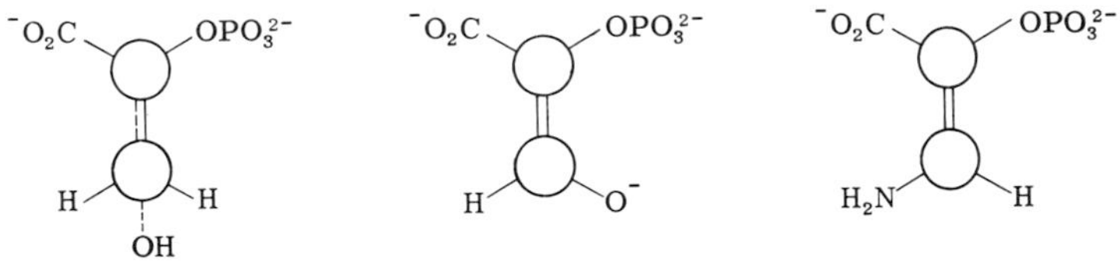

Proposed transition
state structure

Proposed structures in the enzyme-analog
complexes (stereochemistry unknown)

Scheme 1

to titrate the active sites of different enolases; the presence of 2 sites per native enolase dimer in 4 different enolases is based on such titration (see Fig. 1 and the table).[7]

A representation of the complexes of Ttal 2P and Am-ePrvP with the active site of enolase is given in Scheme 1 as a basis for the comparison with the proposed transition state in the reversible dehydration of glycerate 2-phosphate to enolpyruvate phosphate.[8] The observations that the enolase–analog complexes form only in the presence of Mg^{2+} and that both compounds compete with substrate for the active site support the proposition that the two compounds behave as true substrate (transition state) analogs. The absolute requirement for magnesium also suggests that the readily determined spectral properties of the enzyme-Mg^{2+}-analog complexes may provide a simple tool for the investigation of the role of metal ions in the enolase reaction.

[7] T. G. Spring and F. Wold. *Biochemistry* **10**, 4655 (1971).
[8] E. D. Dinovo and P. D. Boyer, *J. Biol. Chem.* **246**, 4586 (1971).

Section IV

Oxidation–Reduction Enzymes

[30] 5-Keto-D-fructose Reductase from *Gluconobacter cerinus*[1]

By SASHA ENGLARD and GAD AVIGAD

Assay Method

Principle. The reduction of 5-keto-D-fructose is followed by measuring the oxidation of NADPH spectrophotometrically at 340 nm.

Reagents

Tris·HCl, 62.5 mM, at pH 7.4

5-Keto-D-fructose, 0.1 M, aqueous solution stored at −20° when not in use. This compound, although occasionally available from commercial sources, is generally isolated from *Gluconobacter cerinus* culture media.[2]

NADPH, 1.25 mM, prepared freshly each day and stored in ice

Procedure. Assays are carried out in cuvettes of 10 mm light path maintained at 30° and containing 0.8 ml of Tris·HCl buffer, 0.1 ml of 5-keto-D-fructose, and 0.1 ml of NADPH. Reactions are initiated by the addition of 10–20 µl of the various enzyme fractions diluted appropriately with 50 mM potassium phosphate, pH 7.4. Changes in absorbancy at 340 nm are measured continuously with a multiple sample absorbance spectrophotometer and recorder.

Definition of Unit. A unit of enzyme activity is defined as the amount of enzyme catalyzing the oxidation of 1 µmole of NADPH per minute under the conditions of assay specified above. Specific activity is the number of units of enzyme activity per milligram of protein as determined by the method of Lowry *et al.*[3] using crystalline bovine serum albumin as a standard.

[1] EC 1.1.1.124, D-fructose: NADP+ 5-oxidoreductase.
[2] G. Avigad and S. Englard, *J. Biol. Chem.* **240**, 2290 (1965). See this Volume [18].
[3] O. H. Lowry, N. J. Rosebrough, A. L. Farr, and R. J. Randall, *J. Biol. Chem.* **193**, 265 (1951). See this series, Vol. 3 [73].

Purification

Several bacterial and yeast extracts have been examined for their capacity to reduce 5-keto-D-fructose in the presence of NADPH and shown to exhibit significant activity, albeit to varying degrees.[4] *G. cerinus* contains by far the most potent 5-keto-D-fructose reductase activity and the enzyme has been extensively purified, characterized, and shown to catalyze the stoichiometric reduction of 5-keto-D-fructose to D-fructose.[5-7] An NADPH:5-keto-D-fructose reductase which produces D-fructose has been obtained from *Gluconobacter albidus*[8] and detected in extracts prepared from *Acetobacter melanogenum*.[9] The method reported for preparation of 5-keto-D-fructose reductase from *G. cerinus*[5] is described below.

Cultivation of Bacteria. The bacterium used is *G. cerinus* subsp. *ammoniacus Asai*, 1FO, 3267, ATCC 19441, kept with frequent transfers (1–2 weeks) on fructose-yeast agar slants.

Erlenmeyer flasks of 2-liter capacity, containing 400 ml of 1% fructose in 0.5% yeast extract medium, are inoculated with 1/20 the volume of a 24-hr bacterial culture suspension. Incubation is carried out at 30° in a rotary shaker at 160 rpm. After about 46 hr, cells are collected by centrifugation (10,000 g or 15 min) and washed twice with cold 25 mM potassium phosphate buffer, pH 7.2. The yield of packed cells is usually 7–9 g per liter of culture medium.

To obtain cells in larger quantities, 35 liters of the same fructose–yeast extract culture medium is incubated at 30° in a 60-liter stainless steel baffled fermenter. The inoculum used is a 24-hr shake flask culture, and its volume is 4% of the total culture medium. Aeration is promoted by constant stirring at 210 rpm and bubbling of air through a bottom sparger at a rate of 1 liter/min. After 24 hr, cells are collected in a Sharples supercentrifuge, washed three times with cold 10 mM phosphate buffer, pH 7.4, and lyophilized. The yield generally ranges between 1.0 and 1.3 g of dry cells per liter of culture medium. Freshly packed cells can also be dried with 10 volumes of acetone at −20°. Lyophilized cells are kept at −20° and can be used as the source of enzyme for more than 2 years without appreciable loss of activity. Unless otherwise indicated, all operations are carried out at 0–3°, and all buffers and ammonium sulfate solutions used contain 1 mM EDTA and 1 mM 2-mercaptoethanol.

Step 1. Preparation of Crude Extract. Lyophilized *G. cerinus* cells, 5 g, are suspended in 75 ml (15 times their weight) of 50 mM potassium

---

[4] S. Englard, G. Kaysen, and G. Avigad, *J. Biol. Chem.* **245**, 1311 (1970).

[5] G. Avigad, S. Englard, and S. Pifko, *J. Biol. Chem.* **241**, 373 (1966).

[6] S. Englard and G. Avigad, *J. Biol. Chem.* **240**, 2297 (1965).

[7] S. Englard, G. Avigad, and L. Prosky, *J. Biol. Chem.* **240**, 2302 (1965).

[8] Y. Yamada, K. Aida, and T. Uemura, *J. Biochem.* (*Tokyo*) **61**, 803 (1967).

[9] K. Sasajima and M. Isono, *Agr. Biol. Chem.* **32**, 161 (1968).

phosphate, pH 7.4, and dispersed by homogenization for 3 min in a Sorvall High Speed Omni-Mixer. The cells are then disrupted by sonic action for 30 min in a Raytheon 10-kc sonic oscillator, and the supension is centrifuged for 1 hr at 32,000 g. The residue is resuspended in 38 ml of the above buffer (7.5 times the original weight of the dried cells), subjected to sonication for an additional 15 min and centrifuged as above. The clear supernatant is combined with the first extract.

Step 2. Streptomycin Fractionation. A 5% aqueous solution of streptomycin sulfate is added dropwise to the combined crude extract to a final concentration of 1%. The suspension is stirred for 15 min and centrifuged for 1 hr at 32,000 g, and the residue is discarded.

Step 3. Ammonium Sulfate Fractionation. Solid ammonium sulfate is added slowly to the streptomycin-treated supernatant to 40% saturation (28.2 g/100 ml) and the mixture is stirred for 15 min. The residue, collected by centrifugation of the suspension at 32,000 g for 1 hr, is discarded. Ammonium sulfate (19.8 g/100 ml of original volume) is added gradually to the clear supernatant to 68% saturation, and the mixture is permitted to equilibrate for 30 min and centrifuged as before. The residue is suspended in 3.63 M ammonium sulfate, the volume used for this washing procedure being equivalent to one-third that of the streptomycin-treated supernatant. The thick suspension is stirred for 15 min and centrifuged as before, and the supernatant is discarded.

Step 4. Extraction of 40–68% Ammonium Sulfate Fraction. The washed 40–68% ammonium sulfate residue is suspended in 32.5 ml (equivalent to one-fourth the volume of the streptomycin-treated supernatant) of 2.78 M ammonium sulfate, and the mixture is stirred mechanically for 30 min. The suspension is centrifuged at 78,000 g for 45 min in a preparative Spinco ultracentrifuge. The supernatant is collected and the residue is reextracted as before with 26 ml of 2.78 M ammonium sulfate. The third and final extraction is carried out in a similar manner with 22 ml of the same ammonium sulfate solution, and the extract is collected by centrifugation at 105,000 g for 50 min. The three extracts are combined before proceeding with the next step.

Step 5. Ethanol Fractionation. The temperature of the combined 2.78 M ammonium sulfate extracts is lowered to $-5°$, and an equal volume of ethanol precooled at dry ice temperature is added dropwise. During the course of ethanol addition, the temperature is further decreased and maintained at $-10°$, and the mixture is stirred for 35 min prior to centrifugation at 5800 g for 30 min. The bulky residue is suspended in 29 ml (equivalent to one-third the volume of the combined ammonium sulfate extracts) of 50 mM potassium phosphate, pH 7.4. The suspension is stirred for 30 min at 0° and centrifuged at 105,000 g for 75 min. The residue is discarded and the supernatant is dialyzed against the above potassium phosphate buffer for 18 hr; several changes of dialyzing medium are made.

Step 6. Fractionation on DEAE Cellulose. DEAE-cellulose is washed successively with potassium phosphate buffers, pH 7.4, of decreasing concentrations starting with 0.5 M. The anion exchange cellulose is finally equilibrated with 50 mM potassium phosphate, pH 7.4. The dialyzed ethanol fraction (114 ml) is passed through such a DEAE-cellulose column (1.2 × 15 cm.). The column is washed first with an equivalent volume of 50 mM potassium phosphate, pH 7.4, followed by an equal volume of 0.3 M potassium phosphate, pH 7.4. These fractions contain little if any enzymic activity. The eluting buffer is changed to 0.5 M potassium phosphate, pH 7.4, and the column effluent is collected in 17.5-ml fractions. The first two fractions containing insignificant NADPH:5-keto-D-fructose reductase activity are discarded. The next three to four fractions, containing the bulk of the enzymic activity, are pooled, and the combined solution is reduced in volume to 1–2 ml by means of pressure dialysis under a stream of N_2. This concentrated enzyme solution is exhaustively dialyzed against 50 mM potassium phosphate and clarified by centrifugation at 32,000 g for 1 hr.

Contaminating NADP$^+$-mannitol dehydrogenase activity is completely removed at the alcohol fractionation step, and a purification of approximately 1200-fold is generally achieved. The highly purified preparation can be stored at 4° for 3–4 weeks without appreciable loss of activity.

A summary of the purification is given in the table.

PURIFICATION OF *Gluconobacter cerinus* NADPH:5-KETO-D-FRUCTOSE REDUCTASE

Steps	Total units	Total protein (mg)	Specific activity (units/mg protein)	Yield (%)
1. Crude extract	952	1803	0.53	100
2. Streptomycin supernatant	946	—	—	99.4
3. 40–68% Saturated ammonium sulfate fraction	927	—	—	97.4
4. 2.78 M Ammonium sulfate extract of fraction 3	691	—	—	72.6
5. Dialyzed ethanol fraction	341	25.8	13.2	35.8
6. Dialyzed DEAE-cellulose fraction	207	0.33	627	21.7

Properties[5,10]

Effect of pH. Maximum rates of NADPH oxidation occur at pH 6.5 in phosphate buffer and at pH 7.4 in Tris · HCl buffer.

Kinetic Properties. The kinetic constants for NADPH:5-keto-D-fructose reductase have been determined at pH 7.4 and 30°. The K_m values are: 5-keto-D-fructose = 4.5 mM; NADPH = 1.8 μM; D-fructose = 70 mM (K_i = 82 mM as an inhibitor to 5-keto-D-fructose reduction); NADP$^+$ = 1.3 mM. The equilibrium constant (K_{eq} = [5-keto-D-fructose] [NADPH] [H$^+$]/[D-fructose] [NADP$^+$]) has been calculated to be 51 pM. The maximal rate of reaction in the direction of NADP$^+$ reduction and 5-keto-D-fructose formation is only 0.7% that of NADPH oxidation and D-fructose formation. The disparity in maximal velocities between the reactions in either direction, the equilibrium constant (which strongly favors 5-keto-D-fructose reduction) and the kinetic parameters (a K_m value for NADPH which is 750-fold lower than the corresponding constant for NADP$^+$) are properties that should enhance the usefulness of this reductase for NADP$^+$ regeneration in coupled enzyme systems.

Specificity. Partially purified reductase, 0.25 unit per milliliter of reaction mixture (ethanol fraction, specific activity, 13–15), incubated at 30° in the presence of 20 mM substrate and 2 mM NADP$^+$, buffered at pH 7.4 with 50 mM Tris·HCl exhibits the following activities (percent) in comparison to 5-keto-D-fructose: fructose 1-phosphate, 0.40; fructose 6-phosphate, 0.13; fructose 1,6-diphosphate, 0.33; sorbose, 0.05; fructose, 0.07; mannose, 0.21; mannose 6-phosphate, 0.30. It is thus clear that the phosphate esters tested are not appreciably oxidized and the trace of activity observed could be due to contaminants. With more purified preparations of 5-keto-D-fructose reductase, significant reduction of NADP$^+$ is not observed when glucose, mannose, or sorbose are present in the reaction mixtures at concentrations of 25 mM or when xylose, xylitol, mannitol, or sorbitol are tested at levels as high as 250 mM. Similarly, NADPH is not oxidized in the presence of dihydroxyacetone, tagatose, sorbose, glucose, mannose and DL-glyceraldehyde at concentrations varying between 25 and 40 mM. Highly purified preparations of reductase catalyze an NADPH-dependent reduction of 5-keto-D-fructose 1-phosphate[10–12] to fructose 1-phosphate. Under standard conditions of assay, a K_m of 4.2 mM was determined with a maximal rate of reduction of the monophosphate ester only 0.16 of that observed for 5-keto-D-fructose. NADH does not substitute for NADPH in the enzymic reduction of 5-keto-D-fructose.

[10] G. Avigad and S. Englard, *J. Biol. Chem.* **243**, 1511 (1968).
[11] S. Englard, I. Berkower, and G. Avigad, *Biochim. Biophys. Acta* **279**, 229 (1972).
[12] See this Volume [18].

[31] 5-Keto-D-fructose Reductase from Yeast[1]

By SASHA ENGLARD and GAD AVIGAD

Assay Method

Principle. Yeast 5-keto-D-fructose reductase is measured spectrophotometrically by the decrease in absorption at 340 nm due to NADPH oxidation in presence of 5-keto-D-fructose.

Reagents

Tris·HCl, 62.5 mM, at pH 7.5
5-Keto-D-fructose, 0.1 M, aqueous solution stored at −20° when not in use. This dicarbonylhexose although occasionally available from commercial sources is generally isolated from *Gluconobacter cerinus* culture media.[2]
NADPH, 1.25 mM, prepared freshly each day and stored in ice

Procedure. Into a 10-mm light path cuvette, add 0.8 ml of Tris·HCl, 0.1 ml of 5-keto-D-fructose, and 0.1 ml of NADPH. The reaction, carried out at 30°, is initiated by addition of 5–10 μl of the various enzyme fractions properly diluted with 50 mM potassium phosphate, pH 7.4. The rate of NADPH oxidation is conveniently followed by measuring the decrease in absorbancy at 340 nm with a multiple sample absorbance recording spectrophotometer.

Definition of Unit. A unit of activity is defined as that amount of enzyme required to convert 1 μmole of NADPH to NADP+ per minute under the assay conditions just described. Specific activity is defined as the number of units per milligram of protein. Protein is measured by the method of Lowry *et al.*,[3] crystalline bovine serum albumin being used as a standard.

[1] EC 1.1.1.123, L-sorbose: NADP+ 5-oxidoreductose.
[2] G. Avigad and S. Englard, *J. Biol. Chem.* **240**, 2290 (1965). See this volume [18].
[3] O. H. Lowry, N. J. Rosebrough, A. L. Farr, and R. J. Randall, *J. Biol. Chem.* **193**, 265 (1951). See this series, Vol. 3 [73].

Purification

An NADPH:5-keto-D-fructose reductase activity was initially detected as a contaminant in highly purified preparations of Boehringer yeast glucose-6-phosphate dehydrogenase.[4] Extracts from various species of yeast (with the notable exception of *Candida tropicalis*) exhibit the following levels of reductase activity[5] (expressed in terms of units per gram of dry cells): brewer's yeast (dried, Anheuser-Busch, St. Louis, Missouri), 16.2; *Torula utilis* (dried, Lake State Company, Rhinelander, Wisconsin), 12.5; Boehringer yeast (dried, C. F. Boehringer and Soehne, GmbH, Mannheim, Werktutzing, Germany), 8.9; baker's yeast (fresh, the Fleischman Company, New York, New York), 5.8 units per gram of wet packed cells. In each case, the product of reduction was identified to be L-sorbose.[5] The procedure described here for the purification of 5-keto-D-fructose reductase, uses a glucose-6-phosphate dehydrogenase-rich strain of dried baker's yeast (a gift from C. F. Boehringer and Soehne, GmbH, Mannheim) as starting material. Unless otherwise indicated, all operations are carried out at 0–3°, and all buffers contain 1 mM EDTA and 1 mM 2-mercaptoethanol.

Step 1. Preparation of Crude Extract. Dried Boehringer yeast, 500 g, is suspended in 3 volumes of 0.1 M NaHCO$_3$ containing 1 mM EDTA and 1 mM 2-mercaptoethanol. The mixture is allowed to autolyze for 2 hr at 37° with continuous and vigorous mechanical stirring. After cooling to approximately 10° in a salt–ice bath, the suspension is centrifuged in a Sorvall RC-2 centrifuge at 17,000 g and the residue is discarded.

Step 2. Heat Inactivation. For each 100 ml of crude extract, 11.1 ml of 1 M L-sorbose are added. The solution is heated rapidly to 50° in a boiling water bath (3.5 min) and maintained between 50° and 52° for 15 min longer. After rapid cooling in a salt–ice bath at −10°, the heat-treated solution is centrifuged for 90 min at 17,000 g and the precipitate is discarded.

Step 3. First Ammonium Sulfate Fractionation. To the supernatant solution from step 2 is added solid ammonium sulfate to 35% saturation (24.7 g/100 ml of solution). The salt is added slowly with constant stirring. Equilibration is continued for an additional 40 min after the last addition of salt, the suspension is centrifuged at 17,000 g for 90 min, and the precipitate is discarded. The clear supernatant solution is then brought to 52% saturation by the further slow addition of solid ammonium sulfate (12.0 g/100 ml of initial heat-treated supernatant solution). Mechanical stirring is continued for an additional 40 min, the

[4] S. Englard, G. Avigad, and L. Prosky, *J. Biol. Chem.* **240**, 2302 (1965).
[5] S. Englard, G. Kaysen, and G. Avigad, *J. Biol. Chem.* **245**, 1311 (1970).

suspension is centrifuged as before but for only 1 hr and the residue is dissolved in a minimum volume of 50 mM potassium phosphate, pH 6.9. A deep amber and somewhat turbid solution (100–115 ml) is obtained.

Step 4. First Gel Filtration on Sephadex G-100. A column (5 × 90 cm) of Sephadex G-100, equilibrated with 50 mM Tris, pH 7.4, containing 0.1 M NaCl and 1 mM EDTA, is prepared with the use of a Sephadex K50/100 chromatographic column (Pharmacia Fine Chemicals, Inc.). The gel bed is firmly positioned between two flow adaptors, thus allowing the column to be operated by the upward flow of eluting solutions. A single-speed 1 rpm peristaltic-type influsion pump (Micro-Flow tubing pump, Model 7119-1, Cole-Parmer Instrument and Equipment Company, Chicago), and the use of Tygon tubing with internal diameter $\frac{1}{16}$ inch and external diameter $\frac{1}{8}$ inch, provides and maintains a constant flow rate of 22–23 ml per hour. The 35–52% ammonium sulfate fraction from step 3 is applied to the bottom of the column, and elution is continued with the above equilibrating buffer. A large fraction (560–570 ml) is first collected, and smaller fractions (13.5 ml) are then collected with an automatic fraction collector. Usually, no enzymic activity is detected in the first ten fractions, but significant NADPH:5-keto-D-fructose reductase activity emerges between fractions 11 and 34. These latter fractions which contain over 50% of the initial total protein and significantly less than half of the total initial units of enzymic activity are discarded. Reductase of increased specific activity emerges between fractions 35 and 50. These fractions are pooled, and the combined solution is subjected to a second ammonium sulfate fractionation.

Step 5. Second Ammonium Sulfate Fractionation. The combined Sephadex G-100 fractionated enzyme solution obtained in step 4 is brought to 45% saturation with salt by the slow addition of solid ammonium sulfate (31.8 g/100 ml of solution). The mixture is stirred for 30 min and then centrifuged for 30 min at 17,000 g; the residue is discarded. Ammonium sulfate (14.1 g/100 ml of initial solution) is added gradually to the clear supernatant to 65% saturation, and the mixture is stirred for 30 min. The precipitate is collected by centrifugation at 17,000 g for 30 min and dissolved in a minimum volume of 50 mM potassium phosphate, pH 6.9. This solution is poured onto a Sephadex G-25 (medium) column (2.5 × 40 cm), equilibrated previously with 50 mM potassium phosphate, pH 6.9, and the protein is eluted from the column by washing with the same buffer. The protein, freed from ammonium sulfate, generally appears in the 75–120 ml fraction of the eluate and is subjected to ion exchange chromatography.

Step 6. DEAE-Sephadex Column Chromatography. DEAE-Sephadex (bead form) is suspended in 20 volumes of 0.2 M potassium phosphate, pH 6.9, stirred mechanically for 1 hr, and allowed to swell overnight. The ion exchange Sephadex gel is then washed successively with potassium phosphate buffer, pH 6.9, of decreasing concentration, and the finer particles which do not settle within 20 min are removed by decantation. The gel is finally equilibrated with 50 mM potassium phosphate, pH 6.9, and a slurry poured into a K25/45 Sephadex laboratory column. The gel is permitted to pack by gravity to a column height of 40 cm. The ammonium sulfate-free solution obtained in step 5 is applied to the top of the DEAE-Sephadex column and this is followed by the passage of 200–210 ml of 50 mM potassium phosphate, pH 6.9, through the gel bed. The column is then connected to a mixing flask containing 150 ml of 50 mM potassium phosphate, pH 6.9, which in turn is connected to a reservoir containing 0.25 M potassium phosphate, pH 6.9. The mixing flask contains a rapidly spinning magnetic bar, and in this manner the concentration of potassium phosphate in solution entering the DEAE-Sephadex is increased gradually and uniformly. A flow rate of 52–54 ml per hour is maintained with a constant-delivery infusion pump of the type described above but using Tygon tubing with internal diameter $\frac{3}{32}$ inch and external diameter $\frac{5}{32}$ inch. Fractions of approximately 10 ml each are collected, and those containing NADPH:5-keto-D-fructose reductase activity in excess of 1.65 units/ml are combined (fractions 26–32).

Step 7. Third Ammonium Sulfate Fractionation. Solid ammonium sulfate (33.5 g/100 ml of solution) is added slowly to the combined DEAE-Sephadex fractionated enzyme obtained in step 6 (47.5% saturation). The mixture is stirred for 30 min and the precipitate is collected by centrifugation at 32,000 g for 30 min and discarded. The supernatant solution is brought to 62.5% saturation by the further gradual addition of solid ammonium sulfate (10.6 g/100 ml of initial solution). The suspension is allowed to equilibrate for 30 min and centrifuged as before.

Step 8. Second Gel Filtration on Sephadex G-100. The 47.5–62.5% saturated ammonium sulfate residue obtained in step 7 is dissolved in a minimum volume of 50 mM Tris, pH 7.4, containing 0.1 M NaCl and 1 mM EDTA. The clear solution of enzyme is loaded on a Sephadex G-100 column (2.5 × 90 cm), equilibrated in the same buffer. The enzyme is eluted with the same buffer, and approximately 5-ml fractions are collected at a constant flow rate of 22–23 ml/hr. The fractions containing the bulk of enzymic activity (70–80%) with a 2-fold increase in specific activity are combined and concentrated by ammonium sulfate precipitation to 80% saturation.

A purification of approximately 300-fold is generally achieved; when

PURIFICATION OF YEAST NADPH:5-KETO-D-FRUCTOSE REDUCTASE

Steps	Total units	Total protein (mg)	Specific activity (units/mg protein)	Yield (%)
1. Crude extract	4480	50,420	0.089	100
2. Heat inactivation	3927	34,080	0.115	87.7
3. 35–52% Ammonium sulfate fraction	3778	14,260	0.265	84.4
4. First Sephadex G-100 fractionation	2155	3,186	0.675	48.1
5. 45–65% Ammonium sulfate fraction followed by Sephadex G-25 treatment	1581	1,546	1.02	35.3
6. DEAE-Sephadex fractionation	1176	165.0	7.11	26.3
7. 47.5–62.5% Ammonium sulfate fraction	1126	80.0	14.1	25.1
8. Second Sephadex G-100 fractionation	781	27.5	28.4	17.4

stored at 4°, the purified preparations are stable for at least 4 months. Recovery and specific activity data of a typical preparation are given in the table.

Properties[5]

Effect of pH. A rather broad range of optimal activity is observed in Tris-acetate buffer between pH 7.4 and 8.5.

Kinetic Properties. The K_m value for 5-keto-D-fructose and NADPH at pH 7.5 and 30° are 1.0 mM and 46 μM, respectively.

Reversibility of Reaction. All attempts to ascertain the oxidation of L-sorbose by NADP[+] in the reverse direction have failed. Although NADP[+] acts as a competitive inhibitor of 5-keto-D-fructose reduction by NADPH (K_i = 0.13 mM), concentrations of L-sorbose up to 500 mM in the presence of only 0.65 mM 5-keto-D-fructose do not affect the rate of NADPH oxidation. In view of these observations, it is evident that thermodynamic considerations[6] are not the only factors accounting for

---

[6] Assuming that the K_{eq} for the reaction catalyzed by the yeast reductase is the same or even significantly lower than the value of approximately 50 pM at pH 7.2 obtained for the reduction of 5-keto-D-fructose to D-fructose catalyzed by the *Gluconobacter cerinus* enzyme.[7]

[7] G. Avigad, S. Englard, and S. Pifko, *J. Biol. Chem.* 241, 373 (1966).

the apparent irreversibility of the yeast reductase-catalyzed reduction of 5-keto-D-fructose to L-sorbose.

Inhibitors. Although iodoacetaminde and N-ethylmaleimide (up to 5 mM) do not inhibit the rate of NADPH oxidation in the presence of 5-keto-D-fructose, p-hydroxymercuribenzoate at concentrations of 0.40 mM and 1 mM inhibit the reductase to the extent of 21 and 41%, respectively.

Specificity. The following carbonyl-containing substances, examined at concentrations of 25–50 mM in the presence of NAD$^+$, NADH, NADP$^+$ and NADPH and partially purified reductase preparations show insignificant activities (none or less than 1% in comparison to 5-keto-D-fructose): D-fructose, D-tagatose, D-glucose, acetone, acetaldehyde, α-ketoglutarate, pyruvate, D-mannose, L-arabinose, D-fucose, 2-deoxy-D-ribose, D-xylose, and L-rhamnose. On the other hand, 5-keto-D-fructose 1-phosphate[8-10] is a substrate for the yeast reductase ($K_m = 0.29$ mM) with a maximal rate of 1.8% that observed for 5-keto-D-fructose. By analogy to the *G. cerinus* reductase-catalyzed reduction of 5-keto-D-fructose 1-phosphate to D-fructose 1-phosphate,[8] the product of reduction by the yeast reductase is expected to be L-sorbose 1-phosphate.

Occurrence of Enzyme. Extracts from the following bacteria contain (in decreasing order, varying over a 20-fold range) significant levels of NADPH:5-keto-D-fructose reductase which produces L-sorbose[5]: *Bacillus subtilis* var. *aterneum, Aerobacter cloacae,*[11] *Bacillus subtilis* 168-2, *Proteus vulgaris,*[11] *Aerobacter aerogenes,*[11] *Bacillus megaterium,* and *Bacillus brevis* 9999. An NADH-dependent reduction of 5-keto-D-fructose to L-sorbose has been reported to occur in extracts prepared from *Acetobacter melanogenum.*[12] In the presence of NADH, rat and sheep liver sorbitol dehydrogenase (L-iditol:NAD$^+$ oxidoreductase, EC 1.1.1.14) catalyze the reduction of 5-keto-D-fructose specifically to L-sorbose.[4] The presence of NADPH:5-keto-D-fructose reductase activity, however, is not necessarily accompanied by potent polyol dehydrogenase activities. Thus, Boehringer yeast from which the reductase has been partially purified is devoid of NADP-linked sorbitol and mannitol dehydrogenase activi-

---

[8] G. Avigad and S. Englard, *J. Biol. Chem.* **243,** 1511 (1968).
[9] S. Englard, I. Berkower, and G. Avigad, *Biochim. Biophys. Acta* **279,** 229 (1972).
[10] See this volume [18].
[11] With the crude extracts of *A. aerogenes, P. vulgaris,* and *A. cloacae,* traces of D-fructose are also detected by the hexokinase-phosphoglucose isomerase and glucose 6-phosphate dehydrogenase assay system. The total amount of D-fructose, however, never exceeds 10% of the total ketose formed from the 5-keto-D-fructose reduced by these extracts. The difference is always quantitatively accounted for as L-sorbose.
[12] K. Sasajima and M. Isono, *Agr. Biol. Chem.* **32,** 161 (1968).

ties. Although the crude extract contains an NAD-linked mannitol and sorbitol dehydrogenase, these activities are significantly lower than that of the 5-keto-D-fructose reductase (4.3 and 1.0 units per gram of dry cells, respectively, compared to 9.0 units per gram of dry cells for the NADPH:5-keto-D-fructose reductase).

[32] D-Mannitol Dehydrogenase from *Leuconostoc mesenteroides*

By KEI YAMANAKA

$$\text{D-Mannitol} + \text{NAD}^+ \rightleftarrows \text{D-fructose} + \text{NADH} + \text{H}^+$$

Crystalline D-mannitol dehydrogenase has been described from *Lactobacillus brevis* by Horecker.[1,2]

Assay Method

Principle. A spectrophotometric assay is employed making use of the D-mannitol dehydrogenase catalyzed consumption of NADH by fructose at 340 nm.

Reagents

Acetate buffer, 50 mM, pH 5.3
D-Fructose, 1 M
NADH, 10 mM keep in a deepfreeze

Procedure. To a microcuvette of 0.5 ml capacity and 1 cm light path are added 0.10 ml of acetate buffer (5 μmoles), 0.006 ml of NADH (0.06 μmole), 0.03 ml of D-fructose (30 μmoles), and enzyme in a total volume of 0.30 ml. The enzyme is sufficiently diluted in acetate buffer to give an absorbance change of less than 0.10 per minute. The reaction is started by addition of substrate, and the temperature is maintained at 30° in a spectrophotometer equipped with thermospacers and a circulating water bath. A blank is run without substrate for an NADH oxidase correction. The reaction is recorded or read every 30 sec for 3–5 min.

Definition of Unit. One unit of enzyme activity oxidizes 1 μmole of NADH per minute per milliliter in the standard assay.

[1] G. Martinez, H. A. Baker, and B. L. Horecker, *J. Biol. Chem.* **238**, 1598 (1963).
[2] B. L. Horecker, See this series, Vol. 9 [28].

Sources[3]

Among lactic acid bacteria, the homofermentative species ferment fructose but do not produce the D-mannitol dehydrogenase in any detectable amounts. Heterofermentative species can be classified into two groups; one ferments fructose to produce D-mannitol, but the other does not. Four strains of heterofermentative species, *Leuconostoc mesenteroides* ATCC 9135, *Lactobacillus brevis* ATCC 347 and 8287, *L. gayonii* ATCC 8289, and *L. pentoaceticus* ATCC 368 produce D-mannitol dehydrogenase from D-fructose medium. In contrast, *L. brevis* IFO[4] 3960 and *Leuconostoc mesenteroides* IFO 3076 and several unnumbered strains do not produce the enzyme from D-fructose medium. The yield is highest from *L. mesenteroides* after 16 hr culture. Fructose is essential for production of the enzyme in *Lactobacillus brevis* ATCC 8287. Fructose can be replaced, however, by glucose or sucrose for the enzyme production in *Leuconostoc mesenteroides* ATCC 9135. Glucose is not a substrate, but is the best carbon source for enzyme production in *L. mesenteroides*. Neither *Lactobacillus brevis* or *Leuconostoc mesenteroides* can dissimilate D-mannitol in static culture, but both can grow on D-mannitol in shaking culture. The dehydrogenase is produced in amounts almost equal to those cultures from D-glucose medium and static culture.

Purification Procedure[5]

Culture. D-Mannitol dehydrogenase may be purified from *Leuconostoc mesenteroides* ATCC 9135 or oher active strain grown on the following medium: 1% peptone, 1% sodium acetate, 0.2% yeast extract, 0.02% $MgSO_4 \cdot 7H_2O$, 0001% NaCl, 0.0002% $MnSO_4 \cdot 4H_2O$, and 1% D-glucose. The organism is inoculated into 8 ml of the medium in a test tube and incubated overnight at 30°. The whole culture is transferred to 400 ml of the same medium and incubated for 24 hr. The entire culture is then transferred to 10 liters of medium and incubated for 16 hr. Cells are harvested by centrifugation and washed with 50 mM phosphate buffer (pH 7.0) containing 1 mM mercaptoethanol.

Preparation of Cell-Free Extract. Washed cells, 94 g obtained from 50 liters of medium are disrupted in small portion by grinding with levigated alumina.[6] The enzyme is extracted with 1 liter of 50 mM potassium

[3] K. Yamanaka and S. Sakai, *Can. J. Microbiol.* **14**, 391 (1968).
[4] IFO = Institute for Fermentation, Osaka. Japan.
[5] S. Sakai and K. Yamanaka, *Biochim. Biophys. Acta* **151**, 684 (1968).
[6] Levigated alumina, about 300 mesh for chromatography, was purchased from Wako Pure Chemicals, Osaka, Japan.

phosphate buffer (pH 7.0) containing 1 mM mercaptoethanol. Cell debris and alumina are removed by centrifugation.

Protamine Precipitation. A solution of protamine sulfate (2%, 46 ml) is added dropwise with slow stirring. After standing for 1 hr, the mixture is centrifuged and the precipitate is discarded.

Ammonium Sulfate Fractionation. Degree of saturation by ammonium sulfate is calculated from the table given by Green and Hughes[7] after temperature correction using the factor 0.92. Solid ammonium sulfate (674 g) is added to the protamine supernatant (950 ml) to give 100% saturation. The precipitate at each step of the purification is dissolved in 50 mM potassium phosphate buffer (pH 7.0) containing 1 mM mercaptoethanol, and the protein solution is dialyzed against the same buffer with stirring. The precipitate is collected, dissolved in 50 mM phosphate buffer (28.5 ml) and dialyzed (ammonium sulfate fraction I). Fraction I (44.6 ml) is treated again with solid ammonium sulfate. The precipitate obtained with 12.8 g of ammonium sulfate is discarded. Further addition of 4.62 g of ammonium sulfate yields the precipitate (50 to 65% saturation). The precipitate is collected, dissolved in 4 ml of phosphate buffer and dialyzed overnight against 1 liter of phosphate buffer as described above.

Acetone Fractionation. The second ammonium sulfate fraction (7.4 ml) is adjusted to pH 6.0 with 0.2 M acetic acid. Acetone (3.4 ml), previously chilled to −20°, is added dropwise with gentle stirring (30% acetone). The precipitate is removed quickly by centrifugation at −10° and discarded. To the supernatant (9.8 ml) is added 4.3 ml of acetone to give a 50% acetone solution. The precipitate is collected and dissolved in 10 ml of buffer. After removing a trace of insoluble matter by centrifugation, the solution (13.2 ml) is fractionated with 6.8 g of ammonium sulfate to give 80% saturation. The precipitate is dissolved in 5.0 ml of phosphate buffer and dialyzed overnight as described above.

Column Chromatography on DEAE-Cellulose. The acetone fractionation can be replaced by column chromatography on DEAE-cellulose. Cellulose is equilibrated against 20 mM phosphate buffer (pH 6.5) and proteins are eluted by a linear gradient of KCl between 0 and 0.6 M at pH 6.5. The dehydrogenase in the active fractions eluted between 0.25 and 0.30 M KCl are pooled and concentrated by precipitation with ammonium sulfate to 80% saturation.

Crystallization. Crystallization can be performed with either the acetone fraction or the DEAE-cellulose fraction. A saturated ammonium sulfate solution is added dropwise until faint turbidity is observed. The

---

[7] A. A. Green and W. L. Hughes, see this series, Vol. 1 [10].

slight turbidity is clarified by centrifugation and the solution is allowed to stand overnight at 5°; saturation of ammonium sulfate is 50%. The first crystals are collected by centrifugation, dissolved 2.0 ml of 50 mM phosphate buffer (pH 7.0) containing 1 mM mercaptoethanol and dialyzed against 100 ml of the same buffer at 2° overnight (first crystals). Recrystallization is carried out by the same procedure. The overall purification is approximately 30-fold in specific activity. The purification procedure is summarized in the table.

Properties[8]

Properties of the crystalline enzyme are closely similar to those of *Lactobacillus brevis.*[1,2]

Purity. Analysis of the recrystallized enzyme preparation shows a single and symmetric peak on ultracentrifugation. A single band of protein migrated toward the cathode on polyacrylamide gel electrophoresis at pH 9.3 and electrophoresis on a cellulose acetate film at pH 9.0. This band coincided with the activity band of D-mannitol dehydrogenase after incubating the gel or cellulose acetate film in dark with D-mannitol, NAD, phenazine methosulfate, and nitroblue tetrazolium at pH 8.0–9.0.

Molecular Weight. The sedimentation coefficient ($s_{20,w}$) is calculated as 7.37 S. Molecular weight is estimated by exclusion chromatography on Sephadex G-200 and sucrose density gradient centrifugation as 132,000 ± 9000.

PURIFICATION OF D-MANNITOL DEHYDROGENASE

Fraction	Total volume (ml)	Total protein (mg)	Total units	Specific activity (μmoles/mg/min)	Yield (%)
Crude extract	930	3810	48,370	12.7	
Protamine fraction	950	1070	33,330	31.1	69
Ammonium sulfate I (100% saturation)	28.5	1130	33,330	29.2	69
Ammonium sulfate II (50–65% saturation)	7.4	612	29,610	48.3	61
Acetone (30–50% precipitate)	7.7	241	27,780	115.2	56
First crystals	(2.9)	86.7	25,660	296	53
Second crystals	(1.1)	43.2	15,050	348	31
Third crystals	(1.5)	34.1	13,410	393	28

[8] S. Sakai and K. Yamanaka, *Agr. Biol. Chem.* **32**, 894 (1968).

Effect of pH. The pH optima for the reduction of D-fructose is 5.3 and for the oxidation of D-mannitol it is at 8.6. These values are identical with *Lactobacillus brevis* enzyme.[1,2]

Stability. Enzyme solution is stable in the pH range from 5.0 to 9.0 at 35° for 1 hr or at pH 6.0 to 8.0 at 35° for 5 hr. A suspension of crystals kept in 50% ammonium sulfate at 5° is not stable. But losses caused by storage for several months are not so critical as to preclude use in assays for fructose.

Substrate Specificity. The enzyme is specific for D-fructose and D-mannitol. The Michaelis constants are 35 mM for D-fructose at pH 5.3 and 20 mM for D-mannitol at pH 8.6. Those of *L. brevis* enzyme are twice or more higher, i.e., 70 mM and 60 mM for D-fructose and D-mannitol, respectively.[1] The following relative rates were obtained with 3 times crystallized enzyme when the substrates were tested at 30 μmoles and with 0.06 μg of enzyme protein for D-mannitol or D-fructose, and with 3.0 μg for other substrates: D-mannitol, 100; D-sorbitol, 4.0; dulcitol, 0.03; ribitol, 0.03; D-arabitol, 0.17; L-arabitol, 0.01; xylitol, 0.06; *i*-erythritol, 0.01; D-fructose, 100; L-sorbose, 1.32; D-tagatose, 0.37; D-ribulose, 0.18; L-ribulose, 0.03; D-xylulose, 0.15. The Michaelis constants are 0.27 mM for NAD and 10 μM for NADH. NADP is completely inactive in the oxidation of D-mannitol and D-sorbitol. However, NADPH is about 20% as active as NADH. The relative activities with 30 μmoles of ketose and NADPH were: D-fructose, 100; L-sorbose, 1.35; D-ribulose, 0.1; and D-xylulose, 0.04. NADP does not inhibit the oxidation of D-mannitol and D-sorbitol with NAD, while NADP was suggested to be inhibitory for *Lactobacillus brevis* enzyme.[1]

[33] Aldohexose Dehydrogenase from *Gluconobacter cerinus*[1]

By GAD AVIGAD and SASHA ENGLARD

$$\text{D-Aldohexose} + \text{NADP}^+ \rightleftharpoons \text{D-hexonic acid-δ-lactone} + \text{NADPH} + \text{H}^+$$

Assay Method

Principle. The assay is based on the spectrophotometric determination of NADPH formed with D-mannose as substrate.

[1] EC 1.1.1.119, D-aldohexose (D-glucose, D-mannose): NADP⁺ 1-oxidoreductase.

Reagents

Tris·HCl, 0.125 M, at pH 8.0
D-Mannose, 0.3 M
NADP+, 2.5 mM

Procedure. The assay system (1.01–1.02 ml), in addition to enzyme, contains: 0.8 ml of Tris·HCl, 0.1 ml of D-mannose and 0.1 ml NADP+. The reaction is initiated by the addition of NADP+ and the changes in absorbance at 340 nm are determined at 30° with a multiple sample absorbance recording spectrophotometry.

Definition of Unit. One unit of activity is defined as that amount of enzyme which under the above specified conditions of assay catalyzed the formation of 1 μmole of NADPH per minute. Specific activity is defined as the number of units per milligram of protein. Protein is determined either colorimetrically[2] with crystalline bovine serum albumin as a standard or by direct ultraviolet spectrophotometry.[3]

Purification Procedure

Cells of *Gluconobacter cerinus* subsp. *ammoniacus Asai*, 1FO 3267, ATCC 19441, are grown on fructose–yeast medium and harvested as described.[4] Washed cells, which are lyophilized and stored at −18°, lose less then 10% of the extractable enzyme activity after one year of storage in the deep freezer. All procedures are carried out at 0–4°.

Step 1. Preparation of Crude Extract. Lyophilized *G. cerinus* cells are suspended in 25 times their weight of 25 mM phosphate buffer, pH 7.4, or 25 mM Tris·HCl, pH 8.0, and dispersed by homogenization for 3 min in a Sorvall high speed Omni-Mixer. The cells are then disrupted by ultrasonic action for 30 min in a Raytheon 10 kc oscillator, and the suspension is centrifuged at 20,000 g for 1 hr.

Step 2. Protamine Sulfate Precipitation. The supernatant solution from step 1 is adjusted to pH 5.7 by the careful addition of dilute acetic acid. A 1.0% protamine solution, freshly prepared and neutralized to pH 7, is then added to a final concentration of 0.25%. After standing for 25 min, the suspension is centrifuged at 20,000 g for 30 min and the sediment is discarded.

Step 3. Ammonium Sulfate Fractionation. The pH of the protamine-treated supernatant from step 3 is adjusted to pH 5.4 and then solid ammonium sulfate is added in a gradual manner. The precipitate which

---

[2] O. H. Lowry, N. J. Rosebrough, A. L. Farr, and R. J. Randall, *J. Biol. Chem.* **193**, 265 (1951).

[3] See this series, Vol. 3 [73].

[4] S. Englard and G. Avigad, *J. Biol. Chem.* **240**, 2287 (1965). See this volume [30].

forms between 38 and 60% saturation (24 g/100 ml followed by an additional 15 g/100 ml of $(NH_4)_2SO_4$, respectively), is collected by centrifugation, dissolved in a small volume (8–10 ml per gram of dried cells initially used) of 25 mM citrate buffer, pH 5.5, and dialyzed for about 18 hr against 100 volumes of the same buffer. A turbid precipitate, which occasionally appears on dialysis, is removed by centrifugation before proceeding with the next step.

Step 4. Ethanol Fractionation. To the dialyzed 38 to 60% saturated ammonium sulfate fraction cooled to $-10°$, cold ethanol is added with vigorous stirring. The fraction precipitating between 25 and 52% ethanol is collected by centrifugation at 20,000 g for 30 min. The sediment is dissolved in 0.1 M phosphate buffer, pH 7.0, with 3 ml for every gram of dry cells used initially in the preparation.

Step 5. DEAE-Cellulose Fractionation. DEAE-cellulose is equilibrated with 0.1 M potassium phosphate buffer, pH 7.0, and a 2.1 $\times$ 5.0 cm column is used for each 3 ml (containing about 6 mg of protein) of the ethanol-fractionated enzyme solution. A concentration gradient ranging from 0.10 to 0.25 M phosphate buffer, pH 7.0, is applied for elution, and the enzyme emerges at phosphate concentrations between 0.14 and 0.17 M. The fraction containing the enzyme is dialyzed against 10 mM potassium phosphate buffer, pH 7.0, then reabsorbed on a small column of DEAE-cellulose, and finally eluted with 0.5 M potassium phosphate, pH 8.0. This concentrated enzyme solution is dialyzed overnight against 5 mM potassium phosphate buffer, pH 7.0. A summary of a typical purification procedure is presented in Table I.

TABLE I

PURIFICATION OF *Gluconobacter cerinus* ALDOHEXOSE DEHYDROGENASE[a]

Fraction	Protein (mg)	Units	Specific activity (units/mg protein)
1. Crude extract	483	50.0	0.10
2. After treatment with protamine sulfate	186	43.4	0.23
3. Fraction between 38 and 60% ammonium sulfate	76	35.0	0.46
4. Fraction between 25 and 52% ethanol	10.2	24.8	2.43
5. After fractionation on DEAE Cellulose	1.6	16.9	10.1

[a] Results represent yields obtained from 1.0 g of dry bacterial cells.

Properties[5]

Effect of pH. A broad range of optimal activity is observed between pH 7.5 and 9.0. The activity as measured in Tris buffer is consistently higher than that observed in phosphate buffers.

Kinetic Properties. In 0.1 M Tris·HCl, at pH 8.0, the K_m values for D-glucose, 2-deoxy-D-glucose, D-mannose, and 2-amino-2-deoxy-D-mannose are 5.3 mM, 39 mM, 1.8 mM, 29 mM, respectively; with corresponding relative maximal velocities of 1.00, 2.27, 1.25, and 0.95, respectively. The K_m for D-mannose, determined in 0.1 M potassium phosphate, pH 6.2, is 5.6 mM. With D-mannose as substrate, K_m values for NADP$^+$ are 12 μM and 32 μM at pH 8.0 and 6.2, respectively. NADPH inhibits the oxidation of aldohexoses in a competitive manner with respect to NADP$^+$ ($K_m = 38$ μM at pH 8.0). Because of the instability of glucono-δ-lactone in solution, an accurate K_m value for this compound as a substrate in the reverse reaction is difficult to obtain.

Substrate Specificity. All sugars listed and assayed at the concentrations indicated in Table II are neither substrates nor inhibitors under the standard conditions of assay for aldohexose dehydrogenase activity. Thus, of all sugar and sugar derivatives examined, only D-glucose, D-mannose, 2-deoxy-D-glucose, and 2-amino-2-deoxy-D-mannose are oxidized. NAD$^+$, in concentrations as high as 50 mM, does not substitute for NADP$^+$ in the reaction, nor does it inhibit the rate of mannose or glucose oxidation in the presence of 50 μM NADP$^+$.

Reversibility and Equilibrium Constant of Reaction. The addition of solid or freshly dissolved glucono-δ-lactone to an incubation mixture containing buffer, aldohexose dehydrogenase, and NADPH results in an immediate decrease in absorbance at 340 nm and to the formation of stoichiometric amounts of D-glucose as assayed with glucose oxidase. The lability of glucono-δ-lactone in aqueous solutions precludes an exact determination of K_{eq} ([NADPH][H$^+$][D-glucono-δ-lactone]/[NADP$^+$][D-glucose]), but a value of about 0.1 μM has been obtained and is no doubt close to the true equilibrium constant of the overall reaction.

Inhibitors. The addition of iodoacetate, N-ethylmaleimide, p-hydroxymercuribenzoate, and 5,5'-dithiobis-(2-nitrobenzoic acid) (in concentrations ranging from 0.5 mM to 20 mM) to the standard aldohexose dehydrogenase assay mixture does not inhibit the rate of hexose oxidation. The presence of 1 mM ZnSO$_4$ or CuSO$_4$, however, results in the complete inactivation of the enzyme.

[5] G. Avigad, Y. Alroy, and S. England. *J. Biol. Chem.* **243**, 1936 (1968).

TABLE II

Sugars Found to Be Neither Substrates nor Inhibitors for the
Gluconobacter cerinus Aldohexose Dehydrogenase

Sugar	Tested as substrate (μM)	Tested as inhibitor (at concentration of D-mannose as substrate) (μM)	(μM)
D-Ribose	40	10	4
D-Xylose[a]	25	10	4
D-Lyxose	25	10	4
D-Arabinose	50	—	—
D-Allose	1	1	1
D-Altrose	25	—	—
D-Gulose	1	1	1
D-Galactose[a]	100	10	4
D-Talose	25	—	—
L-Rhamnose	40	—	—
L-Sorbose	30	—	—
D-Fructose	30	—	—
5-Keto-D-fructose	20	—	—
D-Mannodialdose	1	1	1
3-O-Methyl-D-glucose	30	30	1
D-Glucuronic acid	60	60	3
D-Galacturonic acid	40	—	—
D-Sorbitol	—	10	4
D-Mannitol	50	—	—
D-Glucose 1-phosphate	—	10	1
D-Glucose 6-phosphate	20	10	3
D-Mannose 6-phosphate	1	—	—
2-Amino-2-deoxy-D-galactose	40	—	—
2-Amino-2-deoxy-D-glucose	40	30	2
N-Acetylmannosamine	30	20	4
N-Acetylglucosamine	40	10	4
Trehalose	40	—	—
Maltose	50	20	4
Melibiose	40	10	4
Cellobiose	40	10	4
Isomaltose	40	10	4
Methyl α-D-glucoside	—	10	4
Methyl β-D-glucoside	—	10	4

[a] Some trace activity found with galactose and xylose was due to a glucose contaminant.

Occurrence. The aldohexose dehydrogenase from *G. cerinus* is unusual in its specificity; thus it does not catalyze the oxidation of aldopentoses and is completely inactive with NAD[+]. Furthermore, oxidation of D-mannose is a relatively unusual feature of the aldose dehydrogenases hitherto obtained from various sources. However, NADP⁺-linked oxidation of D-mannose has been observed in extracts derived from two stains of *Acetobacter suboxydans*[6,7] which are indeed closely related to the *Gluconobacter* group.

Stability. Solutions of the purified enzyme retain full activity for at least 8 weeks when stored at 4° provided that the protein concentration is at least 0.5 mg/ml and that the pH is maintained between 5.4 and 8.5. Prolonged storage at 4°, freezing and thawing, and lyophilization of more dilute enzyme solutions results in significant losses of activity. The presence of bovine serum albumin (1 mg/ml) effectively protects the enzyme against these inactivations.

[6] J. DeLey and A. J. Stouthamer, *Biochim. Biophys. Acta* **34**, 171 (1959).
[7] K. Okamoto, *J. Biochem. (Tokyo)* **53**, 348 (1963).

[34] D-Aldohexose Dehydrogenase

By RICHARD L. ANDERSON and A. STEPHEN DAHMS

$$\beta\text{-D-Glucopyranose} + NAD^+ \rightarrow \text{D-glucono-}\delta\text{-lactone} + NADH + H^+$$

This dehydrogenase also catalyzes the oxidation of other D-aldohexoses, and apparently functions in the metabolism of D-fucose and D-glucose in pseudomonad MSU-1.[1]

Assay Method[1]

Principle. The rate of D-glucose-dependent NADH formation is determined by measuring the rate of absorbance increase at 340 nm.

Reagents

Tris·HCl buffer, 0.3 M, pH 8.1
D-Glucose, 0.25 M
NAD⁺, 30 mM

Procedure. The following are added to a microcuvette with a 1.0-cm light path: 0.05 ml of buffer, 0.01 ml of D-glucose, 0.01 ml of NAD⁺, D-aldohexose dehydrogenase, and water to a volume of 0.15 ml. The reaction is initiated by the addition of D-aldohexose dehydrogenase. The reaction rates are conveniently measured with a Gilford multiple-sample ab-

[1] A. S. Dahms and R. L. Anderson, *J. Biol. Chem.* **247**, 2222 (1972).

sorbance recorder. The cuvette compartment should be thermostated at 25°.

Definition of Unit and Specific Activity. One unit of D-aldohexose dehydrogenase is defined as the amount that catalyzes the reduction of 1 μmole and NAD⁺ per minute in the standard assay. Specific activity is in terms of units per milligram of protein. Protein is conveniently measured by the method of Warburg and Christian,[2] but with crude extracts and fractions high in nucleic acid content, the biuret method[3] may be used.

Purification Procedure

Growth of Organism. Pseudomonad MSU-1[4] was grown aerobically in Fernbach flasks containing 1 liter of medium. The flasks were agitated on a rotary shaker at 32°. The medium consisted of: 1.35% $Na_2HPO_4 \cdot 7$ H_2O, 0.15% KH_2PO_4, 0.3% $(NH_4)_2SO_4$, 0.02% $MgSO_4 \cdot 7$ H_2O, 0.0005% $FeSO_4$, and 3.2% D-glucose. The sugar was autoclaved separately and added aseptically to the mineral medium. Stepwise addition of the sugar was necessary because 1% sugar markedly reduced the growth rate. The first portion of sugar (20 ml of a 25% solution) was added immediately after inoculation. Six additional 20-ml portions of 25% sugar were added at 6-hr intervals. The cells were harvested 4 hr after the final addition of the sugar (about 40 hr after inoculation).

Preparation of Cell Extracts. Cells were harvested by centrifugation, washed in water, and suspended in 0.1 M sodium phosphate buffer (pH 7.0). (Alternatively, 0.1 M Bicine buffer, pH 7.4, containing 0.15 mM 2-thioethanol may be used to inhibit NADH oxidase activity in the resulting cell extract.) The cells were broken by treating the suspension for 13 min in a Raytheon 250-W, 10-kHz sonic oscillator cooled with circulating ice water. The supernatant fluid resulting from a 10-min centrifugation at 40,000 g was used as the cell extract.

General. The following procedures were performed at 0° to 4°. A summary of the purification procedure is given in the table.

Protamine Sulfate Treatment. To a cell extract containing 0.2 M ammonium sulfate was added an amount of 2% (w/v) protamine sulfate solution (pH 7.0) to give a final concentration of 0.33%. After 30 min, the precipitate was removed by centrifugation and discarded.

Ammonium Sulfate Fractionation. The protein in the supernatant

[2] O. Warburg and W. Christian, *Biochem. Z.* **310**, 384 (1941).

[3] A. G. Gornall, C. J. Bardawill, and M. M. David, *J. Biol. Chem.* **177**, 751 (1941).

[4] This organism is obtainable from the American Type Culture Collection as strain ATCC 27855.

PURIFICATION OF D-ALDOHEXOSE DEHYDROGENASE

Fraction	Volume (ml)	Total protein (mg)	Total activity (units)	Specific activity (units/mg protein)	$A_{280}:A_{260}$
Cell extract	650	4220	839	0.198	0.63
Protamine sulfate supernatant	780	4060	755	0.186	0.86
$(NH_4)_2SO_4$ precipitate (30–40%)	53	1010	459	0.454	1.17
Sephadex G-200	135	109	243	2.24	1.25
DEAE-cellulose[a]	330	14.1	195	13.8	1.48
Calcium phosphate gel[a]	66	1.7	112	66.0	1.58

[a] The values given for these fractions have been corrected for the proportions of the previous steps not subjected to further purification.

from the protamine sulfate step was fractionated by the addition of crystalline ammonium sulfate. The protein precipitating between 30 and 40% saturation was collected by centrifugation and was dissolved 10 mM sodium phosphate, pH 7.0.

Sephadex G-200 Chromatography. The above fraction was chromatographed on a column (6 × 60 cm) of Sephadex G-200 equilibrated with 10 mM sodium phosphate buffer, pH 7.0. Fractions (15 ml) were collected during elution with the same buffer, and those with the highest specific activity were combined.

DEAE-Cellulose Chromatography. The above fraction was concentrated from 135 ml to 15 ml with a Diaflo ultrafiltration cell (Amicon Corp.) containing a type UM-10 membrane. Two milliliters of this concentrate was placed on a column (3 × 5 cm) of DEAE-cellulose (Sigma, exchange capacity = 0.9 meq/g) which had been pretreated as recommended by Sober *et al.*[5] and equilibrated with 20 mM sodium phosphate buffer, pH 7.0. The column was then eluted with a stepwise gradient of 60-ml volumes of the same buffer containing 0, 0.10, 0.20, 0.30, 0.40, and 0.80 M NaCl. D-Aldohexose dehydrogenase eluted in the 0.20 to 0.30 M NaCl range. Fractions containing most of the activity were combined.

Calcium Phosphate Gel Step. The pooled DEAE-cellulose fractions were adjusted to pH 6.5 with 0.05 N HCl and dialyzed for 24 hr against 10 mM sodium cacodylate buffer (pH 6.5).To a portion (5.0 ml) of the dialyzed protein solution was added 1.0 ml of calcium phosphate gel (containing 6% solids) prepared as described by Wood.[6] The gel suspension was centrifuged in a clinical centrifuge, and the collected solids were suc-

[5] H. A. Sober, F. J. Gutter, M. Wyckoff, and E. A. Peterson, *J. Amer. Chem. Soc.* **78**, 756 (1956).

[6] See this series, Vol. 2 [25].

cessively eluted with 1.0 ml each of 10, 20, 30, 40, and 50 mM sodium phosphate buffer (pH 7.0). About half of the activity was recovered in the 10–20 mM range. The enzyme in this fraction was 335-fold purified with a 13% overall recovery of activity.

Properties[1]

Substrate Specificity. This enzyme is specific for D-aldohexoses, the following of which serve as substrates, in order of decreasing activity: D-glucose, D-galactose, D-mannose, 2-deoxy-D-glucose, 6-deoxy-D-galactose (D-fucose), 2-deoxy-D-galactose, D-altrose, 6-deoxy-D-glucose (D-quinivose), D-allose, and 3,6-dideoxy-D-galactose (abequose). The following derivatives of D-aldohexoses do not serve as substrates: D-glucuronic acid, D-galacturonic acid, D-galactose 6-phosphate, D-glucose 6-phosphate, D-glucosamine, N-acetyl-D-glucosamine, 6-deoxy-D-allose, 6-iodo-6-deoxy-D-galactose, 2-acetamido-6-deoxy-D-allose, and 2-acetamido-6-deoxy-D-altrose. Other classes of compounds that do not serve as substrates are L-aldohexoses, ketohexoses, pentoses, trioses, polyols, and di- and trisaccharides.

The apparent K_m values for the two substrates considered to be of physiological significance are 0.86 mM for D-glucose and 5.8 mM for D-fucose. The β anomer is preferred over the α anomer and since the product is the δ-lactone, the actual substrates must be β-D-glucopyranose and β-D-fucopyranose.

Nucleotide Specificity. NADP is not used. The K_m value for NAD$^+$ at pH 8.1 is 80 mM.

pH Optimum. Activity as a function of pH is maximal at pH 8.0 to 8.5 in Tris·HCl and at pH 9–10 in glycine-NaOH.

Stability. Sephadex G-200 fractions (in 10 mM sodium phosphate buffer, pH 7.0) are stable to freezing for at least 6 months.

[35] L-*arabino*-Aldose Dehydrogenase

By RICHARD L. ANDERSON and A. STEPHEN DAHMS

$$\text{D-Fucofuranose} + \text{NAD(P)}^+ \rightarrow \text{D-fucono-}\gamma\text{-lactone} + \text{NAD(P)H} + \text{H}^+$$

This dehydrogenase also catalyzes the oxidation of other aldoses possessing the L-*arabino* configuration at carbon atoms 2 through 4 (or their deoxy derivatives), and is believed to function in the metabolism of D-fucose, D-galactose, and L-arabinose in pseudomonad MSU-1.[1]

[1] A. S. Dahms and R. L. Anderson, *J. Biol. Chem.* **247**, 2228 (1972).

Assay Method[1]

Principle. The rate of D-galactose-dependent NADPH formation is determined by measuring the rate of absorbance increase at 340 nm.

Reagents.

Tris·HCl buffer, 0.3 M, pH 8.1
D-Galactose, 0.25 M
NADP+, 30 mM

Procedure. The following are added to a microcuvette with a 1.0-cm light path: 0.05 ml of buffer, 0.01 ml of D-galactose, 0.01 ml of NADP+, L-*arabino*-aldose dehydrogenase, and water to a volume of 0.15 ml. The reaction is initiated by the addition of L-*arabino*-aldose dehydrogenase. The reaction rates are conveniently measured with a Gilford multiple-sample absorbance recorder. The cuvette compartment should be thermostated at 25°.

It is important to use NADP+ rather than NAD+ when assays are made on extracts of D-fucose-grown cells in order to prevent interference by D-aldohexose dehydrogenase.[2]

The pH of the assay is lower than the apparent optimum pH because the enzyme exhibits greater affinity for its substrates at the lower pH.

Definition of Unit and Specific Activity. One unit of L-*arabino*-aldose dehydrogenase is defined as the amount that catalyzes the reduction of 1 μmole of NADP+ per minute in the standard assay. Specific activity is in terms of units per milligram of protein. Protein is conveniently measured by the method of Warburg and Christian,[3] but with crude extracts and fractions high in nucleic acid content the biuret method[4] may be used.

Purification Procedure[1]

The enzyme was purified from pseudomonad MSU-1.[5] The growth of this organism and the preparation of extracts was as described in another article,[6] except that the carbon source in the growth medium was D-galactose instead of D-glucose. Except for the heat step, the following procedures were performed at 0° to 4°. A summary of the purification procedure is given in the table.

Protamine Sulfate Treatment. To a cell extract containing 0.2 M am-

[2] A. S. Dahms and R. L. Anderson, *J. Biol. Chem.* **247**, 2222 (1972).
[3] O. Warburg and W. Christian, *Biochem. Z.* **310**, 384 (1941).
[4] A. G. Gornall, C. J. Bardawill, and M. M. David, *J. Biol. Chem.* **177**, 751 (1941).
[5] This organism is obtainable from the American Type Culture Collection as strain ATCC **27855**.
[6] See this volume [34].

PURIFICATION OF L-*arabino*-ALDOSE DEHYDROGENASE

Fraction	Volume (ml)	Total protein (mg)	Total activity (units)	Specific activity (units/mg protein)	$A_{280}:A_{260}$
Cell extract	350	8505	1674	0.197	0.62
Protamine sulfate supernatant	630	7875	1659	0.211	0.89
Heat step	620	3534	1374	0.389	0.96
(NH₄)₂SO₄ precipitate (40–60%)	35	805	933	1.16	1.22
Sephadex G-200	30	55.5	500	9.01	1.31
Calcium phosphate gel[a]	4.0	4.32	242	55.9	1.49

[a] The values given for this fraction have been corrected for the proportion of the previous step not subjected to further purification.

monium sulfate was added an amount of 2% (w/v) protamine sulfate solution (pH 7.0) to give a final concentration of 0.33%. After 30 min, the precipitate was removed by centrifugation and discarded.

Heat Step. The supernatant from the protamine step was immersed in a 60° water bath and was stirred gently until the temperature reached 55°. The solution was heated for 2 min at 55°, cooled in an ice-water bath, and clarified by centrifugation.

Ammonium Sulfate Fractionation. The protein in the supernatant from the heat step was fractionated by the addition of crystalline ammonium sulfate. The protein precipitating between 40 and 60% saturation was collected by centrifugation and was dissolved in 0.10 M sodium phosphate buffer, pH 7.0.

Sephadex G-200 Chromatography. The above fraction was chromatographed on a column (6 × 60 cm) of Sephadex G-200 equilibrated with 10 mM sodium phosphate buffer, pH 7.0. Fractions (15 ml) were collected during elution with the same buffer, and those with the highest specific activity were combined.

Calcium Phosphate Gel Step. The pooled Sephadex G-200 fractions were adjusted to pH 6.5 with 0.05 N HCl and dialyzed for 24 hr against 10 mM sodium cacodylate buffer (pH 6.5). To a portion (5.0 ml) of the dialyzed protein solution was added 1.0 ml of calcium phosphate gel (containing 6% solids). The gel suspension was centrifuged in a clinical centrifuge and the collected solids were successively eluted with 1.0 ml each of 10, 20, 30, 40, and 50 mM sodium phosphate buffer (pH 7.0). About half the activity was recovered in the 10–20 mM range. The enzyme in this fraction was 284-fold purified with a 14% overall recovery of activity.

Properties[1]

Substrate Specificity. Of 55 compounds tested as possible substrates, the only six that are oxidized by this enzyme are those that possess the L-*arabino* configuration and their deoxy derivatives. These are L-arabinose, D-galactose, 6-deoxy-D-galactose (D-fucose), 2-deoxy-D-galactose, 3,6-dideoxy-D-galactose (abequose), and 6-iodo-6-deoxy-D-galactose. Compounds that do not serve as substrates at 33 mM concentrations are: D-glucose, L-glucose, D-mannose, L-mannose, D-altrose, D-allose, L-galactose, L-fucose, L-rhamnose, 2-deoxy-D-glucose, 6-deoxy-D-glucose, 6-deoxy-D-allose, D-glucuronic acid, D-galacturonic acid, D-glucose 6-phosphate, D-galactose 6-phosphate, D-glucosamine, N-acetyl-D-glucosamine, methyl α-D-glucoside, 2-acetamido-D-allose, 2-acetamido-D-altrose, D-fructose, L-fructose, L-sorbose, D-fructose 6-phosphate, D-xylose, L-xylose, D-lyxose, D-ribose, 2-deoxy-D-ribose, D-arabinose, D-xylulose, DL-glyceraldehyde, D-mannitol, D-glucitol, *myo*-inositol, L-arabitol, D-arabitol, xylitol, ribitol, maltose, cellobiose, sucrose, lactose, melezitose, turanose, melibiose, trehalose, and raffinose.

Nucleotide Specificity. Both NAD and NADP serve as coenzymes. K_m values at pH 8.1 are 15 μM for NAD$^+$ and 68 μM for NADP$^+$.

Effect of pH and Substrate Concentration on Reaction Velocity. At high substrate concentrations the reaction velocity is maximal at about pH 9.4 and about 40% maximal at pH 8.1. However, the affinity of the enzyme for its substrate is greater at the lower pH. The respective apparent K_m values for D-fucose, D-galactose, and L-arabinose are 0.50, 0.17, and 0.14 mM at pH 8.1, and 6.6, 2.0, and 0.52 mM at pH 9.4.

Stability. Sephadex G-200 fractions (in 10 mM sodium phosphate buffer, pH 7.0) are stable to freezing for at least 2 months.

[36] Hexopyranoside: Cytochrome c Oxidoreductase from *Agrobacterium*[1]

By J. VAN BEEUMEN and J. DE LEY

Hexopyranoside + acceptor → 3-keto hexopyranoside + reduced acceptor

Assay Method

Principle. Enzyme activity can be followed either manometrically as oxygen consumption, using 5-methyl phenazinium methylsulfate (PMS)

[1] J. Van Beeumen and J. De Ley, *Eur. J. Biochem.* **6**, 311 (1968).

as intermediate electron carrier, or spectrophotometrically, as decrease in optical density at 600 nm using 2,6-dichlorophenolindophenol (DIP) as terminal acceptor. The former assay is laborious but offers the possibility to analyze the reaction products. A brief description is given in another article.[2] The latter assay is quicker and is described below.

Reagents

DIP, 0.885 mM in 0.4 M phosphate buffer, pH 6.0
Lactose, 0.1 M
Water, distilled
Enzyme, 0.002–0.05 unit/ml

Procedure. The reaction can be followed in any spectrophotometer at 600 nm. A cuvette of 1 cm light path receives 2.58 ml of DIP solution, 0.05–0.1 ml of enzyme, and water to a final volume of 2.7 ml. The amount of enzyme used depends on the possibilities of chart speed and scale expansion of the recorder. The reaction mixture is placed in the thermostated cuvette holder at 30° and, after 2 min of thermal equilibration, the endogenous reaction is recorded for 1 min. Enzyme activity is measured as the initial reduction rate of DIP after mixing in quickly 0.3 ml of lactose solution, kept at 30°. When the endogenous activity is too high, a blank is included with 0.3 ml of water instead of the substrate. Since DIP is known to be a very good acceptor with many other enzymes, one should identify the 3-keto sugar (synonym: 3-ulose) formed, as described below, whenever the presence of a 3-ulose-forming enzyme is suspected in organisms other than *Agrobacterium*.

Definition of Unit. One unit of enzyme is the amount that oxidizes 1 μmole of lactose per minute. Our sample of DIP having an extinction coefficient $\epsilon_{mM} = 11.85$ cm^{-1} mmole^{-1} at pH 6.0, this amount of enzyme gives a decrease of 4 optical density units under the conditions of the assay. Specific activity is expressed as number of units per milligram of protein.

Source of Enzyme. The enzyme is inducible and is known to occur only in most strains of *Agrobacterium tumefaciens* and *A. radiobacter*. The enzyme was completely purified from cells of *A. tumefaciens* ATCC 143.[1] Less pure preparations have been reported from *A. tumefaciens* strain B6,[3] BNV6,[3] and IAM 1525.[4] The latter strain is derived from

---

[2] J. Van Beeumen and J. De Ley, this volume [3].
[3] E. E. Grebner, E. Kovach, and D. S. Feingold, this series, Vol. 9 [17].
[4] K. Hayano and S. Fukui, *J. Biol. Chem.* **242**, 3665 (1967).

strain ATCC 4452. The enzyme has also been called D-aldohexopyranoside dehydrogenase[3,5] or D-glucoside 3-dehydrogenase.[6]

Purification Procedure

Culture Conditions. We purified the enzyme starting from a 150-liter culture, but the amounts can be scaled down if required. *Agrobacterium tumefaciens* ATCC 143 is grown for 40 hr at 30° in a culture containing 1% yeast extract, 0.1% KH_2PO_4, 0.2% $(NH_4)_2SO_4$, 0.025% $MgSO_4 \cdot 7H_2O$, and 1% lactose, final pH 6.8. Lactose is autoclaved separately in concentrated solution. Growth is started with a 5% inoculum. Good aeration (8 liters of air per hour liter of medium) and vigorous stirring are necessary. Approximately 2 kg of cells are harvested by centrifugation at the end of the logarithmic growth phase. They are washed three times with 10 mM phosphate buffer, pH 7.0. The following steps should be carried out at a temperature not exceeding 5°.

Step 1. Preparation of Particle-Free Extract. Portions of 35 g of washed cells in 70 ml of 10 mM phosphate buffer, pH 7.0, are disrupted in a Raytheon 10 kc sonic oscillator after replacement of the air in the cup by hydrogen. Unbroken cells and debris, removed by centrifugation, are resuspended, sonicated, and centrifuged again. The particles in the pooled supernatants are spun down for 1 hr at 100,000 *g* for 4 hr at 40,000 *g*.

Step 2. Removal of Nucleic Acids. To the particle-free supernatant, 0.05 volume of a 1 M $MnCl_2$ solution is added under stirring. After 30 min the precipitate is discarded by centrifugation at 10,000 *g*. The supernatant is dialyzed against a 1000-fold excess of 10 mM phosphate buffer, pH 6.9. A heavy precipitate formed during dialysis is removed by centrifugation.

Step 3. Ammonium Sulfate Fractionation. To each liter of supernatant, 288 g of solid ammonium sulfate is added, and the precipitate formed is removed by centrifugation. The supernatant is then brought to 60% saturation by the addition of 61 g of solid ammonium sulfate to each liter. The precipitate is dissolved in 100 ml of 10 mM phosphate buffer, pH 6.9, and the residual salt is removed by three dialysis experiments against 5 liters of the same buffer.

Step 4. DEAE-Cellulose Chromatography. The preceding preparation is chromatographed on a 4.1 × 68 cm column with 100 g of DEAE-cellulose (0.69 meq/g) equilibrated with 10 mM phosphate buffer, pH 6.9.

[5] E. E. Grebner and D. S. Feingold, *Biochem. Biophys. Res. Commun.* **19**, 37 (1965).
[6] M J. Bernaerts and J. De Ley, *J. Gen. Microbiol.* **22**, 137 (1960).

The ion exchange is pretreated in batch experiments with 0.2 N NaOH, washed with distilled water, charged with 0.1 M phosphate buffer, pH 6.9, until the effluent has the same pH, and equilibrated with the same buffer, diluted 10-fold. The enzyme is eluted under 1 meter of hydrostatic pressure in the 0.23–0.28 M KCl fraction of a 3.4-liter linear gradient of 0 to 0.4 M KCl in 10 mM phosphate buffer, pH 6.9. The enzyme-containing fractions are reduced in volume by precipitation with $(NH_4)_2SO_4$ and dialyzed as described at the end of step 3.

Step 5. Column Electrophoresis on Agarose. The data for the electrophoretic purification step in the table were obtained after two subsequent zone electrophoretic runs on a large LKB 5801 Porath column.[1] The following experiment on a smaller LKB 3340 column (3 × 41 cm) gives the same purification factor and a slightly better yield (85%). The column is filled with 0.18% agarose in 33 mM phosphate buffer, pH 7.27. About 30 mg of protein in 3.5 ml is applied. Electrophoresis is continued for 38 hr at 250 V and 37–38 mA. Electrode buffer is renewed after 18 hr. At the end of the run, the gel column is lifted by hydrostatic pressure (15 cm) and collected in 2-ml fractions. Enzyme activity is measured in the presence of agarose. The agarose is centrifuged at 10,000 g for 20 min, resuspended in buffer and centrifuged again. The enzyme is prepared for step 6 by $(NH_4)_2SO_4$ precipitation and dialysis against 30 mM phosphate buffer, pH 7.17.

Step 6. Gel Filtration. About 40 mg of protein in 4 ml of buffer are loaded on top of a 2.5 × 71 cm column of Sephadex G-75. Eluted at 25 ml/hour, the pure enzyme is collected as the first of two well separated protein peaks. The product is pooled, concentrated with $(NH_4)_2SO_4$, dialyzed against 50 mM phosphate buffer, pH 6.9, and stored under nitrogen at —18° when not in use.

The purification procedure is summarized in the table.

Properties

Stability. The enzyme becomes increasingly labile upon purification, perhaps owing to gradual loss of FAD. The activity half-life of pure enzyme is about a week when kept in the deep freezer. The enzyme is stable for 2 weeks under nitrogen at —15°. The best pH range is 6.5–7.4.

Physical and Chemical Properties. The sedimentation coefficient $s_{0.75\%}^{20}$ of the oxidoreductase is 5.1 S. Its molecular weight is 85,000 ± 7700, as determined with Archibald's method, and assuming a partial specific volume $\bar{v}$ of 0.725. The molecular weight from gel filtration on Sephadex G-200 is about 15% smaller, probably because of a slight retention of the enzyme on the substratelike dextran matrix of the gel. FAD is the

PURIFICATION OF HEXOPYRANOSIDE : CYTOCHROME c OXIDOREDUCTASE[a]

Step	Total units	Protein (mg/ml)[b]	Specific activity	Recovery (%)
1. Particle-free extract	1362	17.8	0.0255	100
2. MnCl$_2$ treatment	1309	9.9	0.0306	96
3. Ammonium sulfate	765	40.1	0.152	56
4. DEAE-cellulose	515	12.6	0.585	38
5. Column electrophoresis[c]	333	10.2	3.84	24
6. Gel filtration	293	11.7	7.63	21

[a] Modified from J. Van Beeumen and J. De Ley, *Eur. J. Biochem.* **6**, 311 (1968).

[b] Calculated from OD_{280}/OD_{260}, except for step 1, where protein is determined by the method of Lowry [O. H. Lowry, N. J. Rosebrough, A. L. Farr, and R. J. Randall, *J. Biol. Chem.* **193**, 265 (1951)].

[c] Results obtained with the Porath column, see text.

cofactor of the enzyme. The difference spectrum shows a flavin peak at 455 nm, and a smaller one at 396.5, when the enzyme is reduced with lactose. Less pure enzyme preparations show the participation of cytochrome c in the oxidation of lactose. Good electron acceptors of the enzyme, apart from PMS and DIP, are $K_3Fe(CN)_6$ and heart muscle cytochrome c. The oxidation rate of lactose is optimal at pH 7 with cytochrome c, at pH 6 with DIP, and at pH 8 with PMS as acceptor. Cyanide, azide, CO, EDTA, Ca^{2+}, and atabrine have no inhibitory effect on the enzyme. At 10 mM, cyanide does inhibit enzyme action by 70%.

Specificity. The first number after the name of each substrate mentioned below denotes its relative rate of oxidation. The next numbers are R_f values of the corresponding 3-keto derivatives on cellulose thin-layer plates, developed in the solvent designated by letter in parentheses and identified under *Detection of 3-Uloses*. Migration values larger than 1 are calculated versus glucose; all others versus the solvent front. The data are lactobionate 100, 0.07 (A), 0.22 (B); cellobiose 100, 0.27 (A), 1.3 (A); methyl β-D-glucose 85, 0.67 (A), 0.25 (E), 11.0 (E); methyl α-D-glucose 84, 0.67 (A), 0.51 (C), 11.0 (E); lactose 84, 0.06 (A), 0.23 (B); D-glucose 84, 0.60 (A), 0.49 (E), 19.0 (E); lactulose 76, 0.05 (A), 0.37 (B), 0.40 (D); p-arbutine 76; maltose 70, 0.20 (A); cellobionate 63, 0.26 (A); sucrose 60, 1.05 (A); D-galactose 56; glucose-1-phosphate 56; maltobionate 46, 0.28 (A); trehalose 30, 0.18 (A), 0.49 (B); β-melibiose 25, 0.16 (A), 0.49 (B); 2-deoxy-D-glucose 22; leucrose 12, 1.10 (A); methyl-β-D-thiogalactose 10, 0.45 (A), 0.67 (B), 0.37 (C); melezitose 10; D-mannose 10; raffinose 7, 0.09 (A); D-glucosamine 5; anhydro-1,6-D-glucose 4; 2-deoxy-D-galactose 2. In terms of substrate configuration, a sugar is oxidized by hexopyranoside: cytochrome c oxidoreductase if it occurs in the hexopyranose C-1 chair form, has a hemiacetal oxygen or sulfur atom at C_1, an equatorial OH group at C_3, and a CH_2OH group at C_5. An equatorial configuration of the OH group at C_2 and C_4 is preferential over an axial one. There is no definite proof yet that the enzyme acts only on D-sugars.

K_m values measured with enzyme preparation from step 3 are cellobiose, 0.2 mM; lactobionate, 0.21 mM; maltobionate, 0.36 mM; lactose, 1.7 mM; maltose, 2.8 mM; glucose, 2.9 mM; sucrose 4.1 mM, and galactose, 25 mM.

Detection of 3-Uloses. These compounds can be detected either chromatographically on thin-layer plates, or polarographically as described elsewhere.[2] Solvents used for the chromatography are (A) acetone–acetic acid–water (20:6:5, v/v), (B) water-saturated phenol, (C) methylethylketone–acetone–water (30:10:6, v/v), (D) ethylacetate–pyridine–acetic acid–water (5:5:1:3, v/v), (E) water-saturated methylethylketone. They display a specific gray-violet color at 100° with *o*-phenylenediamine spray reagent,[6] and a less sensitive reddish color with fluorescence after 10 min at 100° with an urea phosphate spray reagent.[7] The first reagent is made by suspending 400 mg of *o*-phenylenediamine in 3 ml of water, adding 0.65 ml of concentrated HCl, and diluting to 20 ml with ethanol. The second reagent contains 3 g of urea in *n*-butanol–ethanol–water–85% phosphoric acid (80:10:10:6, v/v). It allows detection of 15 μg of 3-ulose.

The chemical structure of the 3-uloses of sucrose[8,9] lactose[6,10] lactobionate,[6] maltose,[6,10] maltobionate,[6] trehalose,[10] cellobiose,[11] glucose,[12] methyl-β-D-glucose,[13] and glucose 1-phosphate[14] has been established, and several of them have been crystallized[9,14]. This is not yet the case for the other presumed 3-uloses.

[7] C. S. Wise, F. J. Dimler, H. A. Davis, and C. E. Rist, *Anal. Chem.* **27**, 33 (1965).

[8] M. J. Bernaerts, J. Furnelle, and J. De Ley; *Biochim. Biophys. Acta* **69**, 322 (1963).

[9] S. Fukui, R. M. Hochster, R. Durbin, E. E. Grebner, and D. S. Feingold, *Bull. Res. Counc. Isr. Sect. A* **11**, 262 (1963).

[10] S. Fukui and R. M. Hochster, *Can. J. Biochem. Physiol.* **41**, 2363 (1963).

[11] K. Hayano and S. Fukui, *J. Biochem.* **64**, 901 (1968).

[12] S. Fukui and R. M. Hochster, *J. Amer. Chem. Soc.* **85**, 1697 (1963).

[13] B. Lindberg and O. Theander, *Acta Chem. Scand.* **8**, 1870 (1954).

[14] S. Fukui, *J. Bacteriol.* **97**, 793 (1969).

[37] Aldose Reductases from Mammalian Tissues

By K. H. GABBAY and J. H. KINOSHITA

D-Glucose + NADPH + H$^+$ → sorbitol + NADP$^+$
D-Galactose + NADPH + H$^+$ → galactitol + NADP$^+$
D-Xylose + NADPH + H$^+$ → xylitol + NADP$^+$
DL-Glyceraldehyde + NADPH + H$^+$ → glycerol + NADP$^+$
D-Glucuronate + NADPH + H$^+$ → L-gulonate + NADP$^+$

Aldose reductase[1,2] (alditol:NADP$^+$ 1-oxidoreductase, EC 1.1.1.21) is widely distributed in mammalian tissues and catalyzes the NADP-linked reduction of a broad group of aldoses and alduronic acids. A closely related enzyme, NADP-L-hexonate dehydrogenase[3] (L-gulonate:NADP$^+$ 1-oxidoreductase, EC 1.1.1.19), with similar broad substrate specificities is also present in many tissues. The two enzymes are distinguished by relative substrate specificities, by column chromatography, and immunologically.

Assay Method

Principle. The enzymes are assayed spectrophotometrically by determining the decrease in NADPH concentration at 340 nm. In the following procedure the reaction is carried out in a temperature-controlled (25°) automated compensating double-beam Unicam spectrophotometer equipped with a scale expander and back-off accessory attached to a recorder.[4] Both blank and sample cuvettes contain NADPH, and a metal diaphragm (absorbance of 1.0) is inserted in the sample cuvette light path. The scale is expanded to give a recorder full scale of 0.400, and the decrease in NADPH in the sample cuvette is measured during the reaction. This procedure is used to compensate for occasional nonspecific NADPH utilization in the blank, occurring especially in the more crude enzyme preparations.

Reagents

Assay buffer—Na-K phosphate, 0.135 M, pH 6.2:
Add 18.5 ml of Na$_2$HPO$_4$ (0.135 M) to 81.5 ml of KH$_2$PO$_4$ (0.135 M) and check pH.

---

[1] H. G. Hers, *Biochim. Biophys. Acta* **37**, 120 (1960).
[2] S. Hayman and J. H. Kinoshita, *J. Biol. Chem.* **240**, 877 (1965).
[3] Y. Mano, K. Suzuki, K. Yamada, and N. Shimasono, *J. Biochem.* **49**, 618 (1961).
[4] Philips Electronics Co., Mount Vernon, N.Y.

Substrates

   a—DL-Glyceraldehyde, 0.1 M. Dissolved in distilled water; aliquots may be frozen at —30°. May be refrozen.

   b—Sodium-D-glucuronate, 1 M. Dissolved in distilled water; may be frozen as above.

   c—D-Xylose, 4 M. As above.

NADPH, 0.73 mM. Dissolve 5 mg of the tetrasodium salt (Sigma type I, preweighed vials) in 7.5 ml of distilled water. Make up fresh daily and keep on ice. Determine the exact concentration from the absorbance at 340 nm and the extinction coefficient of 6.25×10^{-3}.

Mercaptoethanol, 50 mM. Prepared immediately before use by diluting 0.35 ml of the reagent to 100 ml with distilled water.

Enzyme: Add sufficient enzyme solution to obtain 0.05–0.1 change in absorbance at 340 nm in 2 min.

Procedure. To two matched microcuvettes of 1 cm light path and 1.0 ml volume, add 0.5 ml of buffer (final concentration 67 mM), 0.1 ml NADPH (final concentration 73 μM), and 0.1 ml mercaptoethanol solution (final concentration 5 mM). Appropriate amounts of enzyme solution are added to both cuvettes, and enough distilled water to a final volume of 1 ml. The reaction is started by the addition of substrate to the sample cuvette. Glyceraldehyde (10 μl) is added to a final concentration of 1 mM. Glucuronate (10 μl) final concentration is 10 mM, and xylose (25 μl) final concentration is 100 mM. Additions are made with appropriate Eppendorf micropipettes.

Definition of Enzyme Unit and Specific Activity. One unit of enzyme is defined as the amount that causes the oxidation of 1 μmole of NADPH per hour at 25° under the conditions of assay described above. The specific activity is expressed as units of enzyme per gram of protein. Protein is determined by the method of Lowry *et al.*[5] using bovine serum albumin as a standard.

Sources of Enzymes. Lens,[6] peripheral nerves (sciatic), and placenta[7] contain aldose reductase only. Retina,[8] brain,[9] spinal cord, kidney, pancreas,[10] and liver contain both aldose reductase and L-hexonate dehydro-

[5] O. H. Lowry, N. J. Rosebrough, A. L. Farr, and R. J. Randall, *J. Biol. Chem.* **193**, 265 (1961).

[6] S. Hayman, M F. Lou, L. O. Merola, and J. H. Kinoshita, *Biochim. Biophys. Acta* **128**, 474 (1966).

[7] R. S. Clements and A. I. Winegrad, *Biochem. Biophys. Res. Commun.* **47**, 1473 (1972).

[8] K. H. Gabbay, *Isr. J. Med. Sci.* **8**, 1626 (1972).

[9] G. I. Moonsammy and M. A. Stewart, *J. Neurochem.* **14**, 1187 (1967).

[10] K. H. Gabbay, and J. Tze, *Proc. Nat. Acad. Sci. U.S.* **69**, 1435 (1972).

genase. A general procedure has been in use in our laboratories to isolate and purify these enzymes from bovine, rat, and human tissues, and the preparation of the beef kidney enzymes will be described in detail. Purification of aldose reductase from the lens[2,11] and the separation of aldose reductase from L-hexonate dehydrogenase from brain[9] have been described.

Purification Procedure

All steps are at 4°.

Step 1. Fresh calf kidneys are obtained from a local slaughterhouse and kept on ice. The kidney calyces are opened, and the papillary and medullary tissue is removed with scissors. The cortical tissue is discarded since aldose reductase is present in the kidney papilla.[12] Sufficient hexonate dehydrogenase activity is present in the papillary and medullary tissue for isolation purposes. Alternatively, whole kidneys may be used with higher yields of hexonate dehydrogenase, and correspondingly lower yields of aldose reductase. The dissected papillary tissue (approximately 180–220 g) is homogenized in a Waring Blendor for 2 min with 5 mM Tris phosphate buffer, pH 7.4, at a 2:1 volume/weight ratio. The homogenate is centrifuged for 30 min at 45,000 g, and the supernatants are collected.

Step 2. Solid $(NH_4)_2SO_4$ (17.5 g/100 ml) is added at 4° to the crude extract to give a saturation of 0.3. After the preparation has stood for 20 min, the precipitate is removed by centrifugation and discarded. Additional solid $(NH_4)_2SO_4$ (31 g/100 ml) is added to yield a saturation of 0.75, and the precipitate is collected by centrifugation after 30 min. The precipitate is dissolved in 5 mM Tris phosphate buffer, pH 7.4, containing 5 mM mercaptoethanol and dialyzed overnight against 4 changes of 20 volumes each of the same buffer.

Step 3. The dialyzate is centrifuged to remove any formed precipitate, and the supernatant fluid (about 200 ml) is pumped onto a 2.5 × 45 cm column of DEAE-cellulose previously equilibrated with 5 mM Tris phosphate 5 mM mercaptoethanol (pH 7.4). The column is then washed by gravity with the same buffer, and the eluate monitored at 280 nm in a flow cell until the protein elution decreases to 0.1–0.2 absorbance. Twenty milliliter fractions are collected, and usually 400–600 ml of buffer is necessary to accomplish this elution. L-Hexonate dehydrogenase is invariably isolated in this wash peak (Fig. 1), and is characterized by high glucuronate:glyceraldehyde activity ratios under the conditions of the assay described above.

[11] J. A. Jedziniak, and J. H. Kinoshita, *Invest. Ophthalmol.* **10**, 357 (1971).
[12] K. H. Gabbay and E. S. Cathcart, *Diabetes* **23**, 460 (1974).

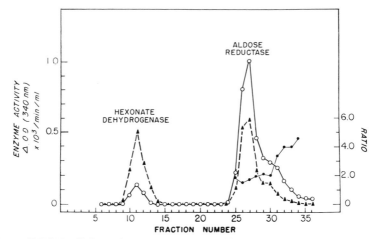

Fig. 1. DEAE-cellulose chromatography of kidney aldose reductase and L-hexonate dehydrogenase. This fractionation was achieved using 40–75% ammonium sulfate fraction of the kidney homogenate. ○———○, Glyceraldehyde; ▲- - -▲, glucuronate; ●———●, ratio glyceraldehyde:glucuronate. (From footnote 12.)

Step 4. The column is subsequently developed with 750 ml of the NaCl-phosphate (pH 7.3 to 6.2) gradient described by Moore and McGregor[13] using a nine-chambered gradient mixer.[14] Twenty-milliliter fractions (elution rate of 20 ml per hour using a pump) are collected and assayed for enzyme activity as described above with all three substrates. Aldose reductase activities emerge at 150 to 200 mM NaCl gradient concentration (Fig. 1); the fractions containing activity are combined and concentrated by pressure filtration through an Amicon UM-10 filter[15] (10,000 MW cutoff), and the buffer is changed to a 5 mM Na-K phosphate–5 mM mercaptoethanol, pH 6.8.

Step 5. The concentrated and dialyzed enzyme is applied to a gravity-packed 1 × 25 cm hydroxyapatite column equilibrated with Na-K phosphate buffer. After emergence of a protein peak in the wash (approximately 60 ml), the activity is eluted with 540 ml of a linear phosphate gradient from 5 to 400 mM (pH 6.8). Ten-milliliter fractions are collected with a flow rate of 8–10 ml per hour. Besides further purification, this procedure resolves two aldose reductase activities (Fig. 2) previously noted on DEAE-cellulose chromatography as a main peak and trailing shoulder (Fig. 1).

[13] B. W. Moore, and D. McGregor, *J. Biol. Chem.* **240**, 1657 (1965).
[14] Buchler Instruments, Fort Lee, New Jersey.
[15] Amicon Corp., Lexington, Massachusetts.

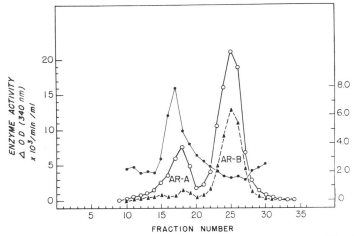

FIG. 2. Hydroxyapatite chromatography of kidney aldose reductase. The fractions recovered from the DEAE chromatography containing aldose reductase activity were combined and subjected to hydroxyapatite column chromatography. ○——○, Glyceraldehyde; ▲---▲, glucuronate; ●——●, ratio glyceraldehyde:glucuronate. [Data from K. H. Gabbay and E. S. Cathcart, *Diabetes* **23**, 460 (1974).]

Step 6. The aldose reductase activities (AR-A and AR-B) are separately concentrated by pressure dialysis as described above and subjected to further purification by chromatography on a 2.5 × 150 cm column of Sephadex G-100 equilibrated with 5 mM Tris phosphate buffer containing 5 mM mercaptoethanol at pH 7.4. Repeated passages (2–3) may be necessary to achieve highest purification.

The purification procedure is summarized in the table.

PURIFICATION OF KIDNEY ALDOSE REDUCTASE

Fraction	Specific activity (units/mg protein)	Total activity (units)	Recovery (%)
Original extract	250	1300	100
40–75% (NH$_4$)$_2$SO$_4$	670	1170	90
DEAE-cellulose (AR peak)	2,490	700	54
Hydroxyapatite			
AR-A	1,810	200	15
AR-B	6,550	420	33
Sephadex G-100			
AR-A	19,040	215	16
AR-B	33,950	445	34

Variations in Procedure. If smaller amounts of tissue are used, column size and elution gradients may be proportionately reduced. In certain cases (e.g., pancreas) where rapid enzyme losses from small samples may occur, it is recommended that the $(NH_4)SO_4$ fraction (step 2) be rapidly desalted by passage through a 2.5×50 cm column of Sephadex G-25 and immediately proceeding to step 3. Elution of the DEAE-cellulose column may also be alternatively attempted with a linear phosphate gradient (5 to 200 mM, pH 7.4).

Properties

Stability. The crude extract is stable for a few days at 5°, while the $(NH_4)_2SO_4$ precipitate loses 90% of the enzyme activity overnight in the cold room. The most highly purified aldose reductase preparations may be stored frozen at $-30°$ *after* extensive dialysis to remove mercaptoethanol, with loss of 50% of the activity in 6 months. Activity may be restored by the addition of mercaptoethanol.

Effect of pH. The optimal pH range for the enzyme with D-glyceraldehyde as a substrate is 5.5–6.8.

Specificity. Aldose reductase has broad substrate specificity reducing D-glyceraldehyde, D-glucuronolactone, D-glucuronate, D-glucose, D-xylose, D-galactose, D-ribose, D-erythrose, and D- and L-arabinose. D-Mannose, L-glucose, L-xylose, and deoxysugars do not serve as substrates.[2] These substrate specificities are essentially the same for aldose reductases isolated from various tissues. AR-A differs from AR-B primarily by its poor ability to reduce D-glucuronate.

Aldose reductases require NADPH as the coenzyme, and show no activity with NADH. NADP inhibits the enzyme.[11] The reaction proceeds in the forward direction with poor reversibility.

Activators and Inhibitors. Lithium sulfate and ammonium sulfate stimulate aldose reductase.[1,2] High concentrations of chloride ions are inhibitory, as are a number of organic anions.[1,2] 3,3-Tetramethyleneglutaric acid (TMG) inhibits lens aldose reductase with a K_i of 10 μM.[11] There are differences in the inhibitory potency of TMG on aldose reductases from various tissue sources and between species. Glutaric acid and 2,2-dimethylglutaric acid have different potencies in inhibiting L-hexonate dehydrogenase and aldose reductase and may be used to differentially inhibit the two enzymes.[10] A description of the type of inhibition with various inhibitors for the lens aldose reductase has been presented.[2,11] The enzymes are totally inhibited by 50 μM p-chloromercuribenzoate, and 1 mM N-ethylmaleimide. Mercaptoethanol protects the enzyme and can restore the activity in stored preparations of aldose reductase.

The basis for some of these interesting effects has been the finding that aldose reductase exists in an equilibrium between an active form and two inactive fragments. NADP, TMG, and N-ethylmaleimide favor the conversion to the inactive forms of the enzyme, while mercaptoethanol favors the formation of the active enzyme.[11]

Immunological Data. Antibodies prepared in rabbits to bovine kidney aldose reductase cross-react with aldose reductase from other bovine tissues. There is no cross reaction with L-hexonate dehydrogenase.[8] Håstein and Velle[16] reported no immunological cross-reaction between the enzymes prepared from sheep seminal vesicle and placenta.

Distribution. Enzymes with similar substrate specificities have been described in *Candida utilis, Candida albicans, Penicillium* sp., *Penicillium chrysogenum, Aspergillus* sp., and *Rhodoturula* sp.

[16] T. Håstein, and W. Velle, *Biochim. Biophys. Acta* **178**, 1 (1969).

[38] Aldose Reductase from Seminal Vesicle and Placenta of Ruminants

By WEIERT VELLE

Glucose + NADPH $\rightleftharpoons$ sorbitol + NADP$^+$ + H$^+$
Glyceraldehyde + NADPH $\rightleftharpoons$ glycerol + NADP$^+$ + H$^+$
Lactaldehyde + NADPH $\rightleftharpoons$ 1,2-propanediol + NADP$^+$ + H$^+$

Assay Method

Principle. Glucose is reduced to sorbitol in the presence of ruminant placental or seminal vesicle aldose reductase and reduced pyridine nucleotide at pH 7.5. The reaction rate is, however, higher using glyceraldehyde or lactaldehyde as substrate. Although the equilibrium of the reaction strongly favors the reduction of the aldehyde, the oxidation reaction takes place when the reaction is carried out at pH 10 using the corresponding alcohols as substrate.

The reactivity is considerably higher for the D-form than for the L-form of the substrates.

For practical purposes, DL-glyceraldehyde is used as substrate in the standard assay system, and the reaction is measured spectrophotometrically by following the decrease in absorption at 340 nm caused by the formation of NADP$^+$.

Reagents

Tris chloride, 0.3 M, pH 7.5

NADPH, 5 mM

DL-Glyceraldehyde, 0.5 M, dedimerized just prior to use by heating at 85° for 10 min and then kept at 37° during periods of assay.

Procedure. Place 1.0 ml of Tris-chloride buffer, 0.1 ml of reduced pyridine nucleotide, 0.1 ml of enzyme solution, and water to a final volume of 2.9 ml in a glass cuvette of 1 cm light path. The reaction is initiated by the addition of 0.1 ml of glyceraldehyde. This order is important. If the enzymes are incubated with the substrate, and the reaction is initiated with coenzyme, lower reaction rates are observed, and reproducibility is poor. The reaction is followed by readings every 10 sec for a period of 1 min, starting 20 sec after the initiation of the reaction. A decrease in reaction rate occurs at an early stage. The reaction rate is therefore determined from the extrapolated values of the three first readings, the change in absorbance per minute being calculated.

Definition of Unit and Specific Activity. One unit of enzyme is defined as the amount of protein catalyzing the transformation of 1 μmole of substrate per minute at 37° under the conditions specified above. Specific activity is expressed as units per milligram of protein determined by absorbancy readings at 280 and 260 nm, applying the correction 1.55 D_{280}–0.76 D_{260}, using cuvettes of 1 cm light path.

Purification Procedure

The following general conditions are used in the purification of the enzymes.[1] All operations are carried out at 0–4°. Fractionations with ammonium sulfate are made by slow addition, with stirring, of the calculated amount of solid ammonium sulfate. The resulting suspensions are kept for 30 min before centrifugation. In the chromatographic procedures, protein in the eluates from the columns is detected by absorbance readings at 280 nm in cuvettes of 1 cm light path, or by continuous scanning at 256 nm.

Step 1. Preparation of Crude Extract. The fetal membranes (cotyledons) or seminal vesicle tissue are cut into small pieces with scissors and homogenized in batches of 100 g of tissue per 300 ml of medium 1 mM in cysteine hydrochloride, 1 mM in EDTA, 10 mM in nicotinamide and 30 mM in sodium bicarbonate. Homogenization is carried out in a Waring Blendor for 2 min at maximal speed. The homogenate is centrifuged for 20 min at 2500 g, and the supernatant fluid is collected.

[1] W. Velle and L. L. Engel, *Endocrinology* **74**, 429 (1964).

Step 2. Microsome Precipitation. To the supernatant fluid is added 1/10 volume of 0.1 M $CaCl_2$ followed by stirring for 1 hr. After standing an additional hour, the microsomes are brought down by centrifugation for 30 min at 2500 g.

Step 3. Ammonium Sulfate Fractionation. The supernatant fluid from the above step is brought to about 50% saturation with solid ammonium sulfate (30 g/100 ml). After centrifugation for 15 min at 4000 g, the precipitate is discarded. The supernatant fluid is brought to about 80% saturation with ammonium sulfate (20 g/100 ml); after centrifugation for 15 min at 4000 g, the precipitate is collected and dissolved in a small amount of 5 mM Tris-phosphate buffer of pH 7.5. Thus prepared the enzymes are stable in the frozen state for months.

Step 4. Desalting of the Protein Solution. This is carried out by passage through a Sephadex G-75 column previously equilibrated with 5 mM Tris-phosphate buffer (pH 7.5) or by dialysis against distilled water and finally against the buffer to be used in subsequent chromatography.

Step 5. Chromatography on DEAE-Sephadex. The solution from the above step in amounts of 20–40 ml (0.5–1.9 g protein) is applied to a column of DEAE-Sephadex (30 × 2.7 cm) that is equilibrated with 5 mM Tris-phosphate buffer of pH 7.5, at a flow rate of approximately 1.2 ml/min. The column is eluted successively with 200 ml of buffer and 200 ml of buffer 0.1 M in NaCl, the eluates being discarded. The enzymes are then eluted by buffer 0.15 M in NaCl (bovine enzymes) or 0.20 M in NaCl (ovine enzymes). At this stage of purification the enzymes are very stable when kept at −15°.

Step 6. Chromatography on Hydroxyapatite. The combined active fractions from the above step may be applied, without desalting, directly to columns of hydroxyapatite prepared according to Anacher and Stoy.[2] Coarse grade sintered-glass filters (4.5 cm in diameter) covered by filter paper are filled to a height of 1.6–2.0 cm with hydroxyapatite from a suspension in 5 mM phosphate buffer, pH 7.0. The capacity of such a column for the proteins in question is of the order of 50 mg, which may be applied in volumes up to at least 200 ml. The fractions are collected in a Thunberg tube, slight suction being applied to give a flow rate of 1.5–2.0 ml/min. The active fraction is eluted with 50 mM phosphate buffer pH 7.0, leaving the bulk of protein on the column. The column is regenerated by elution with 1 M phosphate buffer and may be used repeatedly for long periods.

A summary of the purification data for the ovine enzymes is given in the table.

[2] W. F. Anacher and V. Stoy, *Biochem. Z.* **330,** 141 (1958).

Properties of the Enzymes

Purity. The placental enzyme is not pure after the procedure outlined above. Moving-boundary electrophoresis of the bovine placental enzyme and gel diffusion and immunoelectrophoresis carried out using antiserum against the ovine placental enzyme reveal two protein fractions. The ovine seminal vesicle enzyme, on the other hand, seems to be pure as judged by immunological studies. These studies also revealed that there is no immunological cross-reaction between the ovine placental and seminal vesicle enzymes.[3]

Stability. The enzymes are stable at −15° for months after chromatography on DEAE-Sephadex. After chromatography on hydroxyapatite stability is reduced and repeated freezing and thawing results in marked loss of activity. Heating to 56° for 2 min in the absence or presence of coenzyme leads to a 95–90% loss of activity.

Specificity. In the purified form both the placental and seminal vesicle enzymes are NADP specific. The most reactive substrates are lactaldehyde and glyceraldehyde. The enzymes also act on glucose and galactose. Of the alcohols tested, 1,2-propanediol is the most reactive, followed by glycerol. Sorbitol is also a good substrate. These three alcohols are oxidized more readily in the presence of the placental enzyme than when the seminal vesicle enzyme is used. This applies for both the bovine and ovine enzymes.

With the bovine placental enzyme, the reaction rate using the D-(−)-form of glyceraldehyde and lactaldehyde is about twice the rates found when the DL-forms are used. Similar results are obtained for the D-(−)- and L-(+)-forms of 1.2-propanediol (75% optically pure).

Kinetic Properties. When aldehydes are used as substrates, none of the enzymes follow the Michaelis-Menten equation. The double reciprocal plot covering a 10,000-fold concentration range indicates substrate activation. A valid estimate of Michaelis constants therefore cannot be made.

With alcohols as substrate and a coenzyme concentration of 0.167 mM, both enzymes (from bovine as well as ovine source) follow the Michaelis-Menten equation. For ovine placental enzyme K_m at pH 10.0 and 37° are 1,2-propanediol = 0.7 M; glycerol = 1.4 M; and for ovine seminal vesicle enzyme: 1,2-propanediol = 0.8 M, glycerol = 1.2 M.

With glyceraldehyde as substrate at a concentration of 16.7 mM at pH 7.5 and 37° the K_m for NADPH is 0.17 mM for the ovine placental enzyme and 0.32 mM for the ovine seminal vesicle enzyme.

Inhibitors and Activators. The enzymes are strongly inhibited by sulfhydryl reagents. Rapid and almost complete inhibition is produced by

[3] T. Håstein and W. Velle, *Biochim. Biophys. Acta* **178**, 1 (1969)

PURIFICATION DATA FOR OVINE ALDOSE REDUCTASES[a]

Fraction	Placental enzyme			Seminal vesicle enzyme		
	Specific activity (units $\times$ 10³)	Total units (units $\times$ 10³)	Recovery (%)	Specific activity (units $\times$ 10³)	Total units (units $\times$ 10³)	Recovery (%)
Supernatant fluid after CaCl₂-precipitation	4.68	102.500	100	4.44	42.800	100
50–80% Ammonium sulfate fraction	6.42	52.000	50.7	26.8	5.200	12.2
DEAE-Sephadex eluate	83.8	26.900	26.2	91.3	4.060	9.5
Hydroxyapatite eluate	163.5	16.250	15.9	315	7.800	18.4

[a] From T. Håstein and W. Velle, *Biochim. Biophys. Acta* **178**, 1 (1969).

1 μM parachloromercuribenzoate. Activity may be partly restored by the addition of cysteine.

Both enzymes are sensitive to heavy metal ions. Sulfate ions cause a marked stimulation of the enzyme when glyceraldehyde is used as substrate, while inhibition by sulfate is observed when the substrate is propanediol.

[39] Pyranose Oxidase from *Polyporus obtusus*

By FRANK W. JANSSEN and HANS W. RUELIUS

$$D\text{-Glucopyranose} + O_2 \rightarrow D\text{-glucosone} + H_2O_2 \tag{1}$$
$$D\text{-Xylopyranose} + O_2 \rightarrow D\text{-xylosone} + H_2O_2 \tag{2}$$
$$L\text{-Sorbose} + O_2 \rightarrow 5\text{-keto-}D\text{-fructose} + H_2O_2 \tag{3}$$
$$\begin{array}{ccc} \delta\text{-}D\text{-Gluconolactone} & & 2\text{-keto-}D\text{-gluconic acid} \\ + & \rightarrow & + \\ O_2 & & D\text{-araboascorbic acid} + H_2O_2 \end{array} \tag{4}$$

Pyranose oxidase[1] catalyzes the oxidation of D-glucose, D-xylose, L-sorbose, and δ-D-gluconolactone at the second carbon atom to form 2-keto products and hydrogen peroxide. The purification and properties of the enzyme, isolated from the basidiomycete *Polyporus obtusus*, have been described.[2,3]

Assay Method

Principle. The enzyme is most conveniently assayed by measuring the color produced from H_2O_2 in a coupled reaction utilizing horseradish peroxidase and *o*-dianisidine.

Reagents

Horseradish peroxidase (EC 1.11.1.7) 400 units/mg: 0.1% in distilled water. Store at 4° under toluene.

o-Dianisidine dihydrochloride: 1% solution in 0.25 M HCl. Store at 4°.

D-Glucose: 10 g of β-D-glucopyranose in 100 ml of H_2O. Store at least 24 hr prior to use.

Peroxidase–dianisidine: To 79 ml of distilled water add 10 ml of 0.5 M pH 7.0 sodium phosphate buffer, 10 ml of peroxidase solution, and 1.0 ml of dianisidine solution.

[1] This name has been designated by the I.U.B. Enzyme Commission and the enzyme has been assigned the number 1.1.3.10.
[2] H. W. Ruelius, R. M. Kerwin, and F. W. Janssen, *Biochim. Biophys. Acta* **167**, 493 (1968).
[3] F. W. Janssen and H. W. Ruelius, *Biochim. Biophys. Acta* **167**, 501 (1968).

Glucose–peroxidase–dianisidine: Prepare as described for peroxidase–dianisidine except substitute 10 ml of glucose solution for water.

HCl, 4.0 M

Procedure. Add 1.0 ml of appropriately diluted enzyme preparation to duplicate test tubes in a 25° water bath. At zero time add 4.0 ml glucose–peroxidase–dianisidine reagent. Terminate the reaction at 5 min with 0.2 ml of 4.0 M HCl. Read the color produced in a spectrophotometer at 460 nm or a colorimeter with blue filter (Klett-Summerson No. 42). Correct reading for any blank produced by enzyme + peroxidase–dianisidine reagent. The amount of H_2O_2 produced is determined by comparison to standard solutions of H_2O_2 treated in the same way. 420 Klett units $\cong$ 1 μmole of H_2O_2. The ideal range is 50 to 200 KU.

Definition of Unit and Specific Activity. One unit of enzyme is defined as the amount of enzyme that produces 1 μmole of H_2O_2 per minute in this system. Specific activity is defined as the number of units per milligram of protein.

Production of Pyranose Oxidase

Cultures of *Polyporus obtusus* (Berk) obtained from the Forest Research Institute, Dehradum, India (No. 136) are maintained on Baltimore Biological Laboratories malt agar slants at 25° (Baltimore Biological Laboratories, Baltimore, Maryland) and are used within 2–12 weeks. The growth from one slant is suspended in 10 ml of sterile water by scraping the agar with a sterile needle. The suspension is transferred aseptically to a 250-ml Erlenmeyer flask containing 100 ml of sterile medium composed of 1.0% Phytone (Baltimore Biological Laboratories) and 1.0% dextrose. The flasks, which are closed with a cotton plug, are incubated at 25° on a rotary shaker set at 280 rpm for 7 days. The contents of the flask are transferred aseptically to a sterilized Waring Blendor jar and blended for 15 sec. Five milliliters of this inoculum is transferred aseptically to 500-ml Erlenmeyer flasks containing 150 ml of medium consisting of 20 g of dextrose, 20 g of N Z Amine ET (Sheffield Chemical Co., Norwich, New York), and 1 g of KH_2PO_4 in 1 liter of tap water. The medium is adjusted to pH 6.5 before autoclaving for 15 min at 121°. The flasks are incubated at 25° on a reciprocating shaker set at ninety 4-inch strokes/min. The enzyme content of the mycelium is monitored daily after the fourth day of incubation. An extract is prepared by homogenizing the mycelium from one flask in 150 ml of 50 mM sodium phosphate buffer (pH 6.8) for 5 min in a Waring Blendor. The mycelial extract is filtered and assayed for pyranose oxidase as described above except that the incubation period is 30 min. The maximum enzyme production of about 1.5 units/ml, achieved in 7–13 days, coincides with depletion of glucose and a rise in the pH of the broth. The mycelium is filtered, washed, and stored frozen until fractionated.

Purification Procedure

Step 1. The mycelium is homogenized for three periods of 3 min each in a Waring Blendor with cold 50 mM sodium phosphate, pH 7.0 (7.5 ml/g wet mycelium) with intermittent cooling in an ice bath. All subsequent steps are carried out at room temperature.

Step 2. The turbid supernatant (1 volume) obtained by centrifugation at 5100 g is fractionated by the slow addition of 22.5 g of polyethylene glycol-6000 (PEG) J. T. Baker Chemical Co. for each 100 ml of supernatant while mixing on a magnetic stirrer. The precipitate formed during 30 min is separated by centrifugation at 7000 g for 20 min. The sediment is dissolved with stirring for 30 min in about 0.1 volume 0.2 M NaCl in 50 mM pH 7.0 sodium phosphate, after which the insoluble material is removed by centrifugation at 33,000 g.

Step 3. PEG (0.11 g/ml) is added with stirring to the resulting supernatant as above. After 30 min, the precipitate is separated by centrifugation at 10,300 g for 20 min. The yellow sediment is dissolved in $\frac{1}{12}$ volume of 50 mM, pH 7.0, sodium phosphate buffer. Centrifugation of the opalescent solution at 33,000 g yields a clear yellow supernatant.

Step 4. The enzyme in the supernatant was reprecipitated by the slow addition of PEG (0.11 g/ml) as described above. The precipitate formed during 30 min standing was recovered by centrifugation at 10,300 g for 20 min. The final precipitate is dissolved in $\frac{1}{60}$ volume of 50 mM phosphate buffer (pH 7.0). The results of a typical fractionation are summarized in the table. The enzyme is purified about 17-fold with a recovery of about 50%. Such preparations retain full activity for at least 4 months when stored at 4°. If a dry preparation is desired, the solution is dialyzed overnight against 1 mM pH 7.5 Tris buffer, shell frozen, and lyophilized. About 20–40% of the activity is lost in the preparation of the freeze-dried material, but the remaining enzyme is stable for at least 1 year when stored at 4° over anhydrous silica gel.

Properties

Specificity. In addition to glucose, the enzyme also catalyzes the oxidation of D-xylose, L-sorbose, and δ-D-gluconolactone.[3] No other substrates were found. All the substrates have in common a six-membered ring and the same structural and conformational features on carbons

PURIFICATION OF PYRANOSE OXIDASE BY FRACTIONAL
PRECIPITATION WITH POLYETHYLENE GLYCOL

Enzyme fraction	Volume (ml)	Total activity (units)	Total protein (6.25 × mg N)	Specific activity	Overall recovery (%)
Crude extract	290	1200	1600	0.75	
19% Polyethylene glycol ppt.	32.5	1200	304	3.9	100
11% Polyethylene glycol ppt.	25	1070	106	10.1	89
Second 11% polyethylene glycol ppt.	5	640	50	12.8	53

2, 3, and 4. Gluconic acid (sodium salt), its γ-lactone and the α-anomer of glucose are unreactive. Evidence that the oxidation of the four substrates is catalyzed by a single enzyme was also presented in the previous report.[2,3]

Effect of pH. Pyranose oxidase has a broad pH optimum of 6.0–8.0. It is stable for at least 1 hr at 25° over a pH range of 4.0–9.0. It is precipitated at pH 4.7 with acetic acid. This property may be used for purification.[2]

[40] L-Fucose Dehydrogenase from Sheep Liver

By PATRICK W. MOBLEY, ROBERT P. METZGER,
and ARNE N. WICK

$$\text{L-Fucose} + \text{NAD}^+ \rightleftharpoons (\text{L-fucono-}\delta\text{-lactone}) + \text{NADH} + \text{H}^+$$
$$\text{D-Arabinose} + \text{NAD}^+ \rightleftharpoons (\text{D-arabono-}\delta\text{-lactone}) + \text{NADH} + \text{H}^+$$

At saturating substrate concentrations, sheep liver L-fucose dehydrogenase oxidizes L-fucose and D-arabinose at approximately the same rate. At the pH's required for maximal activity, the lactone products quickly hydrolyze to the corresponding acids. The enzyme is found in the cytosol fraction and preferentially utilizes NAD⁺ as coenzyme. Some physical and kinetic properties of the enzyme have been studied,[1] and it may be

[1] P. W. Mobley, R. P. Metzger, and A. N. Wick, *Arch. Biochem. Biophys.* **139**, 83 (1970).

useful as a model for similar enzymes found in livers of other vertebrate species.[2,3]

Assay Method

Principle. The enzyme is assayed by determining spectrophotometrically the increase in absorption at 340 nm as a consequence of NADH formation.

Reagents

Glycine-NaOH buffer, 0.25 M, pH 10.4
NAD, 10 mM, adjusted to pH 7 before use
L-Fucose, 0.50 M
D-Arabinose, 0.50 M

Procedure. To each of two 1-ml cuvettes are added 0.2 ml of glycine buffer, 0.1 ml of NAD, and 0.1 ml of the selected substrate (L-fucose or D-arabinose). To the sample cuvette is added 0.1 ml of the enzyme preparation. Each cuvette is then adjusted, with water, to 1.0 ml final volume. The contents are stirred well and the cuvettes are immediately placed into a spectrophotometer, in which the change is absorption at 340 nm is followed for 5 min at 25°. Controls containing carbohydrate substrate and NAD are required at high pH values, but a second control, in which 0.1 ml of the appropriately diluted enzyme preparation replaces the substrate sugar in the control cuvette, is recommended for assay of crude enzyme preparations.

Protein levels are measured either by the biuret method or the method of Lowry *et al.*, both as described by Layne.[4]

Purification Procedure

Steps 2 through 5 of the following purification procedure give an enzyme preparation 85-fold purified over that of step 1. This purification can be obtained quickly by doing steps 1 through 4 in one day, applying the sample to the column in step 5 and eluting overnight. Locating the activity from the column will take the morning of the second day. Thus, the one and a half days give a working solution that has four protein

[2] W. Schiwara, W. Domschke, and G. Domagk, *Hoppe-Seyler's Z. Physiol. Chem.* **349**, 1575 and 1582 (1968).
[3] P. W. Mobley, R. P. Metzger, and A. N. Wick, *Comp. Biochem. Physiol.* **43B**, 509 (1972).
[4] See this series, Vol. 3, p. 73.

bands on disc gel analysis and is stable for at least one month at −15°. This preparation is well suited for studying the properties of the enzyme and for determining the fucose content of tissues and fluids.[5]

Source. Fresh frozen sheep liver is almost as good a source of the enzyme as that obtained directly from the slaughterhouse. Livers stored in a refrigerator for extended periods give variable and generally poorer results.

Step 1. Extraction. About 300 g of fresh sheep liver is minced and homogenized for 2 min at low speed in a Waring Blendor in 1200 ml of ice cold 0.25 *M* sucrose. The resulting preparation is centrifuged at 25,000 *g* for 20 min. The pellet is discarded, and the supernatant is poured through four layers of cheesecloth to remove connective tissue and fat. The first assay for activity and protein is performed on the filtered solution.

Step 2. Acid Precipitation. The homogenate is adjusted to pH 5.5 with 1 *M* acetic acid while being stirred on ice. After the pH adjustment, the solution is immediately centrifuged at 10,000 *g* for 20 min. The supernatant from the centrifugation is then adjusted to pH 7.5 with 2 *M* Tris (free base) and assayed. This procedure removes a large amount of microsomal protein.[6]

Step 3. Ammonium Sulfate. Solid ammonium sulfate, 176 g per liter, is added over a 10-min period while the solution is being stirred in an ice bath. After 10 additional minutes the mixture is centrifuged at 10,000 *g* for 20 min. To the supernatant is added 198 g of ammonium sulfate per liter as before. After an identical centrifugation, the pellet is taken up in a volume of 10 m*M* Tris-chloride buffer at pH 8.6 not greater in milliliters than the original weight of the liver in grams. This gives an enzyme solution that has high activity due to the concentration of the enzyme and the stimulatory effect of the high salt content. Assays at this point are not comparable to the other assays throughout the purification procedure. Dialysis of an aliquot for assay will give an indication of the purification obtained in the ammonium sulfate step.

Step 4. Heat Treatment. After ammonium sulfate treatment the enzyme preparation has about 40 mg/ml of protein. A heat treatment lowers the protein content to 5–10 mg/ml and gives a 2–5-fold purification. The solution, while being stirred in a 500-ml Erlenmeyer, is heated to 60° in a boiling water bath. The temperature is carefully maintained at 60° for 10 min, then the flask is placed into an ice bath and stirred continuously until the temperature is less than 10°. The precipitated protein

[5] P. R. Finch, R. Yuen, H. Schachter, and M. A. Moscarello, *Anal. Biochem.* **31**, 296 (1969).
[6] D. M. Zeigler and F. H. Petit, *Biochemistry* **5**, 2932 (1964).

is then removed by centrifugation at 10,000 g for 20 min. The supernatant is retained and assayed.

Step 5. Sephadex G-100 Chromatography. The enzyme solution is concentrated to about 15 ml by use of an Amicon ultrafiltration cell fitted with a UM-10 membrane (retains molecules with a molecular weight greater than 10,000). Aliquots, 5 ml, are applied to a 2.5 × 80 cm Sephadex G-100 column (at this length, best results are obtained with upward flow) that has been equilibrated with 20 mM Tris chloride at pH 8.6. The enzyme is eluted with the same buffer at 17–20 ml/hr and collected in 3-ml aliquots. The enzyme elutes just after a large protein peak at the void volume ($V_e/V_o = 1.2$).

Step 6. DEAE-Cellulose. The peaks from the gel elutions are combined and applied to a DEAE-cellulose column (1 ml DEAE per 10 mg protein) equilibrated with 10 mM Tris chloride, pH 8.6. After washing with one column volume of the same buffer, the column is eluted with buffer containing NaCl in a gradient from 0 to 0.3 N. The enzyme elutes at 0.1 N salt.

Step 7. Sephadex G-200 Chromatography. To desalt and further purify the DEAE eluate it is concentrated (ultrafiltration or solvent uptake by dry gels) and applied to a 1 × 30 cm Sephadex G-200 column. The gel is eluted with the same buffer as in the G-100 step. The purified enzyme as eluted from the Sephadex G-200 column must be concentrated to 1 mg/ml for maximum stability. This preparation is about 300-fold purified over step 1 and gives a single band on disc gel electrophoresis.

Properties[1]

Specificity. Under standard assay conditions (pH 10.4), 2-deoxy-D-ribose is about 10% as active as L-fucose or D-arabinose. No other sugar we tested, including D-galactose, L-arabinose, D-glucose, and 2-deoxy-D-glucose, had greater than 2% activity.

Stability. All preparations greater than 5-fold purified and 1 mg/ml in protein are stable for at least one month in the freezer. The enzyme is also stable to refrigeration (+5°) for a week. The stability to 60° heat is demonstrated in the section on purification procedure.

pH and Ionic Strength. The pH maximum is 10.4 with an ionic strength of 0.05. At $\mu = 1.0$ activity is maximal and the pH curve is broadened, peaking between 9 and 9.5.

Molecular Weight and pI. The molecular weight is approximately 96,000 and the pI is 5.8 (from electrofocusing experiments).

Kinetic Properties. At $\mu = 0.05$ and pH = 8.6, the K_m for L-fucose is 70 μM. The K_m's for D-arabinose and NAD are 0.40 mM and 0.04 mM, respectively. At pH 10.4 the K_m is 1.5 mM for L-fucose, 7.3 mM for D-arabinose, and 10 μM for NAD. Variation of fixed substrate and inhibition of the reaction by NADH give results consistent with an ordered bi-bi mechanism.[7]

[7] W. W. Cleland, *Biochim. Biophys. Acta* **67**, 104 (1963).

[41] Glucose-6-phosphate Dehydrogenase from *Neurospora crassa*

By WILLIAM A. SCOTT

D-Glucose 6-phosphate + NADP + H$_2$O $\rightleftharpoons$
$$\text{D-gluconate 6-phosphate} + \text{NADPH} + \text{H}^+$$

Assay Method

Glucose-6-phosphate dehydrogenase is measured spectrophotometrically by following the rate of NADPH formation at 340 nm in the presence of saturating amounts of glucose 6-phosphate and NADP.

Reagents

Tris·HCl buffer, 0.1 M, pH 7.4 at 25°
Glucose 6-phosphate, 31.6 mM in 0.1 M Tris·HCl buffer, pH 7.4
NADP, 22.8 mM in 1% NaHCO$_3$

Procedure. The assay is carried out at 25° in a recording spectrophotometer. A recorder which has an expanded log scale (0.1 or 0.25 optical density units full scale) and a variable chart speed greatly improves the accuracy of measurements.

The reaction mixture consists of 2.5 ml of buffer, 0.1 ml of glucose 6-phosphate, and 0.03 ml of NADP. The reaction is carried out in a 3.0 ml quartz cell with a 1.0 cm light path and is initiated by the addition of 1–10 μl of enzyme (0.2–2.0 units).[1] The rate of absorbancy change at 340 nm is followed for at least 2 min. Reaction velocities are calculated from initial slopes.

Definition of Enzyme Unit and Specific Activity. The international unit of enzyme activity is employed. The quantity of enzyme which reduces 1 μmole of NADP per minute per milliliter under the above conditions is defined as one unit of activity. Protein is estimated by the method

[1] W. A. Scott and E. L. Tatum, *J. Biol. Chem.* **246**, 6347 (1971).

of Warburg and Christian.[2] Specific activity is defined as units of enzyme activity per milligram of protein.

Materials

Tris·HCl buffers, pH 7.4 and pH 8.3, are prepared as 1 M stock solutions at 25° and are stored at 5°. All buffers with the exception of that used for the extraction of *Neurospora* contain 20 μM NADP and 1 mM β-mercaptoethanol. Buffers are extensively degassed prior to the addition of the reducing reagent.

Before use commercial DEAE-cellulose (250 g) is suspended in enough distilled water to make a thick slurry. This is poured onto a sintered glass funnel (15 cm in diameter) and is washed with 4 liters of distilled water under suction followed by 4 liters of 0.1 N NaOH. With most commercial ion exchange celluloses, varying amounts of a red pigmented material are eluted by the NaOH. A steady vacuum is maintained, and the exposure of the resin to NaOH is kept to a minimum to reduce the possibility of hydrolysis. The damp cake is washed with distilled water until the eluate is neutral to pH paper. This procedure is repeated with 0.1 N HCl–H$_2$O to convert the resin to the chloride form. The resin is stored as a thick slurry under 1 M Tris·HCl buffer, pH 7.4, at 5°. Columns of DEAE-cellulose are packed in 1 M buffer under pressure then are equilibrated with buffer of the appropriate concentration.

Hydroxyapatite (Bio-Rad HTP, Bio-Rad Laboratories) is stored as a dry powder at 5° and is suspended in buffer immediately before use. Fines are removed by decantation before packing the column.

Neurospora crassa wild type RL3-8A (Fungal Genetics Stock Center, California State University, Humboldt, Arcata, California) is conveniently grown at 25° or 30° in 5-gallon Pyrex carboys containing 16 liters of minimal medium[3] plus 2% sucrose. A single 50-ml shake culture, homogenized for 30 sec in a sterile Waring Blendor, is used as an inoculum for each carboy. The contents of the carboys are agitated for 36–48 hr by air forced through a sterile cotton filter. The mycelia are collected on a 24-cm diameter Büchner funnel covered with Miracloth or a coarse filter paper. After washing with cold distilled water, thin sheets of mycelia from approximately one-third of a carboy are pressed between paper towels. The mycelial pad is cut and is rolled to cover the inner walls of 1-liter, wide-mouth, vacuum jar (Thermovac Industries Inc.).

[2] O. Warburg and W. Christian, *Biochem. Z.* **310**, 384 (1942).
[3] H. J. Vogel, *Amer. Natur.* **98**, 435 (1964).

After freeze-drying, the mycelia are ground in a mortar at 5°; they are stored in tightly capped bottles at the same temperature. The yield is approximately 50 g of mycelia per carboy.

Purification Procedure

The purification procedure given here is essentially that described by Scott and Tatum.[1]

Step 1. Extraction of Mycelia. The dried and ground mycelia are mixed with an equal weight of acid-washed sea sand, ground again in a mortar, and suspended in 20 volumes of 0.1 M Tris·HCl buffer, pH 7.4, per gram of tissue. The suspension is stirred with a glass rod, then sand is removed by decantation and washed with buffer until free of tissue. Extraction is carried out by stirring for 2–3 hr, after which the cellular debris is removed by centrifugation at 2500 g for 10 min. The precipitate is resuspended in one-tenth the original volume and is centrifuged again. The combined supernatants form the crude extract. This step and all subsequent steps are carried out at 5°.

Step 2. Ammonium Sulfate Precipitation. Solid ammonium sulfate (209 g per liter) is added slowly to the crude extract with stirring. The precipitate, collected by centrifugation at 12,000 g for 15 min, is discarded. During the addition of ammonium sulfate, the pH of the enzyme mixture is maintained between pH 6.8 and 7.0 by the dropwise addition of 1 N NaOH.

The enzyme is precipitated by adding an additional 164 g of ammonium sulfate per liter of the original crude extract. After centrifugation, the enzyme is redissolved in 0.1 M Tris·HCl buffer, pH 7.4 (one-tenth of the volume of the crude extract) and is dialyzed overnight against 200 volumes of the same.

Step 3. Calcium Phosphate Gel Absorption. Calcium phosphate gel is added to the dialyzed enzyme solution with stirring. A ratio of approximately 3 mg of gel per milligram of protein is necessary for complete absorption. The pH of the mixture is adjusted to 7.0 with 1 M Tris·HCl buffer, pH 7.4, and is allowed to stir gently overnight. After centrifugation, the gel is washed with an equal volume of 0.1 M Tris·HCl buffer, pH 7.4. The wash and gel supernatants are assayed for glucose-6-phosphate dehydrogenase activity and are discarded if they contain less than 1% of the original activity. The enzyme is eluted from the gel by three successive washes with 0.6 ml per gram of gel of 0.1 M Tris buffer, pH 7.4, containing 1.51 M ammonium sulfate. The supernatant fractions are pooled, and 340 g of solid ammonium sulfate per liter of supernatant are added to precipitate the enzyme. The enzyme, collected by centrifugation, is dissolved in a minimum amount of 10 mM Tris buffer, pH 7.4, and is dialyzed overnight against 15 volumes of the same.

Step 4. DEAE-Cellulose Chromatography. A DEAE-cellulose column is packed to provide 1 ml of exchanger per 5–7 mg of protein and is equilibrated with 10 mM Tris·HCl buffer, pH 7.4. The enzyme solution is diluted in buffer to 1 mg of protein per milliliter and is added to the column at a rate of 15 ml/cm² per hour. The enzyme is eluted at a flow rate of 5 ml/cm² per hour by means of a linear gradient of increasing ionic strength formed by mixing 10 mM Tris·HCl buffer, pH 7.4, and the same buffer containing 0.2 M NaCl. The total gradient volume is 8-times the bed volume of the column. Contents of fractions which contain two or more units of enzymic activity are pooled. The enzyme is precipitated by the addition of 472 g of solid ammonium sulfate per liter of solution and is collected by centrifugation. The precipitated enzyme is redissolved in a minimal volume of 0.1 M Tris·HCl buffer, pH 7.4, and is dialyzed overnight against 20 volumes of the same.

Step 5. Hydroxyapatite Chromatography. A hydroxyapatite column is packed to provide 1 ml of adsorbant per 1–2 mg of protein. After equilibration of the column with 0.1 M Tris·HCl buffer, pH 7.4, the enzyme is adsorbed at a flow rate of 1.7 ml/cm² per hour. The chromatogram is developed by means of a linear gradient formed between buffer and buffer containing 0.76 M ammonium sulfate at the same flow rate employed to load the column. The enzyme is pooled and is precipitated with ammonium sulfate as in step 4. After redissolving in a minimum amount of 10 mM Tris buffer, pH 8.3, the enzyme is dialyzed against 100 volumes of the same.

Step 6. DEAE-Cellulose Chromatography II. The chromatography conditions are similar to those described in step 4 except that the pH of the Tris-HCl buffer is increased to pH 8.3 and the protein load is reduced to 1 mg of protein per milliliter of packed exchanger. The enzyme is eluted at a flow rate of 4.7 ml/cm² per hour. Contents of fractions containing enzymic activity are pooled and dialyzed overnight against 100 volumes of 10 mM Tris·HCl buffer, pH 8.3.

Step 7. DEAE-Chromatography III. The chromatography described in step 6 is repeated at a flow rate of 2 ml/cm² per hour. This step is usually necessary for complete removal of protein, which has elution characteristics similar to those of the enzyme. The pooled enzyme is concentrated by ammonium sulfate precipitation as described in step 4 and is redissolved in a minimum volume of 0.1 M Tris·HCl buffer, pH 7.4.

Step 8. Bio-Gel A-1.5m Chromatography. Columns of Bio-Gel A-1.5m are equilibrated with 0.1 M Tris-HCl buffer, pH 7.4. The sample volume should be 1.4% of the column volume or less. Descending chromatog-

PURIFICATION OF *Neurospora* GLUCOSE-6-PHOSPHATE DEHYDROGENASE[a]

Fraction	Total protein (mg)	Total activity (units)	Specific activity	Fold purifi-cation[b]	Recovery (%)
1. Crude extract	142,654	27,675	0.19	—	100
2. Ammonium sulfate	36,100	26,606	0.74	3.8	96.1
3. Calcium phosphate	13,953	20,922	1.50	7.7	75.6
4. pH 7.4 DEAE-cellulose	684	14,861	23	118	53.7
5. Hydroxyapatite	73	7,777	107	552	28.1
6. pH 8.3 DEAE-cellulose II	27	5,155	188	970	18.6
7. pH 8.3 DEAE-cellulose III	8	2,384	298	1,536	8.6
8. Bio-Gel A-1.5m	6	2,820	470	2,423	10.2

[a] Results obtained from the extraction of 400 g of lyophilized mycelia. Data from W. A. Scott and E. L. Tatum, *J. Biol. Chem.* **246** 6347 (1971).
[b] Ratio of the specific activity of the enzyme of the given purification step to that of the crude extract.

raphy is carried out at a flow rate of 5 ml/cm² per hour. Contents of fractions which contain enzymic activity are pooled, and are treated with ammonium sulfate (see step 4). The precipitated enzyme is redissolved in 0.1 M Tris·HCl buffer, pH 7.4, which contains 2% ammonium sulfate to give a final concentration of 2 mg of protein per milliliter.

A summary of averaged data obtained from 10 separate purification schemes is given in the table. The specific activity of the purified enzyme ranges from 250 to 500 and is influenced by the aggregation state of the molecule.

Properties

Purity. Enzyme which has a specific activity of 250–500 appears to be homogeneous by gel filtration and polyacrylamide gel electrophoresis.[1] Preparations of enzyme containing 20 μM NADP consist of a mixture of dimers and tetramers as judged from ultracentrifugal and electron microscopic studies. The ratio of these two aggregation states may account for the 2-fold variation in specific activity[4] (see Physical Properties below).

Stability. The purified enzyme can be stored for 2 weeks in 10 mM

[4] W. A. Scott, *J. Biol. Chem.* **246**, 6353 (1971).

Tris·HCl buffer, pH 7.4, 2% ammonium sulfate if kept in a tightly stoppered evacuated bottle.

Physical Properties. The sedimentation behavior of the enzyme is influenced by the NADP concentration[4] in a manner similar to that reported by Yue *et al.*[5] for the yeast enzyme. Increasing concentrations of NADP cause the enzyme to associate with a concomitant increase in specific activity. The molecular weight of the tetramer is 206,000 and that of the dimer is 104,000. Sedimentation velocity studies in 6 M guanidine·HCl and polyacrylamide gel electrophoresis in sodium dodecyl sulfate indicate the presence of subunits of molecular weight 57,000.

The extinction coefficient of a 1 mg per milliliter of solution of the purified enzyme in 0.1 M Tris·HCl buffer, pH 7.4, at 280 nm is 1.11.[1]

Catalytic Properties. The pH optimum of the glucose-6-phosphate dehydrogenase is broad and ranges from pH 7.4 to pH 8.2. At pH 7.4, the K_m values for glucose 6-phosphate and NADP are 37 μM and 12 μM, respectively. No reaction is detected if NAD is substituted for NADP. Less than 10% activity is observed if 2-deoxyglucose 6-phosphate and galactose 6-phosphate are substituted for glucose 6-phosphate in the reaction mixture.[1]

Genetics. Four unlinked genes in *Neurospora* are known to control the properties of glucose-6-phosphate dehydrogenase.[6-8] Mutations in three of these genes produce gross morphological aberrations in *Neurospora* suggesting that glucose-6-phosphate dehydrogenase is an important determinant of *Neurospora* morphology.[6,7]

Acknowledgments

This work was supported in part by grants from the National Institutes of Health (GM 16224) and the Research Corporation, and a postdoctoral fellowship from the National Institute of Child Development and Human Welfare.

[5] R. H. Yue, E. A. Noltmann, and S. A. Kuby, *J. Biol. Chem.* **244**, 1353 (1969).
[6] S. Brody and E. L. Tatum, *Proc. Nat. Acad. Sci. U.S.* **56**, 1290 (1966).
[7] W. A. Scott and E. L. Tatum, *Proc. Nat. Acad. Sci. U.S.* **66**, 515 (1970).
[8] W. A. Scott and S. Brody, *Biochem. Genet.* **10**, 1024 (1973).

[42] Glucose-6-phosphate Dehydrogenase from Bovine Mammary Gland

By G. R. JULIAN and F. J. REITHEL

D-Glucose 6-phosphate + NADP$^+$ $\rightleftarrows$ 6-phosphoglucono-δ-lactone + NADPH + H$^+$

Assay Method

Principle. The reaction catalyzed results in the formation of NADPH, and the rate of formation is accurately estimated by measuring the increase in absorbancy at 340 nm. Concentrations of substrates must be nonlimiting in order to ensure that the rate corresponds to the limiting concentration of enzyme being assayed.

Reagents

Buffer I: a solution 0.1 M in HCl and 0.1 M in MgCl$_2$ was adjusted to pH 7.2 with Tris. The resultant concentration of Tris was 0.092 M. Water used for this buffer and all other procedures was distilled and deionized.

Glucose 6-phosphate, disodium salt, 0.26 mM. Stock solution in buffer I.

NADP, sodium salt, 0.56 mM. Stock solution in buffer I.

Enzyme: the volume (in μl) of enzyme solution used for assay varied according to its concentrations during purification. In every case, however, sufficient buffer I was added to maintain the total assay volume of 1.0 ml.

Procedure. A recording spectrophotometer was used to measure and record the rate of change of absorbancy at 340 nm. The cuvette compartment was maintained at 25° and 1-ml quartz cuvettes were employed. Appropriate concentrations of substrates were provided by adding a 10-μl aliquot of the NADP and a 20-μl aliquot of the glucose 6-phosphate stock solution to a volume of approximately 1 ml of buffer. The glucose-6-phosphate dehydrogenase reaction was initiated on addition of enzyme. Volumes of enzyme solution over 20 μl were introduced as a droplet on a plastic stick designed as an agitator.

Definition of Unit and Specific Activity. The unit of activity was defined as the amount of enzyme necessary to effect a change of 1.0 unit of absorbance per minute under the conditions specified.

Specific activity considered as activity units per milligram of protein requires estimation of protein. The method used was based on measurements of absorbance at 280 and at 260 nm but corrected for the presence of thioglycolate, when present, as follows.

The absorbance ratios $A_{280}/A_{260} = R_B$ for Tris-thioglycolate (described below) buffer = 0.4. The same ratio, R_P for protein = 1.58. Assuming an extinction coefficient $E_{280}^{1\,\text{mg/ml}} = 1.0$, then the following relationship allows one to calculate a conversion factor K, from experimental values of A_{260}/A_{280}.

$$A_{260}/A_{280} = 1/R_B + K\left[(R_B - R_P)/(E_{280}^{1\,\text{mg/ml}} \cdot R_P \cdot R_B)\right]$$

thus, $A_{280} \cdot K$ = mg protein/ml

Purification Procedure[1]

Cow udders were obtained from the local abattoir soon after slaughter of the animal and cooled to 0–4° with dispatch. The chilled gland was trimmed to remove fat, drained of milk, wrapped in paper or plastic, and frozen at −10°. The frozen gland could then be sliced with a mechanical slicer (a commercial butcher's meat slicer was used) and added to buffer II (10 mM thioglycolate, adjusted to pH 7.2 with solid Tris) sufficient to keep the tissue covered as slices were added. In a typical preparation 12 liters of tissue slices required 8 liters of buffer II for extraction.

Step 1. Preparation of Extract. The sliced gland was allowed to stand in the covering buffer for 2 hr at room temperature with occasional stirring. The first buffer extract was removed by draining through a plastic screen that formed the top of a filter table. A second extract was made with 0.75 times the first volume of buffer and finally a third extract with 0.5 times the first volume of buffer, allowing the slices to soak 30 min each time before draining. The combined extracts were centrifuged to separate the large amounts of fat and cell debris entrained. Careful filtration of the clarified extract through plastic screen and cotton toweling yielded a deep red clear extract.

Step 2. Ethanol Fractionation. The extract was cooled to 1–2°. From a reservoir 95% ethanol, cooled to −20°, was added to the extract with efficient stirring at a rate of about 50 drops per minute through a capillary tipped manifold of nine outlets. Ethanol was added until the final concentration was 24% by volume. The precipitated protein was collected by continuous centrifugation at 4° and the supernatant liquid (containing most of the 6-phosphogluconate dehydrogenase) was discarded. The precipitate was converted to a free-flowing suspension by careful blending or homogenizing in buffer II containing 15% ethanol. About 250 ml of

[1] G. R. Julian, R. G. Wolfe, and F. J. Reithel, *J. Biol. Chem.* **236**, 754 (1961).

buffer was required per 100 g of precipitate. The suspension was finally centrifuged for 30 min at 1.3×10^4 g and 4° to obtain a firmly packed, washed, homogeneous sediment.

Step 3. Extraction with $(NH_4)_2SO_4$ Solution. A stock solution of saturated $(NH_4)_2SO_4$ was prepared by dissolving the salt in 10 mM thioglycolic acid, 2 mM in EDTA, to saturation at room temperature. After adjustment to pH 7.2 with Tris, the solution was filtered, freed of O_2 by bubbling purified N_2 through it, and stored under N_2 in carboys at 4°. This stock was diluted with buffer II to prepare less saturated solutions when needed. In this paper, "saturation" refers to saturation at 4°. When percentage saturation was adjusted by adding saturated $(NH_4)SO_4$ solution. The volume change was ignored. For example, equal volumes of protein solution and saturated $(NH_4)_2SO_4$ solution, when mixed, resulted in a solution considered to be 50% saturated.

The protein precipitate of step 2 was suspended in 40% saturated $(NH_4)_2SO_4$ by low speed homogenization to form a smooth homogeneous slurry. Protein from 1 kg of tissue resulted in a volume of about 500 ml at this point. The slurry was centrifuged at 1.3×10^4 g at 4° for 30 min, the extract was collected, and the residue was extracted twice more by repeating the described procedure.

Step 4. Fractionation by Extraction with $(NH_4)_2SO_4$ Gradient. Celite (Johns-Manville No. 535) was rid of fines by water washing and added to a 9×72 cm Lucite column to within 10 cm of the top. This served to measure accurately the amount of Celite used during this step. It was washed with about 9 liters of 10 mM Tris-thioglycolate, pH 7.2, 10% saturated with $(NH_4)_2SO_4$. Fe(III) contamination when present was evident by the formation of a purple complex resulting from its complex with thioglycolate in basic solution. The column contents were washed with a like volume of distilled water, allowed to drain, removed from the column and stored at 4°.

The extract of step 3 was added to 3 liters of washed Celite and sufficient 40% saturated $(NH_4)_2SO_4$ solution was added to produce a slurry fluid enough for efficient stirring. The protein was precipitated on the Celite particles by raising the $(NH_4)_2SO_4$ concentration to 65% saturation. To achieve this, saturated $(NH_4)_2SO_4$ solution was added through the capillary-tipped manifold, employed in step 2, with efficient stirring.

The supernatant liquid was decanted and the protein-Celite slurry was allowed to settle and compact in the 9×72 cm Lucite column. The bed was washed initially with 1 liter of 65% saturated $(NH_4)_2SO_4$ solution. Protein was then gradually extracted with a descending exponential $(NH_4)_2SO_4$ gradient. A 500-ml filter flask served as a mixing chamber. It was filled with 65% saturated $(NH_4)_2SO_4$ and into it was fed 40%

saturated $(NH_4)_2SO_4$ from a 6-liter reservoir. Fractions of 250 ml were collected and enzymic activity appeared in the eluate after about 3.5 liters of eluate had been collected. Those fractions containing substantial enzymic activity were pooled and made 70% saturated with respect to $(NH_4)_2SO_4$ to precipitate the desired protein.

Step 5. Precipitation of the Enzyme with Zinc Ion. The precipitate from step 4 was collected by centrifugation and dissolved in about 2 liters of buffer II. To this solution was added 80 ml of 0.5 M zinc acetate buffered at pH 7.2 in buffer II. The zinc-containing solution was added very slowly with mild stirring over a period of 6–7 hr. The final concentration of Zn^{2+} was 20 mM. The precipitate was collected by centrifugation without delay. From this precipitate the enzyme protein was extracted with 25 mM EDTA in buffer II at pH 7.2. Again, the protein desired was precipitated from this extract at 70% $(NH_4)_2SO_4$.

Step 6. First Chromatography on Hydroxyapatite. The precipitate of Step 5 was dissolved in 250 ml of buffer II and added to a column of hydroxyapatite 9.5 cm in diameter and 9.0 cm in height. As soon as the solution had entered the bed, 200 ml of buffer was added to remove $(NH_4)_2SO_4$ and unadsorbed protein. Next a quantity of 200 ml 50 mM K phosphate in buffer II (pH 7.2) was applied. The collection of fractions was initiated during the next elution with 200 ml of 0.1 M K phosphate in buffer II. In general, enzyme activity was not eluted from the column until 15 mM K phosphate was used for elution. The enzyme in the eluate was precipitated again at 70% $(NH_4)_2SO_4$ and collected by centrifugation. At this point, the protein was dissolved in a minimum volume of buffer II and desalted by passing the solution through a column of Sephadex G-25.

Step 7. Purification by Use of DEAE-Cellulose. A 2.2 × 19 cm

PURIFICATION OF GLUCOSE-6-PHOSPHATE DEHYDROGENASE (PREPARATION 48)

Fraction	Units of activity	Specific activity	Yield (%)
1. Initial extract	41,300	0.041	
2. Extract of ethanol ppt.	34,200	0.37	83
3. $(NH_4)_2SO_4$ extract	31,800	3.32	77
4. Extract before Zn^{2+} ppt.	19,800	3.36	48
5. Solution after Zn^{2+} removal	16,000	5.3	39
6. Recovery from hydroxyapatite	9,500	8.4	23
7. Recovery from DEAE-cellulose	6,700	70	16
8. Sephadex eluate	2,780	111	6.7
9. Solution of crystallized enzyme	1,180	420	2.9

DEAE-cellulose column was packed and equilibrated with buffer II. To this column was now adsorbed the desalted enzyme solution of step 6 without delay. To the column was attached a 250-ml mixing chamber filled with buffer II. A 500-ml separatory funnel was fitted to the top of the mixing chamber and filled with 0.2 M NaCl in buffer II. With this arrangement the enzyme was eluted with a salt gradient. Five-milliliter fractions were collected at a flow rate of 15–20 ml per hour at 4°. Enzyme usually was present in the eluate after 250 ml had been collected.

Step 8. Second Chromatography on Hydroxyapatite. Since the total amount of protein at this stage was 100–200 mg it was possible to employ a smaller (1.8 $\times$ 10 cm) column. The active fractions of step 7 were added to this column and the enzyme was found to be adsorbed. One column volume of buffer II was percolated through to remove residual NaCl and then gradient elution was started at a concentration of 0.1 M K phosphate, increasing to 0.15 M.

Step 9. Crystallization. The enzyme eluted in step 8 was precipitated with $(NH_4)_2SO_4$, collected by centrifugation and dissolved in a minimal volume of 50 mM Tris-thioglycolate buffer, pH 7.0. This was desalted by passing through a small (1 $\times$ 10 cm) column of Sephadex and the enzyme solution was allowed to stand for a few hours at 4°. Crystals formed, accounting for most of the activity in a solution of 1% protein concentration. Examination by ultracentrifugation showed that at low ionic strength higher molecular weight forms were evident when the enzyme was highly purified.

Comments. The procedure outlined is the result of nearly fifty trials. As described here, our procedure allowed the processing of relatively large amounts of tissue and of protein with reasonable facility. To expedite handling, no attempt was made to obtain a maximum yield at any step, but rather to obtain a moderate yield with dispatch.

In some steps, such as that employing Zn^{2+} precipitation, speed was necessary to avoid aggregation and hence grave losses due to decrease in solubility of the protein. In others, storage or slow collection of fractions in which protein concentration was low resulted in losses at any temperature. Again, during chromatography on DEAE-cellulose, two distinct peaks of protein and activity were discovered but only the most active was collected.

When highly purified enzyme was stored at low ionic strength at any temperature between 4° and 25° the formation of crystals was observed. Repeated crystallization yielded the same activity, but there was increased evidence of aggregation. Storage of crystalline enzyme was accompanied by slow loss of enzyme activity.

The use of ethanol precipitation at an early stage allowed nearly complete removal of 6-phosphogluconate dehydrogenase that would interfere with accurate assay of the glucose-6-phosphate dehydrogenase. Also it should be noted that the use of DEAE-cellulose dictated the use of buffers of low ionic strength, despite an early recognition that the enzyme protein desired had a rather low solubility in such buffers. Losses during dialysis at this step, where the ionic strength was lowered, were substantial and likely involved some aggregation.

Stabilization of the enzyme was achieved partly by use of thioglycolate but principally by using EDTA. Use of Tris-EDTA buffers maintained 90% of the activity for 15 days in one test, 52% of the activity for 76 days in another. However, it was found that EDTA even at concentrations as low as 2 mM interfered with adsorption of the enzyme on hydroxyapatite.

[43] Glucose-6-phosphate Dehydrogenase from Cow Adrenal Cortex[1]

By Kenneth W. McKerns

Glucose 6-phosphate + NADP$^+$ ⇌ 6-phospho-δ-gluconolactone + NADPH + H$^+$

Enzyme Assay

During the purification procedure dehydrogenase activity was measured by monitoring the initial rate of reduction of NADP$^+$ or NAD$^+$. Increase in absorbance at 340 nm was followed using a Beckman spectrophotometer with recorder. It is convenient to have scale expansion to increase the sensitivity of the method. A stop-flow apparatus with oscilloscope recording is desirable for kinetic studies of the enzyme. The routine assay mixture contains the following components at 37°: 0.1 M Tris·HCl buffer (pH 8.0), 10 μM NADP$^+$ or 10 mM NAD$^+$, 0.1 mM glucose 6-phosphate, 6 mM MgCl$_2$, and enzyme. During stages I–IV of enzyme purification, corrections for 6-phosphogluconate dehydrogenase were made by noting the difference in the reaction rate obtained with NADP$^+$, 0.1 mM 6-phosphate gluconate, and 0.1 mM glucose 6-phosphate, and the reaction obtained with NADP$^+$ and 0.1 mM 6-phosphate gluconate. The reactions were initiated with glucose 6-phosphate or 6-phosphate gluconate. All assays were made with the enzyme diluted

[1] (D-Glucose 6-phosphate:NADP$^+$ oxidoreductase, EC 1.1.1.49).

to 1–20 μg protein per 3 ml and were performed under conditions of zero-order kinetics for substrates. Enzyme activity is defined as micromoles of NADP$^+$ or NAD$^+$ reduced per minute at 37°. Specific activity is the number of enzymic activity units per milligram of protein.

Protein assay was by the method of Lowry et al.[2] using crystalline bovine serum albumin as standard.

Materials and Methods

Calcium phosphate gel was prepared according to the method of Tsuboi and Hudson.[3] This consisted of adding 6 ml of concentrated ammonium hydroxide (29% NH$_3$) to 200 ml of 0.5 M Na$_2$HPO$_4$ solution. This was immediately added to 1500 ml of 0.1 M CaCl$_2$. The suspension was allowed to stand for 30 min. The precipitate was centrifuged down and washed three times in deionized water and resuspended in 1 liter of 5 mM phosphate buffer at pH 7.8 containing the three buffer additives listed below.

Ion Exchange and Molecular Sieving Elements. DEAE-Sephadex A50 (3.5 meq/g, 40–120 μm particle size), CM-Sephadex C-50 (4.5 meq/g, 40–120 μm particle size), and Sephadex G-25 were purchased from Pharmacia Fine Chemicals, Piscataway, New Jersey. DEAE-cellulose (DE fibrous) was obtained from Whatman, H. Reeve Angel & Co., Clifton, New Jersey. All materials were treated according to the manufacturer's instructions.

Substrates and Buffers. D-Glucose 6-phosphate and 6-phosphate D-gluconate were obtained from Sigma Chemical Co., St. Louis, Missouri; NADP$^+$ and NAD$^+$ were from P-L Biochemicals, Inc., Milwaukee, Wisconsin. Phosphate buffer was Na$_2$HPO$_4$–NaH$_2$PO$_4$; Tris-HCl buffer was Tris (trizma base) plus HCl added to obtain required pH; acetate buffer was sodium acetate plus acetic acid added to desired pH; Tris–EDTA–boric acid buffer was Tris (trizma base), 5 mM EDTA, plus boric acid to obtain required pH; amine buffer was triethanolamine and 5 mM EDTA. Deionized water was used in all buffers. Chemicals used in the buffers were purchased from Fisher Scientific Co., Fair Lawn, New Jersey: J. T. Baker Chemical Co., Phillipsburg, New Jersey; W. H. Curtin & Co., Jacksonville, Florida;   E. H. Sargent & Co., Springfield, New Jersey.

Buffer Additives and Other Materials. Many of the buffers contained additives: glycerol at 5%; β-mercaptoethanol at 2.7 mM; and EDTA

---

[2] O. H. Lowry, N. J. Rosebrough, A. L. Farr, and R. J. Randall, *J. Biol. Chem.* **193**, 265 (1961).

[3] K. K. Tsuboi and P. B. Hudson, *J. Biol. Chem.* **224**, 879 (1957).

at 0.2 mM. The additives were reagent grade and were purchased from Fisher Scientific Co., Fair Lawn, New Jersey, and Mann Research Laboratories, Inc., New York, New York. The ammonium sulfate used was enzyme grade, from Nutritional Biochemicals Inc., Cleveland, Ohio. A saturated solution of ammonium sulfate was taken as 767 g/liter at 25°. Percentage saturations were derived from this amount. A kit containing proteins of known molecular weight was acquired from Mann Research Laboratories, Inc., New York, New York. Muscle pyruvate kinase (EC 2.7.1.40) was purchased from Worthington Biochemical Corp., Freehold, New Jersey. Materials and equipment for cellulose acetate electrophoresis was purchased from Gelman Instrument Co., Ann Arbor, Michigan. Materials for polyacrylamide gel electrophoresis were obtained from Canalco (Canal Industrial Corp.), Bethesda, Maryland, and the equipment was made locally.

Purification Procedure

The following procedure has been described previously by Criss and McKerns.[4] The method is not only suitable for the preparation of crystalline enzyme from the adrenal cortex, but also from other tissues, such as the corpus luteum of the ovary.

Step 1. Homogenate. The adrenal glands were taken directly from freshly slaughtered cows, placed in ice, and processed within 1 hr. All subsequent procedures were carried out in ice baths or in a cold room (5 ± 1°). The adrenal cortex was separated from the adrenal medulla after cutting thick slices of the gland with a Stadie-Riggs slicer. The cortex was minced with scissors and homogenized in a stainless-steel Sorvall Mini-Mixer with speed control. The homogenate was made up 1 g/2 ml in 10 mM phosphate buffer at pH 6.7 with all three buffer additives.

Step 2. High-Speed Centrifugation. The adrenal homogenate was centrifuged at 200 g for 15 min in a Model V International centrifuge to remove cell debris and unbroken cells. The supernatant was centrifuged at 105,000 g for 1 hr in a Spinco preparative ultracentrifuge. The higher speed supernatant was carefully removed with a syringe fitted with a long needle, to avoid disturbing the large compact layer of lipid at the top of the tubes.

Step 3. Isoelectric Precipitation. The enzyme was precipitated from high-speed supernatant (fraction II) while mixing rapidly, by adding an equimolar amount of 0.1 M acetate buffer at pH 4.5, containing 15%

---

[4] W. E. Criss and K. W. McKerns, *Biochemistry* **7**, 2364 (1968).

ethanol. Rapid mixing was accomplished with the aid of a Fisher Vibro-Mixer and by adding the acetate buffer rapidly through a syringe equipped with a 25-gauge needle. The enzyme precipitate was centrifuged at 25,000 g for 15 min in a Sorvall automatic centrifuge. The precipitate was redissolved in 5 mM phosphate buffer at pH 7.8 with all three additives.

Step 4. Freezing and Thawing. Fraction III was frozen at −30°. The enzyme was stable for up to 3 months at this stage of preparation. It was essential to obtain an extract that would remain stable for short periods because only a limited amount (usually 100–200 g) of adrenal cortical tissue could be obtained and processed daily. Subsequent purification was carried out with several batches of fraction IV, which were prepared from adrenal tissue collected over a 2–3-week period. The frozen fractions were thawed, centrifuged at 25,000 g for 20 min, and pooled for subsequent purification.

Step 5. Calcium Phosphate Gel Adsorption. An equal volume of calcium phosphate gel suspension (see Materials) was added to batches of fraction IV containing approximately 7 mg of protein per milliliter. The suspensions were thoroughly mixed and stirred intermittently for 0.5 hr. The gel was centrifuged at 2000 g for 15 min, and the supernatant was discarded. The gel was resuspended in 5 mM phosphate buffer at pH 7.8 containing ammonium sulfate at 20% saturation and all three buffer additives. This volume of phosphate buffer was equal to approximately one-half the original volume of fraction IV. The mixture was allowed to remain, with intermittent stirring, for 0.5 hr. The gel was centrifuged as before, and the supernatant containing the enzyme was poured off.

Step 6. First Ammonium Sulfate Precipitation. The ammonium sulfate concentration was increased to 40% saturation by the slow addition of ammonium sulfate crystals to fraction IV, which was stirred constantly with a magnetic stirrer. Stirring was continued for 1 hr. The precipitate was centrifuged at 25,000 g for 15 min and discarded. Ammonium sulfate was added (as before) to the supernatant up to a concentration of 55% saturation. The solution was stirred continuously for 1 hr and then centrifuged at 25,000 g for 15 min. This supernatant was discarded, and the precipitated enzyme was redissolved in a small volume of 20 mM phosphate buffer at pH 7.8 with all three additives.

Step 7. DEAE-Sephadex Column Chromatography. Fraction VI was desalted by passing it through an 0.8 × 20 cm column of Sephadex G-25 equilibrated in 20 mM phosphate buffer at pH 7.8 and containing all three buffer additives. NADP+ was added to the desalted fraction VI to give a final concentration of 0.1 mM. This enzyme solution was applied to the top of a 2.5 × 45 cm Pharmacia column containing DEAE-Sephadex A-50 suspended in the phosphate buffer. A volume equal to one-half the column volume was allowed to flow through the column. A NaCl gradient from 0 to 0.2 M was then passed through the column. This gradient was achieved by runnning 0.4 M NaCl in phosphate buffer into a mixing chamber containing 500 ml of phosphate buffer. Flow rate was 18 ml/hr. Fractions of approximately 7 ml/tube were collected throughout the eluting gradient and analyzed for protein (optical density at 280 nm) and enzymic activity. The first tube to contain measurable protein was called tube 1. Tubes 48–62, containing 90% of the glucose-6-phosphate dehydrogenase activity, were pooled.

Step 8. *CM-Sephadex Column Chromatography*. Ammonium sulfate was slowly added to a final concentration of 60% saturation to the pool fractions of step 7. The solution was stirred continuously for 1 hr. The precipitated enzyme was centrifuged at 25,000 g for 15 min and resuspended in a small volume of 20 mM phosphate buffer at pH 6.0 with buffer additives 2 and 3. The enzyme solution was desalted on an 0.8 × 20 cm column of Sephadex G-25 equilibrated with the phosphate buffer. The desalted glucose-6-phosphate dehydrogenase solution was allowed to flow into a Pharmacia reverse flow 2.5 × 45 cm column containing CM Sephadex C50 suspended in the phosphate buffer. One-half column volume was allowed to flow up through the column. A gradient from 0 to 0.4 M NaCl was then passed up through the column. The gradient was established similar to the gradient for fraction 7. Flow rate was 18 ml/hr. Fractions of approximately 8 ml/tube were collected throughout the eluting gradient and analyzed for protein (optical density at 280 nm) and glucose-6-phosphate dehydrogenase activity. The first tube to contain measurable protein was called tube 1. Tubes 58–75 containing 95% of the glucose-6-phosphate dehydrogenase activity were pooled.

Step 9. *Second Ammonium Sulfate Precipitation*. A second ammonium sulfate fractionation was carried out similar to that for fraction VI. The fractionation was performed with the pooled enzyme from fraction VIII. All the glucose-6-phosphate dehydrogenase activity was located between 40 and 55% saturation. The precipitated enzyme was redissolved in a 102-ml volume of 10 mM phosphate buffer at pH 7.8 with buffer additives 2 and 3.

Step 10. *Sephadex G-200 Column Chromatography*. NADP+ was added to fraction IX to give a concentration of 20 μM. This enzyme solution was placed on a 2.5 × 100 cm column containing Sephadex G-200 suspended in 10 mM phosphate buffer at pH 7.8 containing NADP+ at a concentration of 20 μM and additives 2 and 3. Fractions of 8 ml/tube were collected as the buffer was allowed to flow through the column at a rate of 12 ml/hr. The first tube to contain Blue Dextran was called tube 1. The fractions were assayed for protein (optical density at 280 nm) and enzymic activity. Tubes 6–15, containing 95% of the glucose-6-phosphate dehydrogenase activity, were pooled.

Step 11. Third Ammonium Sulfate Precipitation. Crystalline ammonium sulfate was added to fraction X to give a 60% saturation with respect to ammonium sulfate. The solution was continuously stirred for 2 hr and then centrifuged at 25,000 g for 15 min. The precipitate was resuspended in a small volume of 10 mM phosphate buffer at pH 7.8 containing ammonium sulfate at 35% satuaration and with buffer additives 2 and 3. This suspension was allowed to stand overnight. It was then centrifuged at 25,000 g for 30 min. The supernatant containing the enzyme was retained.

Step 12. Crystallization. Fraction XI was dialyzed against 10 mM phosphate buffer at pH 7.8 with buffer additives 2 and 3. Crystals appeared within 24 hr and continued to come out of solution for several days. The crystals were collected by centrifugation at 25,000 g for 15 min, washed (with dialysis buffer), and recentrifuged several times. The crystals were redissolved in 10 mM phosphate buffer at pH 4.0 or in 10 mM phosphate buffer at pH 6.7 including 0.1 mM NADP⁺ (both buffers contained additives 2 and 3). Recrystallizations were accomplished by adding ammonium sulfate to 35% saturation, followed by dialysis as described for the first crystallization.

Comments. Purification of fresh adrenal tissue was carried through to fraction IV, since this fraction could be stored frozen for several weeks with minimum loss of enzyme activity. The pH must be kept between pH 6 and 7 with the early fractions, possibly because of proteases which may be less active at a lower pH. Glycerol helped to stabilize the enzyme at earlier stages but was of little value in the partially purified preparations. Adrenal glucose-6-phosphate dehydrogense was most stable in the presence of either NADP⁺ or glucose 6-phosphate. Since it was least stable in Tris and acetate buffers, phosphate buffers were used during purification.

Purification Summary. A summary of a typical purification of adrenal glucose 6-phosphate dehydrogenase is given in the table. The ten fractionation steps (step 4 was a storage step) produced a 250-fold purification with a 20% yield. The entire procedure usually required a 3-week period of collecting and processing to the fraction IV stage and 1 week between fraction IV and preparation of crystals. The final specific activity was 340 μmoles of NADP⁺ reduced per minute per milligram of protein in 37° in 0.1 M Tris buffer at pH 8.0 with Mg^{2+}.

PURIFICATION OF ADRENAL GLUCOSE-6-PHOSPHATE DEHYDROGENASE

Fraction No.	Purification steps	Total volume (ml)	Total protein (mg)	Total activity (units)	Specific activity[a] (units/mg)	Accumulative purification	Yield (%)
I	Whole homogenate[b]	1227	80,013	9988	0.13		100.0
II	High-speed centrifugation	780	12,437	9879	0.80	6.1	98.9
III	Isoelectric precipitation[b]	493	3,411	8743	2.54	19.5	87.6
IV	Freezing and thawing	487	3,412	8691	2.55	19.6	87.0
V	CaP gel adsorption	280	1,334	8428	6.27	48.3	84.4
VI	First ammonium sulfate	14	532	7075	13.30	102.3	70.9
VII	DEAE-Sephadex column chromatography	140	140	5726	40.90	314.8	57.3
VIII	CM-Sephadex column chromatography	145	70.1	5146	73.42	564.8	51.5
IX	Second ammonium sulfate precipitation	1.5	39.8	3958	99.45	765.3	39.6
X	Sephadex G-200 column chromatography	130	28.6	3425	119.77	921.3	34.3
XI	Third ammonium sulfate precipitation	1.6	17.1	3001	175.48	1348.3	30.00
XII	Crystallization I	2.1	6.6	2271	344.09	2646.8	22.7
	Crystallization II	2.0	6.3	2090	331.70	2551.5	20.9
	Crystallization III	2.0	6.1	1989	326.07	2508.2	19.9

[a] All activity measurements and therefore specific activity calculations were performed with protein concentrations at 1–20 µg/3 ml, in 0.1 M Tris · HCl buffer at pH 8.0, with 1 mM MgCl$_2$, 10 mM NADP$^+$, and 0.1 mM glucose 6-phosphate.

[b] Values represent pooled assay values of 13 daily collections of adrenal glands which were processed up to fraction IV.

Properties

The crystalline enzyme showed only one protein band and one enzyme band when examined by DEAE-cellulose column chromatography, cellulose acetate electrophoresis, and vertical disc electrophoresis in polyacrylamide gel.[4]

Ultracentrifugation of the crystallized enzyme was carried out in a Spinco Model E ultracentrifuge. When a standard centrifuge cell is used, the schlieren patterns reveal a single, symmetrical peak. The molecular weight was calculated from an $s_{20,w}$ of 9.8 S to be 238,700. Molecular sieve analysis on a calibrated column of Sephadex G-200 gave a calculated molecular weight of 235,000.[4]

The enzyme was crystallized three consecutive times. The specific activity remained constant.

The optimum pH was in the range 8–9, and Mg^{2+} increased the activity about 20% in this pH range. There was an approximate doubling of activity between pH 7.0 and 8.0 both with and without Mg^{2+}.

The K_m for $NADP^+$ was calculated to be 5.6 μM and 47 mM for NAD^+. For glucose 6-phosphate the K_m for the $NADP^+$-linked reaction was 42 μM compared to 0.19 mM for the NAD^+-linked reaction.

The enzyme was inhibited by metal chelating agents and especially by sulfhydryl-inhibiting agents such as p-chloromercuribenzoate (PCMB).[5] Compounds having reduced sulfhydryl groups protected the enzyme against PCMB inactivation. $NADP^+$ and glucose 6-phosphate completely protected the enzyme from PCMB. ACTH markedly increased the rate of inactivation of the enzyme by PCMB. This inactivation was prevented by $NADP^+$. Two sulfhydryl groups are present at the $NADP^+$-binding center. The amino acid composition of adrenal glucose-6-phosphate dehydrogenase differs considerably from other glucose-6-phosphate dehydrogenases that have been purified. These results are given in detail elsewhere.[5]

A number of steroids such as estrogen, androgens and progesterone analogs are inhibitors of the catalytic activity of glucose-6-phosphate dehydrogenase prepared from a number of endocrine tissues. Inhibition by steroids involves competition with $NADP^+$ for the $NADP^+$-binding site on the enzymes.[6] On the other hand, adrenal glucose-6-phosphate dehydrogenase is specifically activated by ACTH[7,8] and glucose-6-phosphate dehydrogenase prepared from the corpus luteum of the ovary is

[5] W. E. Criss and K. W. McKerns, *Arch. Biochem. Biophys.* **135**, 118 (1969).
[6] W. E. Criss and K. W. McKerns, *Biochim. Biophys. Acta* **184**, 486 (1969).
[7] K. W. McKerns, *Biochim. Biophys. Acta* **90**, 357 (1964).
[8] W. E. Criss and K. W. McKerns, *Biochemistry* **7**, 2364 (1968).

activated by gonadotropins, such as chorionic gonadotropin and luteinizing hormone.[9] It has been postulated that tropic hormones regulate function and growth of target cells by stimulating unique species of glucose-6-phosphate dehydrogenase in their target tissues[9,10] Stimulation of the rate-limiting enzyme of the pentose phosphate pathway provides reducing equivalents as NADPH for steroidogenesis, fatty acid synthesis, and other processes. The pentose phosphate pathway also provides a net synthesis of ribose-P required for purine and pyrimidine biosynthesis, and thus is a regulatory mechanism in the rate of synthesis of RNA in the target cells.[11–13]

Acknowledgment

The experimental work in this article was supported by grants from the National Institutes of Health.

Adapted from *Biochemistry* [**7**, 126–128 (1968)]. Copyright 1968 by the American Chemical Society; reprinted by permission of the copyright owner.

[9] K. W. McKerns, *in* "The Gonads" (K. W. McKerns, ed.), p. 155. Appleton-Century-Crofts, New York, 1969.
[10] K. W. McKerns, *in* "Functions of the Adrenal Cortex" (K. W. McKerns, ed.), Vol. 1, Chap. 12. Appleton-Century-Crofts, New York, 1968.
[11] K. W. McKerns, *Biochim. Biophys. Acta* **192**, 318 (1969).
[12] K. W. McKerns and W. Ryschkewitsch, *Arch. Biochem. Biophys.* **154**, 341 (1973).
[13] K. W. McKerns, *Biochemistry* **12**, 5206 (1973).

[44] Glucose-6-phosphate Dehydrogenase from *Leuconostoc mesenteroides*

By Charles Olive and H. Richard Levy

D-Glucose 6-phosphate + NAD⁺(NADP⁺) ⇌
                      6-phosphoglucono-δ-lactone + NADH(NADPH) + H⁺

A partial purification of this enzyme was reported in an earlier volume of this series.[1]

Assay Method

Principle. The rate of production of reduced coenzyme is determined from measurements of absorbancy at 340 nm with time in the presence of saturating levels of substrate and coenzyme.

[1] R. D. DeMoss, Vol. 1, p. 328.

Reagents

Tris-chloride buffer, pH 7.8, 0.1 M
Glucose 6-phosphate, sodium salt, 50 mM
NADP$^+$, 2.5 mM
NAD$^+$, 25 mM, freshly neutralized to pH 7

Procedure. Enzyme assays are conducted at 25° in a spectrophotometer with a thermostatted cell compartment. Assay mixtures contain the following components in a final volume of 3.0 ml: 1.0 ml of the Tris-chloride buffer; 0.2 ml of glucose 6-phosphate; and either 0.2 ml of NADP$^+$ or 0.3 ml of NAD$^+$. Reactions are usually initiated by the addition of glucose 6-phosphate or enzyme.

Units. One unit of enzymic activity is that amount which catalyzes the reduction of 1 μmole of NADP$^+$ per minute at 25°. Protein is measured by the ratio of absorbancies at 280 and 260 nm[2] or by a modification of the biuret test[3] when necessary. For pure enzyme, the extinction coefficient (see below) can be used.

Materials

Leuconostoc mesenteroides, ATCC 12291, is obtained from the American Type Culture Collection. The nicotinamide-adenine dinucleotides, glucose 6-phosphate, and protamine sulfate (Salmine) are purchased from Sigma Chemical Co. Hydroxyapatite is prepared by the method of Anacker and Stoy.[4] For reproducible results and good flow rates it is essential to use this procedure, to stir gently and as little as possible, to decant all the "fines" and to equilibrate the column to the correct pH before applying the enzyme.

Purification Procedure

The purification procedure given here is that described by Olive and Levy.[5]

The pregrowth conditions and medium used are those described by DeMoss.[1] The cells are routinely gram-stained and inoculated to a 5% sucrose-AC broth tube prior to the initiation of large-scale growth. The appearance of chains of uniformly gram-positive cocci which produce dextran within 24 hr of growth at 30° in sucrose is considered to be sufficient evidence of a pure culture of *Leuconostoc mesenteroides*.

[2] O. Warburg and W. Christian, *Biochem. Z.* **310**, 384 (1941).
[3] A. G. Gornall, C. J. Bardawill, and M. M. David, *J. Biol. Chem.* **177**, 751 (1949).
[4] W. F. Anacker and V. Stoy, *Biochem. Z.* **330**, 141 (1958).
[5] C. Olive and H. R. Levy, *Biochemistry* **6**, 730 (1967).

The cells are grown for 22 hr at 30° in five tightly stoppered carboys each containing 10 liters of modified AC medium.[1] Cells are collected by centrifugation in a Sharples supercentrifuge at a bowl speed of 45,000 rpm and at a feed rate of 15–20 liters per hour. The cell paste is resuspended in 0.1 M NaHCO$_3$ and the cells are washed in the same solution.

Step 1. Disruption of Cells. The cells are suspended in 0.1 M NaHCO$_3$ such that a 1:100 dilution gives an OD$_{660}$ of 0.4. Fifty-milliliter aliquots of this suspension are placed in a Branson rosette cell, which is cooled by immersion in a stirred ice-water bath and sonicated for 40 min using a 20 kHz Branson sonifier at the highest power setting.

The cellular debris is removed by centrifugation at 20,000 g for 30 min. The supernatant is decanted and saved.

All subsequent operations are performed at approximately 4°, except where noted.

Step 2. Ammonium Sulfate Fractionation. A saturated (4°), neutralized solution of ammonium sulfate (1.5 volumes) is added to the crude supernatant fraction (60% saturation), and the resulting precipitate is removed by centrifugation and discarded. Solid ammonium sulfate (28.3 g/100 ml) is added slowly to the supernatant until the solution is saturated. The solution is stirred for 1 hr, and the precipitate is collected by centrifugation. The precipitate is dissolved in a small volume of 10 mM Tris chloride, pH 7.8 containing 1 mM EDTA and dialyzed for 16–18 hr against 100 volumes of the same buffer.

Step 3. Protamine Sulfate Treatment. After dialysis the extract is diluted to 5 mg/ml with twice distilled water and adjusted to pH 6.1 by the dropwise addition of 1.0 M acetic acid to the stirred solution.

A 1.7% solution of protamine sulfate is prepared and adjusted to pH 4.6–4.7 by the slow addition of 1.0 M NaOH with stirring. Protamine sulfate is added to a small aliquot of the enzyme solution in a centrifuge tube until the ratio of absorbancies at 280 and 260 nm of the supernatant after centrifugation reaches 0.9–1.0. These conditions are then extrapolated to the bulk of the extract. After centrifugation to remove the protamine–nucleic acid complex, the pH of the supernatant is increased to 7.2, and any precipitate which forms is removed by centrifugation.

Step 4. Ammonium Sulfate Fractionation. The protein solution is concentrated to approximately 8 mg/ml and solid ammonium sulfate is added to 80% of saturation (52.6 g/100 ml). The precipitate which forms is discarded and the supernatant brought to ammonium sulfate saturation (14.1 g/100 ml). The saturated ammonium sulfate precipitate is dissolved in a minimum volume of 1 mM K$_2$HPO$_4$–KH$_2$PO$_4$, pH 6.8 and dialyzed overnight against the same buffer.

Step 5. Hydroxyapatite Chromatography. The dialyzed solution is applied to a column (4 × 24 cm) of hydroxyapatite equilibrated with K_2HPO_4–KH_2PO_4, pH 6.8. The sample size was usually 1–2 mg of protein per milliliter of column volume. Elution was in steps with phosphate buffer at room temperature. Aliquots of buffer are used which are 1.5 times the column void volume and of increasing concentration beginning with 1 mM and using increments of 30 mM up to a final concentration of 0.12 M, as suggested by Levin.[6] The enzyme is eluted in 90 mM buffer. The fractions containing enzyme of high specific activity are concentrated by placing a filled dialysis bag in a graduated cylinder containing dry Sephadex G-200 and waiting for several hours. The enzyme was then precipitated at ammonium sulfate saturation (70.7 g/100 ml).

Step 6. Crystallization. The resulting precipitate is dissolved in 0.5 ml of a solution containing 0.2 M NaCl, 1 mM EDTA, 35 mM potassium phosphate, pH 7.2, 50 μM NADP$^+$, and ammonium sulfate at 30% of saturation. The enzyme is placed in a small gauze-covered tube, and a saturated solution of ammonium sulfate is added (1 or 2 drops per day) over a period of a few days. Crystals form when the ammonium sulfate concentration reaches approximately 70–80% of saturation. The solution is allowed to stand for a few days while the turbidity of the crystalline suspension increases. The suspension is then centrifuged for 1 hr at 5000 g, which accomplishes the sedimentation of the crystalline material away from a small amount of amorphous precipitate. The crystals appear as very fine needles. The process is repeated until the specific activity of the crystalline material remains constant. After adsorption of the bound NADP$^+$ with acid-washed charcoal the specific activity is found to be between 290 and 315. A summary of the data from a typical preparation is given in the table.

The enzyme is now commercially available from a number of companies. Generally the commercial preparations are about 50% pure and can be purified by the application of steps 5 and 6 above.

Properties

Purity. The recrystallized enzyme appears to be homogeneous in sedimentation velocity experiments at concentrations of 2–10 mg/ml and in sedimentation equilibrium experiments. Disc gel electrophoresis experiments at pH 9.3 show that less than 1% of the protein present is not enzymically active.

Stability. The enzyme is quite stable during all stages of purification. A partially purified preparation was stored at −20° in 60% ammonium

---

[6] O. Levin, this series, Vol. 5, p. 27.

PURIFICATION OF GLUCOSE-6-PHOSPHATE DEHYDROGENASE FROM
Leuconostoc mesenteroides[a]

Step	Fraction	Volume (ml)	Protein (mg)	Total activity (units $\times$ 10^{-3})[b]	Specific activity (units/mg of protein)	Yield (%)
1	Sonicated cell supernatant	1030	11,700	27.4	2.33	100
2	60–100% (NH₄)₂SO₄ precipitate	71	1,460	24.7	16.9	90
3	Protamine supernatant	345	828	22.7	27.4	83
4	80–100% (NH₄)₂SO₄ fraction	24.8	234	20.8	93.9	76
5	Hydroxyapatite column eluate	83	45.9	9.76	220	36
6	First crystals	0.75	22.9	6.10	266	22
	Second crystals	0.68	21.8	5.13	256[c]	18.6

[a] Reproduced from C. Olive and H. R. Levy, *Biochemistry* **6**, 730 (1967).

[b] Activities refer to the activity with $NADP^+$ under the conditions given in the text.

[c] After adsorption of bound $NADP^+$ with acid-washed charcoal, specific activity was 290 units/mg.

sulfate for nearly 2 months with a loss of 7% of its initial activity. The purified enzyme was stored for 1 month at 4° and at a protein concentration of 2 mg/ml with a loss of 15% of its enzymic activity.

Physical Properties.[7] The sedimentation coefficient as determined by the moving boundary method is 6.01 S. The partial specific volume is 0.717, and the molecular weight is 103,700 from both high- and low-speed equilibrium techniques performed at pH 4.6, which is the approximate isoelectric point. The subunit molecular weight, determined by SDS-gel electrophoresis,[8] is 54,800.

The extinction coefficient of a 0.1% protein solution in 0.09 M Trischloride, pH 7.2, is 1.15 at the 280.5 nm absorption maximum. The ratio of absorbancies at 280 nm to 260 nm is 1.88.

Kinetic Properties.[9] The NAD^+:$NADP^+$ activity ratio remains constant over the course of the purification. Mixing experiments suggest that both coenzymes bind at a common site which displays a rather broad specificity. At pH 7.8 the K_m for $NADP^+$ is 7 μM; the K_m for NAD^+ is 0.12 mM; the K_m for G-6-P varies slightly depending upon the coenzyme

[7] C. Olive and H. R. Levy, *J. Biol. Chem.* **246**, 2043 (1971).

[8] K. Weber and M. Osborn. *J. Biol. Chem.* **244**, 4406 (1969).

[9] C. Olive, M. E. Geroch, and H. R. Levy, *J. Biol. Chem.* **246**, 2047 (1971).

used, but is approximately 60 μM. The molecular activity with NADP⁺ as coenzyme is 3.2×10^4 molecules per minute per molecule of enzyme.

The NADP-linked reaction occurs via an ordered, sequential path. The mechanism of the NAD-linked reaction also is ordered and sequential but appears to involve an isomerization of free enzyme.

The enzyme is inhibited by ATP, but this inhibition is reversed by physiological concentrations of Mg^{2+}. Acetyl-CoA and CoA also inhibit the enzyme, and this inhibition persists in the presence of Mg^{2+}. Mg^{2+} alone has little effect on the enzyme. The inhibition studies taken together with the relevant kinetic constants and known physiological concentrations of NAD⁺ and NADP⁺ in *L. mesenteroides* suggest that CoA and acetyl-CoA are involved in controlling the NAD-linked activity of the enzyme.[10]

Enzyme Structure.[10,11] The amino acid composition of this enzyme differs most significantly from that of other glucose-6-phospate dehydrogenases in the complete absence of cysteine or cystine. Two lysine residues (per 103,700 MW unit) appear to be involved in binding glucose 6-phosphate.

[10] A. Ishaque, unpublished results; H. R. Levy and A. Ishaque, *Fed. Proc., Fed. Amer. Soc. Exp. Biol.* **30**, 1059 (1971).
[11] M. Milhausen, unpublished results.

[45] Glucose-6-phosphate Dehydrogenase from *Penicillium duponti*

By MAXWELL G. SHEPHERD

$$\text{D-Glucose 6-phosphate} + \text{NADP}^+ \rightleftharpoons \text{6-phosphogluconate} + \text{NADPH} + \text{H}^+$$

Assay Method

Principle. The reduction of NADP⁺ to NADPH results in the appearance of an absorption band at 340 nm. Glucose-6-phosphate dehydrogenase (D-glucose-6-phosphate:NADP oxidoreductase, EC 1.1.1.49) is generally measured by determining the amount of NADPH produced at 340 nm. In the presence of saturating concentrations of G-6-P and NADP⁺, the rate of change of absorbancy at this wavelength is proportional to the enzyme concentration.

Determination of Enzymic Activity. The enzyme activity is determined in a 1.5 ml quartz cuvette using a thermostated (25°) recording spectrophotometer. The assay mixture contains 50 μl of 50 mM glucose 6-phosphate; 50 μl of 50 mM NADP$^+$; 880 μl of Tris·HCl buffer pH 8.0, $I = 0.1$, 10 mM MgCl$_2$ and 20 μl of an appropriate dilution of enzyme. Whenever the concentration of substrate or coenzyme is varied, a check must be made to ensure that the enzyme is still saturated.

Definition of Enzyme Unit and Specific Activity. One unit of enzyme is defined as the amount of enzyme that will reduce 1 micromole of NADP$^+$ per minute at 25°. Specific activity is defined as the number of units of enzyme per milligram of protein.

Protein Determination. Protein concentration is measured by a modification[1] of the method of Lowry et al.,[2] with sodium citrate being used instead of sodium tartrate. When there is interference with this method, protein is estimated by the method of Warburg and Christian.[3]

Buffers. Constant-ionic-strength buffers are prepared as described by Long (1961).[4] The buffer used for assay of glucose-6-phosphate dehydrogenase is Tris·HCl, pH 8.0 (25°), $I = 0.1$, containing 10 mM MgCl$_2$. The phosphate buffer (KH$_2$PO$_4$–Na$_2$HPO$_4$) used in the purification of glucose-6-phosphate dehydrogenase, up to and including the Sephadex column step, was pH 6.8, $I = 0.05$, containing 1 mM EDTA; the sodium citrate buffer was pH 5.4, 10 mM, containing 0.1 mM EDTA. The Sephadex G-200 column was equilibrated with phosphate buffer (KH$_2$PO$_4$–Na$_2$HPO$_4$), pH 7.0, $I = 0.01$, containing 0.2 M ammonium sulfate.

Purification Procedure

The purification procedure given here is essentially that described by Malcolm and Shepherd.[5]

Step 1. Preparation of Cell-Free Extract. Submerged cultures of *Penicillium duponti*, ATCC No. 26229, are prepared as described by Broad and Shepherd.[6] The cells are harvested by filtration and washed twice with distilled water. The mycelial mat is resuspended in 5 volumes of ice cold phosphate buffer pH 6.8, $I = 0.05$, 1 mM EDTA. This suspension is sonicated (300 ml at a time) in a fluted vessel (20 min is required for complete breakage in an MSE 100 W Ultrasonic Disintegrator). The

[1] M. Eggstein and F. H. Kreutz, *in* "Techniques in Protein Chemistry" (J. L. Bailey, ed.), p. 340. Elsevier, Amsterdam, 1967.

[2] O. H. Lowry, N. J. Rosebrough, A. L. Farr, and R. J. Randall, *J. Biol. Chem.* **193**, 265 (1951). See also E. Layne, this series, Vol. 3. p. 448.

[3] O. Warburg and W. Christian, *Biochem. Z.* **310**, 384 (1941). See also E. Layne, this series, Vol. 3, p. 451.

[4] Long C. (ed.), "Biochemists' Handbook," p. 29. Spon, London, 1961.

[5] A. Anne Malcolm and M. G. Shepherd, *Biochem. J.* **128**, 817 (1972).

[6] T. E. Broad and M. G. Shepherd, *Biochim. Biophys. Acta* **198**, 407 (1970).

temperature of the sonicating vessel can be maintained below 8° by an ice–salt bath. The sonicate is centrifuged at 20,000 g for 30 min, and the resulting pellet is discarded.

Step 2. Protamine Sulfate. Protamine sulfate (0.1 g per 50 g wet weight of cells) is dissolved in 5–10 ml of the phosphate buffer and added slowly to the cell-free extract; after stirring for 10 min, the suspension is centrifuged for 10 min at 20,000 g. Note that streptomycin sulfate (0.1 g per 50 g wet weight of cells) or $MnCl_2$ (5–10 mM final concentration) may be used instead of protamine sulfate.

Step 3. Ammonium Sulfate Fractionation. The supernatant is adjusted to 45% saturation with solid $(NH_4)_2SO_4$ (277 g/liter of supernatant) and centrifuged (10 min, 20,000 g); the pellet is discarded. The $(NH_4)_2SO_4$ concentration of the supernatant is now increased to 60% saturation [by adding 99 g of solid $(NH_4)_2SO_4$ per liter to the 45% saturated $(NH_4)_2SO_4$ supernatant], and the pellet formed after centrifuging (10 min at 20,000 g) is suspended in 3–5 ml of the phosphate buffer.

Step 4. Sephadex G-100 Chromatography. The enzyme is now layered onto a Sephadex G-100 column (2.5 × 40 cm), which has been equilibrated with phosphate buffer, and the column is eluted at a rate of 30 ml/hr. Fractions (5 ml) are collected with a fraction collector, and those fractions with more than 50% of the activity of the peak fraction are pooled (total volume 40 or 45 ml) and concentrated to about 5 ml in an Amicon Ultrafiltration Cell with a PM-10 membrane. The concentrated solution is diluted to 50 ml with citrate buffer and reconcentrated to 5 ml; this dilution and concentration procedure is repeated and the final volume of concentrate after rinsing the apparatus is about 10 ml.

Step 5. CM-Cellulose Chromatography. The enzyme solution is now applied to a CM-cellulose column (1.6 × 44 cm), which has been equilibrated with citrate buffer, and the column is eluted at a rate of 25 ml/hr with a 500-ml linear gradient of 0 to 80 mM $(NH_4)_2SO_4$ in citrate buffer. Two peaks of glucose-6-phosphate dehydrogenase activity are obtained.

Step 6. Concentration by Ultrafiltration. The active fractions from both peaks are pooled together (120 ml) and concentrated by ultrafiltration to 10 ml, diluted to 50 ml with citrate buffer, and reconcentrated to 30 ml.

Step 7. CM-Cellulose Chromatography. The enzyme is now rechromatographed on CM-cellulose under the conditions described in step 5. From this second column only one region of activity is recovered, and the fractions containing more than 50% of the activity of the peak fraction are pooled (30 ml), concentrated by ultrafiltration to 5–10 ml, and stored at 4°. A summary of this procedure is shown in the table.

PURIFICATION OF *Penicillium duponti* GLUCOSE-6-PHOSPHATE DEHYDROGENASE[a,b]

Step[b,c] and fraction	Volume (ml)	Total activity (units)	Total protein (mg)	Specific activity (units/mg)	Yield (%)
1. Cell-free extract	600	348	2820	0.12	—
2. Protamine sulfate supernatant	600	354	1680	0.21	100
3. 45% saturated $(NH_4)_2SO_4$ supernatant	635	375	1460	0.26	100
3. 60% saturated $(NH_4)_2SO_4$ supernatant	20	406	390	1.04	100
4. Sephadex G-100 eluate	40	404	158	2.56	100
5. First CM-cellulose eluate	120	260	51	5.1	79
6. Concentrated eluate	30	261	21	12.5	79
7. Second CM-cellulose eluate (concentrated)	7	111	1.68	66	34

[a] Reproduced from A. A. Malcolm and M. G. Shepherd, *Biochem. J.* **128**, 817 (1972).

[b] For details see the text.

[c] The number of each step corresponds to that in the text.

Properties

Stability. The purified enzyme is very unstable (half-life of 24 hr at 4° in citrate buffer). Glucose-6-phosphate (0.1 mM final concentration) increases the half-life to 10 days at 4°. Storage at −20° or −40° or freeze drying did not alter the rate of loss of activity. The enzyme is more stable in an impure state. The 60% saturated $(NH_4)_2SO_4$ pellet obtained at step 3 can be stored for over a week in citrate buffer at −20° or −40° without loss of activity; at 4° or room temperature for 24 hr, this fraction loses 30 and 50% of its activity, respectively. $(NH_4)_2SO_4$ (10%) gives almost complete stability for 1–2 days at 4° when the specific activity of the fraction is below about 10 units/mg of protein.

When the stability and activity of the 60% saturated $(NH_4)_2SO_4$ pellet fraction is studied as a function of pH in buffers of constant ionic strength, stability is found to be maximal between pH values of 5.4 and 7.0 (citrate and phosphate buffers), in contrast to a pH-activity optimum in the range pH 8–9.

Molecular Weight. Gel filtration on Sephadex G-200 yields a molecular weight of 126,000 and a Stokes radius of 4.4×10^{-7} cm. A sedimentation coefficient of 6.2 S is obtained from sucrose density gradient centrifugation. If the apparent partial specific volume of 0.744 ml/g derived for yeast glucose-6-phosphate dehydrogenase[7] is used in the formula $M = 6\pi N \eta a s / (1 - \bar{v}\rho)$, where M = molecular weight, N = Avogadro's number, η = viscosity of the medium, a = Stokes radius, s = sedimentation coefficient, $\bar{v}$ = partial specific volume, and ρ = density of the medium, a molecular weight of 121,000 is obtained. Yeast glucose-6-phosphate dehydrogenase is eluted from the Sephadex G-200 column slightly after the *P. duponti* enzyme, but sediments in a sucrose density gradient at the same rate as the fungal enzyme.

The addition of either glucose 6-phosphate or NADP+ to the gradients (0.1–1 mM final concentration) did not alter the sedimentation rate of *P. duponti* glucose-6-phosphate dehydrogenase.

Kinetic Properties. In Tris·HCl buffer, pH 8.0, 25°, $I = 0.09$, the following kinetic constants were obtained; K_m (NADP+) = 43 μM and K_m (G-6-P) = 0.16 mM. Inhibition by NADPH is competitive for both NADP+ and G-6-P, with K_i NADPH of 22 μM variable NADP+ and 30 μM variable G-6-P.

NAD+ is not reduced by *P. duponti* glucose-6-phosphate dehydrogenase, and neither NAD+ nor NADH inhibit its activity; the activation energy of the enzyme is 40.2 kJ·mole⁻¹.

[7] R. H. Yue, E. A. Noltmann, and S. A. Kuby, *J. Biol. Chem.* **244**, 1353 (1969).

[46] Glucose-6-phosphate Dehydrogenase from *Candida utilis*

By G. F. Domagk and R. Chilla

$$\text{D-Glucose 6-phosphate} + \text{NADP}^+ \leftrightarrow \text{6-phospho-D-gluconolactone} + \text{NADPH} + \text{H}^+$$

Assay Method[1]

Principle. Glucose-6-phosphate (G6P) is oxidized in the presence of NADP+ to give 6-phosphogluconate (6P6A) and NADPH. The reaction rate is measured by the increase of optical density at 340 nm.

Reagents

G6P, sodium salt, 20 mM
6PGA, sodium salt, 20 mM

[1] G. F. Domagk, R. Chilla, W. Domschke, H. J. Engel, and N. Soerensen, *Hoppe-Seyler's Z. Physiol. Chem.* **350**, 626 (1969).

NADP, sodium salt, 20 mM in 1 mM HCl

MgCl$_2$, 1 M

Triethanolamine buffer, 0.1 M, pH 7.6, containing 1 mM EDTA

Assay mixture: Mix 9.2 ml of buffer with 0.1 ml of MgCl$_2$, 0.2 ml of NADP, and 0.5 ml of G6P. Keep in ice on the day of use, freeze overnight.

Enzyme. Prior to assay dilute in cold deionized water so that the addition of 10 μl to cuvette will produce a change of optical density between 0.01 and 0.1 per minute.

Procedure. Place 0.99 ml of assay mixture into a thermostatted (25°) glass cuvette with a light path of 1.0 cm. After sufficient warming of the solution, the reaction is initiated by the addition of 0.01 ml of enzyme and thorough mixing. Take readings at 340 nm at 1-min intervals.

In the early stages of purification the values must be corrected for the NADPH formation due to phosphogluconate dehydrogenase. For this, separate readings are taken employing an assay mixture containing 6PGA instead of G6P. This reading is subtracted from the readings obtained with G6P.

Definition of Unit and Specific Activity. One unit of enzyme is defined as that amount which causes an initial change of absorbance (ΔA_{340}) of 6.23 (international enzyme unit = turnover of 1 μmole of substrate per minute). Specific activity is expressed as units per milligram of protein as determined by the biuret method or by the optical densities at 280 and 260 nm, respectively.

Purification Procedure[2]

Step 1. Preparation of Crude Extract. One hundred grams of dried yeast, *Candida utilis* (Sigma Chemical Company, St. Louis, Missouri), are autolyzed in 500 ml of 5 mM EDTA pH 7.5 containing 0.5 mM mercaptoethanol at 37°. After 4 hr the suspension is centrifuged for 10 min at 20,000 g; the residue is discarded.

Step 2. Removal of Nucleic Acids. Protamine sulfate, 90 mg per 100 g of yeast, is dissolved in water at 2% concentration and added dropwise under mechanical stirring. After 5 min in an ice bath, the turbid solution is centrifuged for 10 min and the precipitate is discarded. The supernatant can, without loss of G6PDH activity, be filtered over a column of DEAE cellulose, which will separate the pentose phosphate isomerase.[3]

[2] R. Chilla, K. M. Doering, G. F. Domagk, and M. Rippa, *Arch. Biochem. Biophys.* **159**, 235 (1973).

[3] G. F. Domagk and K. M. Doering, see this volume [90].

Step 3. Chromatography on Phosphocellulose. The supernatant from the protamine precipitate, adjusted to pH 6.2, is mixed with 150 g of phosphocellulose equilibrated with 0.25 M sodium acetate, pH 6.3. The slurry is then poured into a column (55 × 220 mm) and washed with about 3 liters of 0.25 M sodium acetate, pH 6.3. When the absorbance at 280 nm of the effluent falls below 0.1 a fraction collector is connected and further elution is done by 0.2 M phosphate buffer pH 6.3 with 1 mM EDTA and 0.5 mM 2-mercaptoethanol. This step will elute 6PGADH first,[4] followed by G6PDH activity. All fractions containing G6PDH are combined and dialyzed against 10 mM malonate pH 6.0/1 mM EDTA/0.5 mM 2-mercaptoethanol.

Step 4. Chromatography on CM-Sephadex. The dialyzed enzyme is passed through a column (55 × 120 mm) of CM-Sephadex C50 equilibrated against the same buffer. After adsorption the column is washed with a linear gradient of 1 liter of malonate buffer to 1 liter of 0.3 M NaCl in malonate buffer. After this washing a fraction collector is attached, and the elution of the remaining 6PGADH followed by G6PDH is carried out using 0.4 M NaCl in malonate buffer. All fractions containing G6PDH are combined and dialyzed against 0.25 M sodium acetate pH 6.3.

Step 5. Substrate Elution from Phosphocellulose. All the G6PDH activity can now be adsorbed to a small column (12 × 15 mm) of phosphocellulose in acetate buffer. After washing with 200 ml of acetate buffer the G6PDH is eluted quantitatively by 25 ml of a solution of 1% glucose 6-phosphate in 0.25 M sodium acetate pH 6.3.

Step 6. Crystallization. Concentration of the dilute enzyme solution to about 2 ml is achieved by ultrafiltration (Ultrahülse of Membranfiltergesellschaft, Göttingen, Germany, or Diaflo membrane PM 30 in a "stirred cell" of Amicon, Oosterhout, Holland). Crystallization of G6PDH will occur when solid ammonium sulfate is added until the first turbidity is observed.

A summary of the purification procedure is given in the table.

Properties

Physical Constants. The kinetic data (K_m values, pH optimum, temperature optimum) determined for the enzyme preparation reported here do not differ from the values reported earlier.[1] The molecular weight of G6PDH was found to be 110,000 and 220,000, depending on the pH and the Mg^{2+} concentration.

[4] M. Rippa, M. Signorini, and C. Picco, *Ital. J. Biochem.* **19**, 361 (1970).

Step and fraction	Volume (ml)	Total units	Protein (mg)	Specific activity	Yield (%)
1. Crude extract	315	3000[c]	7700	0.39	100
2. Protamine sulfate supernatant	340	2990[c]	5200	0.58	100
3. Eluate from phosphocellulose	226	2290[c]	178	12.8	76
4. Eluate from CM-Sephadex	75	1930	23	83	64
5. Eluate from substrate elution from phosphocellulose	32	1455	2.7	533	49
6. First crystallization fraction	1.3	1400	2.7	510	46

[a] Reproduced from Domagk et al., Arch. Biochem. Biophys. **159**, 237 (1973) by permission of Academic Press.

[b] 100 g of dried yeast were used in this preparation.

[c] Values were corrected for blank given in incubation of enzyme with 6PGA.

Stability. In the presence of 50% saturated ammonium sulfate, the enzyme could be stored in the refrigerator for several months without loss of activity.

Activators and Inhibitors. For full activity G6PDH needs the addition of 10 mM Mg^{2+}. The presence of various sulfhydryl reagents, of NADPH, of phosphate ions, of pyridoxal phosphate, and several adenosine nucleotides, respectively, is inhibitory to the enzyme. NADPH, erythrose 4-phosphate, and glyceraldehyde 3-phosphate have been discussed as physiological regulators of G6PDH.

[47] Glucose-6-phosphate Dehydrogenase from Human Erythrocytes

By PHILIP COHEN and MICHAEL A. ROSEMEYER

Glucose 6-phosphate + NADP $\rightleftharpoons$ 6-phosphogluconolactone + NADPH + H$^+$

Glucose 6-phosphate (D-glucose-6-phosphate:NADP oxidoreductase, EC 1.1.1.49) was first isolated in a homogeneous form from human red cells by Yoshida.[1] The method given below was subsequently developed to provide a good yield of enzyme in a comparatively short time using straightforward procedures. The details of the procedure have been re-

[1] A. Yoshida, J. Biol. Chem. **241**, 4966 (1966).

ported previously in the *European Journal of Biochemistry*.[2] The subunit structure of the enzyme was also reported in the same journal.[3]

Enzyme Assay

Principle. The activity of glucose-6-phosphate dehydrogenase is measured in the presence of excess glucose 6-phosphate and NADP by the increase in absorbance at 340 nm, which corresponds to the formation of NADPH. The reaction goes to completion since the product 6-phosphogluconolactone is rapidly hydrolyzed to 6-phosphogluconate. This hydrolysis is not a rate-limiting step in the standard assay. The maximum reaction velocity and K_m are dependent on the ionic strength of the buffer and are influenced by the particular ions present.[2]

Reagents

Tris·HCl buffer pH 9.0 $I = 0.1$ containing 92.32 g of Tris and 0.1 equivalent of HCl per liter (2.4 ml)
NADP, 1.0 mM (0.3 ml)
Glucose 6-phosphate, 10.0 mM (0.3 ml)

Procedure. The reaction is started by the addition of 0.01–0.05 ml, containing less than 5 units of activity per milliliter. For a total volume of 3 ml in the cuvette the activity of the enzyme solution (units/ml) is given by:

$$\text{Units/ml} = \Delta A_{340}/2.07\ V$$

where ΔA_{340} is the change per minute in absorbance at 340 nm; V is the volume (ml) of the enzyme sample added; the factor 2.07 derives from the millimolar extinction coefficient[4] of NADPH of 6.22 divided by the volume of the assay solution.

Definition of Unit. A unit of enzyme activity is defined as the amount of enzyme which reduces 1 μmole of NADP per minute at 25°.

Materials

Buffers. The following buffers are used repeatedly in the procedure.
(1) Phosphate pH 6.5, 5 mM (1.0 mM EDTA) containing per liter: KH_2PO_4, 0.524 g; K_2HPO_4, 0.20 g; $K_2EDTA·2H_2O$, 0.404 g
(2) Phosphate pH 5.8, 5 mM (1.0 mM EDTA) containing per liter: KH_2PO_4, 0.626 g; K_2HPO_4, 0.0696 g; $K_2EDTA·2H_2O$, 0.404 g

[2] P. Cohen and M. A. Rosemeyer, *Eur. J. Biochem.* **8**, 1 (1969).
[3] P. Cohen and M. A. Rosemeyer, *Eur. J. Biochem.* **8**, 8 (1969).
[4] B. L. Horecker and A. Kornberg, *J. Biol. Chem.* **175**, 385 (1948).

(3) Phosphate pH 7.0, 50 mM (1 mM EDTA) containing per liter: KH$_2$PO$_4$, 2.66 g; K$_2$HPO$_4$, 5.31 g; K$_2$EDTA·2H$_2$O, 0.404 g

(4) Acetate pH 6.0, 0.1 I (1 mM EDTA) containing per liter: CH$_3$COO·Na, 8.20 g; CH$_3$COOH, 4.5 meq; K$_2$EDTA·2H$_2$O, 0.404 g. NADP (10 μM) and 2-mercaptoethanol (0.1% v/v, i.e., 14 mM) are included in the acetate buffer.

Purification Procedure

Blood is stored at 0–4° in acid-citrate-dextrose to prevent clotting. Starting with 60 pints (30 liters) of outdated human blood, the residual plasma is sucked off and the cells washed 3 times with an equal volume of 0.15 M KCl + 1 mM EDTA. The cells are centrifuged after each wash at 3000 g for 15 min, and the supernatant is sucked off. The washed cells are approximately 12 liters in volume.

Step 1. Hemolysate. The cells are lysed by vigorous shaking for 10–20 min with an equal volume of water, containing 1.0 mM EDTA, 0.1% (v/v) 2-mercaptoethanol, with the addition of 5% (v/v) toluene. After centrifugation at 3000 g for 20 min the upper layer containing toluene and a pad of stroma is sucked off and the clear intermediate layer is decanted. One volume of phosphate pH 6.5 5 mM is added to 2 volumes of hemolysate, producing an overall 3-fold dilution of the red cell components and decreasing the chloride concentration to 50 mM.

Step 2. DEAE-Sephadex (Büchner). Each quarter of the hemolysate is poured through a 24 cm Büchner funnel containing 30 g of DEAE-Sephadex (A-50) equilibrated at pH 6.5 with phosphate pH 6.5, 5 mM, with 50 mM KCl. The flow rate is 2–3 liters per hour under mild suction. The Sephadex is washed with 3 liters of phosphate, pH 6.5, 5 mM, with 50 mM KCl, and eluted with 2.5 liters of phosphate pH 5.8, 5 mM, with 0.3 M KCl. In the final stages of the elution the Sephadex is stirred and sucked dry. The eluate is made 2 μM in NADP, 0.1% in mercaptoethanol, and 2 mM in ϵ-aminocaproic acid.

Step 3. CM-Sephadex (Büchner). The combined eluates from the four Büchner funnels are diluted 3-fold (to 30 liters) with phosphate pH 5.8 5 mM (to decrease the KCl concentration to 0.1 M) and the pH adjusted to between 5.7 and 5.8 with 0.5 M acetic acid. Each half of the solution is poured through a 24-cm Büchner funnel containing 25 g of CM-Sephadex (C-50) previously equilibrated with phosphate pH 5.8, 5 mM, with 0.1 M KCl. The exchanger is washed with 2 liters of the same buffer and eluted with 2 liters of phosphate pH 7, 50 mM, stirring and sucking dry the Sephadex in the final stages. The eluate is made 2 μM in NADP, 0.1% (v/v) in mercaptoethanol and 1 mM in ϵ-aminocaproic acid.

Step 4. Ammonium Sulfate Fractionation. In each liter of solution, 350 g of ammonium sulfate is dissolved, and the solution is stored overnight. The suspension is centrifuged at 20,000 g for 15 min; the supernatant is discarded. The precipitate is dissolved in phosphate pH 7, 50 mM, to give a volume of approximately 80 ml and made 20 μM in NADP, 0.1% (v/v) in mercaptoethanol. All subsequent buffers contain this concentration of mercaptoethanol. The solution is then dialyzed overnight against 25 volumes (2 liters) of phosphate pH 7, 50 mM, containing 5 μM NADP. The dialysis is repeated twice.

Step 5. DEAE-Sephadex (Column). After centrifugation at 20,000 g for 15 min to remove the residual precipitate, the solution is applied to a 40 × 2 cm DEAE-Sephadex column which has been equilibrated with phosphate pH 7.0, 50 mM. A linear salt gradient is applied using two 250-ml vessels with phosphate pH 7.0, 50 mM, as the initial solution and the same buffer with 0.3 M KCl as the final solution. Both buffers also contain 5 μM NADP. Material corresponding to 90% of the enzyme which emerges at 0.2–0.25 M KCl is pooled and concentrated by vacuum dialysis against phosphate pH 5.8, 5 mM with 10 μM NADP.

Step 6. Dialysis and CM-Sephadex (Column). After concentration to approximately 40 ml, the protein solution is further dialyzed using three changes of the same buffer. An appropriate concentration of the enzyme for this step is 120–150 units/ml when less than 10% of the enzyme is in the precipitate, which contains certain protein contaminants that are difficult to remove without this dialysis at low ionic strength.

After centrifugation at 20,000 g to remove the precipitate, the supernatant is made 0.1 M in KCl and applied to a 40 × 1.6 cm column of CM-Sephadex, equilibrated with phosphate pH 5.8, 5 mM, with 0.1 M KCl. A linear gradient is applied using two 500-ml vessels containing phosphate pH 5.8, 5 mM, with 10 μM NADP. The initial solution is 0.1 M in KCl, and the final solution is 0.6 M in KCl. Material corresponding to 90% of the enzyme, which emerges at 0.25–0.30 M KCl, are pooled and vacuum dialyzed against acetate pH 6.0, 0.1 I.

Step 7. Ammonium Sulfate Fractionation. After concentration to approximately 5 ml, the enzyme is precipitated by adding an equal volume of acetate buffer containing 580 g of ammonium sulfate per liter of buffer. The suspension is allowed to stand for 3 hr at room temperature, then centrifuged; the supernatant with less than 1% of the activity is discarded. The precipitate is redissolved in the acetate buffer to give a total volume of 1 ml and dialyzed against the same buffer.

Step 8. Gel Filtration. After centrifugation to remove any residual precipitate, the protein solution is applied to a 150 × 1.0 cm Sephadex G-200 column, equilibrated with acetate buffer, and eluted with the same buffer. Fractions containing 90% of the enzyme activity are pooled and vacuum dialyzed against acetate buffer pH 6.0, 0.1 I.

Step 9. Ammonium Sulfate Fractionation. After concentration to between 4 and 5 ml, acetate buffer containing ammonium sulfate (600 g per liter of buffer) is added until the solution just shows a definite turbidity. This requires about 0.8 volume of the ammonium sulfate solution. After 3 hr at room temperature, the suspension is centrifuged and the supernatant with less than 3% of the enzyme activity is discarded. The precipitate is redissolved in acetate pH 6.0, 0.1 I, and dialyzed twice against 50 volumes of the same buffer. The protein is stored in this solution at a concentration between 5 and 10 mg/ml.

A summary of the purification is shown in Table I.

Properties of the Enzyme

Table I shows a 72,000-fold purification of the enzyme. On correction, using the extinction coefficient determined for the enzyme by the method

TABLE I
SUMMARY OF PURIFICATION PROCEDURE[a]

Step	Protein (mg)	Activity (units)	Specific activity (units/mg)	Purification (fold)	Yield (%)
1. Hemolysate[b]	4.3×10^6	10,700	0.0025	1	100
2. DEAE-Sephadex, pH 6.5	—	9,600	—	—	90
3. CMSephadex, pH 5.8	2.9×10^3	7,500	2.6	1,040	70
4. Ammonium sulfate, 350 g/liter	1.6×10^3	7,200	4.5	1,800	67
5. DEAE-Sephadex, pH 7.0	340	6,100	18	7,200	57
6. Dialysis and CM-Sephadex, pH 5.8	70	4,600	66	26,000	43
7. Ammonium sulfate, 290 g/liter	50	4,500	90	36,000	42
8. Sephadex G-200	22	3,600	160	60,000	34
9. Ammonium sulfate, 260–270 g/liter	19	3,400	180	72,000	32

[a] Taken from P. Cohen and M. A. Rosemeyer, *Eur. J. Biochem.* **8**, 1 (1969).
[b] Protein concentration was derived from the concentration of hemoglobin. The activity of glucose-6-phosphate dehydrogenase was taken to be 75% of the total estimated, allowing for the contribution by 6-phosphogluconate dehydrogenase.

[5] J. Babul and E. Stellwagen, *Anal. Biochem.* **28**, 216 (1969).

TABLE II

PROPERTIES OF HUMAN ERYTHROCYTE GLUCOSE-6-PHOSPHATE DEHYDROGENASE

Property[a]	Value
$A_{280}^{1\%}$	12.2 ± 0.4
Specific activity	220 units/mg
Turnover number	$190\ s^{-1}$ (molecules substrate converted per second per subunit)
K_m (glucose 6-phosphate)	$10^{-4}\ M$
K_m (NADP)	$10^{-5}\ M$
Partial specific volume[b]	0.734
Subunit molecular weight	53,000
Molecular weight (pH 6.0, $I = 0.55$)	210,000
$s_{20,w}$ (pH 6.0, $I = 0.55$)	9.0 S

[a] The kinetic properties were measured in the standard assay with pH 9.0, $I = 0.1$.
[b] Calculated from the amino acid composition (Table III).

of Babul and Stellwagen,[5] the freshly prepared enzyme has a specific activity of 220 units/mg. This value declines to 150 units/mg over a period of 2 weeks. This activity does not decrease appreciably on storage of the enzyme for several months at 4° in acetate or phosphate buffers of pH 6.0, $I = 0.1$, containing 10 μM NADP and 0.1% (v/v) mercaptoethanol.

The preparation is homogeneous by the criteria of electrophoresis on starch gel at pH 8.6,[2] and on cellulose acetate at pH 7.5,[2] and on polyacrylamide gels in the presence of sodium dodecyl sulfate. This last method and also sedimentation equilibrium measurements on the maleylated enzyme[3] indicate that the protein is composed of subunits of 53,000 MW. The subunits are also identical in charge, as shown by hybridization of electrophoretic variants of the enzyme[6] and by starch gel electrophoresis in the presence of 8 M urea.[7]

In the pH range 7–8 and ionic strength 0.1, the native enzyme exists in a rapid equilibrium between a tetramer of MW 210,000 and $s_{20,w} = 9.0$ S and a dimer of molecular weight 105,000 and $s_{20,w} = 5.6$ S.[3] The discrete tetrameric species is observed at pH 6.0 and ionic strength 0.55, but at this pH and lower ionic strengths, the tetramers associate to form larger aggregates as shown by an increase in observed molecular weights and sedimentation coefficients.[3]

Some general properties of the enzyme are summarized in Table II and the amino acid composition is given in Table III.

[6] A. Yoshida, L. Steinmann, and P. Harbert, Nature (London) 216, 275 (1967).
[7] A. Yoshida, Biochem. Genet. 2, 237 (1968).

TABLE III

AMINO ACID COMPOSITION OF HUMAN ERYTHROCYTE
GLUCOSE-6-PHOSPHATE DEHYDROGENASE

Amino acid	Residues per subunit (molecular weight 53,000)
Aspartic acid	49.2
Threonine[a]	19.4
Serine[a]	23.9
Glutamic acid	55.5
Proline	22.8
Glycine	32.6
Alanine	29.8
Valine[b]	31.5
Methionine	12.0
Isoleucine[b]	25.0
Leucine	42.2
Tyrosine	17.0
Phenylalanine	23.1
Histidine	11.5
Lysine	27.3
Arginine	28.1
Cysteine[c]	7.2
Tryptophan[d]	6.9
Total	465.0

[a] Extrapolated to zero time from 24, 48, and 72-hr hydrolyses.
[b] Final value after 72-hr hydrolysis.
[c] Determined as cysteic acid.
[d] Determined by the method of W. L. Bencze and K. Schmid, *Anal. Chem.* **29**, 1193 (1957).

[48] 6-Phospho-D-gluconate Dehydrogenase from Sheep Liver

By MICHAEL SILVERBERG and KEITH DALZIEL

6-Phospho-D-gluconate + NADP $\rightleftharpoons$ D-ribulose 5-phosphate + CO_2 + NADPH

6-Phosphogluconate dehydrogenase was first highly purified from a mammalian tissue by Villet and Dalziel,[1] who obtained an apparently homogeneous preparation from sheep liver. The isolation procedure described here is an extensive modification[2] of that method, which gives a preparation of higher specific activity from which the enzyme has been

[1] R. H. Villet and K. Dalziel, *Biochem. J.* **115**, 639 (1969).
[2] M. Silverberg and K. Dalziel, *Eur. J. Biochem.* **38**, 229 (1973).

crystallized. Other modifications have been published recently for the isolation of the enzyme from sheep liver[3] and rat liver.[4]

Assay Method

Reagents

Glycine-NaOH buffer, 0.17 M, pH 8.9
MgCl$_2$, 0.1 M
Sodium 6-phosphogluconate, 20 mM
NADP, 2 mM

Procedure. The initial rate of reduction of NADP at 25° is determined by absorbance measurements at 340 nm with a reaction mixture comprising 1.8 ml of buffer, 0.6 ml of MgCl$_2$, 0.3 ml of 6-phosphogluconate and 0.3 ml of NADP, reaction being initiated by the addition of 5 or 10 μl of enzyme solution containing about 0.15 μg of enzyme. A unit of enzyme is that amount which causes an initial rate of formation of 1 μmole of NADPH per minute.

Determination of Protein and Specific Activity. The specific activity (units per milligram of protein) was calculated from protein determinations by the method of Warburg and Christian[5] in the tissue extract after acidification and at subsequent stages of the purification procedure from absorbance measurements at 280 nm, assuming $E_{1\,cm}^{1\%} = 10.0$, on which basis the specific activity of the pure enzyme is 19. From dry weight determinations,[2] the extinction coefficient of the pure enzyme at 280 nm is $E_{1\,cm}^{1\%} = 11.4$, and on this basis the specific activity is 21.

Purification Procedure

All operations are performed at 0–5°, and all buffers are made up in glass-distilled water and contain 1 mM EDTA.

Step 1. Extraction and Acid Treatment. Fresh, minced sheep liver, 800 g, is soaked for 20 hr in 1.6 liters of 0.125 M potassium phosphate buffer, pH 7.0, and filtered through muslin. The filtered extract is brought to pH 5.0 by the gradual addition of 20% acetic acid and, after standing for 10 min, is centrifuged at 14,000 g for 45 min. The decanted supernatant is immediately brought to pH 6.0 with 25% ammonia solution. The acid treatment results in a 2- to 3-fold increase of specific activity, with

[3] J. E. D. Dyson, R. E. D'Orazio, and W. H. Hanson, *Arch. Biochem. Biophys.* **154**, 623 (1973).
[4] D. Procsal and D. Holten, *Biochemistry* **11**, 1310 (1972).
[5] O. Warburg and W. Christian, *Biochem. Z.* **310**, 384 (1942).

a yield of 90% if carefully carried out. However, the initial extract before acid treatment is difficult to clarify by centrifugation and assays at this stage are not reliable. Values for the yield in the subsequent stages are therefore referred to the supernatant from the acid treatment.

Step 2. Ammonium Sulfate Fractionation at pH 6.0. Ammonium sulfate (analytical grade) is added piecemeal to the stirred supernatant from step 1, to a final concentration of 300 g/liter. The pH is monitored and kept at 6.0 by addition of 25% ammonia solution. After 1.5 hr the precipitate is removed by centrifugation at 14,000 *g* for 40 min and additional 90 g of ammonium sulfate per liter is added to the supernatant. The solution is left overnight, and the precipitate is collected by centrifugation as before. The precipitate is resuspended in the minimum volume of 0.125 *M* phosphate buffer, pH 7.0, containing 1 m*M* 2-mercaptoethanol. The suspension is dialyzed against 10 m*M* phosphate buffer, pH 7.0, containing 1 m*M* mercaptoethanol for about 30 min, to decrease the salt concentration and dissolve the protein without a large increase in volume.

Step 3. Gel Filtration. The protein solution is chromatographed on a column of Sephadex G-100 (4.5 cm diameter × 90 cm) equilibrated with 10 m*M* phosphate buffer, pH 7.0, containing 1 m*M* 2-mercaptoethanol. The protein is eluted with the same buffer at a flow rate of 30 ml/hr. The enzyme-containing fractions are pooled and brought to pH 8.0 with 25% ammonia solution.

Step 4. Chromatography on DEAE-Sephadex. The product from the previous step is loaded onto a column (3.75 cm diameter × 20 cm) of Sephadex A-50 equilibrated with 10 m*M* phosphate buffer, pH 8.0. The enzyme is eluted at a flow rate of 30 ml/hour with a concave gradient made from 1 liter of 35 m*M* phosphate buffer and 1 liter of 10 m*M* phosphate buffer, pH 8.0. The position of elution of the enzyme in the gradient is somewhat variable, presumably because of the difficulty of exactly reproducing the properties of the ion-exchanger in dilute phosphate buffer.

Step 5. Ammonium Sulfate Fractionation at pH 8.0. The pooled enzyme-containing fractions from the previous step are concentrated by ultrafiltration under oxygen-free nitrogen (Amicon system, membrane PM30) until the enzyme concentration is 1.5 units/ml. The latter value is fairly critical; with smaller values, the yield in the ammonium sulfate fractionation is low, and with higher values colored impurity is not completely removed in either this or the following step. Ammonium sulfate is added to a concentration of 350 g/liter and the precipitate is removed after 30 min by centrifugation at 23,000 *g* for 30 min. A further 30 g/liter ammonium sulfate is added, and the suspension is left overnight. The precipitate is removed as before and 100 g of ammonium sulfate per liter is added to the supernatant. After 48 hr, the enzyme-containing precipitate is isolated by centrifugation and dissolved in 0.1 M potassium phosphate buffer, pH 7.0, to give a total volume of 10 ml.

Step 6. Chromatography on CM-Sephadex. The solution is passed through a column of Sephadex G-25 (2.5 cm diameter × 40 cm) equilibrated with 10 mM phosphate buffer, pH 7.0, containing 50 mM KCl. The eluate is loaded onto a column of CM-Sephadex (C-50) of the same dimensions and preequilibrated with the same buffer-KCl solution as the G-25. The enzyme is eluted with a linear gradient of 1 liter of 0.25 M KCl into 1 liter of 0.05 M KCl in 10 mM phosphate buffer, pH 7.0, at a flow rate of 60 ml/hr. The eluates containing the pure enzyme are concentrated by ultrafiltration with the addition of 0.125 M phosphate buffer, pH 7.0. A small precipitate sometimes forms during this process, and the specific activity increases slightly, presumably indicating removal of inactive and denatured enzyme. The product may be stored for some weeks at 2° without loss of activity.

The results obtained in a typical preparation are given in the table. The isolation procedure has been scaled up to 10 kg tissue without significant modification.

Crystallization. Large crystals are readily obtained[2] by dialysis of the pure enzyme solution, 8 mg/ml in 50 mM phosphate buffer, pH 6.2–7.8, against 55% saturated ammonium sulfate solution for 2 days. The specific activity of the redissolved crystalline enzyme is usually about 20% smaller than that of the original enzyme preparation. Crystals over 1 mm long, suitable for X-ray diffraction studies, have also been obtained.[6,7]

SUMMARY OF PURIFICATION PROCEDURE

Step	Volume (ml)	Protein (g)	Activity (units/ml)	Specific activity (units/mg)	Yield (%)
1. Acidified extract	1430	42	1.32	0.045	100
2. (NH₄)₂SO₄, pH 6.0	70	10.5	16.6	0.110	62
3. Sephadex G-100	440	5.3	2.75	0.23	64
4. DEAE-Sephadex	500	0.64	1.72	1.35	46
5. (NH₄)₂SO₄, pH 8.0	10	0.22	80.4	3.68	43
6. CM-Sephadex, ultra-filtration	3	0.030	185.0	18.7	29

[6] M. Silverberg, K. Dalziel, and M. J. Adams, *Biochem. Soc. Trans.* **1**, 1132 (1973).
[7] M. J. Adams, unpublished work, 1973.

Properties

Chemical and Physical. The enzyme does not contain significant amounts of Zn^{2+}, Mn^{2+}, or Mg^{2+}, nor are added bivalent metal ions essential for activity.[8] The complete amino acid composition has been reported, and of the seven thiol groups per subunit two react rapidly with Ellman's reagent and *p*-hydroxymercuribenzoate and appear to be essential for activity.[2] The extinction coefficient for the protein at 280 nm calculated from the tryptophan and tyrosine contents (7 and 11 per subunit, respectively) is $E_{1\,cm}^{1\%} = 11.4$, identical with that estimated directly from dry weight determinations.[2]

The partial specific volume calculated from the amino acid composition is 0.741. The molecular weight from gel filtration, polyacrylamide gel electrophoresis, and sedimentation equilibrium studies by the meniscus depletion method is $94,000 \pm 2000$, and there are two subunits in the molecule as shown by SDS gel electrophoresis and studies of NADPH binding.[2] Similar molecular weights and subunit composition have been reported for the enzymes isolated from *Candida utilis*,[9] *Bacillus stearothermophilus*,[10] rat liver[4] and human erythrocytes.[11] The higher value of 129,000 for the molecular weight of the sheep liver enzyme earlier reported by Villet and Dalziel[1] must be attributed to the presence of impurities.

Preliminary Crystallographic Data. X-Ray diffraction studies of the enzyme crystals by Adams[6,7] show the crystals to be face-centered orthorhombic, space group $C222_1$. The asymmetric unit is probably the subunit (47,000 daltons), of which there are 8 in the unit cell of dimensions 72.7 Å $\times$ 149.0 Å $\times$ 103.8 Å.

Thermodynamics. The enzyme-catalyzed reaction is readily reversible. Dissolved CO_2, not bicarbonate ion, is the immediate substrate or product.[12] The equilibrium constant for the reaction

$$\text{NADP} + \text{6-phosphogluconate} \rightleftharpoons \text{NADPH} + \text{ribulose 5-phosphate} + CO_2 \text{ (gas)} \quad (1)$$

is $K_p = 2.62$ atmospheres, at 25°, and is independent of pH and ionic strength.[12] This refers to a standard state of 1 atm pressure of CO_2. For a standard state of the ideal molar solution of CO_2, the equilibrium constant is $K_c = 79$ mM. For reaction (1), $\Delta G°$ at 25° is —0.51 kcal/mole, and $\Delta H°$ is 13.75 kcal/mole.

Coenzyme. The enzyme from sheep liver does not utilize NAD as co-

[8] R. H. Villet and K. Dalziel, *Eur. J. Biochem.* **27**, 251 (1972).

[9] M. Rippa, M. Signorini, and S. Pontremoli, *Eur. J. Biochem.* **1**, 170 (1967).

[10] B. M. F. Pearse, unpublished work, 1973.

[11] B. M. F. Pearse and M. A. Rosemeyer, this volume [49].

[12] R. H. Villet and K. Dalziel, *Biochem. J.* **115**, 633 (1969)

enzyme.[13] The binding of the active coenzyme NADPH to the enzyme results in increased nucleotide fluorescence,[8,13] which is further augmented in the presence of 6-phosphogluconate, indicating the formation of an abortive ternary complex. From fluorescence titrations,[13] the presence of one binding site for NADPH on each of the two subunits of the enzyme has been established, and the dissociation constant of the binary enzyme–NADPH complex is 5.7 μM in 0.1 M phosphate buffer, pH 7.0, at 25°. From kinetic studies[13] of the inhibition of the oxidative decarboxylation reaction by NADPH, K_i is 9 μM. From the initial rate parameters for the oxidative decarboxylation reaction at pH 7.0, dissociation constants for the enzyme–NAD compound are estimated indirectly to be 2.8 μM in 0.13 M triethanolamine buffer and 26 μM in 0.1 M phosphate buffer[8]

Kinetic Constants. For the oxidative decarboxylation reaction in glycine buffer,[8] with saturating substrate concentrations, the pH optimum is 8.9. In Tris-acetate buffer, the pH optimum was found to increase with increasing substrate concentrations up to pH 9.1, above which the enzyme was unstable.[3]

In 0.13 M triethanolamine buffer, pH 7.0, at 25°, K_m values, with saturating concentrations of the other substrate(s), are [8,14] 6.8 μM for NADP 6.9 μM for 6-phosphogluconate, 0.22 μM for NADPH, 20 μM for ribulose 5-phosphate and 34 mM for dissolved CO_2. In 0.1 M phosphate buffer, pH 7.0, at 25°, K_m values are 30 μM for NADP and 290 μM for 6-phosphogluconate, in both cases considerably larger than in triethanolamine buffer. The maximum specific rate is also 6 times greater in phosphate than in triethanolamine,[8] however, so that phosphate does not simply act as a competitive inhibitor with respect to NADP and 6-phosphogluconate. Also in 0.1 M phosphate buffer, the effect of 20 mM Mg^{2+} on the initial rate parameters for the oxidative deamination is small, the main effect being a small decrease of the K_m for NADP, while higher Mg^{2+} concentrations cause a small inhibition.[8]

At pH 7.7 and 30°, in 50 mM Tris-acetate-KCl buffer, ionic strength 0.1, K_m values are 6.8 μM for NADP and 16 μM for 6-phosphogluconate.[3] In this medium the addition of 20 mM Mg^{2+} causes inhibition by a 3-fold increase of the K_m for 6-phosphogluconate, while at lower ionic strengths and Mg^{2+} concentrations, there is a small activation; it appears that the relatively small effects of increasing ionic strength, Mg^{2+}, Ca^{2+}, and Mn^{2+} ion concentrations are similar and interchangeable, and mainly due to effects on the K_m for 6-phosphogluconate.[3]

[13] M. Silverberg, D. Phil. Thesis, Oxford University, 1973.
[14] R. H. Villett and K. Dalziel, *Eur. J. Biochem.* **27,** 244 (1972).

Initial rate parameters for the oxidative decarboxylation reaction have been estimated at other pH values also.[3,8] From the effects of temperature on the maximum rate at pH 7.7, the activation energy[3] is 15.5 kcal/mole.

Inhibitors. Dyson and D'Orazio[15,16] have studied a range of potential metabolic inhibitors. Nucleoside 5'-triphosphates and 5'-diphosphates inhibit competitively with respect to both NADP and 6-phosphogluconate, with K_i values ranging from 0.2 mM for GTP to 1.3 mM for ATP in Tris-acetate buffer, pH 7.0, at 30°. The 5'-monophosphates and 3'-monophosphates have larger K_i values (2–11 mM). For 2'-AMP and 2'-GMP, the K_i values are 0.36 mM and 0.57 mM, respectively. Fructose-1,6-diphosphate is a strong inhibitor ($K_i = 0.07$ mM), the effect of which is counteracted by EDTA. Other inhibitors with K_i values in the range 3–9 mM include fructose 1-phosphate, fructose 6-phosphate, glucose 6-phosphate, orthophosphate, pyrophosphate, oxaloacetate, and citrate. The sugar phosphates are competitive with respect to 6-phosphogluconate and noncompetitive with respect to NADP.

[15] J. E. D. Dyson and R. E. D'Orazio, *J Biol. Chem.* **248,** 5428 (1973).
[16] J. E. D. Dyson and R. E. D'Orazio, *Biochem. Biophys. Res. Commun.* **43,** 183 (1971).

[49] 6-Phosphogluconate Dehydrogenase from Human Erythrocytes

By BARBARA M. F. PEARSE and MICHAEL A. ROSEMEYER

6-Phosphogluconate + NADP $\rightleftharpoons$ ribulose 5-phosphate + CO$_2$ + NADPH

The details of the following purification of 6-phosphogluconate dehydrogenase [6-phospho-D-gluconate:NADP$^+$ oxidoreductase (decarboxylating), EC 1.1.1.44] have been reported in the *European Journal of Biochemistry.*[1] The subunit structure of the enzyme has been reported in the same journal.[2]

Enzyme Assay

Principle. The activity of 6-phosphogluconate dehydrogenase is measured in the presence of excess 6-phosphogluconate and NADP by the

[1] B. M. F. Pearse and M. A. Rosemeyer, *Eur. J. Biochem.* **42,** 213 (1974).
[2] B. M. F. Pearse and M. A. Rosemeyer, *Eur. J. Biochem.* **42,** 225 (1974).

the ionic strength of the buffer. The maximal activity is observed at an increase in absorbance at 340 nm, which corresponds to the formation of NADPH. The maximum velocity and K_m show some dependence on ionic strength of 0.1 and pH 8.3, which gives a 30% increase in activity over than observed in the standard assay solution given below, which is similar to that used for glucose-6-phosphate dehydrogenase.[3]

Reagents

2.4 ml Tris·HCl, pH 9.0, I 0.1, containing 92.32 g of Tris and 0.1 equivalent of HCl per liter
NADP, 1.0 mM (0.3 ml)
6-Phosphogluconate, 10.0 mM (0.3 ml)

Procedure. The reaction is started by the addition of 0.01–0.05 ml, containing less than 5 units of activity per milliliter. For a total volume of 3 ml in the cuvette, the activity of the enzyme in units per milliliter is given by:

$$\text{Units per ml} = \Delta A_{340}/2.07\ V$$

where ΔA_{340} is the change per minute in absorbance at 340 nm and V is the volume in ml of the enzyme sample.[3]

Definition of Unit. A unit of enzyme activity is defined as the amount of enzyme which reduces 1 μmole of NADP per minute at 25°.

Materials

The following buffers are used repeatedly in the procedure:

Phosphate pH 7.0, $I = 0.1$ (1 mM EDTA) containing per liter: KH_2PO_4, 2.79 g; K_2HPO_4, 4.61 g; $K_2EDTA \cdot 2H_2O$, 0.404 g. This buffer is also used at a 10-fold dilution with an ionic strength of 0.01.

Acetate pH 5.8, $I = 0.05$ (5 mM EDTA) containing per liter: anhydrous sodium acetate, 4.10 g; glacial acetic acid, 0.217 ml; $K_2EDTA \cdot 2H_2O$, 2.02 g. NaCl 14.61 g is included to give a total ionic strength 0.30. The buffer is also used at a 5-fold dilution with an ionic strength of 0.06.

Phosphate pH 6.0, $I = 0.1$ (1 mM EDTA) containing per liter: KH_2PO_4, 9.81 g; K_2HPO_4, 1.62 g; $K_2EDTA \cdot 2H_2O$, 0.404 g. The

---

[3] P. Cohen and M. A. Rosemeyer, this volume [47].

solution also contains 10 μM NADP and 0.1% (v/v) 2-mercaptoethanol.

Acetate pH 6.0, $I = 0.1$ (1 mM EDTA) containing per liter: anhydrous sodium acetate, 8.2 g; glacial acetic acid, 0.257 ml; K$_2$EDTA·2H$_2$O, 0.404 g. The solution also contains 10 μM NADP and 0.1% (v/v) 2-mercaptoethanol.

Purification Procedure

Blood is stored at 0–4° in acid-citrate-dextrose to prevent clotting. Starting with 70 pints (35 liters) of outdated human blood, the residual plasma is sucked off. The cells are washed twice with an equal volume of 0.15 M NaCl containing 1 mM EDTA. Between these saline washes the cells are suspended for an hour in 0.15 M NaNO$_2$ containing 1 mM EDTA, which converts the hemoglobin to methemoglobin. After each treatment with the salt solutions, the cells are centrifuged at 2500 g for 10 min and the supernatant is sucked off. The washed cells are approximately 14 liters in volume.

Step 1. Hemolysate. The cells are lysed by vigorous shaking for 10–20 min with an equal volume of water, with the addition of 5% (v/v) of toluene. After centrifugation at 2500 g for 30 min the upper stromal layer is sucked off and the clear intermediate layer is decanted. About 24 liters of hemolysate are obtained.

Step 2. DEAE-Sephadex (Büchner). The hemolysate is diluted with 4 volumes of water, thus reducing the chloride concentration to 15 mM. The pH is adjusted to pH 7.0 with 0.5 M K$_2$HPO$_4$. Each quarter of the diluted hemolysate is poured through a 24-cm Büchner funnel containing a 2.5-liter bed of DEAE-Sephadex (A-50), which has been equilibrated with phosphate buffer pH 7.0, $I = 0.01$. The flow rate is 7–10 liters per hour under mild suction. The exchanger is washed with equilibrating buffer until the filtrate is colorless, and eluted with 2.5 liters of acetate buffer pH 5.8, $I = 0.3$. In the final stages of the elution the exchanger is stirred and sucked dry. The combined eluates (12 liters) are made 2 μM in NADP, 0.1% in mercaptoethanol, and 2 mM in ε-aminocaproic acid.

Step 3. CM-Sephadex (Büchner). The previous eluate is diluted with 4 volumes of water, and the pH is adjusted to pH 5.7–5 8 with 1 M acetic acid. Each half of the solution is poured through a 24-cm Büchner funnel, containing a 3-liter bed of CM-Sephadex (C-50) previously equilibrated with acetate buffer pH 5.8, $I = 0.06$. The exchanger is washed with 6 liters of the same buffer and is eluted with 3 liters of phosphate buffer pH 7.0, $I = 0.1$, containing 0.3 M NaCl. The exchanger is stirred and sucked dry in the final stages of the elution. The combined eluates (8 liters) are made 2 μM in NADP, 0.1% in mercaptoethanol, and 2 μM in ε-aminocaproic acid.

Step 4. *Ammonium Sulfate Fractionation.* To each liter of the eluate, 350 g of ammonium sulfate is added to give a 55% saturation. The protein is allowed to precipitate for a minimum of 2 hr. After centrifugation the precipitate is removed (and may be used for the isolation of glucose-6-phosphate dehydrogenase). A further 103 g of ammonium sulfate is dissolved in each liter of the supernatant to give a 70% saturation. After overnight storage, the suspension is centrifuged at 10,000 g for 15 min; the supernatant is discarded. The precipitate is suspended in 50 ml of phosphate pH 7.0, $I = 0.1$, made 20 μM in NADP and 0.1% in mercaptoethanol. All subsequent buffers contain the same concentration of mercaptoethanol. The suspension is dialyzed three times against 20 volumes of phosphate pH 7.0, $I = 0.01$, containing 5 μM NADP.

Step 5. *DEAE-Sephadex (Column).* The precipitate remaining after dialysis is removed by centrifugation and the supernatant is applied to a 30 × 2.5 cm column of DEAE-Sephadex equilibrated with phosphate buffer pH 7.0, $I = 0.01$. The proteins are eluted with a linear salt gradient formed by two 250-ml vessels containing phosphate pH 7.0, $I = 0.01$, as the initial solution, and the same buffer + 0.3 M NaCl as the final solution. Both solutions also contain 5 μM NADP. The 6-phosphogluconate dehydrogenase elutes near 0.1 M NaCl. The active fractions are pooled and dialyzed overnight against 20 volumes of acetate pH 5.8, $I = 0.06$, with 5 μM NADP.

Step 6. *CM-Sephadex (Column).* The dialyzed solution (90 ml) is adjusted to pH 5.7–5.8 with 1 M acetic acid and applied to a 30 × 2.5 cm column of CM-Sephadex, equilibrated with acetate pH 5.8, $I = 0.06$, 5 μM in NADP. A linear salt gradient is applied using two 250-ml vessels containing acetate pH 5.8, $I = 0.06$, and this buffer + 0.4 M NaCl as the initial and final solutions, respectively. Both solutions contain 5 μM NADP. The 6-phosphogluconate dehydrogenase elutes near 0.25 M NaCl.

Step 7. *Ammonium Sulfate Fractionation.* The pooled fractions (150 ml) are made 50% saturated in ammonium sulfate (313 g/liter). The precipitate is discarded after centrifugation at 10,000 g for 15 min. The supernatant is made 70% saturated in ammonium sulfate 137 g/liter, allowed to stand for at least 2 hr and centrifuged; the supernatant is discarded. The precipitate is suspended in acetate pH 6.0, $I = 0.1$, to give a total volume of 5 ml. The solution is dialyzed overnight against 20 volumes of the acetate buffer.

Step 8. *Gel Filtration.* The dialyzate (5 ml) is applied to a 150 × 2

TABLE I
SUMMARY OF PURIFICATION PROCEDURE[a]

Step	Volume	Protein	Activity (units)	Specific activity (units/mg)	Purification (fold)	Yield (%)
1. Hemolysate	24 l	4 kg	8000	0.002	1	100
2. DEAE-Sephadex	12 l	25 g	6000	0.24	120	75
3. CM-Sephadex	8 l	15 g	4500	0.30	150	56
4. Ammonium sulfate, 55–70% saturated	80 ml	3.6 g	4000	1.1	550	50
5. DEAE-Sephadex column	90 ml	800 mg	2300	2.9	1500	29
6. CM-Sephadex column	150 ml	300 mg	1100	3.7	1900	14
7. Ammonium sulfate 50–70% saturated	5 ml	140 mg	900	6.4	3200	11
8. Sephadex G-200	50 ml	90 mg	700	7.8	3900	9
9. Ammonium sulfate 65–75% saturated	5 ml	50 mg[b]	500[b]	10.0	5000	6[b]

[a] Taken from B. M. F. Pearse and M. A. Rosemeyer, *Eur. J. Biochem.* **42**, 213 (1974).

[b] By including 10% (v/v) glycerol in all buffers from step 4 onward the final amounts obtained are 200 mg protein, 2000 units activity, and an overall yield of 25%.

cm column of Sephadex G-200 equilibrated with acetate pH 6.0, $I = 0.1$, and eluted with the same buffer. Fractions containing enzyme activity are pooled (50 ml) and vacuum dialyzed[4] against acetate pH 6.0, $I = 0.1$.

Step 9. Ammonium Sulfate Fractionation. After concentration of the protein solution to approximately 5 ml, the acetate buffer saturated with ammonium sulfate is added to give a final saturation of 65% at 25°. The preparation is allowed to stand overnight, then the suspension is centrifuged. The precipitate with less than 10% of the activity is discarded. The supernatant is made 75% saturated in ammonium sulfate, left overnight, and centrifuged; the supernatant is discarded. The precipitate is dissolved in phosphate pH 6.0, $I = 0.1$ (containing EDTA, NADP, and mercaptoethanol) to give a protein concentration of 5–10 mg/ml.

A summary of the purification is shown in Table I.

[4] Vacuum dialysis: The narrow (¼ inch) stem of a glass column is inserted into a length of (9/32 inch) Visking tubing. The Visking tubing is then pulled through a hole in a rubber stopper until the glass stem forms a tight fit. The lower end of the dialysis tubing is knotted. The stopper is placed in a Büchner flask which has been partially filled with buffer solution. The protein solution is pipetted into the glass column and dialysis tubing. The sidearm of the Büchner flask has a short

Properties of the Enzyme

Table I shows a 5000-fold purification of the enzyme. On correction, using the extinction coefficient determined for the enzyme by the method of Babul and Stellwagen,[5] the specific activity of the freshly prepared enzyme is 12.5 units/mg at pH 9 and 15 units/mg at the optimal pH of 8.3. The activity is retained on storage at $4°$ in the phosphate buffer pH 6.0, $I = 0.1$, with ammonium sulfate at 20% saturation.

The enzyme activity varies with ionic strength, being maximal in the range 0.1–0.2. In this range, V is a maximum and the K_m toward 6-phosphogluconate is a minimum. Phosphate ions show competitive inhibition with respect to 6-phosphogluconate. NADPH inhibits competitively with respect to NADP.[1]

On starch-gel electrophoresis at pH 8.6 there is a correspondence between the protein and enzyme stains, indicating that the enzyme is essentially free of other contaminating proteins.[1] Electrophoresis on polyacrylamide in the presence of sodium dodecyl sulfate shows that the enzyme is composed of subunits of 52,000 molecular weight,[2] which is also

TABLE II

PROPERTIES OF HUMAN ERYTHROCYTE 6-PHOSPHOGLUCONATE DEHYDROGENASE

Property[a]	Value
$A^{1\%}_{280}$	12.5 ± 0.5
Specific activity	15.0 units/mg
Turnover number	13 sec^{-1} (molecules substrate converted per second per subunit)
K_m (6-phosphogluconate)	20 μM
K_m (NADP)	30 μM
K_I (NADPH)	30 μM
Partial specific volume[b]	0.737
Subunit molecular weight	52,000
Molecular weight (pH 6.0, $I = 0.1$)	104,000
$s_{20,w}$ (pH 6.0, $I = 0.1$)	5.8 S
$D_{20,w}$ (pH 6.0, $I = 0.1$)	51 μm^2/sec

[a] The kinetic properties were measured near the optimum with pH 8.0, $I = 0.1$.
[b] Calculated from the amino acid composition (Table III).

length of rubber tubing, through which the flask is evacuated on a water pump. The rubber tubing is then closed with a screw clamp and left overnight. In this manner 50 ml of protein solution can be reduced to 5 ml overnight. For larger volumes a number of glass columns (each with attached dialysis tubing) can be inserted through several holes in the same stopper.

the size of the subunit inferred from sedimentation equilibrium measurements of the maleylated enzyme.[2]

Sedimentation equilibrium of the freshly prepared enzyme in the pH range 6–7 gives a molecular weight of 104,000 indicating that the native enzyme is a dimer.[2] However, storage for long periods or exposure to higher pH promotes the formation of larger aggregates.[2]

Some properties of the enzyme are summarized in Table II, and the amino acid composition is given in Table III.

TABLE III
AMINO ACID COMPOSITION OF HUMAN ERYTHROCYTE
6-PHOSPHOGLUCONATE DEHYDROGENASE

Amino acid	Residues per subunit (molecular weight 52,000)
Aspartic acid	49.0
Threonine[a]	20.2
Serine[a]	25.3
Glutamic acid	44.8
Proline	13.4
Glycine	47.1
Alanine	42.0
Valine	29.0
Methionine[a]	11.9
Isoleucine[b]	27.2
Leucine	41.7
Tyrosine[a]	11.5
Phenylalanine	23.6
Histidine	8.8
Lysine	38.0
Arginine	20.2
Cysteine[c]	10.7
Tryptophan[d]	7.6
Total	472.0

[a] Extrapolated to zero time from 20-, 44-, and 68-hr hydrolyses.
[b] Final value after 68-hr hydrolysis.
[c] Determined as cysteic acid.
[d] Determined spectrophotometrically according to G. H. Beaven and E. R. Holiday, *Advan. Protein Chem.* **7,** 319 (1952).

[5] J. Babul and E. Stellwagen, *Anal. Biochem.* **28,** 216 (1969).

[50] 6-Phosphogluconate Dehydrogenase from *Neurospora crassa*

By WILLIAM A. SCOTT and TESSA ABRAMSKY

D-Gluconate 6-phosphate + NADP $\rightleftharpoons$
D-ribulose 5-phosphate + CO_2 + NADPH + H^+

Assay Method

6-Phosphogluconate dehydrogenase is measured spectrophotometrically by following the rate of NADPH formation at 340 nm during the conversion of 6-phosphogluconate to ribulose 5-phosphate and CO_2.

Reagents

Tris·HCl buffer, 0.1 M, pH 8.0 at 25°
NADP, 24.3 mM in 1% $NaHCO_3$
6-Phosphogluconate, 26.5 mM in 0.1 M Tris·HCl buffer, pH 8.0

Procedure. Reactions are conducted at 25° in a 3.0-ml quartz cell with a 1.0 cm light path. The incubation mixture consists of 2.5 ml of buffer, 0.1 ml of 6-phosphogluconate, and 0.03 ml of NADP. The reaction is initiated by the addition of 0.3–1.3 units of enzyme (1–10 μl), and the rate of absorbancy change at 340 nm is followed in a recording spectrophotometer. Reaction velocities are calculated from initial slopes. Enzyme activity is expressed in international units, i.e., micromoles of NADP reduced per minute per milliliter under the above conditions. Specific activity is defined as units of enzyme activity per milligram of protein. Protein is determined by the method of Lowry *et al.*[1] as described below.

Protein Determination

Most buffers used in the purification procedure contain dithiothreitol (DTT), which interferes with standard methods of protein determination. For this reason, enzyme samples are dialyzed against either distilled water or buffers without the reducing reagent. Protein is then determined by the method of Lowry *et al.*[1] with bovine serum albumin as a standard.

[1] O. H. Lowry, N. J. Rosebrough, A. L. Farr, and R. J. Randall, *J. Biol. Chem.* **193,** 265 (1951).

Materials

Tris·HCl buffers, pH 7.4 and pH 8.0, are prepared as 1 M stock solutions at 25° and are stored at 5°. Prior to use, these stocks are diluted to the appropriate concentration and are degassed extensively.

Commercial DEAE-cellulose is cycled through H_2O–NaOH–H_2O followed by HCl–H_2O and is stored under 1 M Tris·HCl buffer, pH 7.4, at 5°.[2] Columns of DEAE-cellulose are packed in 1 M buffer under pressure. Equilibration with buffer is obtained by passage of 5–20 column volumes of the desired buffer through the packed cellulose.

Hydroxyapatite is purchased as Bio-Rad HTP from Bio-Rad Laboratories and is suspended in the appropriate buffer immediately before use.

Conditions for growth, freeze-drying, and storage of large amounts of *Neurospora* mycelia are identical to those previously described.[2]

Purification Procedure

The method given here is essentially that described by Scott and Abramsky.[3] All steps are performed at 5° unless stated otherwise.

Step 1. Extraction. Freeze-dried mycelia (20–200 g) are ground in a mortar with an equal weight of acid-washed sea sand. The mycelia and sand are mixed with 20 volumes of 0.1 M Tris·HCl buffer, pH 7.4, per gram of tissue. After standing, the sand is removed by decantation and the mycelia are extracted for 3 hr with stirring. The supernatant fraction obtained after centrifugation at 2500 g for 10 min is designated as the crude extract.

Step 2. Ammonium Sulfate Fractionation. Solid ammonium sulfate (313 g/liter) is added to the crude extract with stirring. The pH of the extract is maintained between pH 6.8 and 7.0 by the dropwise addition of 1 N NaOH. The precipitate, collected by centrifugation after 1 hr of stirring, is discarded and an additional 137 g of solid ammonium sulfate per liter of the original crude extract is added to the supernatant. The precipitated enzyme is collected by centrifugation and is redissolved in 2 ml of 0.1 M Tris·HCl buffer, pH 7.4, 1 mM DTT per gram of extracted mycelia. The enzyme is dialyzed overnight against 10 volumes of the same buffer.

Step 3. Heat Fractionation. The protein concentration of the enzyme solution is adjusted to between 10 and 23 mg/ml with the buffer used

---

[2] W. A. Scott, this volume [41].

[3] W. A. Scott and T. Abramsky, *J. Biol. Chem.* **248**, 3535 (1973)

in step 2. After heating for 50 min in a 50° water bath with occasional stirring, the precipitate is removed by centrifugation and discarded.

Step 4. Hydroxyapatite Chromatography. A hydroxyapatite column is packed to provide 1 ml of adsorbent per 5 mg of protein and is equilibrated with 0.1 M Tris·HCl buffer, pH 7.4, 1 mM DTT. The enzyme is diluted to 3 mg of protein per milliliter and is loaded onto the column at a flow rate of 2.3 ml/cm² per hour. The enzyme is eluted by means of a linear salt gradient. The gradient, which should be 10 times the volume of the column, is formed by mixing buffer and buffer containing 0.76 M ammonium sulfate. Contents of fractions which contain 6-phosphogluconate dehydrogenase activity are pooled and are dialyzed overnight against 16 volumes of 5 mM Tris·HCl buffer, pH 7.4, 1 mM DTT.

Step 5. DEAE-Cellulose Chromatography. The DEAE-cellulose column should provide 1 ml of packed exchanger per milligram of protein. Both the enzyme solution and exchanger must be completely equilibrated with 5 mM Tris·HCl buffer, pH 7.4, 1 mM DTT since adsorption of the enzyme to DEAE-cellulose at pH 7.4 is weak. It should be noted that chromatography on DEAE-cellulose at higher pH values, which would increase the adsorption of the enzyme results in a complete loss of enzyme activity.

A salt gradient sufficient to develop the chromatogram is formed at a flow rate of 4.5 ml/cm² per hour between 5 mM Tris·HCl buffer, pH 7.4, 1 mM DTT and the same buffer plus 0.16 M NaCl. The gradient volume should be 10 times the bed volume of the column. Usually two partially resolved peaks of 6-phosphogluconate dehydrogenase activity are obtained. Fractions containing the eluted enzyme are pooled and are extensively degassed. The DTT concentration of the solution is raised to 10 mM by the addition of the solid reducing reagent.

Step 6. Concentration. The concentration of the pooled enzyme from the DEAE-cellulose column is low (<0.08 mg of protein per milliliter). In order to avoid storing large volumes, the enzyme solution is concentrated. This is conveniently achieved with sucrose. Dialysis tubing containing the enzyme are packed in finely powdered sucrose in shallow Pyrex dishes (22 × 34 cm). The sucrose is changed periodically before the sugar completely dissolves; usually 3 changes of sucrose are necessary to achieve a 5-fold concentration in about 5 hr.

The concentrated enzyme is filtered through Whatman No. 1 filter paper by gravity, and is again extensively degassed. The enzyme is stored in an Erlenmeyer flask fitted with a ground-glass stopper.

Alternatively, the enzyme can be concentrated by passing the pooled DEAE-cellulose fractions through a small hydroxyapatite column and eluting the enzyme with 0.76 M ammonium sulfate. However, high con-

PURIFICATION OF *Neurospora* 6-PHOSPHOGLUCONATE DEHYDROGENASE[a]

Fraction	Total protein (mg)	Total activity (units)	Specific activity	Purification[b] (fold)	Recovery (%)
1. Crude extract	20,590	3350	0.16	—	100
2. Ammonium sulfate	3,589	3145	0.87	5.3	93.7
3. Heat	1,980	2895	1.46	9.0	86.4
4. Hydroxyapatite chromatography	415	2548	6.14	37.7	76.1
5. DEAE-cellulose chromatography and concentration	43.3	1334	30.8	192.5	39.8

[a] Results obtained from the extraction of 80 g of lyophilized mycelia. Data are from W. A. Scott and T. Abramsky, *J. Biol. Chem.* **248**, 3535 (1973).

[b] Ratio of the specific activity of the enzyme of the given purification step to that of the crude extract.

centrations of sucrose increase the stability of the enzyme on storage. For this reason, concentration of the enzyme with sucrose is preferred. Attempts to concentrate the purified enzyme by ammonium sulfate precipitation lead to large losses of activity due to the insolubility of the precipitate.

Results from a typical purification are given in the table.

Properties

Purity. The purified enzyme contains no detectable glucose 6-phosphate dehydrogenase or phosphoglucomutase activity. The specific activity of the purified 6-phosphogluconate dehydrogenase is comparable to that reported for the yeast enzyme.[4] Preparations of the *Neurospora* enzyme, which have a specific activity of 30 to 35, appear to be homogeneous by gel filtration. On polyacrylamide gel electrophoresis, the enzyme in 10 mM DTT also migrates as a single coincident band of protein and enzymic activity. Multiple bands of protein are obtained, however, if the DTT concentration of the enzyme solution is reduced to 1 mM. Presumably these extraneous bands are aggregates of enzyme formed on decreasing the concentration of the reducing agent.

Stability. The purified enzyme is stable up to 3 months at 5° if care is taken to exclude molecular oyxgen from the enzyme solution. On storage, the enzyme slowly flocculates. Partial reactivation can be obtained by degassing the solution and adding fresh DTT (10 mM).

[4] M. Rippa, M. Signorini, and S. Pontremoli, *Eur. J. Biochem.* **1**, 170 (1967).

Catalytic Properties. The pH profile of the enzymic activity is broad and ranges from pH 7.8 to pH 9.0. At pH 8.0, the K_m values for 6-phosphogluconate and NADP are 10 μM and 30 μM, respectively. No metal requirements for activity have been found. The enzyme has an absolute specificity for NADP; NAD is completely ineffective as a hydrogen acceptor. Neither gluconic acid nor glucose 6-phosphate can replace 6-phosphogluconate as the substrate.

Inhibitors. NADPH is a potent inhibitor of the enzymic activity as are iodoacetamide and inorganic phosphate. The activity is also destroyed by exposure of the enzyme to light in the presence of low concentrations of Rose Bengal. 6-Phosphogluconate protects the enzyme against inactivation by iodoacetamide and Rose Bengal.

Physical Properties. The enzyme has an apparent molecular weight of 110,000 to 120,000 as determined by gel filtration on columns of Bio-Gel A-0.5 m calibrated with proteins of known molecular weight. Subunits of molecular weight 57,000 have been detected by sodium dodecyl sulfate (SDS) polyacrylamide gel electrophoresis after denaturation of the enzyme in SDS. It has been assumed from this information that the enzyme is a dimer composed of subunits of identical molecular weight.

Two peaks are observed in sedimentation velocity studies of the purified enzyme. Assuming the molecule is spherical, the sedimentation coefficients of the two components, 6.0 S and 2.8 S, correspond to the dimeric and monomeric forms of the enzyme respectively.

Genetics. Mutations in two unlinked genes, *col-10* and *col-3*, of *Neurospora* lead to morphological abnormalities and aberrant 6-phosphogluconate dehydrogenases.[5] The mutant enzymes differ from the wild-type enzyme in kinetic parameters, isoelectric points, and thermolability. Correlations between the properties of the enzyme and the morphology of the *col-3* and *col-10* heterokaryon suggest that the altered 6-phosphogluconate dehydrogenases are responsible for the altered morphologies of the two mutants.

Acknowledgments

This work was supported in part by a grant from the National Institutes of Health (GM 16224) and a grant-in-aid from the Research Corporation.

[5] W. A. Scott and T. Abramsky, *J. Biol. Chem.* **248**, 3542 (1973).

[51] 6-Phosphogluconate Dehydrogenase from *Streptococcus faecalis*

By RAYMOND B. BRIDGES and CHARLES L. WITTENBERGER

D-Gluconate 6-phosphate + NADP$^+$ $\rightleftharpoons$
$\quad\quad\quad\quad$ D-ribulose 5-phosphate + CO_2 + NADPH + H$^+$

Assay Method

Principle. The reaction is followed spectrophotometrically by measuring the rate of increase in absorbancy at 340 nm due to the enzyme-catalyzed reduction of NADP by 6-phosphogluconate.

Reagents

Tris-hydrochloride buffer, 1 M, pH 7.4.
D-Gluconate 6-phosphate, 2.0 mM. Dissolve 7.6 mg of the trisodium salt of 6-phosphogluconate dihydrate (Sigma) in 10 ml of water.
NADP, 2.5 mM. Dissolve 20.0 mg of the monosodium salt of NADP dihydrate (Sigma) in 10 ml of water.
All buffers and reagents are prepared in glass-distilled water.

Procedure. The 6-phosphogluconate dehydrogenase activity is measured in a recording spectrophotometer. The assay mixture contains: 0.05 ml of buffer, 0.1 ml of D-gluconate 6-phosphate, 0.1 ml of NADP, a quantity of an enzyme preparation that contains less than 0.1 enzyme unit, and sufficient water to give a final volume of 1.0 ml. The reaction is initiated by the addition of enzyme to the mixed components of the assay system. The velocity of the reaction during the first 10 sec after the addition of enzyme is taken to represent the initial velocity. One unit of 6-phosphogluconate dehydrogenase activity is the amount of enzyme that catalyzes the reduction of 1 μmole of NADP per minute under the above assay conditions. Protein is measured by the colorimetric method of Lowry et al.,[1] using bovine serum albumin as a standard. Specific activity is expressed as enzyme units per milligram of protein.

Purification Procedure

Growth of the Organism. Streptococcus faecalis, ATCC 27792 (formerly designated as strain MR[2]), is grown in a complex medium contain-

---

[1] O. H. Lowry, N. J. Rosebrough, A. L. Farr, and R. J. Randall, *J. Biol. Chem.* **193**, 265 (1951).
[2] J. London and E. Meyer, *J. Bacteriol.* **102**, 130 (1970)

ing the following per liter: dibasic potassium phosphate, 10 g; yeast extract (Difco), 2.5 g; tryptone (Difco), 2.5 g; and dextrose, 5 g (added as a separate sterile solution). A 20-liter culture incubated at 37° for 8 hr is used to inoculate 400 liters of medium in a fermentor. The large culture is incubated anerobically (under an atmosphere of helium) at 37° for 18 hr. Cells are then harvested, washed 3 times with 50 mM Tris·HCl buffer, pH 7.5, and stored as a wet cell paste at—60° until further use. The average cell yield from a 400-liter culture is about 600 g wet weight under the above conditions. For smaller runs, the organism can also be cultivated in 20-liter carboys filled completely with medium.

General Considerations. The results below are those from a typical fractionation procedure. Some minor variations have been observed with respect to the appearance of the enzyme in the specified column effluent fractions. Unless otherwise indicated, all steps of the purification procedure are carried out at 0–5°.

Step 1. Preparation of Cell Extract. The frozen cells (150 g) are suspended in 3 times their volume of 50 mM Tris·HCl buffer, pH 7.5, containing 1 mM 2-mercaptoethanol and are disrupted by treatment for 30 min in a Branson 185 W Sonifier. Unbroken cells and cellular debris are removed by centrifugation at 37,000 g for 30 min, and the yellow, opalescent, supernatant fluid is collected by decantation. Smaller particulate material is removed by further centrifugation at 105,000 g for 60 min.

Step 2. Acid Precipitation. The 105,000 g supernatant fluid (500 ml) is dialyzed overnight against 14 liters of 50 mM sodium acetate buffer, pH 4.5, during which time a precipitate forms. The acid precipitate is collected by centrifugation at 15,000 g for 45 min and resuspended in about 500 ml of 50 mM Tris-acetate buffer, pH 6.0. Sodium hydroxide (0.1 N) is slowly added to the turbid suspension with stirring to adjust the pH to 6.0, and the Tris-acetate buffer, pH 6.0, is added to bring the total volume to about 1 liter. At this point the turbid suspension is clarified and MgCl$_2$ is then added to give a final concentration of 5.0 mM.

Step 3. Cellulose Phosphate Column Chromatography. Cellulose phosphate (Sigma, coarse mesh) is treated prior to use by washing successively with 0.25 N NaOH, distilled water, 0.2 N HCl, distilled water, and then equilibrated with 10 mM Tris-acetate buffer, pH 6.0, containing 5 mM MgCl$_2$. A cellulose phosphate column (6 × 36 cm) is prepared, and the enzyme solution from step 2 is applied. The column is then washed with 10 mM Tris-acetate buffer, pH 6.0, containing 5 mM MgCl$_2$ until the absorbancy of the eluate at 280 nm is less than 0.1. A linear gradient extending between zero and 1.0 M KCl in a total volume of 2 liters of 10 mM Tris-acetate buffer, pH 6.0, containing 5 mM MgCl$_2$

is next applied to the column with a peristaltic action pump, and 20-ml effluent fractions are collected at a rate of one fraction every 3–4 min. The enzyme is eluted at approximately 0.55 M KCl, and the appropriate fractions (numbers 45–59), having a specific activity greater than 2.0 are pooled. The enzyme solution is dialyzed overnight against several changes of 50 mM Tris·HCl buffer, pH 7.5, containing 1 mM 2-mercapto-ethanol and 5 mM MgCl$_2$. Prompt dialysis after elution from the cellulose phosphate column is recommended, since considerable inactivation of the enzyme occurs in the presence of the high salt concentration and because of the absence of a sulfhydryl reducing agent. Magnesium chloride is included in all buffers for this chromatographic step because it appears to facilitate binding of the enzyme to the cellulose phosphate column. It is important to note that the presence of certain metal complexing agents (such as 2-mercaptoethanol or EDTA) in these buffers prevents binding of the enzyme to the column.

Step 4. DEAE-Cellulose Column Chromatography. DEAE-cellulose (Whatman, DE-52, microgranular, preswollen) is used directly after re-moval of the "fines" and subsequent equilibration with the buffer used in the chromatography. The dialyzed enzyme from step 3 is applied to a DEAE-cellulose column (2.5 $\times$ 30 cm) and is eluted with a linear gra-dient extending between zero and 0.5 M KCl in 1 liter of 50 mM Tris·HCl buffer, pH 7.5, containing 1 mM 2-mercaptoethanol. Fractions of 6.0 ml each are collected at a rate of one fraction every 3 min and the enzyme is eluted at approximately 0.22 M KCl. The peak fractions that have a specific activity greater than 4.0 (numbers 80–88 in a typical run) are pooled and concentrated to about 20 ml by ultrafiltration using an Amicon ultrafiltration cell (Model 401) equipped with a UM 10 membrane.

Step 5. Preparative Polyacrylamide Gel Electrophoresis. The appara-tus used is that manufactured by Buchler Instruments (Poly-Prep), which incorporates the original design of Jovin et al.[3] The procedures for preparing the anionic gel column as well as the upper and lower buffers are as described in the Buchler preparative polyacrylamide gel electro-phoresis instruction manual (pH 9.3 system). The resolving gel portion of the column is 6.0 cm in length and the concentrating gel portion is 1.0 cm long. A current of 37 mA is applied to the column for 1 hr prior to the application of the step 4 fraction. As much as 125 mg of protein from the step 4 fraction (about 19 ml) can be layered onto the column. A current of 37 mA (approximately 160–180 V) is again applied to the column, and the elution buffer (0.1 M Tris·HCl buffer, pH 8.1, containing

[3] T. Jovin, A. Chrambach, and M. A. Naughton, *Anal. Biochem.* **9**, 351 (1964).

10 mM dithiothreitol) is pumped through the elution chamber at a constant rate of 0.8 ml/min. Fractions of 3.7 ml each are collected, and enzyme activity begins to appear in the effluent fractions after about 13 hr. In order to determine which fractions are to be pooled, samples from the peak fractions (numbers 145–165) that have a specific activity of about 14 are placed on each of two polyacrylamide gel columns (0.6 × 10.0 cm). The columns are prepared and electrophoresis is conducted as described in the Buchler Polyanalyst instruction manual (pH 9.3 system). After electrophoresis, one column is stained for protein using Coomassie blue,[4] and the other is stained for enzyme activity by incubation in the following solution: disodium-6-phosphogluconate, 1.03 mM; NADP, 0.24 mM; phenazine methosulfate, 0.078 mM; nitro blue tetrazolium, 0.39 mM.[5] Those fractions containing a single protein and corresponding activity band are pooled, concentrated in an Amicon ultrafiltration cell as described previously, and then dialyzed overnight against 50 mM Tris·HCl buffer, pH 7.5, containing 10 mM dithiothreitol. The total recovery of enzyme activity from the preparative polyacrylamide gel column is always greater than 96%.

A summary of the results from a typical purification procedure are given in the table. This procedure gives the enzyme in a yield of greater than 12% with an overall purification of about 180-fold. The enzyme is homogeneous as demonstrated by polyacrylamide gel electrophoresis,[5] immunoelectrophoresis, and sedimentation equilibrium.[6]

Properties

Stability and pH Optimum. The purified enzyme (step 5 fraction) is stable to storage at 5° in 50 mM Tris·HCl buffer, pH 7.5, containing 10 mM dithiothreitol and 20% glycerol.[7] No significant decrease in activity occurs over a 2-week period under the above storage conditions. The enzyme is labile to storage in the absence of a sulfhydryl reducing agent and a single freezing and thawing results in as much as a 50% loss of activity. The pH optimum for catalytic activity is between 7.6 and 8.0.

[4] A. Chrambach, R. A. Reisfeld, M. Wyckoff, and J. Zaccari, *Anal. Biochem.* **20**, 150 (1967).

[5] C. L. Wittenberger, M. P. Palumbo, R. B. Bridges, and A. T. Brown, *J. Dent. Res.* **50**, 1094 (1971).

[6] R. B. Bridges and C. L. Wittenberger, *Federation Proc.* **30**, 1059 (1971).

[7] For routine daily studies, a portion of the enzyme is passed through a Sephadex G-50 column previously equilibrated with 50 mM Tris·HCl buffer, pH 7.5, in order to remove the dithiothreitol and glycerol. As will be noted later, it is particularly important to remove the reducing agent before undertaking studies on fructose 1,6-diphosphate inhibition of the enzyme.

PURIFICATION OF D-GLUCONATE-6-PHOSPHATE DEHYDROGENASE

Step	Fraction	Activity (total units)	Protein (total mg)	Specific activity (units/mg protein)	Recovery (%)
1	Crude extract	1115	13,847	0.08	100
2	Acid precipitate	938	10,200	0.09	84.1
3	Cellulose phosphate eluate, pooled	320	148.9	2.15	28.7
4	DEAE-Cellulose eluate, pooled	204	43.6	4.68	18.3
5	Preparative polyacrylamide gel electrophoresis eluate, pooled	136	9.4	14.47	12.2

Kinetic Properties and Coenzyme Specificity. The reaction rate is a hyperbolic function of the coenzyme concentration at saturating levels of substrate and is also a hyperbolic function of the substrate concentration in the presence of saturating levels of the coenzyme. The respective K_m values for NADP and 6-phosphogluconate are 0.015 mM and 0.024 mM. The enzyme is specific for NADP; no activity is observed when NAD is substituted for NADP. It should be noted that high levels of 6-phosphogluconate dehydrogenase activity are observed with NAD in crude extracts of gluconate-grown cells. This activity, however, has been shown to be due to a separate and distinct NAD-specific enzyme, which is induced during growth of the organism on gluconate.[8]

Physical and Chemical Properties. The enzyme has a molecular weight of 108,000, as determined by sedimentation equilibrium, and is composed of two subunits of identical molecular weight, as determined by sodium dodecyl sulfate polyacrylamide gel electrophoresis.[6] Its amino acid composition is: ASP_{85} Thr_{49} Ser_{51} Glu_{137} Pro_{36} Ala_{94} ½ Cys_5 Val_{53} Met_{26} Ile_{66} Leu_{80} Tyr_{43} Phe_{36} His_{11} Lys_{69} Arg_{37} Trp_{17}.[9]

Activators and Inhibitors. Divalent cations, such as Mg^{2+}, Mn^{2+}, or Ca^{2+}, have no effect on catalytic activity at concentrations between 1.0 and 10.0 mM. Concentrations of $MgCl_2$ greater than 50 mM are inhibitory. The enzyme is strongly inhibited by certain heavy metal ions (zinc, cadmium, copper) and by p-chloromercuribenzoate.

Fructose 1,6-diphosphate is a potent and specific physiological inhibitor of the enzyme.[6,10] This inhibition is a complex phenomenon, however, and any one of a number of variations in the assay procedure affect

[8] A. T. Brown and C. L. Wittenberger, *J. Bacteriol.* **109**, 106 (1972).
[9] R. B. Bridges and C. L. Wittenberger, unpublished observation, 1972.
[10] A. T. Brown and C. L. Wittenberger, *J. Bacteriol.* **106**, 456 (1971).

both the kinetics and the degree of inhibition observed at a constant of this negative effector. Maximum inhibition, with a linear initial reaction rate, is observed under the following conditions: add to the assay cuvette (in the order shown) buffer, 6-phosphogluconate, fructose 1,6-diphosphate, and enzyme. Mix the components and incubate at room temperature for 3–5 min. Initiate the reaction by adding NADP. The K_i for fructose 1,6-diphosphate is about 0.053 mM under these conditions.[9] It is important to note that the inhibition is reversed, either partially or completely, by 2-mercaptoethanol,[10] EDTA, and certain other metal complexing agents.[6] Such compounds, therefore, must be carefully excluded from the assay system when studying fructose 1,6-diphosphate inhibition of the enzyme.[7] None of the above compounds have any effect on catalytic activity when included in the standard assay system. It has been found recently that substituting various buffers with metal-binding capacity for Tris·HCl in the assay also markedly decreases the degree of inhibition by fructose 1,6-diphosphate. These buffers include: imidazole, glycylglycine, Bicine, and histidine.[11]

[11] C. L. Wittenberger, unpublished observation, 1973.

[52] 6-Phosphogluconate Dehydrogenase from *Candida utilis*

By Mario Rippa and Marco Signorini

D-Gluconate 6-phosphate + NADP →
$$\text{D-ribulose 5-phosphate} + CO_2 + NADPH + H^+$$

Assay Method

Principle. A spectrophotometric assay based on the reduction of NADP by 6-phosphogluconate is employed. The formation of NADPH is followed by the change in absorbance at 340 nm.

Reagents

Tris·HCl buffer, 50 mM, pH 8.0, containing 1 mM EDTA
NADP, sodium salt, 15 mM
6-Phosphogluconate, sodium salt, 50 mM
Enzyme: dilute enzyme solution in the Tris buffer to obtain 1–0.5 unit of enzyme per milliliter.

Procedure. One milliliter of the reaction mixture contains: 0.940 ml Tris buffer, 20 μl of NADP, and 20 μl of 6-phosphogluconate. The reaction is started by the addition of 20 μl of enzyme and readings at 340 nm are taken at 10-sec intervals for 1 min. The assay is carried out at 20°.

Definition of Enzyme Activity and Specific Activity. A unit of enzyme is defined as that amount which catalyzes the formation of 1 μmole of NADPH per minute under the conditions specified above.

The specific activity is expressed as units of enzyme per milligram of protein. The protein concentration is measured from the absorbancy at 280 nm, based on dry weight determination. A solution containing 1 mg of purified enzyme per milliliter has an absorbancy of 1.270 OD at 280 nm. Before the column chromatography a turbidimetric method[1] for the determination of the protein concentration is used.

Source of Enzyme. Candida utilis dried at low temperature can be purchased from Sigma Chemical Co., St. Louis, Missouri.

Purification Procedure[2]

Crude Extract and Protamine Treatment. All operations are performed at 0–4°, unless otherwise stated. All solutions contain 1 mM EDTA.

The dry *Candida utilis* (200 g) is suspended in 1 liter of glass-distilled water containing 1 mM EDTA and kept with occasional stirring at 37° for 4 hr. The autolyzed suspension is centrifuged for 30 min at 20,000 g in a refrigerated centrifuge. The supernatant (crude extract) is treated with 2 g of protamine sulfate previously dissolved in 100 ml of water. The suspension is adjusted to pH 6.2 with 0.1 M NaOH and centrifuged as above. The supernatant (protamine fraction) is subjected to column chromatography.

Phosphocellulose Column Chromatography. This column chromatography does not require a previous equilibration of the phosphocellulose with buffer. The protamine fraction is applied to a phosphocellulose column (4.5 × 40 cm) which had been previously washed with water. The exchanger is washed overnight with 2 liters of 0.25 M acetate buffer, pH 6.3. This washing removes most of the proteins from the resin. The enzyme is then eluted from the resin with 0.20 M phosphate buffer, pH 6.2, as a sharp peak in the first protein fractions.

The elution is carried out at a flow rate of 1 ml/min and fractions of 5 ml each are collected. The enzymically active fractions are pooled (column eluate).

[1] T. Bücher, *Biochim. Biophys. Acta* **1**, 292 (1947).
[2] M. Rippa, M. Signorini, and S. Pontremoli, *Ital. J. Biochem.* **19**, 361 (1970).

PURIFICATION PROCEDURE FOR 6-PHOSPHOGLUCONATE DEHYDROGENASE

Fraction	Total volume (ml)	Activity (units/ml)	Protein (mg/ml)	Specific activity (units/mg)	Recovery (%)	Purification (fold)
Crude extract	650	5.7	48	0.12	100	1
Protamine fraction	740	5.0	19	0.25	98	2.1
Column eluate	68	47.3	4.8	9.65	85	80
$(NH_4)_2SO_4$ fraction	72	41.6	4.2	10.8	81	90
First crystals	10	130	6.2	20.9	35	173
Second crystals	9	138	4.3	32.1	33	266
Third crystals	8	148	3.5	41.8	32	346

Ammonium Sulfate Precipitation and Crystallization. The column eluate is brought to 50% saturation with solid ammonium sulfate (29.1 g/100 ml of solution), keeping the pH at 6.2. The proteins precipitated are discarded by centrifugation. To the supernatant (ammonium sulfate fraction) a cold saturated ammonium sulfate solution is added dropwise, slowly, under constant stirring, until a slight turbidity appears. The solution is then kept at 4°. The formation of the crystals begins in a few minutes and is complete in 24 hr.

The crystalline suspension is centrifuged at 20,000 g for 10 min. The supernatant, which contains type II 6-phosphogluconate dehydrogenase,[3] is discarded and the precipitate is dissolved in 5 ml of 50 mM phosphate buffer, pH 6.2 (first crystals). The enzyme is crystallized twice as described above and is stored as a crystalline suspension in 60% saturated ammonium sulfate, pH 6.2. The second crystallization and the following ones require only 5–6 hr to be complete. The enzyme crystallizes in long needles.

The purification procedure is summarized in the table.

Properties

Stability and pH Optimum. The enzyme can be stored as a crystalline suspension in ammonium sulfate for several months without any loss of enzyme activity. The pH optimum for the catalytic activity is 8.0.[2]

Affinity Constants. The Michaelis constants measured at pH 8.0 is 54 μM for 6-phosphogluconate and 20 μM for NADP.[2] The order of addition of the substrate and the coenzyme is not compulsory.[4]

[3] M. Rippa, M. Signorini, and S. Pontremoli, *Eur. J. Biochem.* **1**, 170 (1967).
[4] S. Pontremoli, E. Grazi, A. De Flora, and G. Mangiarotti, *Arch. Sci. Biol.* **46**, 83 (1962).

Activators. No metal requirement for the activity is observed. The activity of the enzyme is not enhanced by magnesium ions.

Physical and Chemical Properties. The enzyme has a molecular weight of 100,000 and is composed of two apparently equal subunits.[5]

The crystalline enzyme catalyzes also the oxidation of 2-deoxy-6-phosphogluconate to 3-keto-2-deoxy-6-phosphogluconate,[6] the decarboxylation of this last product,[6] and a tritium exchange reaction between tritiated ribulose 5-phosphate and water.[7] The last two reactions require the presence of NADPH,[6,7] but in these reactions the reduced coenzyme does not play a redox role.[7–9] The presence at the active site of the enzyme of residues of lysine,[10] cysteine,[11,12] tryosine,[13] and histidine[14–16] has been detected.

[5] M. Rippa, M. Signorini, and S. Pontremoli, *Ital. J. Biochem.* **18**, 174 (1969).
[6] M. Rippa, M. Signorini, and F. Dallocchio, *J. Biol. Chem.* **248**, 4920 (1973).
[7] G. E. Lienhard and I. A. Rose, *Biochemistry* **3**, 190 (1964).
[8] M. Rippa, M. Signorini, and F. Dallocchio, *Biochem. Biophys. Res. Commun.* **48**, 764 (1972).
[9] M. Rippa, M. Signorini, and F. Dallocchio, *FEBS Lett.* **36**, 168 (1973).
[10] M. Rippa, L. Spanio, and S. Pontremoli, *Arch. Biochem. Biophys.* **118**, 49 (1967).
[11] E. Grazi, M. Rippa, and S. Pontremoli, *J. Biol. Chem.* **240**, 234 (1965).
[12] M. Rippa, E. Grazi, and S. Pontremoli, *J. Biol. Chem.* **241**, 1632 (1966).
[13] M. Rippa, C. Picco, M. Signorini, and S. Pontremoli, *Arch. Biochem. Biophys.* **147**, 487 (1971).
[14] M. Rippa, M. Signorini, and S. Pontremoli, *Arch. Biochem. Biophys.* **150**, 503 (1972).
[15] M. Rippa and S. Pontremoli, *Arch. Biochem. Biophys.* **133**, 112 (1969).
[16] M. Rippa, C. Picco, and S. Pontremoli, *J. Biol. Chem.* **245**, 4977 (1970).

[53] Glycerol-3-phosphate Dehydrogenase from the Honey Bee

By STEVEN C. FINK and RONALD W. BROSEMER

Dihydroxyacetone phosphate + NADH + H$^+$ = L-α-glycerophosphate + NAD$^+$

Extramitochondrial glycerol-3-phosphate dehydrogenase (EC 1.1.1.8) plays a central role in insect flight metabolism. It is one of the two enzymic members of the glycero-P cycle, which accounts for rapid reoxidation of cytoplasmic DPNH during glycolysis.[1] The activity of this dehydrogenase is very high in the flight muscles of insects, such as honey bees, which utilize carbohydrate as fuel.[2]

[1] R. G. Hansford and B. Sacktor, *in* "Chemical Zoology" (M. Florkin and B. T. Scheer, eds.), Vol. 6, p. 213. Academic Press, New York, 1971.
[2] A. M. T. Beenakkers, *J. Insect Physiol.* **15**, 353 (1969).

Assay Method

Principle. The method is based upon the oxidation of NADH by dihydroxyacetone-P. The reverse reaction can also be used, but it often does not show reproducible kinetics.

Reagents

2-(N-Morpholino)-propane sulfonic acid (MOPS), 75 mM, pH 6.6
NADH, 1.4 mg/ml of 1% (w/v) NaHCO$_3$, made up fresh each day
Dihydroxyacetone-P, 3 mM, prepared from the dimethylketal dimonocyclohexylamine salt according to the directions supplied by the commercial distributors.

Procedure. Mix 2.0 ml of MOPS buffer, 0.2 ml of NADH, and sufficient water to give a final volume of 3.0 ml. After reaching temperature equilibrium at 30°, a suitable aliquot of glycerol-3-P dehydrogenase is added, followed immediately by 0.3 ml of dihydroxyacetone-P.

Definition of Units and Specific Activity. One unit of enzyme activity catalyzes the disappearance of 1 μmole of substrate per minute in the above assay. Specific activity is the number of units of activity per milligram of protein.

In the early stages of purification, protein is determined by the biuret reaction.[3] The protein is first precipitated by addition of trichloroacetic acid to a final concentration of 10% (w/v), washed once with 10% (w/v) trichloroacetic acid, and once with ether. The precipitation of the protein eliminates interference of (NH$_4$)$_2$SO$_4$ in the biuret reaction. Bovine serum albumin is used as protein standard.

After the initial crystallization of the enzyme, concentrations are measured by absorbance at 280 nm using an extinction coefficient[4] of 0.48 ml mg^{-1} cm^{-1}.

Purification Procedure

The preparation of thoraces and of the homogenate is identical to that for the isolation of triose-P dehydrogenase.[5] The 2.2 M (NH$_4$)$_2$SO$_4$ supernatant described below can be used to isolate triose-P dehydrogenase, arginine kinase, and cytochrome c.[4]

Preparation of Thoraces. Honey bees (*Apis mellifera*) are readily available from most beekeepers; the bees are stored frozen. The thoraces

---

[3] E. Layne, see this series Vol. 3 [73].

[4] C. W. Carlson, S. C. Fink, and R. W. Brosemer, *Arch. Biochem. Biophys.* **144**, 107 (1971).

[5] R. W. Brosemer, this volume [60].

(including wings and legs) are removed from the bees, care being taken that the bees and thoraces remain frozen. It is convenient to place the material on dry ice and use tweezers to snap off the heads and abdomens. It requires about 6 man-hours of labor to prepare 100 g of thoraces.

Ammonium Sulfate Fractionation. Unless otherwise noted, all subsequent steps are carried out at 0–2°. The thoraces (100 g) are homogenized 1 min in a Waring Blendor with 300 ml of 0.1 M Tris, 10 mM EDTA, pH 7.5 (pH measured at 23°). The homogenate is centrifuged 30 min at 12,000 g. The pellet is reextracted with 90 ml of the 0.1 M Tris, 10 mM EDTA buffer and again centrifuged. The two extracts are combined, and solid $(NH_4)_2SO_4$ is added over a 30-min period to a final concentration of 1.4 M [19.6 g of $(NH_4)_2SO_4$/100 ml of extract]. The suspension is centrifuged 1 hr at 12,000 g; the pellet is discarded. The supernatant can be filtered through nonwoven plastic cloth, Miracloth (Chicopee Mills, Inc., New York), at this point in order to remove fat and other large particles; the filtration is not nesessary. To the supernatant is added solid $(NH_4)_2SO_4$ over a 30-min period to a final concentration of 2.2 M [14.2 g of $(NH_4)SO_4$ per 100 ml of supernatant]. The suspension is centrifuged 1 hr at 12,000 g.

pH Precipitation. The pellet is suspended in 75 ml of 5 mM K-phosphate, 1 mM EDTA, pH 6.5, and dialyzed overnight against the same buffer. The suspension is centrifuged 60 min at 25,000 g and the pellet extracted with 10 ml of the phosphate-EDTA buffer. Solid $(NH_4)_2SO_4$ is added to the combined supernatants to a final concentration of 1.9 M. After centrifugation at 25,000 g for 60 min, the pellet is suspended in 12 ml of 0.1 M histidine, 0.1 M Tris, 10 mM EDTA, pH 5.8 (measured at 23°). After standing for 4 days, the suspension is centrifuged. The crystalline pellet is extracted by stirring for 30 min at room temperature with 6 ml of 0.1 M Tris, 10 mM EDTA, pH 8.5 (measured at 23°). The suspension is centrifuged and reextracted twice at room temperature with 4 ml of the same buffer. The clear supernatants are combined, cooled

PURIFICATION OF GLYCEROL-3-PHOSPHATE DEHYDROGENASE FROM HONEY BEES

Fraction	Volume (ml)	Protein (mg)	Total units	Specific activity
1. Extract	360	9700	33,000	3.4
2. Dialyzed ammonium sulfate-precipitated fraction	11	1820	19,200	10.5
3. First crystallization	1.5	46	12,900	242
4. Second crystallization	1.5	40	13,400	335
5. Third crystallization	0.9	33	11,200	340

to 0°, and adjusted with 1 N acetic acid to pH 6.5 (measured at 0°). Crystals in the shape of needles form overnight.

Recrystallization. The crystals are dissolved as described above by extraction 3 times at room temperature with 0.1 M Tris, 10 mM EDTA, pH 8.5. The enzyme is then crystallized either by lowering the pH to 6.5 with 1 N acetic acid as described above or by precipitating with $(NH_4)_2SO_4$. In the latter method, solid $(NH_4)_2SO_4$ is added slowly with stirring at 0° until a slight definite cloudiness appears. The solution is centrifuged, and a small quantity of $(NH_4)_2SO_4$ added until the cloudiness is fairly intense. The suspension is stored at 2° and small quantities of solid $(NH_4)_2SO_4$ added over a period of 1–4 days until most of the protein has crystallized.

The ammonium sulfate method is prefered for crystallization if the protein concentration is less than 1–2 mg/ml.

Modification of the Purification Procedure. The dialysis of the ammonium sulfate sediment against 5 mM K-phosphate, 1 mM EDTA is not a necessary step. The ammonium sulfate sediment can be directly suspended in 30 ml of 50 mM histidine, 5 mM EDTA, pH 5.8 (measured at 23°). The pH of the suspension is then adjusted to pH 6.4 (measured at 0°) with 1 N acetic acid or 1 N Tris. Glycerol-3-P dehydrogenase usually crystallizes from the suspension over a 4-day period at 2°; on rare occasions the enzyme does not crystallize at this point. The enzyme isolated by this method must be recrystallized three or four times before reaching maximum specific activity; only two recrystallizations are necessary with the isolation procedure which includes the dialysis step. It is not clear why dialysis against dilute phosphate-EDTA results in a cleaner first crystal crop.

Slight modifications of the isolation procedure sometimes result in crystals with a reddish tinge that cannot be removed by repeated recrystallizations. The red contaminant can be removed by dissolving the crystals in 80 mM Tris, 8 mM EDTA, pH 8.5 (measured at 23°) and applying the sample at 2° to a DEAE-cellulose column equilibrated with the same buffer. Glycerol-3-P dehydrogenase emerges with the void volume while the red color adheres to the DEAE-cellulose.

Properties

Stability. The crystals can be stored without detectable loss in activity at 2° in 0.1 M Tris, 10 mM EDTA, 2 mM dithiothreitol, pH 7.6 to which $(NH_4)_2SO_4$ has been added to a final concentration of 1.7 M. It is best to harvest the crystals by centrifugation every month, discard the supernatant, and add fresh buffer–$(NH_4)_2SO_4$.

The stability of the enzyme crystals at 2° and pH 6.5 has not been systemically checked. There is no loss of activity after a month under these conditions.

The enzyme is stable for 15 min at 21° from pH 4.8 to 9.9. In Tris–histidine–EDTA–bovine albumin buffer, glycerol-3-P dehydrogenase is stable for 5 min at 55° and completely inactivated at 61°.[6]

Purity. The enzyme preparation is homogeneous by the following criteria: (a) purification to a constant specific activity, (b) electrophoresis on cellulose acetate at several pH values,[7] (c) disc gel electrophoresis,[4] (d) gel electrophoresis in the presence of sodium dodecylsulfate,[8] (e) sedimentation in the ultracentrifuge,[7] (f) reaction of rabbit antibodies prepared against the enzyme with honey bee extracts.[9]

Other Insect Species. A modification of the above procedures has been used to purify glycerol-3-P dehydrogenase from three bumble bee species and from yellow jackets.[10,11]

Kinetic Properties. The value of K_m for dihydroxyacetone-P is 0.23 mM in MOPS buffer and 0.33 mM in Tris-histidine.[11] V_m is also higher in Tris-histidine than in MOPS. The K_m in MOPS is independent of temperature in the range of 16–47°. As is typical with many dehydrogenases, higher levels of substrate (in this case dihydroxyacetone-P) result in inhibition.[6] The value of K_m for NADH is less than 10 μM.[6]

The enzyme exhibits a broad pH optimum around pH 6.6.[6]

The relative reaction rates with NAD$^+$, deamino-NAD$^+$, and 3-acetyl-pyridine-NAD$^+$ are 100, 68, and 1.5, respectively, in the direction of pyridine nucleotide reduction.[6]

Inhibition. The enzyme is inhbited over 60% by 0.1 μM p-chloromercuribenzoate and 0.1 mM N-ethylmaleimide but is not inhibited by 1 mM iodoacetate.[6] The bumble bee enzyme is inhibited competitively by phosphate and noncompetitively by L-α-glycero-P.[11]

Molecular Weight. The molecular weight of the honey bee enzyme has been reported to be 76,000 based upon Sephadex chromatography of crude extracts,[12] 73,000 based upon amino acid composition,[11] and 74,000–79,000 based upon SDS-gel electrophoresis.[8]

[6] R. W. Brosemer and R. R. Marquardt, *Biochim. Biophys. Acta* **128**, 464 (1966).
[7] R. R. Marquardt and R. W. Brosemer, *Biochim. Biophys. Acta* **128**, 454 (1966).
[8] Y. Tomimatsu and R. W. Brosemer, *Comp. Biochem. Physiol.* **43B**, 403 (1972).
[9] R. W. Brosemer, D. S. Grosso, G. Estes, and C. W. Carlson, *J. Insect Physiol.* **13**, 1757 (1967).
[10] S. C. Fink, C. W. Carlson, S. Gurusiddaiah, and R. W. Brosemer, *J. Biol. Chem.* **245**, 6525 (1970).
[11] S. C. Fink and R. W. Brosemer, *Arch. Biochem. Biophys.* **158**, 19 (1973).
[12] H. B. White, III, *Arch. Biochem. Biophys.* **147**, 123 (1971).

Recent extensive analysis of the molecular weight by Robert Dyson (personal communication) using the Yphantis sedimentation equilibrium method shows a molecular weight of 79,500 ± 1900. Exhaustive dialysis often results in apparent dissociation into monomers. In high concentrations of guanidine-HCl the enzyme dissociates into units with molecular weight of 39,000. SDS-gel electrophoresis experiments by Dyson also support a dimeric molecular weight of around 79,000.

Other Properties. There is only one isozyme present in adult honey bee flight muscle; six to eight isozymes occur in bumble bee flight muscle.[10,11]

The amino acid composition[11] and peptide map of the honey bee enzyme have been reported. Both subunits have a C-terminal alanine and a blocked N-terminal residue.[13] Glycerol-3-P dehydrogenase binds 2 moles of NADH per mole of enzyme.[13] The optical rotatory dispersion spectrum for the protein suggests an alpha helical content of about 30%.[13]

The honey bee enzyme binds to honey bee actin, as do some of the bumble bee isozymes.[11] Glycerol-3-P dehydrogenase was localized in honey bee flight muscle with immunofluorescent techniques; it is located at or near the Z-band cross striations.[14]

[13] R. W. Brosemer and R. W. Kuhn, *Biochemistry* **8**, 2095 (1969).
[14] R. W. Brosemer, *J. Histochem. Cytochem.* **20**, 266 (1972).

[54] Glycerol Phosphate Dehydrogenase of Chicken Breast Muscle

By Harold B. White, III

Glycerol 3-phosphate + NAD^+ ⇄ dihydroxyacetone phosphate + $NADH$ + H^+

Assay Method

Principle. The assay is based on the spectrophotometric determination of the dihydroxyacetone phosphate-dependent oxidation of NADH.

Reagents

Dihydroxyacetone phosphate (DHAP), 30 mM. A 1:1 mixture of DHAP and glyceraldehyde 3-phosphate can be prepared enzymically from fructose 1,6-diphosphate,[1] or pure DHAP can be gener-

[1] G. Beisenherz, T. Bücher, and K.-H. Garbade, this series, Vol. I [58].

ated chemically by the hydrolysis of the commercially available dimethyl ketal.[2]

NADH, 10 mM

Tris·HCl, 50 mM, pH 7.85 (25°)

Procedure. To a 3-ml quartz cuvette (1-cm light path), add 2.90 ml of Tris buffer, 50 µl of DHAP, and 30 µl of NADH. Initiate the reaction with 10–50 µl of glycerol phosphate dehydrogenase solution (≤0.1 unit of activity), and record the decrease in absorbance at 340 nm for 1 min. This assay is applicable at all stages of purification.

Definition of Unit and Specific Activity. One unit of activity is defined as the amount of enzyme which catalyzes the disappearance of 1 µmole of NADH per minute. One unit is equivalent to an absorbance change of 2.08 per minute in the standard assay. Specific activity is expressed as units per milligram of protein.

Purification Procedure

General. All steps in this purification except the heat step are carried out at 0–5°. pH measurements are made at 0–5°, since there is a marked temperature dependence for the pK of Tris.[3] Dialysis tubing is boiled at least twice in 50 mM EDTA (pH 7.4) before use. All buffers and ammonium sulfate solutions contain 1 mM 2-mercaptoethanol and 1 mM EDTA.

Extraction. Approximately 6 kg of chicken breast muscle, obtained from 30 lb of chicken breasts, are passed through an electric meat grinder and then extracted for 1 hr in 3 liters of 1 mM EDTA and 1 mM 2-mercaptoethanol (pH 7.4) per kilogram of muscle. Large tissue fragments are removed by straining the solution through two layers of cheese cloth. The remaining cellular debris is removed by centrifugation at 13,000 *g* for 20 min. The supernatant is poured through a funnel containing a glass wool plug which removes much of the fat floating on the surface.

Ammonium Sulfate Fractionation. To the above crude extract 431 g of solid ammonium sulfate per liter (65% saturation) is added. No adjustment is made for pH which drops below 6.0. The protein precipitated by 65% ammonium sulfate is collected by centrifugation at 13,000 *g* for 15 min. The supernatant, which contains less than 10% of the glycerol phosphate dehydrogenase activity, can be discarded. (Neutralization of this supernatant fraction yields a gelatinous yellow precipitate of glycer-

---

[2] C. E. Ballou, *Biochem. Prep.* **7**, 45 (1960).

[3] Sigma, Tentative Technical Bulletin No. 106 B, revised August 1967.

aldehyde-3-phosphate dehydrogenase.) The 65% precipitate is suspended in a minimal amount of 5 mM Tris·HCl, 1 mM EDTA, 1 mM 2-mercaptoethanol at pH 7.6. Additional buffer is added until the specific gravity drops to 1.135 (45% saturation). Undissolved and denatured proteins are removed by centrifugation at 13,000 g for 10 min. The resulting supernatant is dialyzed overnight against 10 volumes of 54% (343 g/liter) ammonium sulfate (unneutralized). Centrifugation at 13,000 g for 10 min removes much of the precipitated protein; however, a copious fine precipitate of crystalline aldolase remains suspended and can best be removed by centrifugation at 39,000 g for an hour or more. The aldolase obtained here is recrystallized several times and can be used for the preparation of DHAP.[1]

Heat Step. The supernatant obtained after the removal of aldolase is heated at 68° within 10 min with constant stirring. The solution is then cooled rapidly in an ice bath and the denatured proteins are removed by a brief centrifugation at 13,000 g. A 10-fold purification can be achieved at this step. The addition of NADH does not enhance the thermal stability of this enzyme in ammonium sulfate solutions, as it does in solutions of low ionic strength.[4]

DEAE-Cellulose Column Chromatography. The supernatant protein from the previous step is precipitated by the addition of 60 g of solid ammonium sulfate per liter of solution. The precipitate is collected by a 10-min centrifugation at 13,000 g and then dissolved in a minimum amount of 5 mM Tris·HCl buffer at pH 7.6. The dissolved protein is dialyzed twice against 40 volumes of the above buffer which contains 1 mM EDTA and 1 mM 2-mercaptoethanol. The dialyzed solution, containing 40–60% of the original enzyme activity, is applied to a 2.6 liter (4.5 × 142 cm) DEAE-cellulose column equilibrated with the above buffer. The amount of protein added should not exceed 3 g per liter of packed column volume. A 4-liter linear gradient of NaCl (0 to 0.3 M) in pH 7.6 Tris·HCl is begun shortly before the breakthrough volume has passed. Glycerol phosphate dehydrogenase will elute at the beginning of the gradient in 5–10 mM NaCl. The pH of the step is quite critical; if lower than about pH 7.2, the enzyme may not be retained by the column and if above pH 8.0, less than 10% of the activity is recovered.

Phosphocellulose Column Chromatography. The enzyme eluted from the preceding column is of greater than 90% purity. The remaining contaminants are removed and the glycerol phosphate dehydrogenase is concentrated by adding the pooled fractions from the DEAE-cellulose column directly to a 400 ml (2.8 × 75 cm) phosphocellulose column equili-

[4] W. Rouslin, Ph.D. Thesis, University of Connecticut, 1968.

PURIFICATION PROCEDURE FOR GLYCEROL
PHOSPHATE DEHYDROGENASE

Fraction	Total units	Specific activity[a]
Crude extract	37.7×10^4	2.95
65% $(NH_4)_2SO_4$ precipitate	17.6×10^4	2.43
50% $(NH_4)_2SO_4$ supernatant	14.5×10^4	4.67
68° Supernatant	13.5×10^4	20.0
DEAE-cellulose eluate	13.6×10^4	289
Phosphocellulose eluate	7.75×10^4	310

[a] Protein determinations were made by the method of O. Warburg and W. Christian, *Biochem. Z.* **310**, 384 (1941); see Vol. 3 [73]. This method underestimates the amount of protein in purified preparations by about 50%.

brated with 5 mM sodium phosphate at pH 6.2. Immediately after the sample has been applied, a 2-liter linear NaCl gradient (0 to 1.0 M) is started. The enzyme eluted appears to be homogeneous.

Crystallization. Needlelike crystals of chicken breast muscle glycerol phosphate dehydrogenase form upon dialysis of the active fractions against the above sodium phosphate buffer saturated with ammonium sulfate.

Summary of the Purification. The purification of chicken breast muscle glycerol phosphate dehydrogenase is summarized in the table. This procedure has been repeated successfully six times. The enzyme exhibits a single band on gel electrophoresis, migrates as a single peak in the analytical centrifuge, and behaves as a homogeneous antigen in rabbits.[5]

Properties

Stability. The crystalline enzyme can be kept in ammonium sulfate solution for several months without appreciable loss of activity. On prolonged storage an insoluble amorphous precipitate forms which lacks enzymic activity.

pH Optimum. The pH-rate profile for the reduction of DHAP shows a rather sharp optimum between pH 7.5 and 8.0.

Specificity. There is almost complete discrimination between NADH and NADPH. Reaction with NADPH can be shown only at high enzyme concentration and lower pH values where the additional phosphate group becomes protonated.

Kinetic Properties. Compared with other purified glycerol phosphate dehydrogenases, the enzyme from chicken muscle is unusual with respect

[5] H. B. White, III, and N. O. Kaplan, *J. Biol. Chem.* **244**, 6031 (1969).

to its high K_m values for glycerol 3-phosphate (≥ 2 mM) and NAD$^+$ (0.5 mM).[5] The K_m values for DHAP and NADH are 230 μM and 6 μM, respectively. These distinctive kinetic properties have been linked with the enzyme's role in a tissue which relies heavily on glycolytic energy.[6] A purified isozyme from chicken liver has completely different kinetic properties.[5]

Prosthetic Groups. Adenosine diphosphoribose, which has been found associated with the rabbit muscle enzyme,[7] is not isolated with the chicken muscle enzyme.

Extinction Coefficient. The molar extinction coefficient for chicken muscle glycerol phosphate dehydrogenase at 280 nm is 3.1×10^4 M^{-1} cm^{-1}.

Molecular Structure. Glycerol phosphate dehydrogenase from chicken muscle is a dimeric protein composed of identical subunits. The molecular weight of the native enzyme is estimated to be 68,000;[8] however, this value may be slightly low.[9]

[6] H. B. White, III, and N. O. Kaplan, *J. Mol. Evol.* **1**, 158 (1972).
[7] H. Ankel, T. Bücher, and R. Czok, *Biochem. Z.* **332**, 315 (1960).
[8] H. B. White, III, *Arch. Biochem. Biophys.* **147**, 123 (1971).
[9] R. W. Brosemer and R. W. Kuhn, *Biochemistry* **8**, 2095 (1969).

[55] L-Glycerol-3-phosphate Dehydrogenase from *Escherichia coli*

By DAVID J. SPECTOR and LEWIS I. PIZER

$$
\begin{array}{ccc}
\text{CH}_2\text{OH} & & \text{CH}_2\text{OH} \\
| & & | \\
\text{C=O} & + \text{TPNH} \rightleftharpoons \text{HOCH} & + \text{TPN}^+ \\
| & & | \\
\text{H}_2\text{COPO}_3 & & \text{H}_2\text{COPO}_3
\end{array}
$$

L-Glycerol-3-phosphate dehydrogenase catalyzes the reduction of dihydroxyacetone phosphate (DHAP) to L-glycerol-3-phosphate (G3P). This reaction is the first step in the biosynthesis of phospholipids from glycolytic intermediates and serves to supply the glycerol backbone. The enzyme is subject to product inhibition by L-glycerol-3-phosphate.[1]

Assay Method

The enzyme is assayed routinely by following the disappearance of TPNH spectrophotometrically or the oxidation of G3P by measuring TPNH production fluorometrically.

[1] M. Kito and L. I. Pizer, *J. Biol. Chem.* **244**, 3316 (1969).

Reagents

Triethanolamine·HCl buffer, 0.5 M pII 7.5, (TEA-HCl)

Dihydroxyacetone phosphate, 46 mM (DHAP); the dimethyl ketal is hydrolyzed by incubation in acidic solution as described by Ballou and Fisher[2]

TPNH, 1 mM

Dithiothreitol, 0.1 M (DTT)

Hydrazine, 0.1 M, pH 7.5

TPN, 1 mM

Glycerol phosphate, 20 mM, sodium salt, dissolved in TEA·HCl

Procedures. To assay the reduction of DHAP, the assay mix contains 0.1 ml TEA·HCl buffer, 0.1 ml TPNH, 0.1 ml of DTT and enzyme, and is made up to 0.98 ml total volume with water. The presence of TPNH oxidase in extracts will give a substrate-independent absorbance change that must be subtracted from the total activity observed after addition of substrate. The reaction is initiated by the addition of 20 μl of DHAP, and the absorbancy change occurring at 340 mM is recorded continuously. Because of product inhibition, the initial slope of the curve should be used to calculate enzyme activity.

To assay G3P oxidation, the assay mix contains 0.1 ml of TEA·HCl buffer, 0.1 ml of DTT, 0.1 ml of hydrazine, 0.1 ml of TPN, and enzyme in a total volume of 0.98 ml. The reaction is initiated with 20 μl of G3P. The cuvette is activated at 340 mM, and the fluorescence change is measured at 470 mM. An Aminco-Bowman fluorometer calibrated with standard solutions of TPNH is suitable for these measurements.

Definition of Enzyme Activity. One unit of enzyme activity catalyzes the disappearance or appearance of 1 nmole of TPNH per minute at 25°. Specific activity is expressed as units per milligram of protein.

Purification Procedures

Two procedures are presented: the first has been used to study the properties of the enzyme. The second, an abbreviated procedure, is suitable for studies of the enzyme activity in different bacterial strains by removing the interfering TPNH oxidase background.

Enzyme Purification Procedure

Disruption of Cells. Frozen *Escherichia coli* B cells harvested in the late log phase of growth were purchased from the Grain Processing Com-

[2] C. E. Ballou and H. O. L. Fisher, *Amer. Chem. Soc.* **78**, 1659 (1956).

pany (Muscatine, Iowa). One pound, wet weight, of cells was suspended in 2 liters of 10 mM triethanolamine·HCl buffer, pH 7.5, containing 2 mM dithiothreitol (buffer A), and the cells were disrupted by two passages through a Manton-Gaulin homogenizer. Cell debris was removed by centrifugation for 15 min at 8000 g (all centrifugations were performed in a Sorvall RC2 centrifuge at 0–2°).

Streptomycin and Ammonium Sulfate Precipitation. The nucleic acids were precipitated from the supernatant fluid by the addition of 0.3 volume of 5% (w/v) streptomycin sulfate. After removal of the precipitated material by centrifugation, solid ammonium sulfate was added to the supernatant fluid to give a final concentration of 75% of saturation, 0.57 g/ml.

Ammonium Sulfate Sulfate Fractionation. The precipitated material was removed from the ammonium sulfate solution by a 15-min centrifugation at 8000 g. The precipitate was dissolved in 1 liter of buffer A to give about 1100 ml of solution with an absorbance at 280 nm of approximately 20. The precipitate was assumed to be 0.57 g of ammonium sulfate per milliliter. From the volume increase, the volume of the original pellet and, therefore, the amount of ammonium sulfate in the total solution, was calculated and additional solid ammonium sulfate was added to give a final concentration of 40% of saturation, or 0.30 g/ml.

DEAE-Sephadex Chromatography. The protein precipitated by 40% ammonium sulfate was collected by centrifugation and dissolved in buffer A. The resulting solution was dialyzed against 2 liters of buffer A containing 20 mM ammonium sulfate, changed once, and then applied to a column of DEAE-Sephadex A-50, 4.5 × 36 cm, containing 24 g of ion exchange Sephadex previously equilibrated with buffer A containing 20 mM ammonium sulfate. To the column were applied about 7 g of protein; after washing with 2 column volumes of buffer, the chromatogram was developed with a 2-liter linear gradient of ammonium sulfate ranging from 20 mM to 0.2 M. The enzyme eluted at 85 mM ammonium sulfate. The fractions with high specific activity were pooled, and the protein present was precipitated with 60% ammonium sulfate, 0.46 g/ml.

The precipitate was collected by centrifugation and redissolved in buffer A. This solution was dialyzed against 1 liter of buffer A supplemented with 20 mM ammonium sulfate and 5 mM DL-G3P, changed once. The protein solution was placed on a column that contained 5 g of DEAE-Sephadex previously equilibrated with the same buffer. The column was developed with an 800-ml linear gradient of ammonium sulfate. The eluting buffer was buffer A containing 5 mM DL-G3P and ammonium sulfate from 20 mM to 0.2 M. In the presence of G3P, the enzyme eluted from the column at a higher ammonium sulfate concentration (0.12 M)

than in the initial chromatogram. The fractions with high specific enzyme activity were pooled, and the protein was precipitated by 60% ammonium sulfate.

Sephadex G-150 Chromatography. The precipitate obtained after the second DEAE-Sephadex chromatography was dissolved in 50 mM potassium phosphate buffer, pH 7.1, containing 2 mM dithiothreitol. This solution was chromatographed on a 400-ml column of Sephadex G-150 previously equilibrated with the same buffer. The enzyme activity was relatively slow in eluting from the column when compared with contaminating proteins. The fractions possessing activity were pooled, and the protein was precipitated with 60% ammonium sulfate.

Third DEAE-Sephadex Chromatography. The ammonium sulfate precipitate obtained above was dissolved in Buffer A containing 20 mM ammonium sulfate. This solution was passed through a 58-ml Sephadex G-25 column equilibrated with the same buffer and then added to a column containing 1 g of DEAE-Sephadex. The enzyme was eluted from the column with a 600-ml linear gradient of ammonium sulfate. The salt concentration ranged from 0.02 M to 0.2 M and was essentially the same as used for the development of the first column. The protein in the most active fractions was precipitated with 75% ammonium sulfate. A summary of the purification procedure is shown in the table.

Abbreviated Purification

Growth of Bacteria. Fifteen liters of *Escherichia coli* B were grown on 2 mg/ml glucose and M9 salts[3] to late log phase and harvested by centrifugation. Cells were suspended in 15 ml of buffer A per liter of culture

PURIFICATION PROCEDURE FOR L-GLYCEROL-3-PHOSPHATE
DEHYDROGENASE FROM *Escherichia coli*

Step and fraction	Total enzyme activity	Protein (mg)	Specific activity
1. Crude extract	275,000	55,000	5.0
2. Streptomycin and 75% (NH$_4$)$_2$SO$_4$	380,000	25,000	15.2
3. 40% (NH$_4$)$_2$SO$_4$	335,000	6,900	48.5
4. First DEAE-Sephadex	250,000	870	290
5. Second DEAE-Sephadex	144,000	230	630
6. Sephadex G-150	66,000	47	1,400
7. Third DEAE-Sephadex	24,000	5	4,800

[3] L. I. Pizer, and M. L. Potochny, *J. Bacteriol.* **88**, 611 (1934).

and disrupted by sonic treatment. All debris was removed by centrifugation for 15 min at 8000 g.

Streptomycin Sulfate Precipitation. Nucleic acids were precipitated by the addition of 0.33 volume 5% streptomycin sulfate solution. Addition was dropwise and the suspension was allowed to stand for 45 min with stirring. The precipitate was removed by centrifugation at 8000 g for 15 min.

Ammonium Sulfate Fractionation. The supernatant of 293 ml was brought to 45% ammonium sulfate by the addition of 238 ml of ammonium sulfate solution saturated at 4° and after stirring for 30 min, the precipitate was discarded. The supernatant was then brought to 55% ammonium sulfate with 117 ml of saturated solution; the precipitate was collected and resuspended in 2 ml of buffer A per liter of culture.

DEAE-Sephadex Chromatography. The protein was dialyzed against buffer A containing 20 mM ammonium sulfate and applied to a 1.5 mm $\times$ 30 cm column of 20 ml of DEAE-Sephadex equilibrated with buffer A containing 20 mM ammonium sulfate. The column was washed with 2 column volumes of buffer and developed with a 600-ml linear gradient of ammonium sulfate of 20 mM to 0.2 M. Fractions containing high specific activity eluted at 85 mM ammonium sulfate. The pooled fractions had a specific activity of approximately 90 units/mg, were free of TPNH-DPNH oxidase activity, and were suitable for kinetic studies. The protein was precipitated with 60% ammonium sulfate and stored frozen at −20°.

Properties

Purity of the Enzyme. The enzyme was purified approximately 1000-fold; however, a homogeneous preparation was not obtained, as zonal electrophoresis indicated at 3 bands. The purification procedure was found to have copurified triose-phosphate isomerase.

Molecular Properties. A modification of the zonal centrifugation technique of Rosenbloom *et al.*[4] gave an s_{20} of 5.2 S for the fastest sedimenting enzymically active species. This sedimentation constant suggests a molecular weight of 65,000–80,000, and this value was supported by Sephadex chromatography I.

Kinetic Properties. The enzyme is optimally active at pH 7.5. The inhibition by G3P was most pronounced at or below this pH, and dropped off rapidly at pH values higher than 7.5. The kinetic constants were: K_m(DHAP) = 170 μM, K_m (G3P) = 210 μM, K_m (TPNH) = 10 μM, K_m (DPNH) = 10 μM. TPNH gave the highest activity, and deamino-

[4] J. Rosenbloom, E. Sugimoto, and L. I. Pizer, *J. Biol. Chem.* **243**, 2099 (1968).

DPNH was equally effective. DPHN had 50% of maximum activity, and 3-acetylpyridine DPNH showed no activity.

Inhibition by Glycerol-P. A test of the effect of G3P on enzyme activity showed that this compound inhibited DHAP reduction. When the extent of inhibition occurring at different G3P concentration was determined at 25°, it was found that 35 μM inhibited 50% and the inhibition curve was sigmoid. At 37°, the concentration of G3P needed to give 50% inhibition, or I_{50}, was 50 μM and the maximum inhibition obtained was 60% (D. Spector, unpublished data).

Inhibition and Substrate Analogs. Several substrate analogs failed to inhibit DHAP reduction,[1] but of the inhibitor analogs tested, only dihydroxybutyl 1-phosphonate, an isostere of G3P, inhibited 65% at 900 μM.

Stability. The enzyme was unstable in dilute salt solution (buffer A) and was inhibited in solutions of high ionic strength.[1] Ammonium sulfate precipitates were routinely resuspended in buffer A and desalted with Sephadex G-25 columns before assay. The enzyme was found to be rapidly inactivated at 52°, a temperature which also inactivates TPNH oxidase.

Distribution. The enzyme was not detected in two *E. coli* strains that have a nutritional requirement for glycerol or G3P (strain GR-1[5] and strain E5 isolated by L. Mindich). Enzymes with similar properties have been studied from mammalian and avian sources. They have been purified from muscle and liver tissue of the rat, rabbit, and chicken.[6-10] The rabbit liver enzyme appears to be an allosteric protein.[8]

[5] C. C. Hsu and C. F. Fox, *J. Bacteriol.* **103**, 410 (1970).
[6] T. P. Fondy, J. Solomon, and C. R. Ross, *Arch. Biochem. Biophys.* **145**, 604 (1971).
[7] H. B. White, III, and N. O. Kaplan, *J. Biol. Chem.* **244**, 6031 (1969).
[8] Y. P. Lee and J. E. Crane, *J. Biol. Chem.* **246**, 7616 (1971).
[9] T. P. Fondy, C. R. Ross, and S. J. Sollohub, *J. Biol. Chem.* **244**, 1631 (1969).
[10] H. L. Young and N. Pace, *Arch. Biochem. Biophys.* **75**, 125 (1958).

[56] L-3-Glycerophosphate Dehydrogenase from Pig Brain Mitochondria

By A. P. DAWSON and C. J. R. THORNE

L-3-Glycerophosphate + acceptor $\rightleftharpoons$
dihydroxyacetone phosphate + reduced acceptor

Assay Method

General Considerations. The enzyme reacts with a wide variety of electron acceptors. 2,6-Dichlorophenolindophenol (DCIP)[1,2] phenazine methosulfate,[3] ferricyanide, and long- and short-chain ubiquinone[2] derivatives have all been used. Of these, DCIP is the most convenient, although it does have the disadvantage that the velocity is rather dependent on the dye concentration. The combined use of phenazine methosulfate, with DCIP as terminal acceptor, has been advocated to overcome this problem[4] but, particularly for kinetic measurements, the competing direct reduction of DCIP is a complicating feature. In addition, the use of a direct assay is to be preferred to a coupled assay.

Reagents

Potassium phosphate, 0.1 M, pH 7.6
DCIP, 0.475 mM
DL-3-Glycerophosphate, 0.5 M

Procedure. Activity is measured spectrophotometrically at 600 nm in the following reaction medium at 38°; 1.0 ml phosphate, 0.2 ml DCIP, 0.2 ml DL-3-glycerophosphate, water, and finally enzyme to a total volume of 2.0 ml. The decrease in absorbance is measured in a recording spectrophotometer. The initial reaction velocity is proportional to enzyme concentration up to an absorbance change of at least 0.2/min and the change in absorbance is linear with time for the first minute or more throughout this range. The concentration of DCIP is arranged so that the initial absorbance is 1.0. The extinction of 1.0 mM DCIP at pH 7.6 is 21. In our hands the apparent K_m for DCIP does not change during the purification procedure. However, it has been reported[4] that, under somewhat different conditions, a change in K_m for DCIP can occur on solubilization, in which case extrapolation to infinite dye concentration would be necessary. There are no side reactions when the assay is applied to acetone powder suspensions or to solubilized preparations.

Definition of Unit and Specific Activity. One unit is defined as the amount of enzyme that catalyzes an initial rate of reduction of DCIP of 1.0 μmole per minute under the above assay conditions. Specific activity is expressed as units of enzyme per milligram of protein, measured by the microbiuret method, using bovine serum albumin as standard.

Purification Procedure

As with all respiratory chain-linked enzymes which are tightly bound to the mitochondrial inner membrane, the central problem of the purifica-

---

[1] K. H. Ling, S. H. Wu, S. M. Ting, and T. C. Tung, *Proc. Int. Symp. Enzyme Chem. 1957*, p. 260.
[2] A. P Dawson and C. J. R. Thorne, *Biochem. J.* **111**, 27 (1969).
[3] R. L. Ringler and T. P. Singer, *J. Biol. Chem.* **234**, 2211 (1959).
[4] J. I. Salach and A. J. Bednarz, *Arch. Biochem. Biophys.* **157**, 133 (1973).

tion procedure is to solubilize the enzyme without greatly modifying its properties. It is possible to extract the enzyme with deoxycholate[1] or with phospholipase A,[5] but the method which seems to give least modification of activity and greatest ease of purification uses Triton X-100 as the solubilizing agent.[2] The degree of purification achieved depends to some extent on the specific activity of the mitochondrial acetone powder, so that contamination of the mitochondrial fraction with synaptosomes should be avoided as far as possible.

All steps are carried out at 0–4°, unless otherwise stated.

Preparation of Mitochondrial Fraction. This method is an adaptation of that described by Brody and Bain.[6] Pig brains (1 kg) are obtained within 1.5 hr of the death of the animal and packed in ice. The cerebral hemispheres and cerebellum are collected by removal from the basal structures. As much white matter as possible is scraped away and discarded, then the remaining tissue is homogenized for 1 min at medium speed in a Waring Blendor with 9 liters of 0.25 M sucrose, previously adjusted to pH 7.6 with 1.0 M K_2HPO_4. The homogenate is readjusted to pH 7.6 with more 1.0 M K_2HPO_4 and centrifuged at 700 g for 20 min. The supernatant is centrifuged in a continuous-flow rotor at 18,000 rpm with a flow rate of 35 ml/min (M.S.E. 18 centrifuge). This approximates to 20,000 g for 20 min in discontinuous centrifugation. The pellet consists of a white fluffy layer and a darker heavy layer. As much as possible of the white upper layer of the pellet should be removed and discarded, and the dark mitochondrial layer is resuspended in 2.5 liters of 0.25 M sucrose, pH 7.6. This is then centrifuged at 20,000 g for 20 min; the pellet is collected and finally suspended in 0.25 M sucrose, pH 7.6, to give a final volume of about 100 ml.

Acetone-Dried Mitochondria. The mitochondrial suspension is added drop by drop to 3 5 liters of dry acetone at −15° with vigorous stirring, which is continued for 5 min after all the suspension has been added. The acetone-dried powder is filtered off under suction through two thicknesses of Whatman No. 1 filter paper, and the resulting moist cake is resuspended in 2.5 liters of dry acetone at −15°. After stirring for 10 min the suspension is filtered as before and the process is repeated with a further 2 5 liters of acetone. The precipitate is sucked dry on the filter funnel, dried for 1 hr at room temperature in a stream of air, and finally left overnight in a vacuum desiccator containing paraffin wax chippings and solid KOH. The yield is approximately 10 g.

Phosphate Extraction. This step is in common with the procedure of Ringler.[5] Mitochondrial acetone-dried powder, 1.8 g, is suspended in 60

[5] R. L. Ringler, *J. Biol. Chem.* **236**, 1192 (1961).
[6] T. M. Brody and J. A. Bain, *J. Biol. Chem.* **195**, 685 (1952).

ml of 30 mM KH$_2$PO$_4$–NaOH buffer, pH 7.6, with a nylon in glass homogenizer. After stirring for 20 min, the suspension is centrifuged at 38,000 g for 15 min, and the supernatant is discarded. The pellets are extracted with 60 ml of the same buffer and centrifuged as before.

Solubilization with Triton X-100. The pellet from the preceding step is homogenized in 57 ml of 30 mM KH$_2$PO$_4$–NaOH buffer, pH 7.6. Then 2.7 ml of 6% (w/v in water) Triton X-100 is added, and, after stirring for 20 min, the suspension is centrifuged at 105,000 g_{av} for 45 min. The pellet is discarded.

Treatment with Hydroxyapatite. Bio-Gel HTP (Bio-Rad Laboratories, Richmond, California), 6 g dry weight, is washed with 100 ml of water and sedimented by centrifugation. The washing water is decanted off, and the clear supernatant from the previous step is poured onto the pellet of Bio-Gel HTP. The mixture is stirred for 1 hr, then centrifuged at 2000 g for 10 min, and the pellet is discarded.

Concentration. The enzyme activity in the supernatant from the previous step is rather dilute for most purposes. The simplest method of concentration is to reduce the volume 10-fold by rotary evaporation at 30°. Care must be taken to avoid exaporation to dryness. An alternative procedure, which is also satisfactory, is partial freeze drying. The resulting solution is stored at −20°. Concentration of the enzyme by ammonium sulfate precipitation is not recommended, since a flotate is obtained, resulting in large mechanical losses.

A summary of the procedure is given in the table.

PURIFICATION PROCEDURE FOR L-3-GLYCEROPHOSPHATE DEHYDROGENASE FROM PIG BRAIN MITOCHONDRIA[a]

Stage	Volume (ml)	Activity (units/ml)	Protein (mg/ml)	Specific activity (units/mg)	Purification (fold)	Yield (%)
1. Acetone-dried powder	65	0.57	8.35	0.068	1	100
2. Phosphate-extracted acetone-dried powder	60	0.57	7.60	0.075	1.1	92
3. Whole Triton X-100 extract	56	0.60	6.84	0.088	1.3	91
4. Triton X-100 supernatant	54	0.38	0.70	0.54	8.0	55
5. Hydroxyapatite supernatant	56	0.16	0.14	1.15	17	24
6. Concentrated supernatant	7.3	1.05	0.93	1.15	17	21

[a] Data taken from A. P. Dawson and C. J. R. Thorne, *Biochem. J.* **111,** 27 (1969), and reproduced by kind permission of the publisher.

Comments on the Purification Procedure. The enzyme activity in the acetone powder is very much more stable than that in the solubilized extract (see below). Since the purification procedure can be readily completed in 1 day, starting from the acetone powder, it is convenient to work up small batches as required.

Apart from the specific activity of the starting material, the other important factor in determining the degree of purification achieved is the ratio of the Triton-X-100 concentration to the protein concentration. Any major deviation from the ratio described here will result in an extract of lower specific activity.

Properties

Stability. In the form of the acetone powder, the enzyme shows no loss of activity after storage for 4 months at −20°. After solubilization, the enzyme can be stored frozen at −20° for 1 week without significant loss of activity if repeated freezing and thawing is avoided.

Molecular Properties. At a specific activity of 1.15, the preparation contains 1 mole of acid-liberatable FAD/5.5×10^6 g of protein, and negligible quantities of FMN. The partially purified preparation of Ringler,[5] which has a somewhat lower specific activity, contained 1 mole of FAD/2.1×10^6 g of protein. The enzyme also contains 1 g-atom of acid liberatable iron/3.1×10^5 g of protein, which is in good agreement with the iron content of Ringler's preparation.

Specificity. As far as has been tested,[2] the enzyme is specific for L-3-glycerophosphate as electron donor. No activity was observed with D-3-glycerophosphate, DL-glycerol 1-nitrate, glyceric acid, glycerol, DL-propane-1,2-diol 1-phosphate, D-3-phosphoglyceric acid, and DL-glyceraldehyde 3-phosphate. The enzyme shows very broad specificity for electron acceptors.[7] Extrapolated to infinite electron acceptor concentration, the highest rate is obtained with DCIP, phenazine methosulfate and ubiquinones (coenzyme Q) 0, 1, 2, and 6, all of which are equally active; 39% of this rate is attainable with Q_{10} and 78% with menadione.

Inhibitors. The following are competitive inhibitors with respect to L-3-glycerophosphate: L-3-glyceraldehyde phosphate ($K_i = 35$ μM), D-3-glyceraldehyde phosphate ($K_i = 0.5$ mM), D-3-phosphoglyceric acid ($K_i = 14$ mM), and D-3-glycerophosphate ($K_i = 40$ mM).

The chelating agents 1,2-dihydroxybenzene 3,5-disulfonic acid (Tiron) and 1-(2-thenoyl)-3,3,3-trifluoroacetone (TTA) are both reversible inhibitors of the enzyme and behave as though they are competitive with

[7] A. P. Dawson and C. J. R. Thorne, *Biochem. J.* **114**, 35 (1969).

respect to both substrates. For Tiron, K_i with respect to L-3-glycerophosphate is 3.4 mM, and with respect to DCIP, K_i is 10.4 mM. For TTA, K_i with respect to L-3-glycerophosphate is 1.0 mM, and with respect to DCIP, it is 3.6 mM.[8]

In the presence of 30 μM p-chloromercuribenzoate, the enzyme loses 80% of its activity in 30 min at 20°. Under these conditions the presence of L-3-glycerophosphate potentiates the inhibition. Activity is restored by incubation with mercaptoethanol.[8]

Kinetic Properties. The enzyme shows parallel lines kinetics, indicating the presence of a free, modified enzyme intermediate.[7] K_m values are constants only when extrapolated to infinite concentration of the other substrate. Under these conditions, K_m for L-3-glycerophosphate is 10 mM with DCIP, phenazine methosulfate, Q_0, Q_1, Q_2, or Q_6 as electron acceptor. With Q_{10} as acceptor, K_m for L-3-glycerophosphate is 6.2 mM. For the electron acceptors, K_m values at infinite L-3-glycerophosphate are as follows: DCIP, 0.125 mM; Q_0, 0.125 mM; Q_1, 52 μM; Q_2, 23 μM; Q_6, 0.11 mM; Q_{10}, 50 μM.

[8] A. P. Dawson, Ph.D. Thesis, Cambridge Univ., 1967.

[57] L-α-Glycerophosphate Dehydrogenase from Beef Liver

By HECTOR GONZALEZ-CEREZO and KEITH DALZIEL

L-α-Glycerophosphate + NAD$^+$ ⇌ dihydroxyacetone phosphate + NADH + H$^+$

α-Glycerophosphate dehydrogenase has been isolated from skeletal muscle and liver of rabbit,[1–3] chicken,[4] and rat,[5,6] and the properties of the isoenzymes were compared. The muscle enzymes and the enzyme from rabbit liver,[3] were crystallized. We describe here a procedure[7] for isolat-

[1] T. Baranowski, *J. Biol. Chem.* **180**, 535 (1949).
[2] J. van Eys, J. Judd, J. Ford, and W. B. Womack, *Biochemistry* **3**, 1755 (1964).
[3] J. Otto, A. Raggi, W. Machleidt, and T. Bücher, *Hoppe-Seyler's Z. Physiol. Chem.* **353**, 332 (1972).
[4] H. B. White and N. O. Kaplan, *J. Biol. Chem.* **244**, 6031 (1969).
[5] T. P. Fondy, L. Levin, S. J. Sollohub, and C. R. Ross, *J. Biol. Chem.* **243**, 3148 (1968).
[6] T. P. Fondy, K. J. Herwig, S. J. Sollohub, and D. B. Rutherford, *Arch. Biochem. Biophys.* **145**, 583 (1971).
[7] H. Gonzalez-Cerezo, D. Phil. Thesis, Oxford University, 1973.

ing a pure, crystalline liver enzyme in high yield from a convenient bulk source, beef liver, and some of its chemical, physical, and catalytic properties.

Assay Method

Principle. Earlier workers have assayed α-glycerophosphate dehydrogenase by spectrophotometric measurements of the rate of oxidation of NADH by dihydroxyacetone phosphate.[3,4] Both these substrates are expensive and unstable. A more convenient assay is provided by the reduction of NAD by α-glycerophosphate at alkaline pH.

Reagents

0.1 M glycine–NaOH buffer, pH 10.0, containing 1 mM EDTA
2-Mercaptoethanol, 10 mM
NAD, 27 mM, Boehringer Grade II, 20 mg/ml
DL-α-Glycerophosphate, 0.124 M. Sigma Grade X, disodium salt (6H$_2$O), 43 mg/ml. By enzymic assay,[8] this solution contains 62 mM L-α-glycerophosphate.

Procedure. The initial rate of reduction of NAD at 25° is determined by absorbance measurements at 340 nm with a reaction mixture composed of 2.3 ml of glycine buffer, 0.1 ml of mercaptoethanol, 0.2 ml of NAD, 0.25 ml of DL-α-glycerophosphate, and 0.15 ml of distilled water. Reaction is started by the addition 10 μl (about 0.2 μg) of enzyme. A unit of enzyme is that amount which causes an initial rate of formation of 1 μmole of NADH per minute. Under the above conditions, the progress curve is linear for 1 min or more if the absorbance change (1 cm path) does not exceed 0.1 per minute. In the absence of EDTA or mercaptoethanol, enzyme inactivation causes nonlinearity.

Determination of Protein and Specific Activity. During the isolation procedure, the protein concentration and the specific activity (units per milligram protein) are calculated from absorbance measurements at 280 nm, assuming that an absorbancy of 1.0 (1 cm path) represents 1 mg of protein per milliliter. For the pure enzyme, the absorbancy of a 1 mg/ml solution is 0.56.

Purification Procedure

All steps are carried out at 0–4°, and all buffers are made up in glass-distilled water and contain 1 mM EDTA and 1 mM 2-mercaptoethanol.

[8] H. J. Hohorst, *in* "Methods of Enzymatic Analysis (H. U. Bergmeyer, ed.), p. 215. Academic Press, New York, 1965.

Step 1. Extraction. Livers are transported in ice and used within 2 hr of slaughter. Minced tissue, 1000 g, is suspended in 2 liters of 0.125 *M* phosphate buffer, pH 7.0, for 2 hr, and then filtered through several layers of muslin.

Step 2. Acid Denaturation. The extract is brought to pH 5.0 by drop-wise addition of 20% (v/v) acetic acid with continuous stirring, and after 15 min is centrifuged at 23,000 *g* for 20 min. The decanted supernatant is immediately brought to pH 7.0 by gradual addition of 20% (v/v) ammonium hydroxide.

Step 3. Ammonium Sulfate Fractionation at pH 7.0. The salt (analytical grade) is added gradually with stirring to a concentration of 230 g/liter (38% saturation) and pH 7.0 maintained by addition of ammonia solution. After 2 hr, the precipitate is removed by centrifugation at 23,000 *g* for 20 min. The ammonium sulfate concentration of the supernatant is raised to 55% saturation by addition of a further 110 g/liter, and the preparation is left overnight. The enzyme-containing precipitate is isolated by centrifugation as before, and resuspended in about 500 ml of 35% saturated ammonium sulfate solution (211 g/liter) in 0.125 *M* phosphate buffer, pH 7.0 by stirring for 30 min. The insoluble protein is removed by centrifugation. To the supernatant, 100 g/liter ammonium sulfate is added; the precipitate formed after 2 hr is separated by centrifugation. The precipitate is suspended in the minimum volume of 0.125 *M* phosphate buffer, pH 7.0, for transfer to a dialysis sac, and dialyzed for 30 min against 1 liter of 5 m*M* phosphate buffer, pH 7.7, to dissolve the protein.

Step 4. Heat Denaturation. The solution is quickly warmed to 49° in a water bath and after 10 min is chilled in ice and centrifuged at 27,000 *g* for 10 min. The supernatant is decanted, the precipitate is resuspended in its own volume of 5 m*M* phosphate buffer, pH 7.7, and after centrifugation the second supernatant is added to the first.

Step 5. Gel Filtration. The preparation is passed into a column of Sephadex G-100 of bed volume 1.5 liters (5.5 cm diameter × 75 cm) which is first swollen and equilibrated with 5 m*M* phosphate buffer, pH 7.7. The enzyme is eluted with the latter buffer at a flow rate of 60 ml/hr. The pH and conductivity of the combined enzyme-containing eluates are measured to confirm that equilibration with the eluting buffer has occurred.

Step 6. Chromatography on DEAE-Cellulose. The ion-exchanger (Whatman DE23) is equilibrated with 5 m*M* phosphate buffer, pH 7.7, in a column (2.4 cm diameter × 75 cm) after precycling according to the manufacturer's directions. The enzyme solution is run in, the column is washed with 700 ml of the equilibration buffer, and the enzyme is then eluted with a linear gradient of 0 to 0.1 M NaCl in the same buffer.

Step 7. Heat Denaturation. The enzyme is precipitated from the combined eluates from step 6 by the addition of 430 g of ammonium sulfate per liter. After 24 hr, the precipitate is collected by centrifugation at 27,000 g for 20 min, dissolved in about 20 ml of 10 mM phosphate buffer, pH 7.7, and subjected to heat treatment as in step 4.

Step 8. Chromatography on CM-Sephadex. The combined supernatant and washing from the heat denaturation is dialyzed against 10 mM phosphate buffer, pH 6.7, for 24 hr (4 × 750 ml buffer), and then loaded onto a column (2.5 cm diameter × 20 cm) of CM-50 equilibrated with the same buffer. The column is then washed with 100 ml of the latter, and the enzyme is eluted with a linear gradient of 0 to 0.2 M NaCl in the same buffer.

Step 9. First Crystallization. The combined enzyme-containing eluates from the CM-Sephadex are concentrated by ultrafiltration under oxygen-free nitrogen (Amicon system, PM30 membrane) to a final volume of 2–3 ml and a protein concentration of 10–15 mg/ml. The solution is dialyzed against 250 ml of 1.8 M ammonium sulfate in 0.1 M phosphate buffer, pH 7.0, for 24 hr. The precipitate containing 80–85% of the total enzyme is separated by centrifugation, and the pale yellow supernatant is discarded. The precipitate, which consists of needlelike enzyme crystals and some amorphous material, is dissolved in about 2 ml of 0.1 M phosphate buffer, and the insoluble material is removed by centrifuging.

Step 10. Second Crystallization. The crystallization is repeated exactly as in step 9.

The results obtained in a typical preparation are shown in the table. Further recrystallization results in no change of specific activity, which is 270 ± 10 (6 preparations) on the basis of absorbancy at 280 nm, and 150 units/mg on the basis of the true protein concentration, since $E_{1cm}^{0.1\%} = 0.56$ for the pure enzyme.

Properties

Stability. The enzyme is unstable in the absence of mercaptoethanol. In solution, 1 mg/ml in 0.1 M phosphate buffer, pH 7.0, containing 1 mM EDTA and 1 mM 2-mercaptoethanol, its activity declines by 5–10% in 1 week at 4°. A crystalline enzyme suspension in 1.8 M ammonium sulfate solution in the same buffer, stored at 4° for 6 months, retains 70% of its original activity.

Chemical and Physical. The complete amino acid composition has been determined. There are 8 thiol groups, 2 tryptophan, and 3 tyrosine

PURIFICATION PROCEDURE FOR L-α-GLYCEROPHOSPHATE
DEHYDROGENASE FROM BEEF LIVER

Step	Volume (ml)	Protein (g)	Activity (units/ml)	Specific activity (units/mg)	Yield (%)
1. Extraction	1800	121	4.4	0.066	100
2. Acid denaturation	1780	45	4.5	0.176	101
3. (NH₄)₂SO₄, pH 7.0	88	10.0	52	0.46	57
4. Heat denaturation	88	8.3	55	0.58	57
5. Sephadex G-100	350	3.1	15.4	1.76	68
6. DEAE-cellulose	480	0.41	7.0	8.15	42
7. Heat denaturation	51	0.20	58	15.0	37
8. CM-Sephadex	154	0.031	17.0	85	33
9. First crystallization	2.55	0.0093	860	235	27
10. Second crystallization	1.90	0.0081	1140	270	27

residues in the subunit of 31,000 daltons, in very close agreement with published data for the enzymes from both liver and muscle of rabbit,[2,3] chicken,[4] and rat.[5,6] In accordance with the low aromatic amino acid content, the extinction coefficient for the pure enzyme at 280 nm, based on protein estimation by refractometry,[9] is $E_{1cm}^{0.1\%} = 0.56$. The ratio of absorbancy $A_{280}:A_{260}$ is 1.55.

The partial specific volume calculated from the amino acid composition is 0.743. From gel filtration, sedimentation equilibrium by the meniscus depletion method, and sodium dodecyl sulfate gel electrophoresis, the molecular weight is 62,000 ± 2000 and there are 2 subunits in the molecule. Similar values have been reported for the molecular weights of the liver and muscle enzymes of rat[5,6] and chicken,[4] but higher values have also been reported for the chicken enzymes[10] (68,000) and for the rabbit muscle enzyme[2] (78,000).

Kinetics. Initial rate studies of the reaction in both directions by fluorometry establish a ternary complex mechanism. Provisional estimates of the K_m values at pH 7.0 in 0.11 M phosphate buffer at 25° are 46 μM for NAD, 820 μM for L-α-glycerophosphate, 15 μM for NADH, and 600 μM for dihydroxyacetone phosphate. The maximum turnover numbers per subunit are 9 sec⁻¹ for L-α-glycerophosphate oxidation and 140 sec⁻¹ for dihydroxyacetone phosphate reduction. The dissociation constants for the enzyme–coenzyme compounds estimated from the initial rate parameters are 250 μM for E·NAD and 1.0 μM for E·NADH. The

[9] J. Babul and E. Stellwagen, *Anal. Biochem.* **28**, 216 (1969).
[10] H. B. White, *Arch. Biochem. Biophys.* **147**, 123 (1971).

binding of NADH to the enzyme is accompanied by a 6-fold enhancement of the coenzyme fluorescence emission at 460 nm, and fluorescence titrations indicate the binding of 1 NADH molecule to each subunit of 31,000 daltons, with a dissociation constant of about 0.3 μM.

[58] Glyceraldehyde-3-phosphate Dehydrogenase from Rabbit Muscle[1]

By REMI E. AMELUNXEN and DANIEL O. CARR

D-Glyceraldehyde 3-phosphate + NAD$^+$ + P$_i$ $\rightleftharpoons$
$$1,3\text{-diphosphoglycerate} + \text{NADH} + \text{H}^+ \quad (1)$$
D-Glyceraldehyde 3-phosphate + NAD$^+$ $\xrightarrow{\text{HAsO}_4{}^{2-}}$
$$3\text{-phosphoglycerate} + \text{NADH} + \text{H}^+ \quad (2)$$

Assay Method

Principle. The reversible physiological reaction, Eq. (1), is not often used for assaying the enzyme; instead, the irreversible reaction using arsenate, Eq. (2), is employed. The enzyme is assayed spectrophotometrically at 340 nm by the method of Krebs[2] except that Tris buffer is substituted for the pyrophosphate buffer.

Reagents

DL-Glyceraldehyde 3-phosphate (GAP) solution prepared from the barium salt of the diethyl acetal by the method of Sigma Chemical Co.[3] The content of the D-isomer is determined by the assay procedure given below except that the cysteine is omitted, and 20 μg of crystalline enzyme are added. The standardized solution is diluted to 7.6 mM with respect to the D-isomer.

NAD$^+$, 7.6 mM, is prepared from commercially available reagent and standardized by the absorbance at 260 nm (extinction coefficient = 1.8×10^4).[4]

Tris-acetate buffer, 30 mM, pH 8.6, which is 6 mM with respect to cysteine. Prepare immediately before use: 0.3 M cysteine·HCl is

[1] R. E. Amelunxen and D. O. Carr, *Biochim. Biophys. Acta* **132,** 256 (1967).
[2] E. G. Krebs, this series, Vol. I, p. 407.
[3] Sigma Technical Bulletin No. 10, September 1961. Sigma Chemical Co., St. Louis, Missouri.
[4] A. Kornberg and W. E. Pricer, Jr., *in* "Biochemical Preparations" (E. E. Snell, ed.), Vol. 3, p. 20. Wiley, New York, 1953.

adjusted to pH 8.6 with 4 N NaOH and diluted 1:50 with buffer. Sodium arsenate, 0.17 M

Procedure. A representative assay mixture of 3.0 ml in a 1-cm light path cuvette is prepared as described by Krebs[2] and is as follows: 2.7 ml of Tris-acetate buffer (containing cysteine), 0.1 ml of NAD[+], and 1 μl of the enzyme are added to the cuvette and mixed. Incubate at 30° for 7–10 min to ensure activation of the enzyme by cysteine. The initial absorbance at 340 nm is recorded, and the reaction is initiated by the addition of 0.2 ml of a 1:1 mixture of GAP and arsenate. Absorbance is recorded for the earliest 30-sec interval after initiating the reaction. The initial rate may also be obtained using a recorder. As isolated below, the enzyme has 80% of the total activity in the absence of cysteine.

Definition of Enzyme Units and Specific Activity. One enzyme unit is defined as the production of 1 μmole of NADH per minute. The rate of the reaction is not linear with time, but maximal linearity of the assay procedure is observed during the earliest 30-sec period after initiating the reaction, if the absorbance does not exceed 0.200.

$$\text{Units} = (A_{30} \times 2)/2.07$$

where A_{30} is the change in absorbance during the earliest 30-sec interval (corrected for the initial absorbance), and 2.07 is the absorbance of 1 μmole of NADH per 3 ml.[4] Protein determinations are carried out using a semimicro biuret method.[5] The specific activity is expressed as units per milligram of protein.

This assay may not be useful in other crude systems where a glycerophosphate dehydrogenase,[6] triose isomerase,[6] NADH oxidase,[7] or alcohol dehydrogenase is present. Although the effect of the first three is obvious, it should be recognized that the solution of GAP contains 4 moles of ethanol for every mole of reactive GAP.

Purification Procedure

General Considerations. The frozen rabbit muscle can be obtained from Pel-Freeze or a local vendor. All steps in the purification scheme are carried out at 0–4°. In order to minimize the inactivating effects of heavy metal ions, the saturated ammonium sulfate solutions should be prepared with enzyme grade reagent. In addition, EDTA, KCN, and dithioerythritol (DTE) or dithiothreitol (DTT) are used throughout

[5] L. C. Mokrasch and R. W. McGilvery, *J. Biol. Chem.* **221**, 909 (1956).
[6] W. S. Allison, this series, Vol. 9, p. 210.
[7] R. E. Amelunxen, *Biochim. Biophys. Acta* **122**, 175 (1966).

purification to protect the sulfhydryl groups of the enzyme. The percent saturation of the ammonium sulfate solutions is calculated by the formula of Kunitz.[8] The pH of the ammonium sulfate solutions is determined with a pH meter after diluting 1:20. All other reagents are commercially available and used as received. Modification of this procedure has been used for the isolation of the enzyme from microorganisms[7] and has been employed by other laboratories.[9] The method described below is a more rapid procedure based on the work of Cori et al.[10,11]

Step 1. Preparation of the Extract. A total of 100 g of frozen rabbit muscle is broken into small pieces and added slowly into a Waring Blendor containing 250 ml of 30 mM KOH–1 mM EDTA. Once the muscle is dispersed, the final mixture is blenderized at high speed for a total of 4 min. The homogenate is centrifuged at 20,000 g for 20 min; all subsequent centrifugations are carried out in like manner. The supernatant fluid (extract 1) is collected, and the pellet is resuspended in 50 ml of 30 mM KOH–1 mM EDTA. The pellet is extracted by stirring for a few minutes and then centrifuged; the supernatant fluid (extract 2) is collected. Extracts 1 and 2 are combined and referred to as crude extract.

Step 2. Fractionation with Ammonium Sulfate. The crude extract is brought to 52% of saturation by the addition of a saturated solution of ammonium sulfate, pH 7.5. The preparation is centrifuged, and the supernatant fluid is brought to 70% of saturation by the addition of solid ammonium sulfate. The preparation is allowed to stand 30 min at 0–4°. After centrifugation the supernatant fluid is brought to 90% of saturation by the addition of solid ammonium sulfate and allowed to stand 30 min. After centrifugation, the precipitate is taken up in about 40 ml of deionized water containing EDTA, KCN, and DTE, each being 1 mM. The supernatant fluid contains no detectable activity.

Step 3. Crystallization. Two volumes of a saturated solution of ammonium sulfate, pH 8.4, are added slowly to the resuspended precipitate of step 2. This solution is allowed to stand at 4°; crystals begin to form within a few hours and crystallization is essentially complete in 2–3 days.

Step 4. Recrystallization. In recrystallizing the enzyme, the crystals are centrifuged and dissolved in about 20 ml of deionized water (containing EDTA, KCN, and DTE), after which 2 volumes of saturated ammonium sulfate, pH 8.4, are added. The purification procedure as summarized in the table was repeated several times with only slight variation.

[8] M. Kunitz, *J. Gen. Physiol.* **35**, 423 (1952).
[9] G. G. Hammes, P. J. Lillford, and J. Simplicio, *Biochemistry* **10**, 3686 (1971).
[10] G. T. Cori, M. W. Slein, and C. F. Cori, *J. Biol. Chem.* **173**, 605 (1948).
[11] S. F. Velick, this series, Vol. 1, p. 401.

PURIFICATION PROCEDURE FOR GLYCERALDEHYDE-3-PHOSPHATE
DEHYDROGENASE FROM RABBIT MUSCLE

Fraction	Volume (ml)	Units/ml	Protein (mg/ml)	Specific activity
1. Crude extract	194	208	15.6	13.3
2. Supernatant fluid at 52% $(NH_4)_2SO_4$	380	105	4.8	21.4
3. Supernatant fluid at 70% $(NH_4)_2SO_4$	388	102	2.3	44.1
4. Dissolved precipitate of 90% $(NH_4)_2SO_4$ fraction	46	832	13.0	64.0
5. First crystals	124	234	2.8	83.5
6. Second crystals	63	356	3.9	91.3
7. Third crystals	63	311	3.2	97.2
8. Fourth crystals	64	303	3.1	97.9

Properties

The crystalline suspension in 67% saturated ammonium sulfate containing EDTA, KCN, and DTE is stable for many months at 4°. Others have reported similar stabilities when only EDTA is added.[12] The ratio of the A_{280}/A_{260} absorbance is about 1.0, suggesting that the enzyme has its full complement of NAD^+.[13] However, in another laboratory, this procedure has produced enzyme with a ratio of 1.2–1.3, consistent with 2–3 moles of NAD^+ per mole of enzyme.[9] The various chemical and physical properties of the enzyme have been reviewed.[11,14] More recently, the enzyme has been reported to have a molecular weight of 145,000 and to be composed of four identical subunits.[15] After specific inhibition of the rabbit muscle enzyme with [1-^{14}C]iodoacetic acid and tryptic digestion, a single radioactive peptide is obtained which contains [1-C^{14}]carboxymethylcysteine.[16] The amino acid sequence of the "active-site" peptide is similar to those of the enzyme isolated from other sources.[6]

[12] W. Bloch, R. A. MacQuarrie, and S. A. Bernhard, *J. Biol. Chem.* **246**, 780 (1971).
[13] R. Singleton, Jr., J. R. Kimmel, and R. E. Amelunxen, *J. Biol. Chem.* **244**, 1623 (1969).
[14] S. F. Velick and C. Furfine, *in* "The Enzymes" (P. D. Boyer, H. Lardy, and K. Myrbäck, eds.), Vol. 7, p. 243. Academic Press, New York, 1963.
[15] W. F. Harrington and G. M. Karr, *J. Mol. Biol.* **13**, 885 (1965).
[16] J. I. Harris, B. P. Meriwether, and J. H. Park, *Nature (London)* **198**, 154 (1963).

[59] Glyceraldehyde-3 phosphate Dehydrogenase from *Bacillus stearothermophilus*[1]

By REMI E. AMELUNXEN

$$\text{D-Glyceraldehyde 3-phosphate} + \text{NAD}^+ + \text{P}_i \rightleftharpoons$$
$$\text{1,3-diphosphoglycerate} + \text{NADH} + \text{H}^+ \quad (1)$$

$$\text{D-Glyceraldehyde 3-phosphate} + \text{NAD}^+ \xrightarrow{\text{HAsO}_4{}^{2-}}$$
$$\text{3-phosphoglycerate} + \text{NADH} + \text{H}^+ \quad (2)$$

Assay Method

Reagents

Glycine-NaOH buffer, 0.10 M, pH 10.0, which is 6 mM with respect to cysteine. Prepare immediately before use: 0.3 M cysteine-HCl is adjusted to pH 10.0 with 4 N NaOH and diluted 1:50 with buffer.

NAD$^+$, 7.6 mM; standardized using the extinction coefficient at 260 nm[2]

Glyceraldehyde-3-phosphate (as the diethyl acetal), 7.6 mM. A solution of DL-glyceraldehyde 3-phosphate (GAP) is prepared from the barium salt of the diethyl acetal according to the method of Sigma Chemical Co.[3] The concentration of the D-isomer is determined as indicated,[3] or by the assay procedure described below for the thermophilic enzyme (except that cysteine is omitted and 15 μg of crystalline enzyme are added).

Sodium arsenate, 0.17 M

Procedure. The assay for activity is carried out spectrophotometrically essentially as described by Krebs.[4] The irreversible reaction using arsenate, Eq. (2) above, is employed. An incomplete reaction mixture containing the following components is prepared in a 3-ml cuvette with a 1-cm light path: 2.7 ml of glycine–NaOH buffer (containing cysteine), 0.1 ml of NAD$^+$, and 1 μl of a solution of the enzyme. The mixture is incubated at 30° for 10 min to ensure temperature equilibration, and for

---

[1] R. E. Amelunxen, *Biochim. Biophys. Acta* **122**, 175 (1966).

[2] A. Kornberg and W. E. Pricer, Jr., *in* "Biochemical Preparations" (E. E. Snell, ed.), Vol. 3, p. 20. Wiley, New York, 1953.

[3] Sigma Technical Bulletin No. 10, September 1961. Sigma Chemical Co., St. Louis, Missouri.

[4] E. G. Krebs, this series, Vol. 1, p. 407.

complete activation of the enzyme by cysteine (freshly prepared enzyme at any stage of purification is fully active without cysteine). After recording the zero time absorbance at 340 nm, 0.2 ml of a 1:1 mixture of GAP and arsenate is added with rapid mixing to initiate the reaction, and the increase in absorbance is recorded for the earliest 30-sec interval. Alternatively, initial rates can be obtained with a recorder. Maximal linearity in the assay procedure is observed within 30 sec after initiating the reaction with substrate-acceptor, if the absorbance does not exceed 0.200.

Definition of Enzyme Units and Specific Activity. One unit is defined as the amount of enzyme which reduces 1 μmole of NAD^+ per minute at 340 nm, and is calculated from the extinction coefficient of 6.22×10^3 for NADH.[2] Protein determinations are carried out using a semimicro biuret method.[5]

$$SA = (A_{30} \times 2 \div 2.07)/\text{mg protein} = \text{units/mg protein}$$

where A_{30} is the change in absorbance during the earliest 30-sec interval (corrected for the zero time absorbance), and 2.07 is the absorbance of 1 μmole of NADH per 3 ml.

Crude extracts of thermophilic cells contain an active NADH oxidase giving an apparent lower activity for glyceraldehyde-3-phosphate dehydrogenase. It is impossible to make corrections with any degree of accuracy. After heat treatment (step 2 below), the oxidase is essentially removed. Although every reactive mole of GAP contains 4 moles of ethanol, interference by alcohol dehydrogenase in the assay system is minimal at pH 10.

Purification Procedure

General Considerations. In the purification scheme originally reported,[1] enzymic assays were determined at pH 8.6 rather than the optimal pH of 10. The scheme to be described is representative of data obtained after several years of working with this enzyme, and is based on optimal conditions.

All steps in the purification scheme are carried out at 0–4°. Enzyme grade ammonium sulfate is used to reduce the amount of heavy-metal ions added; the formula of Kunitz[6] is employed in all calculations of percentage saturation. Throughout purification, EDTA is added to reduce heavy-metal contamination, and dithiothreitol (DTT) to diminish oxidation of sulfhydryl groups; each reagent is added to 1 mM.

[5] L. C. Mokrasch and R. W. McGilvery, *J. Biol. Chem.* **221**, 909 (1956).
[6] M. Kunitz, *J. Gen. Physiol.* **35**, 423 (1952).

Step 1. Preparation of Extract by Sonication. Thirty-five liters of log phase cells of *B. stearothermophilus* strain 1503 (ATCC 7954) grown at 60° in the medium originally described[1] (or in Difco Antibiotic Medium 3) are concentrated, washed in 0.15 M NaCl, and resuspended in 60–70 ml of deionized water containing EDTA-DTT. The cells are ruptured by sonication in the Branson Sonifier, Model S-125 (or an equivalent instrument) for a period of 3–4 min (using alternating cycles with the sonifier on for 30 sec, and off for 45 sec); the cells are cooled during sonication in an ice-water bath and the final temperature of the preparation after sonication is about 15°. The sonicate is centrifuged at 27,000 g for 30 min, and the supernatant fluid is filtered through glass wool to remove any nonsedimentable material.

Step 2. Heat Treatment. After adjusting the protein concentration to 30–34 mg/ml, the brownish lysate is heated at 80° for 5 min with constant agitation, rapidly cooled, and centrifuged at 12,000 g for 15 min. Since heat treatment essentially eliminates the NADH oxidase, and the glyceraldehyde-3-phosphate dehydrogenase shows greater total enzyme units, the recovery at this stage of purification is taken as 100%.

Step 3. Fractionation with Ammonium Sulfate. The supernatant fluid from step 2 is brought to 72% of saturation with solid ammonium sulfate; the preparation is allowed to stand overnight at 4°. After centrifugation at 12,000 g for 15 min, the supernatant fluid is brought to 90% of saturation with solid ammonium sulfate (slow addition of the salt) and stored overnight at 4°.

Step 4. Crystallization. The preparation at 90% of saturation is centrifuged at 12,000 g for 15 min, and the precipitate dissolved in deionized water (containing EDTA-DTT) to a protein concentration of 12 mg/ml; any turbidity is clarified by centrifugation. Two volumes of saturated ammonium sulfate, pH 8.4, are added very slowly to the dissolved precipitate with constant stirring; EDTA and DTT are then added to give a final concentration of 1 mM for each reagent. Crystals begin to form within a few hours, and crystallization is complete within 2–3 days. If the protein concentration of the dissolved precipitate is in excess of 12 mg/ml, amorphous precipitation results; if less than 10 mg/ml, crystallization is much slower.

Step 5. Recrystallization. The first crystals are harvested by centrifugation at 12,000 g and dissolved in deionized water (containing EDTA-DTT) to 12 mg of protein per milliliter. Two volumes of saturated ammonium sulfate, pH 8.4, are added as in step 4.

Morphologically the crystals are needles, and similar to those of the crystalline yeast glyceraldehyde-3-phosphate dehydrogenase. Crystalline suspensions are stored routinely in 67% saturated ammonium sulfate at

PURIFICATION PROCEDURE FOR GLYCERALDEHYDE-3-PHOSPHATE
DEHYDROGENASE FROM *Bacillus stearothermophilus*

Fraction	Volume (ml)	Units/ml	Protein (mg/ml)	Specific activity
1. Sonic extract	140	102	30.0	3.4
2. Supernatant fluid after heat step	112	135	11.8	11.4
3. Supernatant fluid at 72% $(NH_4)_2SO_4$	124	87	2.0	43.5
4. Dissolved precipitate of 90% $(NH_4)_2SO_4$ fraction	12	730	12.0	61.0
5. First crystals	36	140	1.3	108.0
6. Second crystals	12	407	3.6	113.0
7. Third crystals	11	396	3.5	113.0

4°. The purification scheme is summarized in the table. Suzuki and Harris[7] have described another procedure for purification of this enzyme from *B. stearothermophilus* 1503.

Properties

Stability of the Enzyme. One of the unique properties of obligate thermophilic bacteria is the remarkable thermostability of their enzymes. When crystals of glyceraldehyde-3-phosphate dehydrogenase are dissolved in deionized water to a protein concentration of 10 mg/ml, little or no inactivation can be demonstrated after heat treatment at 80° for 10 min when assayed without cysteine. After 10 min at 90°, there is about 10% inactivation if assayed without cysteine, but total activity can be restored after preincubation with cysteine in the assay reaction mixture. By contrast, nonthermophilic glyceraldehyde-3-phosphate dehydrogenases are extensively and irreversibly inactivated after 10 min at about 70°.

When the crystalline enzyme is stored at 4° in 67% saturated ammonium sulfate (containing EDTA-DTT), maximal activity can be maintained for months. If assayed without cysteine, about 10–15% of the activity is lost after several months of storage, but preincubation with cysteine restores the activity to its original value.

Physical and Chemical Properties. The homogeneity of thermophilic glyceraldehyde-3-phosphate dehydrogenase has been demonstrated using several physical methods.[1,8,9] Like the rabbit muscle enzyme,[10] there are

[7] K. Suzuki and J. I. Harris, *FEBS Lett.* **13**, 217 (1971).
[8] R. Singleton, Jr., J. R. Kimmel, and R. E. Amelunxen, *J. Biol. Chem.* **244**, 1623 (1969).
[9] R. E. Amelunxen, M. Noelken, and R. Singleton, Jr., *Arch. Biochem. Biophys.* **141**, 447 (1970).
[10] A. L. Murdock and O. J. Koeppe, *J. Biol. Chem.* **239**, 1983 (1964).

4 moles of firmly bound NAD⁺ per mole of enzyme.[8] A procedure is described[11] for crystallization of the thermophilic enzyme after removal of bound NAD⁺, the thermostability of the apoenzyme is essentially the same as that of the holoenzyme. The optimal pH[12] for enzymic activity is 10, in contrast to other glyceraldehyde-3-phosphate dehydrogenases, which have optimal pH values ranging from 8.4 to 8.6.

Using the sedimentation equilibrium method,[9] the average molecular weight of the thermophilic enzyme in 0.1 M KCl is 148,700; in 5.0 M guanidine-HCl, the molecular weight is 36,000 ± 1200, indicating that the native enzyme has a tetrameric structure and dissociates into four subunits. The molecular weight for both dissociated and undissociated enzyme are consistent with those obtained for rabbit muscle, lobster muscle, and pig muscle enzymes.[13–15] The $s_{20,w}^0$ and $D_{20,w}^0$ values for the thermophilic enzyme[8] are 7.17×10^{-13} sec, and 3.95×10^{-7} cm²/sec, respectively.

There are no striking differences in the amino acid composition of the thermophilic enzyme[8] when compared to other glyceraldehyde-3-phosphate dehydrogenases. However, the total content of aspartic and glutamic acids is elevated, and the acidic nature of the protein is reflected by a low pI of 4.6. Using 3 parameters for analysis of the degree of hydrophobicity,[8] the thermophilic enzyme showed no differences when compared to counterparts from other sources. Titration for sulfhydryl groups using 5,5′-dithiobis-(2-nitrobenzoic acid) indicates that there are 4 sulfhydryl groups per mole of enzyme,[8] and unpublished data[16] indicate that all sulfhydryl groups are critical based on stoichiometric inhibition by iodoacetic acid. Analyses for cysteic acid show 12 half-cystines per mole of enzyme, suggestive of 1 disulfide bond and 1 sulfhydryl group per subunit.[8] Inhibition of the thermophilic enzyme by p-hydroxymercuribenzoate[12] is markedly different from that of the rabbit muscle enzyme. The irreversible inactivation of the thermophilic enzyme by NADH[12] is similar to the rabbit muscle enzyme,[17] except that thermophilic temperatures are required.

The kinetics of inactivation[9] of the thermophilic enzyme in 8.0 M

[11] R. E. Amelunxen and J. Clark, *Biochim. Biophys. Acta* **221**, 650 (1970).

[12] R. E. Amelunxen, *Biochim. Biophys. Acta* **139**, 24 (1967).

[13] W. F. Harrington and G. M. Karr, *J. Mol. Biol.* **13**, 885 (1965).

[14] B. E. Davidson, M. Sajgo, H. F. Noller, and J. I. Harris, *Nature (London)* **216**, 1181 (1967).

[15] J. I. Harris and R. N. Perham, *Nature (London)* **219**, 1025 (1968).

[16] R. E. Amelunxen, 1968.

[17] R. Amelunxen and S. Grisolia, *J. Biol. Chem.* **237**, 3240 (1962).

urea show it to be unusually resistant to this denaturant at 30°, 40°, and 50°. However, at thermophilic temperatures (55° and 60°), inactivation is rapid but some reversibility can be demonstrated. In 5.0 M guanidine-HCl, enzymic inactivation is rapid and irreversible at 30°. Based on optical rotatory dispersion studies,[9] the enzyme is extensively unfolded in both 8.0 M urea and 5.0 M guanidine-HCl at 30° and above. Optical rotatory changes of the enzyme in water at various temperatures suggest that a change in secondary structure occurs between 55° and 60°.

Immunochemical data show that the activity of the thermophilic enzyme is inhibited by 96% after reaction with homologous antiserum.[18] No cross-inhibition occurs with the rabbit muscle and yeast enzymes. Using immunodiffusion, data involving the effect of denaturants indicate that there are marked differences between the thermophilic enzyme and the yeast enzyme.

Tryptic digestion of the thermophilic enzyme after specific inhibition with [2-¹⁴C]iodoacetic acid, yields a single radioactive peptide,[19] which contains the active-site cysteine. In most cases, active-site tryptic peptides from glyceraldehyde-3-phosphate dehydrogenases consist of 17 residues[20,21] comprising positions 4–20 of the 20-residue peptide from the thermophilic enzyme. The thermophilic peptide shows considerable homology with its counterparts from many other sources, but differs by the addition of Ala-His-His at the N-terminus, and by the substitution of phenylalanine for leucine in position 18. The presence of two adjacent histidines is a novel feature of the active-site sequence.

[18] R. L. Sauvan, O. J. Mira, and R. E. Amelunxen, *Biochim. Biophys. Acta* **263**, 794 (1972).
[19] J. Bridgen, J. I. Harris, P. W. McDonald, R. E. Amelunxen, and J. R. Kimmel, *J. Bacteriol.* **111**, 797 (1972).
[20] J. I. Harris and R. N. Perham, *J. Mol. Biol.* **13**, 876 (1965).
[21] G. M. T. Jones and J. I. Harris, *FEBS Lett.* **22**, 185 (1972).

[60] Triosephosphate Dehydrogenase from the Honey Bee

By RONALD W. BROSEMER

D-Glyceraldehyde 3-phosphate + NAD⁺ + HPO₄²⁻
= D-1,3-diphosphoglycerate + NADH + H⁺

The flight muscles of many insect species are capable of supporting the highest metabolic rates of any known tissue. Insects that utilize mainly carbohydrate for flight fuel have very high levels of the glycolytic enzymes in flight muscle.[1] This tissue is thus an especially good source for the purification of triose-phosphate dehydrogenase (D-glyceraldehyde 3-phosphate:NAD⁺ oxidoreductase (phosphorylating), EC 1.2.1.12).

Assay Method

Principle. The method is based upon the reduction of NAD^+ by glyceraldehyde-3-P in the presence of arsenate.[2]

This assay is invalid in crude extracts of the flight muscle of many insects, since high levels of triose-P isomerase and glycerol-3-P dehydrogenase result in rapid reoxidation of NADH by dihydroxyacetone-P. If it is desirable to establish the activity of triose-P dehydrogenase in such crude extracts, an assay based upon the reverse reaction can be used.[3]

Reagents

Buffer: 44 mM Na pyrophosphate–20 mM Na arsenate–3 mM D-fructose-1,6-diP–8 mM dithiothreitol, pH 9.0
NAD^+, 3 mM
Rabbit muscle aldolase, suspension of 10 mg/ml of 3.2 M $(NH_4)_2SO_4$

Procedure. Mix 2.0 ml of buffer, 0.3 ml of NAD^+, about 2 μl of aldolose, and sufficient water to give a final volume of 3.0 ml. After reaching temperature equilibrium (usually at 30°), the aliquot to be assayed is added. The absorbance increase between 15 and 45 sec is used for calculating enzyme activity.

Definition of Unit and Specific Activity. One unit of enzyme activity is defined as that amount of enzyme catalyzing the disappearance of 1 μmole of substrate per minute in the above assay at 30°. Specific activity is expressed in terms of units of activity per milligram of protein.

In the early stages of purification, protein is determined by the biuret reaction.[4] The protein is first precipitated by addition of trichloracetic acid to a final concentration of 10% (w/v). The centrifuged pellet is washed once with 10% trichloroacetic acid and once with ether. This protein precipitation eliminates interference of $(NH_4)_2SO_4$ in the biuret reaction. Bovine serum albumin is used as protein standard.

After the initial crystallization of triose-P dehydrogenase, concentrations are measued by absorbance at 280 nm using an extinction coefficient of 1.00 ml mg⁻¹ cm⁻¹ (Fox and Dandliker[5]).

[1] A. M. T. Beenakkers, *J. Insect Physiol.* **15**, 353 (1969).
[2] S. F. Velick, see this series, Vol. 1 [60].
[3] R. Wu and E. Racker, *J. Biol. Chem.* **234**, 1029 (1959).
[4] E. Layne, see this series, Vol. 3 [73].
[5] J. B. Fox, Jr., and W. B. Dandliker, *J. Biol. Chem.* **221**, 1005 (1956).

Purification Procedure

The preparation of thoraces and of the homogenate described below is identical to that for the isolation of glycerol-3-P dehydrogenase.[6] The 2.2 M $(NH_4)_2SO_4$ supernatant from the glycerol-3-P dehydrogenase isolation can be used to isolate triose-P dehydrogenase. In addition, slight modification of the procedure below allows purification of arginine kinase and of cytochrome c.[7] Thus four proteins can be isolated from one bee extract.

Preparation of Thoraces. Honey bees (*Apis mellifera*) are readily available from most beekeepers; the bees are stored frozen. The thoraces (including wings and legs) are removed from the bees, care being taken that the bees and thoraces remain frozen. It is convenient to place the material on dry ice and use tweezers to snap off the heads and abdomens. It requires about 6-man-hours of labor to prepare 100 g of thoraces.

Ammonium Sulfate Fractionation. Unless otherwise noted, all subsequent steps are carried out at 0–2°. The thoraces (100 g) are homogenized 1 min in a Waring Blendor with 300 ml of 0.1 M Tris, 10 mM EDTA, pH 7.5 (pH measured at 23°). The homogenate is centrifuged 30 min at 12,000 g. The pellet is reextracted with 90 ml of the 0.1 Tris, 10 mM EDTA buffer and again centrifuged. The two extracts are combined and solid $(NH_4)_2SO_4$ added over a 30-min period to a final concentration of 2.2 M (36.0 g of $(NH_4)_2SO_4$ per 100 ml of extract). The suspension is centrifuged 1 hr at 12,000 g and the pellet discarded. To the supernatant is added solid $(NH_4)_2SO_4$ until the solution is almost saturated (about 34 g of $(NH_4)_2SO_4$ per 100 ml of supernatant). The suspension is allowed to stand either for 1 hr or overnight and then is centrifuged 70 min at 12,000 g.

Ion Exchange Chromatography. The pellet is suspended in about 30 ml of 5 mM K phosphate–1 mM EDTA–1 mM dithiothreitol, pH 6.5 (termed phosphate buffer) and then dialyzed overnight against three changes of the same buffer. Denatured protein is removed by centrifugation.

The dialyzed supernatant is applied to a 2.5 × 15-cm column of DEAE-cellulose that is equilibrated with the phosphate buffer. The column is flushed with the same buffer. Triose-P dehydrogenase emerges in the void volume and these fractions are immediately applied to a 2.5 × 15-cm column of CM-cellulose equilibrated with the phosphate buffer. After eluting with phosphate buffer, the enzyme again emerges in the void volume.

---

[6] S. C. Fink and R. W. Brosemer, this volume [53].
[7] C. W. Carlson, S. C. Fink, and R. W. Brosemer, *Arch. Biochem. Biophys.* **144**, 107 (1971).

Purification of Triosephosphate Dehydrogenase from Honey Bees

Fraction	Volume (ml)	Protein (mg)	Total units	Specific activity
1. Extract	365	9700	ca. 13,000[a]	ca. 1.3
2. Dialyzed ammonium sulfate-precipitated fraction	90	1560	12,000	7.7
3. First crystallization	23	390	11,000	28
4. Fifth crystallization	10	170	10,100	60

[a] The enzyme assay in the extract is at best a rough estimate due to interference by high levels of glycerol-3-P dehydrogenase.

The two columns can be set up in series so that only fractions emerging from both columns need be monitored.

Crystallization. The protein eluted from the second column is precipitated by adding solid $(NH_4)_2SO_4$ to a final concentration of 3.7 M (6.6 g $(NH_4)_2SO_4$ per 10 ml). The yellow sediment is collected by centrifugation and suspended in 10 ml of 0.1 M Tris–10 mM EDTA-1 mM dithiothreitol, pH 7.5. Insoluble material is removed by centrifugation and suspended in a second 10 ml of the same buffer. After centrifugation, the two supernatants are combined. Solid $(NH_4)_2SO_4$ is added until a definite turbidity appears. Crystals form within a time of 2 sec to several hours. More solid $(NH_4)_2SO_4$ is added in small amounts over the next 4 days until essentially all the protein has crystallized.

For recrystallization, the crystals are harvested by centrifugation, dissolved in 0.1 M Tris–10 mM EDTA-1 mM dithiothreitol, pH 7.5, and crystallized as above. Three to five crystallizations are required in order to reach maximum specific activity.

Other Insect Species. The above procedure has been used to purify triosephosphate dehydrogenase from the following insects: four species of bumble bees (*Bombus nevadensis, B. occidentalis, B. appositus, Psithyrus suckleyi*), leafcutting bees, flesh flies, and screwworm flies.[8] The isolation procedure is so reproducible and reliable that the enzyme is usually not assayed during the fractionation, even when using an insect species for the first time.

Properties

Stability. The enzyme can be stored without loss of activity for several months at 2° as a crystalline suspension in 3.5 M ammonium sulfate, if a sulfhydryl compound is present in order to protect the enzyme from

[8] C. W. Carlson and R. W. Brosemer, *Biochemistry* **10**, 2113 (1971).

air oxidation. A solution containing 10 mM dithiothreitol is satisfactory if it is replaced with fresh solution every 10 days.

Kinetic Properties. The catalytic properties of the insect enzymes have not been as extensively investigated as those of the mammalian muscle or yeast enzymes. The effect of glyceraldehyde-3-P and NAD$^+$ concentrations on the reaction velocity of the honey bee enzyme has been reported.[9] In the presence of 4.3 mM DL-glyceraldehyde-3-P, a kinetic transition is observed when the NAD$^+$ concentration is raised from 0.16 to 0.18 mM. In the presence of 1 mM NAD$^+$, a similar transition occurs when D-glyceraldehyde-3-P is increased from 0.62 to 0.72 mM. Lowering either the NAD$^+$ or glyceraldehyde-3-P concentration eliminates this transition.

Half-maximal velocity of the honey bee enzyme is attained with about 40 μM NAD$^+$ or 200 μM D-glyceraldehyde-3-P.[9] The pH optimum of the honey bee and bumble bee enzymes is 9.1.[10]

Inhibition. Iodoacetate, N-ethylmaleimide, ATP, ADP, AMP, and 3′,5′-AMP inhibit the honey bee enzyme.[9,11]

Molecular Weight. Both the sedimentation coefficient[8] and molecular weight estimation with Sephadex gel chromatography[9] indicate that the molecular weight of the honey bee enzyme is identical to that of the enzyme from other sources,[12-17] i.e., 140,000.

The honey bee enzyme binds 4 moles of NAD$^+$.[9] Tryptic peptide maps of the insect enzymes suggest a tetrameric structure composed of identical subunits.[8] Other studies indicate that triose-phosphate dehydrogenases from mammals and yeast consist of four identical subunits, each of which binds 1 mole of NAD$^+$.[12-18]

Other Properties. The enzyme is yellow due to the absorbance of bound NAD$^+$ at 360 nm. The NAD$^+$ can be removed with charcoal without altering electrophoretic mobility or enzyme specific activity.[10]

The dissociation constants of the four NAD$^+$-binding sites were deter-

[9] W. G. Gelb, E. J. Oliver, J. F. Brandts, and J. H. Nordin, *Biochemistry* **9**, 3228 (1970).
[10] R. R. Marquardt, C. W. Carlson, and R. W. Brosemer, *J. Insect Physiol.* **14**, 317 (1968).
[11] W. G. Gelb, E. J. Oliver, J. F. Brandts, and J. H. Nordin, *Cryobiology* **8**, 474 (1971).
[12] R. N. Perham and J. I. Harris, *J. Mol. Biol.* **7**, 316 (1963).
[13] W. S. Allison, and N. O. Kaplan, *J. Biol. Chem.* **239**, 2140 (1964).
[14] J. I. Harris and R. N. Perham, *J. Mol. Biol.* **13**, 876 (1965).
[15] W. Harrington and G. Karr, *J. Mol. Biol.* **13**, 885 (1965).
[16] B. E. Davidson, M. Sajgo, H. F. Noller, and J. I. Harris, *Nature (London)* **216**, 1181 (1967).
[17] J. I. Harris and R. N. Perham, *Nature (London)* **219**, 1025 (1968).
[18] A. Conway and D. E. Koshland, Jr., *Biochemistry* **7**, 4011 (1968).

mined by equilibrium dialysis to be: 1×10^{-6}, 3×10^{-5}, 3×10^{-5}, 6×10^{-4} M^{-1} for the first through fourth sites, respectively.[9]

The amino acid composition and peptide maps of most of the insect triosephosphate dehydrogenases that have been isolated have been reported.[8]

The enzyme was localized in honey bee flight muscle with immunofluorescent techniques; it is located at or near the Z-band cross-striations.[19]

[19] R. W. Brosemer, *J. Histochem. Cytochem.* **20**, 266 (1972).

[61] 3-Phosphoglycerate Dehydrogenase from Seedlings of *Pisum sativum*

By J. C. SLAUGHTER and D. D. DAVIES

3-Phosphoglycerate $+ NAD^+ \rightleftharpoons$ phosphohydroxypyruvate $+ NADH + H^+$

Higher plants contain enzymes which can in principle synthesize serine from glycerate. However, the physiological significance of these enzymes may lie in the synthesis of carbohydrates from serine during photosynthesis.[1] Isotopic evidence[2] indicates that plants can synthesize serine from 3-phosphoglycerate and the enzymes involved have been demonstrated in pea seedlings.[3] The first enzyme in this sequence, 3-phosphoglycerate dehydrogenase, has been purified from pea seedlings.[4] The allosteric properties of the enzyme are consistent with current ideas concerning feedback inhibition of the first enzyme in a metabolic sequence.[5] However, the allosteric properties are unstable and are lost during purification.[4,6] Glycerol and other polyhydroxy compounds protect the enzyme.

Assay Method

Principle. 3-Phosphoglycerate dehydrogenase activity can be determined, in either the forward or reverse reaction, by measuring the initial rates of reduction or oxidation of NAD and NADH, respectively. The measurement is carried out in a spectrophotometer equipped with a constant temperature cell holder maintained at 25° at wavelength 340 nm, using silica cells of 10-mm light path.

[1] N. E. Tolbert, *Symp. Soc. Exp. Biol.* **27**, 215 (1973).
[2] J. Hanford and D. D. Davies, *Nature (London)* **182**, (1958).
[3] J. C. Slaughter and D. D. Davies, *J. Exp. Bot.* **20**, 451 (1969).
[4] J. C. Slaughter, and D. D. Davies, *Biochem. J.* **109**, 743 (1968).
[5] J. C. Slaughter and D. D. Davies, *Biochem. J.* **109**, 749 (1968).
[6] I. Y. Rosenblum and H. J. Sallach, *Arch. Biochem. Biophys.* **137**, 91 (1970).

3-Phosphoglycerate Oxidation

Reagents

Sodium glycinate buffer, 100 mM, pH 9.5
Tricyclohexylammonium D-3-phosphoglycerate, 35 mM
NAD⁺, 60 mM

The reactants are dissolved in the sodium glycinate buffer.

Procedure. The reaction mixture contains 0.10 ml of 3-phospho-glycerate, 0.10 ml of NAD⁺, 2.65 ml sodium glycinate buffer, 0.05 ml of water, and 0.10 ml of enzyme. The reaction is started by addition of the 3-phosphoglycerate. Because of the unfavorable chemical equilibrium, it is essential to use scale expansion.

3-Phosphohydroxypyruvate Reduction

Reagents

Sodium succinate buffer, 100 mM, pH 5.5
Potassium phosphohydroxypyruvate, 9.0 mM[7]
NADH, 100 mM

The NADH is dissolved in potassium phosphate buffer 100 mM, pH 7.5

Procedure. The reaction mixture contains 0.02 ml of phosphohydroxy-pyruvate, 0.02 ml of NADH, 0.95 ml of sodium succinate buffer, and 0.01 ml of enzyme. The reaction is started by addition of the phosphohydroxypyruvate.

Units. One unit of enzyme is defined as the amount that causes the oxidation of 1 μmole NADH per minute under the stated conditions. Specific activity is defined as the number of units per milligram of protein. Protein is measured by the method of Warburg and Christian,[9] but for very low concentrations the method of Murphy and Kies[10] was substituted.

Purification Procedure

Pea seeds (var. Alaska) are soaked overnight in running water and planted thickly in moistened vermiculite. The trays are stored in the dark

---

[7] Potassium phosphohydroxypruvate was prepared by the method of Ballou and Hesse[8] from phosphoyhydroxypyruvic acid dimethylketal tricyclohexylammonium salt obtained from Calbiochem. (Los Angeles, California).

[8] C. E. Ballou and R. Hesse, *J. Amer. Chem. Soc.* **78**, 3718 (1956).

[9] O. Warburg and W. Christian, *Biochem. Z.* **310**, 384 (1942).

[10] J. B. Murphy and M. W. Kies, *Biochim. Biophys. Acta* **45**, 382 (1960).

at 25°, and the epicotyls are harvested after 8–10 days. All operations are carried out at 2°. Shoots, 200 g, are blended with 200 ml of potassium phosphate buffer (pII 6.5, 10 mM) containing glycerol (2.5 M) for 30 sec in an Ato-Mix blender. The homogenate is strained through fine nylon mesh and centrifuged at 15,000 g for 20 min to give a clear supernatant—the crude extract.

Step 1. Ten grams of Sigma alumina C_γ gel (8% solids) is added, removing about 90% of the activity. The gel is separated by centrifugation at 3000 g for 10 min, then washed with five 50-ml portions of potassium phosphate buffer (pH 6.5, 100 mM) containing glycerol (2.5 M). Washes two to five inclusive, are combined.

Step 2. Protein is precipitated with $(NH_4)_2SO_4$ and the fraction that precipitates between 28 and 49 g per 100 ml is collected by centrifugation at 5000 g for 10 min. The precipitate is dissolved in the minimum volume of potassium phosphate buffer (pH 6.5, 50 mM) containing glycerol (2.5 M).

Step 3. The solution is passed through a column of Bio-Gel P-10 previously equilibrated with potassium phosphate buffer (pH 6.5, 50 mM) containing glycerol (2.5 M). The eluate is assayed for enzyme activity, and the active fractions are combined.

Step 4. The active fractions are adsorbed on a column of DEAE-cellulose (120 × 12 mm) previously equilibrated with potassium phosphate buffer (pH 6.5, 50 mM) containings glycerol (2.5 M). The column is washed with 30 ml of the same buffer containing glycerol (2.5 M), and the enzyme is eluted with 100 mM potassium phosphate buffer pH 6.5 containing glycerol (2.5 M). By using air pressure, a high flow rate (50 ml per hour) is obtained. This is desirable as the enzyme is unstable while adsorbed.

Step 5. The active fractions were passed through a column of Sephadex G-200 (400 × 28 mm) with upward flow of potassium phosphate buffer (pH 6.5, 10 mM) containing glycerol (2.5 M).

The purification procedure is summarized in the table.

Properties

Specificity. NADPH gave about 25% of the rate with NADH, but NADP+ could not replace NAD+ at the concentration used in the standard assay. The enzyme shows A stereospecificity for NADH.

Stability. The activity of the purified enzyme declined slowly and steadily at 2°, some 60% remained after 7 days.

pH Optimum. In the direction of NADH oxidation, the enzyme had optimum activity at pH 5.5. In the direction of NAD+ reduction an opti-

PURIFICATION OF 3-PHOSPHOGLYCERATE DEHYDROGENASE FROM PEAS[a]

Fraction	Total activity (units)	Total protein (mg)	Specific activity (units/mg)	Purification (fold)	Yield (%)	Inhibition by serine (%)
Crude extract	59.2	2320	0.025	1	100	83
Alumina C_γ eluate	48.0	312	0.154	6	81	25
(NH₄)₂SO₄ ppt.	31.0	101	0.307	12	52	15
Bio-Gel eluate	23.5	36	0.65	25	40	16
DEAE-cellulose	16.1	4.0	3.35	131	27	0
Sephadex G-200 eluate	5.9	0.5	11.8	470	10	0

[a] The initial extraction buffer was potassium phosphate, pH 6.5 (10 mM), containing glycerol (2.5 M), and throughout the procedure all buffers contained glycerol (2.5 M). All operations were carried out at 2°. The results for the inhibition due to serine were obtained with 1.0 mM L-serine in the assay mixture.

mum at pH 9.5 was accompanied by a sharp fall in activity at higher pH values.

K_m *Values.* The following values were obtained: NAD⁺ at pH 9.5, 125 μM; 3-phosphoglycerate at pH 9.5, 50 μM; NADH at pH 5.5, 10 μM; phosphohydroxypyruvate at pH 5.5, 50 μM.

Inhibitors. p-Hydroxymercuribenzoate, iodacetate, and N-ethylmaleimide inhibit the enzyme. Substrate protection by 3-phosphoglycerate, but not by NAD⁺, was observed only with the weakest of the inhibitors—N-ethylmaleimide.

When a freshly prepared extract is examined for its response to L-serine at pH 6.5, the kinetics are those of "classical" noncompetitive inhibition, but on storage at 2°, sigmoid kinetics of inhibition are observed. Storage at 2° reduces the inhibition produced by serine, but sensitivity to serine can be partially recovered by incubating at 25° for 4 hr before assay.[5]

The enzyme is inhibited by ATP, GTP, ADP, and GDP, but not by AMP or GMP.[11] The inhibition produced by the purine nucleotides is competitive with respect to NADH, and ATP is the most effective inhibitor.

Activators. L-Methionine activates the enzyme, and the extent of activation increases with the time of incubation up to a period of 8–10 min at 25°. The concentration of methionine necessary for maximum activation is 10 mM.[12]

[11] J. C. Slaughter, *Phytochemistry* **12**, 2627 (1973).
[12] J. C. Slaughter, *FEBS Lett.* **7**, 245 (1970).

[62] D-3-Phosphoglycerate Dehydrogenase from Hog Spinal Cord[1]

By Ronald D. Feld and H. J. Sallach

D-3-Phosphoglycerate $+$ NAD$^+$ $\leftrightarrow$ phosphohydroxypyruvate $+$ NADH $+$ H$^+$

Assay Method

The assay method is identical to that described for the enzyme from wheat germ (see this volume [63]) with respect to reagents, procedure, definition of unit, and specific activity.

Purification Procedure

The following general conditions are used unless otherwise stated. All buffers contain 1 mM EDTA and 5 mM 2-mercaptoethanol. In column chromatographic procedures, the resins used are equlibrated with the same buffer as used for dialysis of the enzyme preparation for the appropriate step. All other general conditions are those outlined for the wheat germ enzyme [63]. A typical enzyme preparation is described below.

Step 1. Preparation of Crude Extract. Hog spinal cords, which are collected on ice at the local slaughterhouse, are stripped of outer membranes. The tissue (1500 g) is homogenized for 1 min with 2 volumes (w/v) of 0.154 M KCl containing 10 mM potassium phosphate buffer, pH 7.4. After centrifugation, the resulting supernatant solution is decanted through glass wool to remove insoluble lipids.

Step 2. Treatment with Protamine Sulfate. A concentrated solution of protamine sulfate (3 g/100 ml of buffer) is added slowly, with stirring, to the filtered supernatant solution (2000 ml) in a volume equivalent to the addition of 1.5 mg of protamine sulfate per milliliter of supernatant solution.[2] After equilibration for 15 min, the precipitate is removed by centrifugation and discarded.

Step 3. Ammonium Sulfate Fractionation. The enzyme is precipitated

---

[1] This work was supported by National Institutes of Health Grant No. NS10287 and by Research Contract No. AT(11-1)-1631 from the U.S. Atomic Energy Commission.

[2] Protamine sulfate is obtained from Sigma (Grade I). It is advisable to carry out a pilot run to determine how much of the particular preparation (lot) of protamine sulfate can be added before precipitation of enzyme occurs. Streptomycin sulfate is not as effective at this step.

from the above supernatant solution by the slow addition, with stirring, of 38 g of solid ammonium sulfate per 100 ml of solution. After centrifugation, the 0–38% ammonium sulfate residue is resuspended in a 1.64 M ammonium sulfate solution containing 0.1 M potassium phosphate buffer, pH 7.4.[3] The suspension is equilibrated with stirring for 15 min and centrifuged; the supernatant solution is discarded. The residue is then resuspended in a 0.82 M ammonium sulfate solution containing 0.1 M potassium phosphate buffer, pH 7.4.[3] After equilibration and centrifugation as above, the residue is discarded and, to the supernatant solution, is added 15 g of solid ammonium sulfate per 100 ml. After equilibration, the ammonium sulfate precipitate is recovered by centrifugation and retained.

Step 4. First CM-Sephadex Chromatography. The ammonium sulfate precipitate from the above step is dissolved in 15 ml of 10 mM potassium phosphate buffer, pH 6.5, and dialyzed against 1 liter of the same buffer, with 30-min changes, for 5 hr. Insoluble material is removed by centrifugation, and the resulting supernatant solution is applied to a CM-Sephadex column (5 × 12 cm). After the sample has been applied, the column is washed with 100 ml of the same buffer and then with 650 ml of 35 mM potassium phosphate buffer, pH 6.5; these fractions are discarded. The enzyme is eluted with 35 mM potassium phosphate buffer, pH 6.5, containing 50 mM KCl. The flow rate of the column is 60–70 ml per hour. Fractions containing enzyme activity are pooled and solid ammonium sulfate (60 g/100 ml) is added to the resulting solution. The 0–60% ammonium sulfate precipitate is recovered by centrifugation and retained.

Step 5. DEAE-Sephadex Chromatography. The 0–60% ammonium sulfate fraction is dissolved in a minimal volume of 18 mM potassium phosphate buffer, pH 7.4, and dialyzed against the same buffer until the dialyzate is free of ammonium ions. The solution is then applied to a DEAE-Sephadex A-50 column (1.5 × 18 cm), and the column is then washed with 100 ml of the same buffer prior to the initiation of a linear gradient of 35 mM to 0.1 M potassium phosphate buffer, pH 6.5. Each vessel contains 150 ml of the appropriate buffer, and the flow rate of the column is 25 ml per hour. Fractions (5 ml) containing enzyme activity are pooled, and the pH of the resulting solution is adjusted to 6.5 with 1 N acetic acid. The solution is then applied to a hydroxyapatite column (1 × 6 cm) that has been equilibrated with 35 mM potassium phosphate buffer, pH 6.5, containing 50 mM KCl. The enzyme is eluted from the column with 0.2 M potassium phosphate buffer, pH 7.4, containing 50

---

[3] The volume of solution used for the resuspension is 0.05 of that from which the 0–38% ammonium sulfate precipitate was prepared. The resuspension of the residues is facilitated by homogenization in a Potter-Elvehjem glass homogenizer.

mM KCl. Fractions containing enzyme activity are pooled[4] and the enzyme is precipitated by the addition of 60 g of ammonium sulfate per 100 ml of solution. The 0–60% ammonium sulfate residue is recovered by centrifugation, and this fraction is stable to storage at −15° for several weeks. In practice, a second 1500-g sample of spinal cord is carried through the above five steps. The 0–60% ammonium sulfate fractions obtained at this step are then combined and utilized in the next step.

Step 6. Second CM-Sephadex Chromatography. The 0–60% ammonium sulfate residues from the above step are dissolved in a minimal volume of 10 mM potassium phosphate buffer, pH 6.5, and dialyzed against the same buffer until the dialyzate is free of ammonium ions. The solution is then applied to a CM-Sephadex column (1.5 × 12 cm), the column is washed with one bed volume of application buffer, and then a linear gradient is applied. The mixing vessel contains 100 ml of 0.01 M potassium phosphate buffer, pH 6.5, and the reservoir contains 100 ml of 35 mM potassium phosphate buffer, pH 6.5, in 50 mM KCl. Fractions (3 ml) containing enzyme activity are pooled and the enzyme is concentrated by passing the solution through a hydroxyapatite column (1 × 2 cm) that has been equilibrated with 35 mM potassium phosphate buffer, pH 6.5, containing 50 mM KCl. The enzyme is eluted with 0.2 M potassium phosphate buffer, pH 7.4, containing 50 mM KCl. Fractions containing enzyme activity are pooled and the enzyme is precipitated by the addition of 60 g of ammonium sulfate per 100 ml of solution. The 0–60% ammonium sulfate residue is recovered by centrifugation and is stable to storage at −15° for several weeks.

At this stage of purification, the enzyme preparation is free of both D-glycerate and L-lactate dehydrogenases. The results of a typical purification are shown in the table.

Properties

The properties of the enzyme resemble those of the purified enzyme from chicken liver[5] in that NADP[+] is inactive as a coenzyme but certain analogs of NAD[+] are utilized. With the activity of NAD[+] as 100%, the relative activities of the analogs are deamino-NAD[+] (220%); 3-acetyl-pyridinedeamino-NAD[+] (167%); 3-acetylpyridine-NAD[+] (125%); and thio-NAD[+] (17.5%).

The enzyme is widely distributed in different mammalian species and occurs in most tissues including liver, kidney, brain, adrenals, adipocyte,

---

[4] Chromatography on hydroxyapatite is used as a means of concentrating the enzyme; little purification is achieved under these conditions.

[5] D. A. Walsh and H. J. Sallach, *Biochim. Biophys. Acta* **146**, 26 (1967).

PURIFICATION DATA FOR D-3-PHOSPHOGLYCERATE
DEHYDROGENASE FROM HOG SPINAL CORD

Purification step	Total units	Specific activity	Yield (%)
1. Crude extract	95,000	5	—
2. Protamine sulfate	95,000	16	100
3. Ammonium sulfate fractionation	89,000	56	93
4. First CM-Sephadex	51,000	390	53
5. DEAE-Sephadex	31,000	1750	32
6. Second CM-Sephadex	15,500	3100	16

ovary, prostate, and testis.[6-8] The level of activity of the hepatic enzyme has been shown to be altered under different dietary and hormonal conditions.[9-12]

[6] J. E. Willis and H. J. Sallach, *Biochim. Biophys. Acta* **81**, 39 (1964).
[7] D. A. Walsh and H. J. Sallach, *J. Biol. Chem.* **241**, 4068 (1966).
[8] H. J. Sallach, unpublished observations.
[9] H. J. Fallon, E. J. Hackney, and W. L. Byrne, *J. Biol. Chem.* **241**, 4157 (1966).
[10] H. J. Fallon, *Advan. Enzyme Regul.* **5**, 107 (1967).
[11] G. P. Cheung, J. P. Cotropia, and H. J. Sallach, *Arch. Biochem. Biophys.* **129**, 672 (1969).
[12] H. J. Sallach, R. A. Sanborn, and W. J. Bruin, *Endocrinology* **91**, 1054 (1972).

[63] D-3-Phosphoglycerate Dehydrogenase from Wheat Germ[1]

By I. Y. ROSENBLUM and H. J. SALLACH

D-3-Phosphoglycerate + NAD$^+$ ↔ phosphohydroxypyruvate + NADH + H$^+$

Assay Method[2]

Principle. Either the oxidation of 3-phosphoglycerate (3-PGA) or the reduction of phosphohydroxypyruvate with the concomitant formation or utilization of NADH may be used as an assay. However, due to the greater stability and availability of 3-PGA, as compared to phospho-

[1] This work was supported by National Institutes of Health Grant No. NS10287 and by Research Contract No. AT(11-1)-1631 from the U.S. Atomic Energy Commission.
[2] J. E. Willis and H. J. Sallach, *Biochim. Biophys. Acta* **81**, 39 (1964).

hydroxypyruvate, the most convenient assay is the spectrophotometric determination of the rate of NADH formation at 340 nm in the presence of 3-PGA and dehydrogenase. Although the equilibrium of the reaction strongly favors the reduction of phosphohydroxypyruvate, it is shifted to the right by carrying out the reaction at pH 9 and in the presence of the carbonyl-trapping reagent, hydrazine.

Reagents

3-PGA, potassium salt, pH 7.0

NAD^+, 15 mM

Glutathione, 50 mM, adjusted to pH 6 with 1 N KOH

Stock buffer solution. Mix 5 volumes of 1.0 M Tris chloride, pH 9.0, 4 volumes of 1.0 M hydrazine acetate, pH 9.0, and 1 volume of 0.25 M EDTA, pH 9.0, on day of use.

Enzyme, prior to assay, dilute in 0.1 M potassium phosphate buffer, pH 7.4, containing 1 mM EDTA and 0.1 mM dithiothreitol, so that 0.1 ml will contain 10–20 units as defined below.

Procedure. Place 1.0 ml of buffer solution, 0.1 ml of glutathione, 0.25 ml of 3-PGA (omit from control and blank), 0.1 ml of enzyme, and water to a final volume of 2.25 ml in each of three quartz cells with 1.0-cm light paths. The reaction, which is carried out at 25°, is initiated by the addition of 0.1 ml of NAD (replaced by water in the blank) and is most conveniently followed by measuring the change in absorbancy at 340 nm with a Beckman DU-Gilford 2000 recording spectrophotometer.

Definition of Unit and Specific Activity. One unit of activity is defined as that amount of enzyme that produces an increase in absorbancy of 0.001 per minute under the above conditions which is equivalent to the formation of 1.61×10^{-4} μmoles of NADH per minute. Specific activity is defined as the number of units per milligram of protein as determined by the method of Lowry et al.[3] with bovine serum albumin as the standard or from the absorbancy at 280 nm according to the method of Warburg and Christian.[4]

Purification Procedure[5]

Unless otherwise stated, the following general conditions are used. All operations are carried out at 0–4°. Ammonium sulfate fractionations are

[3] O. H. Lowry, N. J. Rosebrough, A. L. Farr, and R. J. Randall, *J. Biol. Chem.* **193**, 265 (1951).

[4] O. Warburg and W. Christian, *Biochem. Z.* **310**, 384 (1942). See also this series, Vol. 3 [73].

[5] I. Y. Rosenblum and H. J. Sallach, *Arch. Biochem. Biophys.* **137**, 91 (1970).

made by the slow addition, with stirring, of the calculated amount of solid salt. The resulting suspensions are equilibrated with stirring for 30 min prior to centrifugation. Centrifugations are carried out at 14,000 g for 30 min. Protein in eluates from column chromatographic procedures is detected by absorbancy at 280 nm. All buffers are prepared from the potassium salts and contain 3 mM EDTA and 2.5 mM 2-mercaptoethanol. A typical enzyme preparation is described below.

Step 1. Preparation of Crude Extract. Wheat germ acetone powder[6] (200 g) is extracted by stirring for 30 min with 10 volumes (w/v) of 0.1 M Tris-chloride buffer, pH 8.0. After centrifugation, the resulting supernatant solution is filtered through glass wool to remove insoluble lipid material.

Step 2. Streptomycin Treatment. To the above filtered solution (1850 ml) is added 25 ml of a solution of streptomycin sulfate (400 mg/ml). The resulting suspension is equilibrated with stirring for 30 min prior to the removal of the precipitate by centrifugation.

Step 3. Ammonium Sulfate Fractionation. The supernatant solution from the above step is fractionated with ammonium sulfate (34 g/100 ml) and the 0–34% precipitate is removed by centrifugation. To the resulting supernatant solution, ammonium sulfate (16 g/100 ml) is added, and the precipitate is recovered by centrifugation. This 34–50% ammonium sulfate fraction may be stored at −15° for several weeks without loss of activity.

Step 4. Gel Filtration and Ammonium Sulfate Fractionation. The above precipitate is dissolved in 100 ml of 0.1 M potassium phosphate buffer, pH 7.4, and passed through a Sephadex G-25 column (4.8 × 26 cm) which is equilibrated with the same buffer. This step effectively removes contaminating yellow pigments. Enzyme activity is precipitated from the protein eluate by the addition of 41 g of ammonium sulfate per 100 ml of solution. The 0–41% ammonium sulfate fraction is recovered by centrifugation and is stable to storage at −15° for several weeks.

Step 5. DEAE-Sephadex A-50 Chromatography. One-fourth of the above ammonium sulfate fraction is dissolved in 10 ml of 0.1 M potassium phosphate buffer, pH 7.4, and dialyzed against the same buffer with changes until the dialyzate is free of ammonium ions as determined with Nessler's reagent. The dialyzed enzyme solution is applied to a DEAE-Sephadex A-50 column (3 × 27 cm) which is equilibrated with 50 mM

---

[6] See this series Vol. 1 [8] Method 3. Preliminary studies with raw wheat germ (purchased from General Mills, Inc., Minneapolis, Minnesota) showed that purification procedures are complicated by the large amounts of lipid in the raw tissue which is removed without loss of enzyme activity in the preparation of the acetone powder.

potassium phosphate buffer, pH 7.4. The column is washed with the same buffer (flow rate = 30 ml/hr) until the eluate is essentially free of protein as determined by absorbancy at 280 nm. The enzyme is eluted from the resin with 0.2 M potassium phosphate buffer, pH 7.4, in the next protein fraction. To this fraction, ammonium sulfate (41 g/100 ml) is added and the precipitate is recovered by centrifugation at 35,000 g for 20 min. This 0–41% ammonium sulfate fraction is stable to storage at −15° for 1 month.

Step 6. SE-Sephadex C-50 Chromatography. One-half of the above ammonium sulfate precipitate is dissolved in 1.0 ml of 50 mM potassium phosphate buffer, pH 7.4, and dialyzed against 50 mM potassium phosphate buffer, pH 6.5, until free of ammonium ions. The dialyzed solution is then applied to a SE-Sephadex C-50 column (3.0 × 24 cm) equilibrated with the latter buffer. The column is washed with the same buffer (flow rate = 20–30 ml/hr) and 40 ml of enzyme solution is collected immediately after the void volume. The above eluate is adjusted to pH 7.4 by the slow addition of 1 M potassium phosphate (dibasic) and concentrated to one-half volume by ultrafiltration (Diaflo ultrafiltration apparatus Amicon Corp., Lexington, Massachusetts). Ammonium sulfate (48 g/100 ml) is then added to the concentrated enzyme solution. During the fractionation, the solution is maintained at pH 7.4 by the addition of 1 M potassium phosphate (dibasic). The 0–48% ammonium sulfate precipitate is recovered by centrifugation and is stored at −15°. Zone electrophoresis of this fraction followed by protein staining gives three major protein bands only one of which stains for phosphoglycerate dehydrogenase activity.

A summary of the purification data is given in the table.

PURIFICATION DATA FOR D-3-PHOSPHOGLYCERATE
DEHYDROGENASE FROM WHEAT GERM

Step and fraction	Specific activity	Recovery (%)
1. Acetone powder extract	5	—
2. Streptomycin-treated supernatant	6	100
3. 34–50% Ammonium sulfate fraction	30	80
4. 0–41% Ammonium sulfate precipitate from Sephadex G-25	40	80
5. 0–41% Ammonium sulfate fraction from DEAE-Sephadex A-50 eluate	500	58
6. 0–48% Ammonium sulfate fraction of concentrated eluate from SE-Sephadex C-50	1250	38

Properties[5]

The enzyme is susceptible to inactivation, particularly at high dilutions. Addition of dithiothreitol increases the stability. Sulfhydryl reagents are inhibitory.

Although NADP[+] is essentially inactive, several analogs of NAD[+] are effective. Relative activities as compared to NAD[+] (100%) are 3-acetylpyridine-NAD[+] (199%); 3-acetylpyridinedeamino-NAD[+] (193%); deamino-NAD[+] (81%); thio-NAD[+] (66%); and NADP[+] (6.5%). Kinetic constants for wheat germ phosphoglycerate dehydrogenase have been determined at pH 8 and 25°. The K_m values are NAD[+] = 0.50 ± 0.06 mM; 3-PGA = 1.35 ± 0.23 mM; NADH = 12.0 ± 1.5 μM; and phosphohydroxypyruvate = 40.2 ± 6.9 μM.

Other Plant Sources

The enzyme is widely distributed in plants and, in general, 3-phosphoglycerate dehydrogenase activity is found primarily in seeds and/or embryonic tissue with little or no activity in the green leaf.[7] The enzyme has been partially purified from pea epicotyl.[8]

[7] G. P. Cheung, I. Y. Rosenblum, and H. J. Sallach, *Plant Physiol.* **43**, 1813 (1968).
[8] J. C. Slaughter and D. D. Davies, *Biochem. J.* **109**, 743 (1968).

[64] D-Glycerate Dehydrogenase from Hog Spinal Cord[1]

By RONALD D. FELD and H. J. SALLACH

$$\text{D-Glycerate} + \text{NAD}^+ \leftrightarrow \text{hydroxypyruvate} + \text{NADH} + \text{H}^+$$

Assay Method[2]

Principle. The reaction is measured in the direction of D-glycerate (D-GA) oxidation by the increase in absorbancy at 340 nm due to the formation of NADH. The reverse reaction (reduction of hydroxypyruvate) cannot be used in crude extracts since hydroxypyruvate is reduced to L-glycerate in the presence of mammalian L-lactate dehydrogenase and reduced pyridine nucleotides.[3,4]

[1] This work was supported by National Institutes of Health Grant No. NS10287 and by Research Contract No. AT(11-1)-1631 from the U.S. Atomic Energy Commission.
[2] J. E. Willis and H. J. Sallach, *J. Biol. Chem.* **237**, 910 (1962).

Reagents

D-GA, potassium salt, 0.1 *M*, pH 7.0[5]

NAD, 15 m*M*

Glutathione, 50 m*M*, adjusted to pH 6 with 1 *N* KOH

Stock buffer solution. Mix 5 volumes of 1.0 *M* Tris chloride, pH 9.0, 4 volumes of 1.0 *M* hydrazine acetate, pH 9.0, 1 volume of 0.25 *M* EDTA, pH 9.0, on the day of use.

Enzyme. Prior to assay, dilute in 10 m*M* phosphate buffer, pH 7.0, so that 0.1 ml will contain approximately 20 units as defined below.

Procedure. Place 1.0 ml of buffer solution, 0.1 ml of D-GA (omit from control and blank), 0.1 ml of enzyme, and water to a final volume of 2.5 ml in each of three quartz cells with a light path of 1.0 cm. The reaction is initiated by the addition of 0.1 ml of NAD (replaced by water in blank) and is followed by measuring the change in absorbancy at 340 nm with a Beckman DU-Gilford 2000 recording spectrophotometer.

Definition of Unit and Specific Activity. One unit of activity is defined as that amount of enzyme that produces an increase in absorbancy of 0.001 per minute under the above conditions; this is equivalent to the formation of 1.61×10^{-4} μmole of NADH per minute. Specific activity is defined as the number of units per milligram of protein as determined by the method of Lowry *et al.*[6] with bovine serum albumin as the standard or from the absorbancy at 280 nm according to the method of Warburg and Christian.[7]

Purification Procedure[8]

The following general conditions are followed unless otherwise specified. All steps are carried out at 0–4°. Fractionations with ammonium sulfate are made by the slow addition, with stirring, of the calculated amount of solid salt. The resulting suspensions are equilibrated with stirring for 30 min prior to centrifugation which is at 14,000 *g* for 30 min. Ammonium sulfate precipitates are dissolved in the appropriate buffer

[3] A. Meister, *J. Biol. Chem.* **197**, 309 (1952).

[4] S. R. Anderson J. R. Florini, and C. S. Vestling, *J. Biol. Chem.* **239**, 2991 (1964).

[5] E. Baer, J. M. Grosheintz, and H. O. L. Fisher, *J. Amer. Chem. Soc.* **61**, 2607 (1939).

[6] O. H. Lowry, N. J. Rosebrough, A. L. Farr, and R. J. Randall, *J. Biol. Chem.* **193**, 265 (1951).

[7] O. Warburg and W. Christian, *Biochem. Z.* **310**, 384 (1942). See also this series, Vol. 3 [73].

[8] R. D. Feld and H. J. Sallach, unpublished observations.

for the given purification step and are dialyzed with changes against the same buffer until the dialyzate is free of ammonium ions as determined with Nessler's reagent. In chromatographic procedures, gels are equilibrated with the appropriate buffer prior to pouring of the columns, which then are washed with several column volumes of buffer prior to use. Protein in effluent solutions from columns is detected by absorbancy at 280 nm. Enzymic solutions are concentrated by ultrafiltration (Diaflo ultrafiltration cell, Amicon Corp., Lexington, Massachusetts). All buffers contain 4 mM 2-mercaptoethanol and 1 mM EDTA.

Step 1. Crude Extract. Hog spinal cords are collected in ice at the slaughterhouse, and then the outer membranes are removed. The tissue (1500 g) is homogenized in a Waring Blendor for 60 sec with 22 volumes (w/v) of homogenizing medium consisting of 0.154 M KCl containing 10 mM potassium phosphate buffer, pH 7.4. After centrifugation, the resulting supernatant solution is filtered through glass wool to remove insoluble lipids.

Step 2. Protamine Sulfate Treatment. To the filtered solution (2029 ml) from the above step is added slowly, with stirring, a concentrated solution of protamine sulfate; the volume of the latter used is equivalent to the addition of 1.5 mg of protamine sulfate per milliliter of supernatant solution.[9] The resulting suspension is allowed to equilibrate for 15 min prior to removal of the precipitate by centrifugation.

Step 3. Ammonium Sulfate Fractionation. The supernatant solution from the above step is fractionated with ammonium sulfate (35 g/100 ml) and the 0–35% ammonium sulfate precipitate is retained.

Step 4. CM-Sephadex Chromatography. The above precipitate is dissolved in a minimal volume of 10 mM potassium phosphate buffer, pH 6.5, dialyzed against the same buffer and then applied to a CM-Sephadex column (5 $\times$ 18 cm) previously equilibrated against the same buffer. The column is washed with 600 to 700 ml of buffer, and the wash is discarded. The column is eluted with 35 mM potassium phosphate buffer, pH 6.5, at a flow rate of 60–65 ml per hour, and the enzyme appears in the next protein peak. Fractions containing activity are pooled and concentrated by ultrafiltration (1–2 mg of protein per milliliter).

Step 5. Calcium Phosphate Gel Treatment. Calcium phosphate gel[10] (25 mg/ml), previously equilibrated with 50 mM potassium phosphate buffer, pH 6.0, is added to the above concentrated enzyme solution to

[9] A pilot run is advisable to determine how much protamine sulfate may be added before precipitation of enzyme occurs. Streptomycin sulfate is not as effective as protamine sulfate at this step.

[10] D. Keilin and E. F. Hartree, *Proc. Roy. Soc. Ser. B.* **124,** 397 (1938). See also this series, Vol. 1 [11].

give a final gel-protein ratio of 4:1 (w/w). The slurry is allowed to equilibrate for 10 min before centrifugation at 3000 g for 5 min. The recovered supernatant solution is concentrated to a final protein concentration of 1–2 mg/ml. To this solution, solid ammonium sulfate (60 g/100 ml) is added, and the 0–60% precipitate is retained. During this fractionation, the pH is maintained at pH 6.0 by the addition of dibasic potassium phosphate.

Step 6. DEAE-Sephadex Chromatography. The 0–60% ammonium sulfate fraction from the above step is dissolved in a minimal volume of 0.018 M potassium phosphate buffer, pH 7.4, dialyzed against the same buffer until free of ammonium ions and then applied to a DEAE-Sephadex column (2 × 45 cm). The column is washed with the equilibrating buffer (rate = 25 ml/hr) until the eluate is free of protein. The enzyme is eluted with 50 mM potassium phosphate buffer, pH 7.4 in the next protein peak. Fractions containing enzyme activity are pooled and concentrated (1–2 mg of protein per milliliter).

Step 7. Second CM-Sephadex Chromatography. The concentrated enzyme solution from the above step is dialyzed against 50 volumes of 10 mM potassium phosphate buffer, pH 6.5, for 4 hr with 30 min changes prior to its application to a CM-Sephadex column (1.6 × 39 cm) which has been equilibrated with the same buffer. For elution of the enzyme, a two-chamber linear gradient apparatus containing 150 ml of solution per chamber is used. The mixing vessel contains 10 mM potassium phosphate buffer, pH 6.5, and the reservoir contains 50 mM potassium phosphate buffer, pH 6.5. Fractions containing enzyme activity are pooled and concentrated to a final protein concentration of 1–2 mg/ml. This solution can be stored at −15° for several weeks without appreciable loss of enzyme activity. At this point, the enzyme preparation is free of both 3-phosphoglycerate and lactate dehydrogenases.

The results of a typical purification are shown in the table.

Properties of the Enzyme[8]

The enzyme is active with the oxidized and reduced forms of both pyridine nucleotide coenzymes. In the forward direction (oxidation of D-glycerate), the relative rates with NAD$^+$ and NADP$^+$ do not differ significantly. On the other hand, marked differences are observed in the reverse direction (reduction of hydroxypyruvate). For example, at 50 μM hydroxypyruvate, the rate of oxidation of NADH is about eight times faster than that of NADPH under standard assay conditions with reduced coenzymes at 0.1 mM. Although the enzyme utilizes glyoxylate as well as hydroxypyruvate, the former is a much poorer substrate in

PURIFICATION DATA FOR D-GLYCERATE DEHYDROGENASE FROM
HOG SPINAL CORD

Purification step	Total units	Specific activity	Yield (%)
1. Homogenate	31,400	1.8	—
2. Protamine sulfate	30,900	4.3	98
3. (NH₄)₂SO₄ precipitation	33,990	31.6	108
4. First CM-Sephadex	24,810	199	79
5. CaPO₄ gel	24,560	403	78
6. DEAE-Sephadex	17,190	3900	55
7. Second CM-Sephadex	12,380	4630	40

that at 50 μM glyoxylate there is no detectable activity with either NADH or NADPH. At higher substrate concentrations (67 or 83 μM), the ratio of activities of hydroxypyruvate to glyoxylate is 15 with 0.1 mM NADH. These and other properties of the enzyme resemble those of D-glycerate dehydrogenase purified from beef liver.[11-13]

[11] I. Y. Rosenblum, D. H. Antkowiak, H. J. Sallach, L. E. Flanders, and L. A. Fahien, *Arch. Biochem. Biophys.* **144**, 375 (1971).
[12] E. Sugimoto, Y. Kitagawa, K. Nakanishi, and H. Chiba, *J. Biochem.* (*Tokyo*) **72**, 1307 (1972).
[13] E. Sugimoto, Y. Kitagawa, M. Hirose, and H. Chiba, *J. Biochem.* (*Tokyo*) **72**, 1317 (1972).

[65] D(−)-Lactate Dehydrogenases[1] from Fungi

By H. B. LeJohn and Roselynn M. Stevenson

$$D(-)\text{-Lactate} + NAD^+ \rightleftharpoons \text{pyruvate} + NADH + H^+$$

Lactate dehydrogenases are key enzymes in anaerobic energy metabolism. Two forms of the enzyme are recognized depending upon the isomer of lactic acid formed or oxidized. The L(+)-lactate dehydrogenases have been well studied in animals,[2] plants,[3] and microorganisms.[4-6] The en-

[1] D(−)-Lactate:NAD⁺ oxidoreductase, EC 1.1.1.28.
[2] N. O. Kaplan, *in* "Evolving Genes and Proteins" (H. J. Vogel, and V. Bryson, eds.), Academic Press, New York, p. 243, 1965.
[3] J. D'Auzac and J.-L. Jacob, *Bull. Soc. Chim. Biol.* **50**, 143 (1968).
[4] A. Yoshida, and E. Freese, *Biochim. Biophys. Acta* **99**, 56 (1965).
[5] N. Haugaard, *Biochim. Biophys. Acta* **31**, 66 (1959).
[6] A. Obayashi, H. Yorifuji, T. Yamagata, T. Ijichi, and M. Kanie, *Agr. Biol. Chem.* **30**, 717 (1966).

zymes that catalyze D(—)-lactate to pyruvate are also widespread in microorganisms.[7-10] The coenzyme used by both types of enzyme is NAD+. Less frequently the D(—)-lactic dehydrogenase is cytochrome-linked.[11]

The D(—)-lactate dehydrogenases of some microorganisms, e.g., the slime mold *Polysphondylium pallidum*,[12] *Streptococcus*,[13] and *Aerobacter aerogenes*,[14] are sensitive to salts,[12] various ions, and urea.[14] The streptococcal enzyme is allosterically regulated by fructose 1,6-diphosphate, and is practically inactive in its absence. The enzyme form from fungi other than yeast are less well studied, as the reports of Gleason *et al.*[9,15,16] demonstrate. Recently, we showed[17] that this enzyme is widely distributed among species of the so-called lower fungi. The enzyme has now been purified to homogeneity in 4 species, representing 3 distinct orders, of these lower fungi. The methods used for the purification and study of their properties are described here.

Fungi of the class Oomycetes contain only the D(—)-lactate dehydrogenase enzyme,[18] which is subject to allosteric control by GTP[17-19] Studies of the electrophoretic patterns and regulation of the isozymes present in 49 species of the Oomycetes support the general conclusion that GTP binds cooperatively to the enzyme in the presence of either NADH or D(—)-lactate but not with respect to NAD+ of pyruvate.[19]

Assay Method: Spectrophotometric Assay

Principle. Activity is determined in cell-free extracts and at various stages of purification by following the rate of NADH oxidation at 25° spectrophotometrically, at 340 nm.

[7] E. M. Tarmy and N. O. Kaplan, *J. Biol. Chem.* **243**, 2579 (1968).
[8] C. L. Wittenberger, *J. Biol. Chem.* **243**, 3067 (1968).
[9] F. H. Gleason, R. A. Nolan, A. C. Wilson, and R. Emerson, *Science* **152**, 1272 (1966).
[10] K. Purohit and G. Turian, *Arch. Mikrobiol.* **84**, 287 (1972).
[11] T. P. Singer, C. Gregolin, and T. Cremona, *in* "Control Mechanisms in Respiration and Fermentation," (B. E. Wright, ed.), p. 47. Ronald, New York, 1963.
[12] R. C. Garland and N. O. Kaplan, *Biochem. Biophys. Res. Commun.* **26**, 279 (1967).
[13] M. J. Wolin, *Science* **146**, 775 (1964).
[14] M.-C. Pascal and F. Pichinoty, *Ann. Inst. Pasteur* **107**, 55 (1964).
[15] F. H. Gleason and J. S. Price, *Mycologia* **61**, 945 (1969).
[16] F. H. Gleason, *Mycologia* **64**, 663 (1972).
[17] H. B. LeJohn, *J. Biol. Chem.* **246**, 2116 (1971).
[18] H. B. LeJohn, *Nature (London)* **231**, 164 (1971).
[19] H. S. Wang and H. B. LeJohn, *Can. J. Microbiol.* **20**, 575 (1974).

Reagents

Tris-acetate buffer, 0.2 M, pH 8.0
NADH, 5 mM
Pyruvate, 0.1 M

Procedure. The reaction solution (in 3-ml cells; d = 1.0 cm) contains 1 ml of buffer, 100 μl of NADH (0.5 μmole), 0.3 ml of pyruvate (30 μmoles), and H_2O to 3-ml volume. Fifty microliters or less of the enzyme protein is added to start the reaction depending upon the stage of purification of the preparation. In studies involving the use of agents that inhibit the activity of the enzyme, such agent(s) is added before bringing the reaction solution to volume with H_2O.

Units. One unit of enzyme activity gives a change of 0.001 OD per minute. This value can be converted to international millienzyme units by multiplying by 0.48. Specific activity is expressed as enzyme units per milligram of protein. Protein concentration is determined either by the phenol method of Lowry *et al.*[20] or by the spectrophotometric A_{280}/A_{260} ratio method of Warburg and Christian[21] on purified preparations.

Purification Procedure

Step 1. Preparation of Cell-Free Extracts. Mycelia from liquid cultures grown as described[22] (flasks, or carboys under forced aeration) are collected by filtering through Whatman No. 1 filter paper, washed with distilled water, and sucked dry by vacuum pressure. The wet weight of the cell mat is determined. At this stage the mats can be stored frozen before further processing. Using the wet weight of the mycelium as one volume, the cells are suspended in four volumes of buffer composed of 10 mM Tris-acetate pH 7, 1 mM potassium phosphate, 1 mM $Na_4 \cdot$EDTA, 0.1 mM dithiothreitol, and placed in a Waring Blendor. The mycelia are disrupted by blending at 4° for 2 min followed by sonication for 15 min in the cup of a Raytheon 10 Hz sonicator at 10°. Cell debris are removed by centrifuging at 48,000 g for 15 min at 10° (Sorvall).

The supernatant solution is saved, and may be stored at —20° overnight then thawed and centrifuged again at 48,000 g for 15 min. This freezing and thawing technique removes insoluble materials, giving a 2-fold purification, and also reduces the contents of undetermined mate-

[20] O. H. Lowry, N. J. Rosebrough, A. L. Farr, and R. J. Randall, *J. Biol. Chem.* **193**, 265 (1951).
[21] O. Warburg and W. Christian, *Biochem. Z.* **310**, 384 (1942).
[22] H. S. Wang and H. B. LeJohn, *Can. J. Microbiol.* **20**, 567 (1974).

rials which interfere with the resolution of the enzyme by DEAE-cellulose chromatography. If this is done, step 2 may prove to be unnecessary.

Step 2. Protamine Sulfate Precipitation. A protamine sulfate precipitation step is added to the procedure to further clarify the cell extract. A 2% solution is added slowly with stirring (1 part protamine sulfate: 2.5 parts cell-free extract) at 0°. The extract is centrifuged at 48,000 g for 10 min at 4°, and the supernatant saved.

Step 3. DEAE-Cellulose Chromatography. The supernatant is adsorbed directly onto a column of DEAE-cellulose (0.93 meq coarse) that has been previously equilibrated with the homogenizing buffer. Fifty milliliters of extract may be absorbed on a 40×2.5 cm column; larger volumes of extract require longer columns. A linear gradient of 0 to 0.5 M KCl in homogenization buffer is used to elute the enzyme, with 300 ml of elution buffer in each chamber of the gradient maker. The enzyme comes off at approximately 0.25 M KCl. For a 60×2.5 cm column, 500 ml was used in each chamber. Fractions containing enzyme activity are pooled.

Step 4. Ammonium Sulfate Fractionation. The pooled fractions of step 3 are fractionated by the slow addition of powdered ammonium sulfate with constant stirring at 4°. The protein material which precipitates between 45 and 60% (28 g/100 ml, and 11 g/100 ml) is collected by centrifugation (48,000 g, 20 min), and the pellet is "redissolved" in a small volume of homogenizing buffer.

Step 5. Sephadex Gel Filtration. The enzyme preparation is then filtered through a Sephadex G-200 column (100×2.5 cm) with the homogenizing buffer. The enzyme is collected in 4.5-ml samples. Fractions containing the enzyme are pooled. The central core of tubes containing enzyme with very high activity (25,000 mE units/ml) are collected separately from the other tubes. The table includes total activity from all fractions.

Step 6. Ammonium Sulfate Fractionation. The pooled fractions are fractionated with the addition of ammonium sulfate as before, collecting the 45–60% (28 g/100 ml, and 11 g/100 ml) precipitate by centrifugation, and redissolving in the elution buffer.

Step 7. Desalting. The enzyme preparation may be filtered through Sephadex G-25 gel in a small column to remove ammonium ions prior to further experimental work. For storage, the presence of ammonium ions improves the stability at 4°.

Data on the purification of 3 isozymes of the Oomycetes representing the Peronosporales and Leptomitales are summarized in the table.

Properties

Polyacrylamide gel electrophoresis of the purified enzyme preparations is carried out at pH 8 using Tris–barbital buffer (1 g of Tris, and 5.52 g of barbital diethylbarbituric acid in 1 liter of H_2O) system. A current of 3–4 mA per tube is applied, and electrophoresis is carried out for 1¾ hr. All purified preparations show single bands of protein that are coincident with activity bands in gels stained with a reaction solution composed of 100 mg NAD⁺, 200 mg of D(—)-lactate, 0.5 mg of phenazine methosulfate, 10 mg of nitroblue tetrazolium in 25 ml of 0.1 M Tris acetate, pH 8. Staining proceeds in the dark for 30 min. Protein stain is achieved with 0.1% solution of Coomassie blue.

The enzymes are generally cathodic in pH 8 gels with marked variation observed between the isozymes of the Peronosporales, Saprolegniales, and Leptomitales[19] of the Oomycetes. The isozyme from *Sapromyces elongatus* is the most cathodic of all isozyme species tested to date. Wang and LeJohn[19] have surveyed 49 species of the Oomycetes for their D(—)-LDH isozyme patterns and representative taxonomic isozyme charts are presented in that report.

Stability. The enzymes are generally stable at 4° for several weeks, and at —20° for at least 2 years. Occasional species (e.g., *Pythium butleri subramaniam*, CBS 164.68) may have an enzyme that loses about 50% of its activity in crude extracts when stored frozen or kept on ice overnight.

Molecular Weight. Molecular weight determinations made on *P. debaryanum* isozyme[17] gave values of 115,000 for Sephadex gel filtration, and 98,000 for sucrose density gradient sedimentation. The molecular weights of the enzymes of the other species have not yet been carefully determined.

Specificity and Kinetics. Oomycetes apparently contain the D(—)-LDH isozyme which utilizes NAD⁺ as coenzyme. The enzymes are all freely reversible. The isozyme from *P. debaryanum*, from kinetic studies, shows a sequential ordered binary-binary[17] mechanism with either NAD⁺ or NADH as the first substrate to bind to the free enzyme before lactate or pyruvate. Similar studies have now been concluded on the isozyme from *Sapromyces elongatus*.[23]

Inhibitors and Regulation. GTP has been found to be an allosteric inhibitor of all D(—)-LDH's of the Oomycetes,[19] increasing the coopera-

---

[23] LeJohn, H. B., (Manuscript in preparation).

PURIFICATION PROCEDURES FOR D(−)-LACTIC DEHYDROGENASES OF
THREE SPECIES OF THE OOMYCETES

Step	Vol (ml)	Total units	Total protein (mg)	Specific activity	Purification (fold)
Peronosporales *Pythium debaryanum* Hesse (CBS 114.19)[a]					
1a. Crude extract	220	2,816,000	1980	1,422	1.00
1b. After freezing	188	2,406,400	1203	2,000	1.40
2. Protamine sulfate	210	2,436,000	1302	1,871	1.31
3. DEAE-cellulose	160	1,888,000	102	18,510	13.01
4. Ammonium sulfate	4.6	1,131,600	39.6	28,605	20.11
5. Sephadex G-200	90	1,008,000	8.7	115,464	81.19
6. Ammonium sulfate	3.6	432,000	2.6	166,153.8	116.85
Peronosporales *Pythium undulatum* Petersen (CBS 323.47)[a]					
1a. Crude extract before freezing	105	1,512,000	871.5	1,740	1.00
1b. Crude after freezing	95	1,444,000	361.0	4,000	2.29
3. DEAE-cellulose	136	1,088,000	49	22,204	12.76
4. Ammonium sulfate	2.3	763,600	7.13	107,097	61.55
5. Sephadex G-200	50	512,000	2.8	182,857	105.09
6. Ammonium sulfate	2.0	152,000	0.44	345,455	198.54
Leptomitales *Sapromyces elongatus* (cornu) Thaxter (CBS 337.39)[a]					
1a. Crude extract before freezing	82	2,934,000	1875	1,565	1.00
1b. Crude after freezing	80	2,560,000	1390	1,882	1.20
3. DEAE-Cellulose	94	1,353,600	256	5,287	3.37
4. Ammonium sulfate	2.4	1,052,000	46	22,721	14.51
5. Sephadex G-200	34	680,000	9.5	71,568	45.73
6. Ammonium sulfate	2.6	624,000	3.9	160,000	102.23

[a] Centraalburean voor Schimmelcultures, Baarn, The Netherlands.

tive binding of NADH and D(−)-lactate, but not of pyruvate and NAD⁺. ATP and ITP act as competitive inhibitors of NADH binding[17,19] but are not allosteric modulators. A summary of the inhibitory effect of GTP on isozymes of representative species is given in the report by Wang and LeJohn.[19]

[66] D(−)-Lactate Dehydrogenase from *Butyribacterium rettgeri*

By Charles L. Wittenberger

$$\text{Pyruvate} + \text{NADH} + \text{H}^+ \rightarrow \text{D}(-)\text{-Lactate} + \text{NAD}^+$$

Assay Method

Principle. Pyruvate reduction to D(−)-lactate is followed spectrophotometrically by measuring the rate of decrease in absorbancy at 340 nm due to the oxidation of NADH. As will be discussed later, the reaction catalyzed by this enzyme is, for all practical purposes, kinetically irreversible when NAD serves as the coenzyme.[1,2]

Reagents

Potassium phosphate buffer, 1.0 M, pH 6.2
NADH, 1.0 mM. Prepare fresh daily.
Potassium pyruvate, 50 mM. Prepare fresh daily.

Procedure. Since crude extracts and some preparations from the early fractionation steps contain low but significant levels of NADH oxidase, this activity must be corrected for and can be done so in a single assay by the following procedure: Into a silica cuvette (1-cm light path; approximately 1.5 ml volume) add 0.1 ml of 1.0 M potassium phosphate buffer, pH 6.2, 0.1 ml of 1.0 mM NADH, and 0.65 ml distilled water. Add 0.05 ml of an appropriate dilution of the enzyme preparation and record the rate of decrease in absorbancy at 340 nm against a water blank for 1–2 min. This gives the rate of NADH oxidation due to NADH oxidase. Next add 0.1 ml of 50 mM potassium pyruvate to the experimental cuvette and again record the rate of decrease in absorbancy at 340 nm. This gives the rate of NADH oxidation due to the combined activities of NADH oxidase and lactate dehydrogenase. By subtracting the reaction rate observed in the absence of pyruvate from that seen with pyruvate a "corrected" value for lactate dehydrogenase is obtained. One unit of lactate dehydrogenase activity is the amount of enzyme that catalyzes the pyruvate-dependent oxidation of 1.0 μmole of NADH per minute. One unit of NADH oxidase is the amount of enzyme that catalyzes the oxidation of 1.0 μmole of NADH per minute in the absence of pyruvate.

[1] C. L. Wittenberger, *Biochem. Biophys. Res. Commun.* **22**, 729 (1966).
[2] C. L. Wittenberger and J. G. Fulco, *J. Biol. Chem.* **242**, 2917 (1967).

Specific activity is expressed as units per milligram of protein, as determined by the biuret method.[3]

Purification Procedure

Growth of the Organism. Butyribacterium rettgeri (ATCC 10825), an obligate anaerobe, was grown in the semisynthetic medium of Kline and Barker[4] with the following modifications: the final concentration of potassium phosphate was increased to 0.11 M, and 0.54% glucose (w/v), which was sterilized separately, was substituted for sodium lactate. Much higher levels of the enzyme are present in glucose-grown cells than in lactate-adapted cultures.[5] Methylene blue was not added to the medium. The organism is maintained by daily transfer into 15×150 mm tubes containing 10 ml of medium. After inoculation, a potassium carbonate-pyrogallic acid seal is applied to produce the required anaerobic conditions. Inocula for 20-liter carboys are grown in 2-liter volumetric flasks with the anaerobic seal and are incubated at 37° for 20 hr. Carboys are not fitted with the anaerobic seal but are filled to the top with medium after inoculation. They are then sealed with a sterilized rubber stopper equipped with a water trap to permit gas outflow and prevent air inflow. Carboys are incubated at 37° for 18–20 hr, after which time the cells are harvested and washed twice with 10 mM potassium phosphate buffer, pH 6.2, containing 0.001% reduced glutathione. The cells are then resuspended in a minimal volume of the same buffer, dried by lyophilization, and stored at −20°.

General Considerations. The principal objective of the procedure given below was to resolve the *B. rettgeri* lactate dehydrogenase from a contaminating NADH oxidase. Unless otherwise indicated, all steps are carried out at 0–4°. Percentage of saturation of various fractions with respect to ammonium sulfate was based on Table I, p. 76 of Vol. 1 of this series, even though the fractions were kept at 0–4°.

Step 1. Crude Extract. Eight to ten grams dry weight of glucose-grown cells are suspended in about 60 ml of cold 10 mM potassium phosphate buffer, pH 6.2, containing 0.001% reduced glutathione. The cell suspension is disrupted by two passages through a French pressure cell or by sonification for 20 min in a Branson 185 W Sonifier. Either method of cellular disruption may be employed, but if sonification is used better breakage is obtained by starting with a more dilute cell suspension (about 5.0 g dry weight of cells in 80 ml of the buffer). After centrifugation of

---

[3] A. G. Gornall, C. J. Bardawill, and M. M. David, *J. Biol. Chem.* **177**, 751 (1949).
[4] L. Kline and H. A. Barker, *J. Bacteriol.* **60**, 349 (1950).
[5] C. L. Wittenberger and A. S. Haaf, *J. Bacteriol.* **88**, 896 (1964).

the disrupted cell suspension for 30 min at 30,000 g, the supernatant fluid is collected by decantation and diluted with the potassium phosphate-reduced glutathione buffer so that the final protein concentration is about 10 mg/ml.

Step 2. First Ammonium Sulfate Treatment. Solid ammonium sulfate is slowly added to the diluted cell extract to 40% saturation, while maintaining the pH between 6.2 and 6.5 with 1.0 M ammonium hydroxide. The turbid mixture is allowed to stir for 15 min after the last ammonium sulfate addition and is then centrifuged for 10 min at 30,000 g. The 0–40% pellet is discarded and the supernatant fluid, which contains virtually all of the lactate dehydrogenase activity, is retained.

Step 3. Second Ammonium Sulfate Treatment. To the supernatant fluid from step 2 is added solid ammonium sulfate to 60% saturation as described previously. The protein precipitate is collected by centrifugation, redissolved in a minimal volume of the potassium phosphate-reduced glutathione buffer, and dialyzed overnight against two changes (4 liters each) of the same buffer.

Step 4. First DEAE-Cellulose Column. The dialyzed fraction (about 25 ml) is placed on a DEAE-cellulose column (2.4 × 42 cm) previously equilibrated with 10 mM potassium phosphate buffer, pH 6.2. The enzyme is eluted with a linear potassium chloride gradient extending between zero and 0.5 M in a total volume of 1 liter of 10 mM potassium phosphate buffer, pH 6.2. The eluting solution is applied to the column with a peristaltic action pump and the flow rate adjusted so that one 10-ml effluent fraction is collected every 3 min. The active fractions are pooled (numbers 64 through 80 in a typical run,[2] although some variations have been encountered with different preparations) and the protein is concentrated by adding solid ammonium sulfate to 80% saturation as described in step 2. Centrifuge at 30,000 g for 20 min, resuspend the pellet in a minimal volume of 10 mM potassium phosphate buffer, pH 6.2, containing 0.001% reduced glutathione, and dialyze overnight against two changes (4 liters each) of the same buffer. After this dialysis step, a precipitate usually forms, which is removed by centrifugation and discarded.

Step 5. Second DEAE-Cellulose Column. The light yellow supernatant fluid (about 15 ml) from step 4, which still contains some NADH oxidase activity, is placed on a second DEAE-cellulose column (1.2 × 38 cm) previously equilibrated with 10 mM potassium phosphate buffer pH 6.2. This column is eluted and effluent fractions are collected exactly as described in step 4. Since this is the step that results in a resolution of the lactate dehydrogenase from the NADH oxidase, each fraction in the range of numbers 14 through 38 should be tested for both activities, as described under *Procedure*. Those fractions containing only lactate dehydrogenase activity are pooled.

Step 6 Calcium Phosphate Gel. To the pooled fractions from step 5 (about 150 ml), add 3.0 ml of a freshly prepared calcium phosphate gel suspension (Sigma) containing 50 mg dry weight of gel per milliliter of distilled water. Allow the mixture to stir for 30 min and remove the gel by centrifugation. Test the supernatant fluid for lactate dehydrogenase activity; if activity is found, add more calcium phosphate gel in increments until all the enzyme has been adsorbed. Wash the gel pellet(s) with about 10 ml of distilled water. The enzyme is eluted from the gel by three successive washes (4.0 ml each) with 0.5 M potassium phosphate buffer, pH 6.2. The supernatant fluids from the gel washes are pooled, dispensed in 1-ml portions into tubes, and stored at −20°.

A summary of the date from a typical preparation of the *B. rettgeri* lactate dehydrogenase is given in the table.

Properties

General. The purified enzyme is quite stable to storage in 0.5 M potassium phosphate buffer, pH 6.2, at 0° or at −20°. Losses of 10–15% in specific activity have been observed over a 2-week period under either storage condition, although the major activity loss generally occurs during the first 48 hr. Repeated freezing and thawing also leads to a loss of activity. Dithiothreitol (1.0 mM) or 2-mercaptoethanol (1.0 mM) have no effect on either the stability of the enzyme to storage or on its catalytic activity. Significant losses of activity occur when the enzyme is stored at a pH above 6.5 or in a medium of low ionic strength. Dilutions of the purified enzyme should also be made in 0.5 M potassium phosphate buffer, pH 6.2, since substantial activity losses occur during the course of a day when dilutions are made in a medium of low ionic strength. The enzyme has a rather broad pH optimum between 5.8 and 6.5, but activity decreases sharply at more alkaline pH values. It has a molecular weight of about 145,000 as determined by Sephadex G-200 chromatography, using ribonuclease A, chymotrypsinogen A, ovalbumin, and aldolase as standards.[6]

Specificity. The enzyme is specific for NADH as coenzyme and for pyruvate as substrate. The reaction rate is a hyperbolic function of the coenzyme concentration, and the K_m for NADH is 45 μM. In contrast, the reaction rate is a sigmoidal function of the substrate concentration, and the concentration of pyruvate that gives one-half maximal velocity is about 0.4 mM.

[6] C. L. Wittenberger, unpublished observation, 1972.

PURIFICATION DATA FOR D(−)-LACTATE DEHYDROGENASE FROM
Butyribacterium rettgeri[a]

Fraction	Total protein (mg)	NADH oxidase (units)	NAD LDH[b] (units)	Specific activity (units/mg)	Recovery (%)
Crude extract	3,300	104.5	369.8	0.11	100
Ammonium sulfate 0–40%, supernatant	2,830	83.6	340.2	0.12	92
Ammonium sulfate 40–60%, precipitate	1,968	23.2	232.2	0.12	62.8
First DEAE-cellulose eluate, pooled	270.7	17.8	201.0	0.74	54.4
Ammonium sulfate 0–80%, precipitate	187.3	8.0	115.8	0.62	31.3
Second DEAE-cellulose eluate, pooled	8.4	0	107.7	12.82	29.1
Calcium phosphate gel eluate, pooled	2.8	0	102.3	36.54	27.6

[a] See footnote 2.
[b] Corrected for NADH oxidase.

As indicated previously, attempts to demonstrate NAD reduction by D(—)-lactate have been unsuccessful. Lactate oxidation can be demonstrated however, if the 3-acetylpyridine analog of NAD serves as the coenzyme under assay conditions described elsewhere.[2] Employing these conditions, only the D(—)-stereoisomer of lactate is oxidized.

Activators and Inhibitors. Enzyme activity is not affected by the divalent cations Zn^{2+} or Mg^{2+}. Neither is the enzyme inhibited to any significant extent by 50 mM EDTA. Mercuric chloride (1.0 μM) and *p*-chloromercuriphenyl sulfonic acid (5.0 μM) inhibit 40 and 73%, respectively, and potassium oxamate (7.0 mM) causes a 50% inhibition. α-Ketobutyrate does not serve as a substrate for the enzyme, but when included with pyruvate in the standard assay, it causes a transposition of the pyruvate saturation curve from sigmoidal to hyperbolic. The substrate analog, at a final concentration of 10 mM, stimulates activity at rate-limiting concentrations of pyruvate and is somewhat inhibitory at saturating concentrations of the substrate.[2]

Among a variety of nucleotides tested as potential physiological modifiers of the D(—)-lactate dehydrogenase, ATP is the most effective inhibitor. The available evidence indicates that ATP exerts its effect by interacting with the enzyme at a site distinct from either the substrate or coenzyme binding sites.[7]

[7] C. L. Wittenberger, *J. Biol. Chem.* **243**, 3067 (1968).

[67] Lactate Dehydrogenase[1] from *Bacillus subtilis*

By AKIRA YOSHIDA and ERNST FREESE

$$\text{L-Lactate} + \text{NAD}^+ \leftrightarrow \text{pyruvate} + \text{NADH} + \text{H}^+$$

Principle. L-Lactate dehydrogenase activity is measured spectrophotometrically by following either the reduction of NAD in the presence of L-lactate or the oxidation of NADH in the presence of pyruvate.

Reagents

For oxidation of L-lactate:
L-Lactate, sodium salt, 0.5 M
NAD, 20 mM
Na_2HPO_4-KH_2PO_4 buffer, 60 mM, pH 7.2, at 25°
For reduction of pyruvate:
Pyruvate, sodium salt, 10 mM
NADH, 10 mM
Na_2HPO_4–KH_2PO_4 buffer, 60 mM, pH 6.0, at 25°
All reagents, except the buffer solutions, should be stored frozen. "Phosphate buffer" = Na_2HPO_4–KH_2PO_4 buffer.

Procedure. The enzyme is dissolved just before assay in 10 mM phosphate buffer, pH 7.0, to obtain a concentration of 0.1–10 units of enzyme per milliliter (see definition below).

OXIDATION OF L-LACTATE. Use per milliliter of reaction mixture: 0.8 ml of 60 mM phosphate buffer, pH 7.2; 0.05 ml of NAD; 0.1 ml of lactate; 5–50 μl of enzyme solution, and water to make the correct final volume in a silica cell with a 1-cm light path. The increase of absorbancy at 340 nm is recorded at 25°, and the maximal (initial) rate is used for calculations.

REDUCTION OF PYRUVATE. Use per milliliter reaction mixture: 0.8 ml of 60 mM phosphate buffer, pH 6.0; 10 μl of NADH; 10 μl of sodium pyruvate; 5–50 μl of enzyme solution, and water to make the correct final volume. The decrease of absorbancy at 340 nm is recorded at 25°. A control is run without pyruvate to measure the NADH oxidation activity.

Units. One unit of enzyme activity reduces NAD (oxidation of lactate) or oxidizes NADH (reduction of pyruvate) at the rate of 1 μmole per minute at 25°. The units of enzyme per milliliter of reaction mixture are calculated from the rate of absorbancy change (Δ OD/min at 340 nm divided by 6.22).

[1] L-Lactate:NAD$^+$ oxidoreductase, EC 1.1.1.27.

Specific activity is the number of enzyme units per milligram of protein.

Purification Procedure

When ordinary *Bacillus subtilis* strains, such as strain 168 of Spizizen or its biochemical mutants, are grown with maximal aeration in minimal glucose medium, they usually produce no detectable amount of lactate dehydrogenase at 30° and very little at 37°. However, when the aeration is drastically reduced or completely stopped, significant enzyme activity is usually observed after growth at 37° and even more after growth at 45°. Since these results are variable, strain 60295, isolated by a plate-assay method, was used for enzyme production because it apparently is constitutive in lactate dehydrogenase as shown by the production of consistently large amounts of the enzyme.[2] The strain can be obtained from E. Freese.

Bacillus subtilis strain 60295 (requiring tryptophan and nicotinic acid for growth) is grown in minimal-glucose medium[3] containing per liter: K_2HPO_4, 14 g; KH_2PO_4, 6 g; sodium citrate, 1 g; ammonium sulfate, 2 g; $MgSO_4 \cdot 7H_2O$, 0.25 g; L-tryptophan, 50 mg; nicotinic acid, 2 mg; and, added after autoclaving, 50 ml 10% glucose. An inoculum of 3×10^5 cells per milliliter produces approximately 10^9 cells per milliliter after 16–18 hr of cultivation with vigorous aeration at 40°. The culture is then kept for 1 hr with low or no aeration and the cells are harvested in a Sharples centrifuge. The cell paste is dried by acetone (cooled at $-20°$). Ten grams of acetone-dried cells contain about 20 mg of the enzyme.

Step 1. Extraction. About 30 g of freshly harvested cells (or 10 g of acetone-dried cells) are well suspended in 200 ml of 50 mM Tris-chloride buffer, pH 8.0, containing 1 mM mercaptoethanol, and digested with lysozyme (0.5 mg/ml) at 37° for 30 min. Subsequently, the cell suspension is digested by 30 min of further incubation in the presence of pancreatic ribonuclease (25 μg/ml), pancreatic deoxyribonuclease (1 μg/ml), and 2 mM $MgCl_2$. The liquefied extract is centrifuged at 30,000 g for 30 min. Addition of 50 mM $MnSO_4$ to the supernatant produces a precipitate that is removed by centrifugation. The yield is about 5×10^4 units (reduction of pyruvate at pH 6.0). All subsequent procedures, including chromatography and dialysis are carried out at 0–4°.

Step 2. Fractionation with Ammonium Sulfate. Ammonium sulfate is added to the supernatant (60 g/100 ml); after 1 hr the precipitate is collected by centrifugation. After removal of material that is soluble

[2] A. Yoshida and E. Freese, *Biochim. Biophys. Acta* **99**, 56 (1965).
[3] C. Anagnostopoulos and J. Spizizen, *J. Bacteriol.* **81**, 741 (1951).

in 65% saturated ammonium sulfate [saturated ammonium sulfate (at 4°): water = 65:35, v/v], the insoluble fraction is extracted with 45% saturated ammonium sulfate. The extraction is repeated once more, and the combined extract (about 120 ml) is reprecipitated by adding solid ammonium sulfate (30 g/100 ml of the extract). The precipitate is suspended in 50 ml of 10 mM phosphate buffer, pH 7.7, and dialyzed against the same buffer.

Step 3. Calcium Phosphate-Gel Chromatography. The dialyzed extract is placed on a calcium phosphate gel (hydroxyapatite prepared by the method of Tiselius, Hjertén, and Levin[4]) column (2.5 × 30 cm) with 10 mM phosphate buffer, pH 6.8. The column is eluted first with 500 ml of 75 mM phosphate buffer, pH 6.8, and then with a gradient of phosphate buffer, pH 6.8, whose concentration increases from 75 mM to 0.17 M. The gradient was produced by continuous addition of 0.2 M phosphate buffer to an airtight mixing chamber which initially contained 500 ml of 75 mM phosphate buffer (fixed buffer volume). Lactate dehydrogenase is eluted at phosphate buffer concentrations ranging from 0.1 to 0.15 M, while malate dehydrogenase, alanine dehydrogenase, and other impurities eluate during the first washing with 0.075 M phosphate buffer. Enzyme activity of the effluent fractions is measured and the bulk of the enzyme (4 × 10⁴ units in 300–350 ml of effluent) is precipitated with ammonium sulfate (60 g/100 ml).

The precipitate is resuspended in about 7 ml of 20 mM phosphate buffer, pH 7.7, and centrifuged. The supernatant is then dialyzed against the same buffer containing 1 mM EDTA.

Step 4. DEAE-Sephadex Column Chromatography. The dialyzed solution is placed on a DEAE-Sephadex column (1 × 30 cm) washed with 20 mM phosphate buffer, pH 7.7 containing 1 mM EDTA, and it is eluted with increasing concentration of NaCl, from 0 to 0.3 M. The gradient was produced by continuous addition of phosphate buffer containing 0.4 M NaCl to an airtight mixing chamber which initially contained 250 ml of phosphate buffer (fixed buffer volume). The enzyme elutes at NaCl concentrations ranging from 0.1 to 0.15 M. The bulk of the enzyme fraction [stationary activity (see under stability) = 3 × 10⁴ units in about 40 ml of effluent] is precipitated with ammonium sulfate (60 g/100 ml) redissolved in several milliliters of 10 mM phosphate buffer, pH 7.7, and dialyzed against the same buffer containing 1 mM EDTA.

Step 5. ECTEOLA-Cellulose Column Chromatography. The dialyzed solution is placed on an ECTEOLA cellulose column (1 × 30 cm), washed with 20 mM phosphate buffer, pH 7.7, containing 1 mM EDTA.

[4] A. Tiselius, S. Hjertén, and D. Levin, *Arch. Biochim. Biophys.* **65,** 132 (1956).

The enzyme is eluted by gradually increasing NaCl from 0 to 0.2 M. The gradient was produced by continuous addition of phosphate buffer containing 0.25 M NaCl to an airtight mixing chamber which initially contained 250 ml of phosphate buffer (fixed buffer volume). The bulk of the enzyme (between 160 ml and 250 ml of effluent) is collected by precipitation with ammonium sulfate (60 g/100 ml).

Step 6. Crystallization. The precipitate is dissolved in about 1 ml of 50 mM phosphate buffer, pH 7.5, containing 1 mM EDTA. Saturated ammonium sulfate is added until the solution becomes slightly turbid. After cold storage overnight, a small precipitate is removed by centrifugation. Additional saturated ammonium sulfate is added until the supernatant becomes turbid. After cold storage overnight, the precipitate is collected by centrifugation and redissolved in the smallest feasible amount of the same buffer. When saturated ammonium sulfate is added until the solution becomes turbid, tiny cubic crystals develop after cold storage overnight. Two or three times recrystallization causes the tiny crystals to assemble as large hexagonal particles.

The yield of the enzyme was 2×10^4 units (stationary activity) measured by reduction of pyruvate at pH 6.0, in 8–9 mg of protein, after three times crystallization.

Properties

Effect of Enzyme Concentration. The specific activity of the enzyme rapidly (2–3 min) declines to a lower value stationary specific activity when the enzyme is diluted below 1 mg/ml. The following characteristics were determined using the enzyme that had been diluted to 10–20 μg/ml in 10 mM phosphate buffer, pH 7.2. Under these conditions, the enzyme kept a stable stationary activity.

Effect of pH. The optimal pH for the oxidation of L-lactate is 7.2 (phosphate buffer); that for the reduction of pyruvate is 6.0 (phosphate buffer).

Effect of Substrate Concentration. The Michaelis constant (K_m) for the primary substrates are: L-lactate 30 mM, NAD 0.9 mM (oxidation of lactate at pH 7.2); pyruvate 0.8 mM, NADH 65 μM (reduction of pyruvate at pH 6.0). When the concentration of pyruvate exceeds 0.8 mM, a slight suppression of activity is observed (85% activity at 8 mM).

Turnover Number. (a) Oxidation of lactate: The approximate maximum specific activity obtained at optimal pH (pH 7.2) and extrapolated to conditions of saturation of the enzyme with substrate is 6.3 at 25°. Since the molecular weight of the enzyme is 146,000, the turnover number is about 920 moles of substrate per minute per mole of enzyme.

(b) Reduction of pyruvate: The approximate maximum specific activity at pH 6.0 is 5200 at 25°. The maximum turnover number, therefore, is about 7.5×10^5 moles of substrate per minute per mole of enzyme.

Specificity. None of the following analogs of L-lactate is oxidized by the enzyme: D-lactate, DL-α-hydroxybutyrate, DL-β-hydroxybutyrate, DL-α-hydroxyisovalerate, mesoxalate, citrate, isocitrate, L(+)-tartrate, D(−)-tartrate, mesotartrate, L-malate, glycerate, glycolate, L- or D-alanine. Various keto acids are reduced by the enzyme at a slower rate than pyruvate. Their K_m and V_{max} (relative to pyruvate) are: glyoxylate, 25 mM, 57%; hydroxypyruvate, 5 mM, 45%; α-ketobutyrate, 1.0 mM, 7.9%; α-ketoisovalerate, 5.9 mM, 0.17%; oxaloacetate, 1.5 mM, 3.5%. NADP and NADPH are inactive as substrates. 3-Acetylpyridine adenine dinucleotide ($K_m = 0.125$ mM, $V_{max} = 8.5$ times that of NAD) and 3-acetylpyridine deaminoadenine dinucleotide ($K_m = 1$ mM, $V_{max} = 8.5$ times that of NAD) are better substrates than NAD ($K_m = 0.9$ mM) itself.

Inhibitors. Competitive inhibitors for the oxidation of lactate are: D-lactate ($K_i = 45$ mM), isocitrate ($K_i = 13$ mM), citrate ($K_i = 30$ mM), L(+)-tartrate ($K_i = 42$ mM), D(−)-tartrate ($K_i = 40$ mM), mesotartrate ($K_i = 47$ mM), glycerate ($K_i = 36$ mM), glycolate ($K_i = 160$ mM), DL-α-hydroxyisocaproate ($K_i = 18$ mM), mesoxalate ($K_i = 1.5$ mM), and L-alanine ($K_i = 48$ mM). Various metal ions (Cu^{2+}, Hg^{2+}, Zu^{2+}, Ag^+) inhibit the enzyme noncompetitively, Ag being most effective (1 μM for complete inhibition). p-Chloromercuribenzoate inhibits the enzyme almost completely at a concentration of 20 μM. The inhibition by metal ions and p-chloromercuribenzoate can be completely reversed by L- or D-cysteine or by mercaptoethanol.

Stability. The enzyme can be stored in crystalline form in partially saturated ammonium sulfate without loss of the activity for more than a year at 4°. When the enzyme is diluted in a buffer of various pH values from a stock crystalline suspension, the specific activity rapidly decreases, reaching a stationary state after several minutes at 0°. The extent of the decrease depends on the pH of the buffer and on the initial concentration of the enzyme. The enzyme is fully active (11,000–11,500 units/mg) at concentrations higher than 2 mg/ml (activity measured immediately after dilution), and it is about 16–18% active (about 2000 units/mg) at concentrations lower than 0.5 mg/ml at pH 7.2. No further decrease of activity occurs during incubation of the diluted enzyme at 50° for 4 hr. In an alkaline buffer (pH 8.8), an inactivation is observed even at high enzyme concentrations (3 mg/ml), and at concentrations lower than 1 mg/ml only 2% enzyme activity remains. The activity reduced by dilution in buffers at pH values ranging from 6.8 to 8.8 is fully restored when the solution is neutralized and the enzyme is precipitated (90–95% saturation) with ammonium sulfate and redissolved in a buffer (pH 7.0–7.5) at a high protein concentration (>3 mg/ml).

Physical and Chemical Properties. The molecular weight of the enzyme is 146,000 at pH 7.2, and 72,000 at pH 8.8 (by sedimentation equilibrium method). The latter form of protein is enzymically inactive. The sedimentation constant ($s_{20,w}$) is 6.7 S at a concentration of 0.4% of the protein. The enzyme is composed of four identical or nearly identical subunits of molecular weight 36,000. The subunits have the following amino acid composition: Asp 39, Thr 17, Ser 18, Glu 33, Pro 11, Gly 31, Ala 32, Cys 4, Val 29, Met 6, Ile 20, Leu 25, Tyr 13, Phe 13, Lys 23, His 9, Arg 6, Trp 2.

[5] A. Yoshida, *Biochim. Biophys. Acta* **99**, 66 (1965).

[68] D-Lactate Dehydrogenase of *Peptostreptococcus elsdenii*

By HOWARD L. BROCKMAN, JR., and W. A. WOOD

D-Lactate + acceptor → pyruvate + reduced acceptor

Assay Method

Principle. The spectrophotometric assay is based on the reduction of ferricyanide and the resultant disappearance of absorbance at 420 nm. The procedure is a modification of that of Symons and Burgoyne.[1]

Reagents

D-Lactate, 3.2 M

Potassium ferricyanide, 10 mM in 0.2 M potassium phosphate buffer, pH 7.0

Procedure. Each microcuvette (0.5 ml volume, 1 cm light path) contained in a 0.2 ml reaction volume 160 μmoles of D-lactate, 0.5 μmole of potassium ferricyanide, 10 μmoles of phosphate buffer, pH 7.0, and lactate dehydrogenase. The temperature was 24°. Anaerobic conditions are not required.

Definition of Unit Activity. The rate was recorded on a Gilford Model 2000 spectrophotometer and converted to international units using an ex-

[1] See this series, Vol. 9.

tinction coefficient of 1040 l mole⁻¹ cm⁻¹ for ferricyanide. Thus, international units (μmoles/min) in the cuvette $= \Delta A/\text{min} \times 0.192$. Protein was determined spectrophotometrically using the absorbance at 280 and 260 nm in conjunction with the extinctions published by Warburg and Christian.[2]

Alternate Assay

In studies of the electron transport system in which D-lactate dehydrogenase functions, activity was also measured spectrophotometrically at 450 nm as the rate of reduction of the flavin moiety of acyl-CoA dehydrogenase present in stoichiometric amount.[3] An electron-transferring flavoprotein[3] was also required in catalytic amounts. The reaction mixture contained in 0.2 ml: acyl-CoA dehydrogenase (4 nmoles), electron-transferring flavoprotein (3 μg), D-lactate (80 μmoles), and potassium phosphate buffer, pH 7.0 (9 μmoles). The ability of oxygen to serve as an alternate electron acceptor necessitates anaerobic conditions. Accordingly, a 3/4-inch long piece of 1/2-inch i.d. gum rubber tubing was fitted over the top 3/8 inch of a standard microcuvette and secured with copper wire. After addition of D-lactate, buffer, D-lactate dehydrogenase, and acyl-CoA dehydrogenase, a septum derived from the upper half of the stopper from a Vacutainer (Becton and Dickinson) was fitted into the rubber tubing. A 20-gauge needle from an evacuation manifold was inserted through the septum into the air space above the sample. The manifold was constructed to afford alternate evacuation and filling with argon. During deaeration, the bottom half of the cuvette can be immersed in water to bring the contents to a defined temperature. Small losses through foaming and bumping occur, but these can be minimized with practice. The reaction was initiated by the addition of electron-transferring flavoprotein from a microsyringe.

Purification[4]

Crude Extract. Peptostreptococcus elsdenii, ATCC No. 17752, was maintained in stock culture as described by Elsden and Lewis[4] with 0.001% resazurin as a redox indicator. Large quantities were grown in 55-gallon drums on the medium of Ladd and Walker[5] and Bryant and Robinson.[6] For 200 l (55 gal) of medium were added: Na$_2$HPO$_4$, 800

---

[2] O. Warburg and W. Christian, *Biochem. Z.* **310**, 384 (1942).

[3] H. L. Brockman, Ph.D. Thesis, Michigan State University, 1971.

[4] S. R. Elsden and D. Lewis, *Biochem. J.* **55**, 183 (1953).

[5] D. J. Walker, *Biochem. J.* **69**, 524 (1958).

[6] M. P. Bryant and I. M. Robinson, *Appl. Microbiol.* **9**, 91 (1961).

g; NH_4Cl, 100 g; $CaCl_2$, 40 g; $MgCl_2$, 40 g; $FeSO_4 \cdot 7 H_2O$, 4 g; $MnSO_4 \cdot 4$ H_2O, 1 g; $Na_2MoO_4 \cdot 2 H_2O$, 1 g; $ZnSO_4$, 1 g; Na DL-lactate, 3300 ml of a 60% solution in H_2O; 1600 ml of corn steep liquor (A. E. Staley Co., Decatur, Illinois), or 800 ml of molasses or 1 lb of yeast extract. Corn steep liquor, when used, was acidified to pH 1 with concentrated HCl, autoclaved, and neutralized to pH 7.5 with NaOH. The precipitate was removed by centrifugation and discarded. Dithionite (200 mg/l) was sterilized separately and added to the medium for the inoculum only. A 10% inoculum was added to the large-scale batch of medium and incubation was carried out anaerobically at 37°. The cells were harvested at the end of log phase (11 hr) with a yield of 300 g wet weight.

After centrifugation, the cells were frozen until used. Frozen cells were thawed by suspending them in an equal volume of distilled water (1 g/ml). Disruption was performed either by sonication for 20 min in a 200-W Raytheon 10 kHz sonic oscillator or by two passages through a laboratory homogenizer (Manton-Gaulin Co., Inc., Everett, Massachusetts) operated at 1600 psi. The latter method was used for 100 g or more of cells. Both procedures were carried out at 0–4°, and DNase was added to homogenized extracts to decrease the viscosity. Extracts were centrifuged at 18,000 g to remove debris, and the supernatant was fractionated immediately.

In the following two purification steps, 300 g of cells, derived from 200 l of culture were used. All purification steps were performed in 1 mM dithiothreitol to reduce activity losses.

DEAE-Cellulose Chromatography. The crude extract (550 ml) was mixed with one-fourth volume of moist DEAE-cellulose, stirred for 10 min at 4°, and filtered under vacuum on Whatman No. 1 filter paper. The exchanger containing the bound lactic dehydrogenase activity was washed twice with 500 ml of 0.3 M potassium phosphate buffer, pH 6.0, and applied to the top of a column (5 × 30 cm) containing the same quantity of moist DEAE-cellulose equilibrated with 0.3 M potassium phosphate buffer, pH 6.0. Lactic dehydrogenase was eluted with 1200 ml (600 ml per chamber) of the same buffer in a linear gradient from 0.3 to 0.8 M. The enzyme was eluted in 70 ml of approximately 0.7 M phosphate buffer and then precipitated by addition of ammonium sulfate (472 g/l, 70% saturation at 4°). The resulting precipitate was dissolved to a volume of 7 ml with water and 1 ml amounts were desalted by passage through a 2.5 × 10 cm column of Sephadex G-25, equilibrated against 0.1 M potassium phosphate buffer (pH 7.0). The elution volume each time was 3 ml.

Hydroxyapatite Chromatography. Of the eluate from the previous step, 2 ml were applied to a column of hydroxyapatite (0.75 × 15 cm) equilibrated against 0.1 M potassium phosphate buffer (pH 7.0). The enzyme was eluted with the same buffer in a volume of 1.4 ml.

The procedure gives a 193-fold purification with 9.4% recovery. Scans of polyacrylamide gels at 280 and 450 nm revealed a single large peak, which coincided at these wavelengths. There was also a minor component at both wavelengths appearing as a shoulder at the trailing edge of the main peak.

Properties

In the presence of an acceptor, the enzyme reduces D-lactate to pyruvate. Only D-lactate and DL-α-hydroxybutyrate show appreciable activity. With ferricyanide, lactate binding appears to be highly negatively cooperative; the Hill n was 0.46 and the K_m was 13 M. With the complete electron transport system and assay for the rate or reduction of acyl-CoA dehydrogenase as described above, the kinetics were hyperbolic and the K_m for D-lactate was 26 mM. Variation of the level of D-lactate dehydrogenase in this system with other parameters held constant gave saturation kinetics. Based on an approximate molecular weight of 125,000 for the enzyme, the apparent equilibrium constant for the interaction of D-lactate dehydrogenase with the electron transferring flavoprotein was 1×10^{-7} M.[3]

Although the dehydrogenase is relatively specific for electron donors, its acceptor specificity is rather broad. Dyes, oxygen, and cytochrome c serve as acceptors to some degree. However, pyridine nucleotides do not serve as acceptor. The apparent K_m for ferricyanide was 0.23 mM at a D-lactate concentration of 50 mM.

Cofactors and Activators. The oxidized form of the enzyme exhibits spectral maxima at 450 and 378 nm which are characteristic of a flavoprotein. Flavin adenine dinucleotide was the sole flavin identified on thin-layer plates. o-Phenanthroline (4 mM) causes 99% inactivation. Kinetic studies of the inactivation showed that it occurs in two stages—a rapid loss of activity followed by a slower pseudo-first-order decay. The non-chelating analog, m-phenanthroline, also causes the first rapid inactivation phase, but not the second, slower one. Zn^{2+} and Co^{2+} cause substantial reactivation whereas Mn^{2+} is inactive. The reaction velocity is linearly dependent on the phosphate ion concentration, but phosphate is not required for activity. Of 27 phosphate compounds tested, only thiamine pyrophosphate and inorganic pyrophosphate showed partial stimulation relative to orthophosphate.

pH optimum was broad between pH 6.0 and 8.0, with a slight maximum at pH 7.9.

[69] D-Lactate Dehydrogenase from the Horseshoe Crab

By GEORGE L. LONG

Assay Method

Principle. The routine analysis of lactate dehydrogenase activity is based on the spectrophotometric loss of 340 nm absorption of NADH when it is oxidized in the coupled reaction with pyruvate reduction to D-lactate. Alternative methods for measuring the oxidation of lactate are reported elsewhere.[1]

Reagents

Potassium phosphate, 0.1 M, pH 7.5
Sodium pyruvate, 0.1 M, stored at —20°
NADH, 14 mM, made fresh daily

Procedure. In a total volume of 3 ml, the following are contained: potassium phosphate, 50 micromoles, pH 7.5; sodium pyruvate, 20 μmoles; NADH, 0.4 μmole; lactate dehydrogenase to give a change in optical density at 340 nm of between 0.1 and 0.2 in 1 min.

The reaction is followed by measuring the optical density of the above mixture in a 3.0 ml cuvette having a 1-cm light path at 15-sec intervals for at least 75 sec at room temperature after the addition of enzyme.

Definition of the Unit and Specific Activity. One unit of enzyme activity is defined as the amount of protein solution that would cause a decrease in optical density of 1.0 or the oxidation of 0.48 μmole of NADH in 60 sec under the above conditions. Specific activity is defined as the number of enzyme units per milligram of protein.

Purification Procedure

All operations are carried out at 4° in the presence of 14 mM β-mercaptoethanol and 1 mM EDTA.

Step 1. Crude Extract. One kilogram of freshly thawed *Limulus* skeletal muscle is homogenized in 2 liters of 50 mM potassium phosphate buffer, pH 6.0, for 60 sec at low speed in a Sears two-speed blender. The homogenate is centrifuged 30 min at 10,000 rpm in a GSA rotor in a Sorvall RC2 refrigerated centrifuge. The sediment is discarded.

Step 2. Ammonium Sulfate Precipitation and Fractionation. Solid am-

---

[1] G. L. Long and N. O. Kaplan, *Arch. Biochem. Biophys.* **154**, 711 (1973).

monium sulfate is slowly added to the above sulernatant, until the solution is at 40% saturation.[2] The suspension is stirred 0.5 hour and centrifuged as in the preceding step. The resulting sediment is discarded. The supernatant is brought to 65% ammonium sulfate, stirred 1 hr, and centrifuged as before. The supernatant contains very little enzyme activity. The sediment is dissolved in a minimum amount of fresh extraction buffer and dialyzed 2 hr in 3 liters of the buffer. A second dialysis of 4 hr duration is then performed. The resulting mixture is centrifuged as above, and the supernatant is saved and assayed.

Step 3. Sephadex G-100 Chromatography. The above sample (235 ml) is applied to a Sephadex G-100 column (7.4 × 104 cm, 4.6 liters) which has been preequilibrated with extraction buffer. Fractions are collected and monitored for enzyme activity and optical density at 280 nm (protein). Tubes containing enzyme activity (800 ml) are pooled and dialyzed in 10 liters of 100% ammonium sulfate containing β-mercaptoethanol and EDTA for 4 hr, and then centrifuged as in previous steps and the sediment collected. No activity is observed in the supernatant. The sediment is dissolved in a minimum amount of 5 mM potassium phosphate buffer, pH 6.0. This product is then dialyzed in 1 liter of fresh buffer for 1, 2, and 3 hr, respectively, centrifuged as above, and the supernatant saved.

Step 4. AE-Cellulose Chromatography.[3] The supernatant is applied to a preequilibrated AE-11 column (4.6 cm × 75 cm, 1.31 liters) and eluted with 200 ml of 5 mM potassium phosphate, pH 6.0. A linear gradient in buffer from zero to 0.3 M NaCl (total volume 4 liters) is then started. Two major enzyme peaks are observed, and these are concentrated separately as in Step 3. The resulting ammonium sulfate sediments are taken up in minimum amounts of 5 mM potassium phosphate buffer, pH 6.9, dialyzed in 3 500-ml amounts of fresh buffer for 1, 1, and 3 hr, respectively, and then centrifuged as above. The two resulting supernatants reveal upon analysis by starch gel electrophoresis a partial separation of the most anodic triad of isoenzymes and an absence of the more cathodic forms of the enzyme (Fig. 1). The first fraction eluted (A) contains primarily the middle and lower forms of the triad. The second fraction eluted (B) contains primarily the upper form of the triad.

Step 5. DEAE-Cellulose Chromatography (pH 6.9).[3] Fraction (B) above is further purified on a preequilibrated DEAE-11 column (6.0

---

[2] Percent saturation is determined from Table 1 of A. A. Green and W. L. Hughes, p. 76 of Vol. 1 of this series. No correction is made for temperature.

[3] The resin was prepared by the method outlined for DEAE-cellulose by A. Peace, R. H. McKay, F. Stolzenbach, R. D. Cahn, and N. O. Kaplan, *J. Biol. Chem.* **239**, 1753 (1964).

cm × 35 cm, 1 liter) and eluted with 240 ml of 5 mM potassium phosphate, pH 6.9. A linear gradient in buffer from zero to 0.3 M NaCl (total volume 4 liters) is then started. Fractions in the main enzyme peak are pooled and concentrated as previously described. The resulting ammonium sulfate sediment is dissolved in a minimum amount of 5 mM potassium phosphate buffer, pH 6.4, dialyzed in 1 liter of buffer twice for 2-hr intervals, and submitted to the next purification step.

Fraction (A) is also submitted to essentially the same step as outlined above. This is done to isolate the two isoenzymic forms for high salt hybridization studies. The electrophoretic patterns of the resulting two fractions are shown in Fig. 1.

Step 6. DEAE-Cellulose Chromatography (pH 6.4). The largest enzyme sample from the previous step is put on a preequilibrated DEAE-11 column (3.0 × 83 cm, 560 ml) and eluted with 45 ml of 5 mM potassium phosphate, pH 6.4. The linear gradient in buffer from zero to 0.2 M NaCl (total volume 5 liters) is then started. Fractions from the main

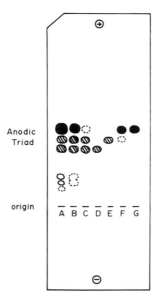

Fig. 1. A tracing of starch gel electrophoresis of *Limulus* lactate dehydrogenase from various purification steps. The intensity of staining is (●) very intense, (◒) strong, (○) clearly distinguishable, (◌) barely visible. The samples run were (A) a crude muscle extract, (B) after Sephadex G-100, (C) fraction A after AE-11 chromatography, (D and E) fractions I and II after DEAE-11 chromatography of fraction A, (F) fraction B after AE-11 chromatography, and (G) fraction after DEAE-11 chromatography of fraction B. Electrophoresis was for 15 hr at 19 mA (~180 V), pH 7.0, 4°. From I. H. Fine and L. Costello, this series, vol. 6, p. 965.

enzyme peak are pooled and concentrated as before, dissolved in a minimum amount of 5 mM potassium phosphate buffer, pH 6.0, and dialyzed
in 1 liter of fresh buffer for 5 and 14 hr, respectively.

Step 7. DEAE-Cellulose Chromatography (pH 6.0). The above sample is put on a preequilibrated DEAE-11 column (2.8 × 45 cm, 300
ml). A linear gradient of 5 mM potassium phosphate, pH 6.0, from zero
to 0.2 M NaCl (total volume 2 liters) is then started. Fractions in the
enzyme peak are pooled and concentrated as above and dialyzed extensively in 5 mM potassium phosphate, pH 6.5. The purified enzyme can
be stored at 4° for several months with little loss of enzymic activity.
The table summarizes the results of the above purification procedure.
The procedure results in the separation of the most anodic triad into
individual isoenzymes and the purification of the most anodic and abun-
dant isoenzymic form to homogeneity as shown in Fig. 1.

Properties of the Purified Isoenzyme

Substrate Specificity. The enzyme is stereospecific for the D-isomer
of lactate and will not catalyze the oxidation of L-lactate. It also stereo-
specifically transfers the 4-α hydrogen of NADH to pyruvate. The enzyme does not oxidize NADPH at a 0.17 mM concentration.[1]

Molecular Weight and Subunit Structure. Limulus D-lactate dehydrogenase has been shown by several criteria to be a dimer with a molecular
weight of approximately 70,000.[4] Sephadex gel filtration and equilibrium
sedimentation yield molecular weights of about 80,000 and 70,000, respectively. The native enzyme has an $s_{20,w}^{0}$ of 3.95.

Kinetic Properties. Limulus muscle lactate dehydrogenase shows only
slight substrate inhibition above 2 mM pyruvate. (The "heart" type
lactate dehydrogenase from *Limulus* shows marked inhibition above 0.2
mM pyruvate.) The apparent K_m's for pyruvate, D-lactate, NADH, and
NAD$^+$ are 70 μM 3.8 mM, 0.21 μM, and 0.44 μM, respectively. The turnover number for the moles of NADH oxidized per mole of enzyme per
minute at 20° is 38,000.[1]

Inhibitors. The purified enzyme is stoichiometrically and reversibly
inhibited by p-hydoxymercuribenzoate. Extrapolation to 2 moles of
pHMB bound to 1 mole of enzyme yields 100% inhibition. Aklylation
and resulting inhibition greater than 50% by iodocetamide or iodoacetate
occurs even in the absence of urea or guanidine-HCl. Inhibition by
oxamate is partially competitive with respect to pyruvate. The apparent
K_i is 0.6 mM. The reduced NAD-pyruvate adduct is a competitive

[4] G. L. Long and N. O. Kaplan, *Arch. Biochem. Biophys.* **154**, 696 (1973).

PURIFICATION PROCEDURE FOR D-LACTATE DEHYDROGENASE FROM *Limulus polyphemus*

Step	Volume (ml)	Total enzyme units	Total protein (mg)	Specific activity	Fold purification	Recovery (%)
Initial extract	2550	134,000	74,000	1.8	1.00	100
40% Ammonium sulfate supernatant	2800	137,000	64,400	2.1	1.17	102
65% Ammonium sulfate precipitate after dialysis	235	128,000	12,100	10.6	5.9	96
Sephadex G-100 column	80.5	107,000	3,380	31.7	17.6	80
AE-11 column: Fraction A	47	34,200	580	58.9	32.7	26
AE-11 column: Fraction B	42	59,400	817	72.7	40.4	44
DEAE-11 column (pH 6.9) on Fraction A: Fraction I	21.4	4,840	192	25.2	14.0	4
DEAE-11 column (pH 6.9) on Fraction A: Fraction II	18.1	29,700	277	107.0	59.5	22
DEAE-11 column (pH 6.9) on Fraction B	19.5	36,800	141	261	145	28
DEAE-11 column (pH 6.4)	23.5	23,700	34	697	387	18
DEAE-11 column (pH 6.0)	16.0	17,500	25.4	689	382	13

inhibitor $(K_i = 6.0\ \mu M)$ with respect to NADH and a noncompetitive inhibitor $(K_i = 12\ \mu M)$ with respect to pyruvate.[1]

Stability. The enzyme is most stable between pH 5.8 and 6.7. Enzyme activity is initially higher in Tris·HCl than in phosphate buffers, but with time the activity drops markedly. EDTA and β-mercaptoethanol have a protective effect on the enzyme. D-lactate and NADH appear to provide no stability to the enzyme. Incubation for 3 min at 50° and 60° resulted in 95% and 10% remaining activity, respectively.[5]

[5] G. L. Long, Doctoral Thesis, Brandeis University, 1971.

[70] Lactate Dehydrogenase-X from Mouse Testes and Spermatozoa

By ERWIN GOLDBERG

$$\text{L-Lactate} + \text{NAD}^+ \rightleftharpoons \text{pyruvate} + \text{NADH} + \text{H}^+$$

The five commonly described isozymes of lactate dehydrogenase are tetramers of A (or M) and B (or H) subunits. A third polypeptide type (C) forms lactate dehydrogenase-X, also a tetramer, and is the product of a separate gene. Lactate dehydrogenase-X is found only in spermatozoa and in mature testes of mammalian and some avian species. This isozyme has been isolated and purified from mouse,[1,2] rat,[3] and bull[4] testes.

Assay Procedure

In a total volume of 3 ml the following are contained:

Sodium phosphate, 150 μmoles, pH 7.0

Sodium pyruvate, 0.75 μmole

NADH, 0.4 μmole

Lactate dehydrogenase-X to give a change in optical density at 340 nm of between 0.1 and 0.2 in 1 min

One unit of enzyme is defined as that amount which will give a change in absorbance of 1.0 in 1 min under the conditions of the assay.

[1] E. Goldberg, *J. Biol. Chem.* **247**, 2044 (1972).

[2] C. Wong, R. Yanez, D. M. Brown, A. Dickey, M. E. Parks, and R. W. McKee, *Arch. Biochem. Biophys.* **146**, 454 (1971).

[3] L. Schatz and H. L. Segal, *J. Biol. Chem.* **244**, 4393 (1969).

[4] E. Kolb, G. A. Fleisher, and J. Larner, *Biochemistry* **9**, 4372 (1970).

Assay of LDH-X in a Crude Extract

LDH-X from mouse testes has a high affinity for other α-keto and α-hydroxy acids. To estimate LDH-X in a mixture of LDH isozymes, pyruvate is replaced in the reaction mixture by 60 nmoles of α-ketovalerate.[5] The ratio of α-ketovalerate/pyruvate activity is a close approximation of the percentage of LDH-X in a testes extract.

Purification Procedure

Crude Extraction. Testes are removed from mature mice after CO_2 euthanasia, and rinsed in 20 mM Tris-chloride buffer, pH 7.4. Approximately 100 g of tissue are homogenized with 5 volumes of cold buffer in a Waring Blendor for 2 min and allowed to stir at 4° for 30 min. The extract is poured through a double layer of cheesecloth and then centrifuged at 13,000 g for 30 min.

Heat Step. The supernatant, in a 4-liter Erlenmeyer flask, is heated to 60° in a 65° water bath, incubated in a 60° bath for 15 min, and cooled in ice. The heated extract is centrifuged at 13,000 g for 40 min; the pellet is discarded.

First Ammonium Sulfate Step. Solid ammonium sulfate is added to the supernatant to give a saturation of 25%. The pH is maintained at 7 with dilute ammonium hydroxide during addition of the salt. After being allowed to stand at 4° for several hours, the suspension is centrifuged at 13,000 g for 45 min.

Second Ammonium Sulfate Step. To the clear supernatant, solid ammonium sulfate is added to 70% saturation. The pH is maintained at 7 as before. The precipitate containing the enzyme activity is collected by centrifugation at 27,000 g for 30 min. The enzyme activity is extracted from the precipitate with 35% saturated ammonium sulfate in 50 mM potassium phosphate buffer at pH 7.4. Usually, all the activity is recovered with two extractions.

Third Ammonium Sulfate Step. The enzymically active fraction in 35% saturated ammonium sulfate is clarified by centrifugation for 20 min at 27,000 g. Solid ammonium sulfate is added to the supernatant to 50% saturation. The pH is maintained at 7.0, as above. After centrifugation for 20 min at 27,000 g, the enzymically active precipitate is dissolved in a minimal volume of 80 mM sodium phosphate buffer at pH 7.0.

Chromatography on DEAE-Sephadex, A-50. DEAE-Sephadex is prepared for use by swelling overnight in distilled water, decanting the fines, and then washing successively with 0.5 M NaOH, distilled water, and

[5] C. O. Hawtrey and E. Goldberg, *J. Exp. Zool.* **174,** 451 (1970).

0.5 M HCl. The resin is rinsed with distilled water to remove excess acid and then equilibrated with 80 mM sodium phosphate at pH 7.0 and containing 1 mM EDTA. A column (40 $\times$ 2.5 cm) containing about 7 g of Sephadex is poured, allowed to settle under gravity to give a bed 30 cm high, and equilibrated with the sodium phosphate–EDTA buffer. The enzymic solution is layered on the column and eluted at room temperature with the equilibration buffer at a flow rate of 1 ml/min. LDH-X appears in the first column volume after the void volume. Those fractions with the highest specific activity are combined.

Final Purification with Ammonium Sulfate. Ammonium sulfate is added to the pooled fractions from the DEAE-Sephadex column until 55% saturation is reached. The solution is allowed to stand overnight at 4° and the precipitate collected by centrifugation at 27,000 g for 30 min.

Crystallization of the Enzyme. The precipitate from the previous step is dissolved in a minimal volume of buffer (80 mM sodium phosphate, pH 7.0). Solid ammonium sulfate is added until a faint turbidity appears. The preparation is kept at 4° for about a week until crystals form. Subsequently, it has been possible to achieve crystallization of LDH-X within 24 hr by seeding with these crystals. The isozyme crystallizes between 30 and 35% saturation and can be stored at 4° as a crystalline suspension in ammonium sulfate. A typical purification is summarized in Table I.

Properties

Purity. The birefringent crystals of LDH-X from mouse testes are thin laths about 0.15 mm thick, 0.1 mm wide, and over 0.4 mm long.[6] The crystalline preparation is homogeneous by ultracentrifugal analysis, polyacrylamide gel electrophoresis, and immunological analysis. The sedimentation velocity pattern at a protein concentration of 6.5 mg/ml reveals a single symmetrical peak with a sedimentation coefficient of 7.0 S. A molecular weight of 140,000 has been calculated from zone velocity centrifugation data.[1]

Stability. The crystalline suspension is stable for several months at 4°.

Substrate Specificity. One of the more striking characteristics of LDH-X is its rather broad substrate specificity. The enzyme catalyzes reversibly, the reduction of α-ketobutyrate and α-ketovalerate[5] as well as α-ketoglutarate,[1,3] in the presence of NADH. When each substrate is assayed at the concentration which gives maximal initial reaction veloci-

[6] A. D. Adams, M. J. Adams, M. G. Rossmann, and E. Goldberg, *J. Mol. Biol.* **78**, 721 (1973).

TABLE I
PURIFICATION PROCEDURE FOR LACTATE DEHYDROGENASE-X[a]

Step	Vol. (ml)	Units per ml	Total units	Protein (mg/ml)	Units per mg protein	Yield[b] (%)
Crude extract	500	40.8	20,400	—	—	—
Heat treatment	424	23.9	10,150	19.4	1.23	100
Ammonium sulfate 0–25% (14.4 g/100 ml)	439	23.9	10,492	3.5	6.8	103
Ammonium sulfate 25–70% (30.7 g/100 ml)	32	279.1	8,932	7.0	39.8	88
Ammonium sulfate 35–50% (9.4 g/100 ml)	2.5	2,802	7,005	51.2	54.7	69
Pooled eluents from DEAE-Sephadex column	30	179	5,370	1.8	99.4	53
Ammonium sulfate 55% saturated (35.1 g/100 ml)	3.0	1,286	3,858	11.6	110.9	38
Crystallization	3.0	1,005	3,015	8.7	116	30

[a] Values shown are for 100 g of mouse testes.

[b] Calculated on the basis of the heated extract, which contains no detectable LDH activity other than LDH-X.

ties, pyruvate is about 30% more active than α-ketoglutarate and 40% more active than α-ketovalerate.

Heat Stability. Although this property of the enzyme shows some species variability, LDH-X is generally less thermolabile than the other isozymes. In the mouse, the half-life of each isozyme at 65° is 9 min for LDH-5, 10 min for LDH-1, and 43 min for LDH-X.

Inhibitors. Oxamate and oxalate are competitive inhibitors of pyruvate reduction and lactate oxidation respectively,[7] catalyzed by LDH-X. Other lactate dehydrogenases are similarly inhibited by these compounds.[8]

Mercurial binding causes 50% inhibition with 4 moles of p-hydroxymercuribenzoate (HMB) bound per mole of LDH-X.[1] In contrast, LDH-1 (B_4) rapidly binds 4 moles of HMB per mole of enzyme with concomitant loss of 90–97% activity.[9]

Both excess substrate and product inhibit LDH-X activity.

Turnover Number. The turnover number of LDH-X with pyruvate

[7] E. Goldberg, *Arch. Biochem. Biophys.* **109,** 134 (1965).

[8] W. B. Novoa, A. D. Winer, A. J. Glaid, and G. W. Schwert, *J. Biol. Chem.* **234,** 1143 (1959).

[9] G. DiSabato, A. Pesce, and N. O. Kaplan, *Biochem. Biophys. Acta* **77,** 135 (1963).

is considerably lower than for the other isozymes. A value of 2980 (moles of NADH oxidized per mole of LDH-X per minute at 25°) has been obtained in comparison to values ranging from 41,500 to 160,000 reported for several lactate dehydrogenases.[10]

Subunit Associations. Catalytically active heterotetramers of C and A as well as C and B polypeptides can be formed *in vitro.*[7,11] In a few species of mammals, C_3A and C_2A_2 tetramers occur *in vivo.*[12]

Purification of LDH-X from Spermatozoa by Affinity Chromatography

It is possible to obtain a good yield of highly purified LDH-X from a relatively small amount of starting material by the technique of affinity chromatography. Sepharose 4B may be substituted with aminohexyl groups after cyanogen bromide activation,[13] or this derivative may be obtained commercially (AGHEXAMINE: P-L Biochemicals, Inc., Milwaukee, Wisconsin). The terminal amino group is condensed with oxalate via an amide bond, to produce the insolubilized oxamate derivative, by the following procedure.[14] To each 5 ml of packed aminohexyl-Sepharose is added 140 mg of potassium oxalate in 2 ml of water, adjusted to pH 4.7 with 1 N HCl. This is followed by dropwise addition of 370 mg of the water-soluble 1-ethyl-3-(3-dimethylaminopropyl) carbodiimide, dissolved in 1 ml of water. The mixture is stirred gently at room temperature for 20 hr, and the Sepharose derivative is washed on a sintered-glass funnel with 1 liter of distilled water. Oxamate is a structural analog which replaces pyruvate in the formation of the ternary complex, LDH:NADH:oxamate.

Chromatography on this material was carried out at room temperature in an 11 mm diameter column (6.5 ml column volume) at a flow rate of 1 ml per minute. The column was equilibrated with 20 mM sodium phosphate, pH 6.8, containing 0.1 M NaCl. NADH at a concentration of 40 μM was added to the irrigant buffer just prior to application of the sample. The sample itself was adjusted to pH 6.8, 0.1 M NaCl and 40 μM NADH. After collecting two column volumes of effluent, NADH was omitted from the irrigating buffer. LDH-X eluted almost exactly one column-volume later, just behind the trailing edge of the NADH, indicating that LDH-X dissociates from the column material as soon

[10] A. Pesce, T. P. Fondy, F. Stolzenbach, F. Castillo, and N. O. Kaplan, *J. Biol. Chem.* **242**, 2151 (1967).

[11] W. H. Zinkham, A. Blanco, and L. Kupchyk, *Science* **142**, 1303 (1963).

[12] E. Goldberg, *J. Exp. Zool.*, **186**, 273 (1973).

[13] P. Cuatrecasas, see this series, Vol. 22 [31].

[14] P. O'Carra and S. Barry, *FEBS Lett.* **21**, 281 (1972).

TABLE II
PURIFICATION OF LACTATE DEHYDROGENASE-X FROM RABBIT SPERMATOZOA

Fraction	Vol. (ml)	Total units	Specific activity (units/mg protein)
Sperm sonicate	22	84	—
DEAE-Sephadex fractions	40	35	5.5
Affinity column eluent	10	24	245

as the excess coenzyme is washed ahead of it.[15] LDH-X is completely unretarded by this column at NADH concentrations greater than 40 μM, and NaCl concentrations higher than 0.1 M.[15]

The results of a typical purification of LDH-X from rabbit spermatozoa are presented in Table II. Combined ejaculates containing 12×10^9 cells were washed free of seminal plasma with phosphate buffered saline, and then suspended in 80 mM sodium phosphate, pH 7.0. The sperm were disrupted with a Bronwill Biosonik at full power output, and centrifuged at 27,000 g. The sonicate was applied to a DEAE-Sephadex column similar to that described above to separate LDH-X from the other isozymes. To the eluent was added NADH and NaCl to 40 μM and 0.1 M final concentrations, respectively. This solution was then applied to the affinity column as described. After two column volumes of the coenzyme-containing buffer were collected, NADH was omitted from the irrigant. LDH-X was then eluted in a single column volume.

[15] E. Goldberg, unpublished results.

[71] D-2-Hydroxy Acid Dehydrogenase from Animal Tissue

By R. CAMMACK

D-2-Hydroxy acid + acceptor → 2-oxo acid + reduced acceptor

Assay Method

Principle. The enzyme-catalyzed reduction of 2,6-dichlorophenol-indophenol with D-lactate as substrate, is measured by decrease in absorption at 600 nm[1,2]. To compensate for any nonenzymic reduction of the dye, by contaminants in the enzyme such as thiols, a reference cell is used containing no lactate.

[1] P. K. Tubbs and G. D. Greville, *Biochem. J.* **81**, 104 (1961).
[2] R. Cammack, *Biochem. J.* **115**, 55 (1969).

Reagents

Potassium D-lactate, 0.5 M
2,6-Dichlorophenolindophenol, 0.8 mM
Tris chloride, 100 mM, pH 8.6

Procedure. Two cells of 1 cm pathlength are placed in the sample and reference positions of a recording split-beam spectrophotometer, maintained at 30°; 0.1 ml of dye solution and 1.0 ml of buffer are added to each cell, and 0.1 ml of D-lactate to the cell in the *reference* position. Both cells are made up to 1.9 ml with water; 0.1 ml of enzyme solution is added to each cell, and they are mixed simultaneously with a double-stirrer arrangement. The change in absorbance at 600 nm is followed, and the initial rate is measured. Since the D-lactate is in the reference cell, the recording will show a net increase in A_{600}. This arrangement is used to prevent the pen from running off the bottom of the chart. For purified enzyme preparations the cell without D-lactate can be omitted, and the cell containing D-lactate is then placed in the sample position.

Definition of Unit and Specific Activity. The unit of enzyme activity is defined as the quantity catalyzing the oxidation of 1 μmole of D-lactate per minute under the above conditions. Thus 0.1 ml of a solution of enzyme containing 1 unit/ml would cause a rate of ΔA_{600} of 1.10 per minute, assuming a molar absorbance for the dye of 22,000.[3]

Application of the Assay to Crude Tissue Extracts. The enzyme is present in a particulate fraction of the cell, and when solubilized it is partially inactive.[4,5] Crude tissue extracts are also likely to contain inhibitors of the enzyme which must be removed. The following procedure can be used to obtain the enzyme in an active form suitable for assay starting from a whole tissue homogenate. The tissue extract (10 ml) is placed in a glass beaker cooled in ice and salt, and subjected to ultrasonic oscillation using a Dawe Soniprobe with 0.5-inch tip at a current of 2.8 A for 1 min.[6] The sample is then centrifuged at 140,000 g for 1 hr. From the center of the tube, 1.0 ml of the clear solution is removed; care must be taken not to disturb the fatty layers at top and bottom of the tube, which usually contain an activity, possibly cytochrome oxidase, that catalyzes the oxidation of leuco-2,6-dichlorophenolindophenol, and thus interferes with the assay. The clear solution is passed through a 10 × 1.1-cm column of Sephadex G-25 equilibrated with 10 mM phosphate, pH 6.3, and the protein fraction (which is usually colored) is col-

---

[3] J. M. Armstrong, *Biochim. Biophys. Acta* **86**, 194 (1964).
[4] P. K. Tubbs and G. D. Greville, *Biochim. Biophys. Acta* **34**, 290 (1959).
[5] R. Cammack, *Biochem. J.* **109**, 46P (1968).
[6] The quantities may be scaled down if a smaller Soniprobe tip is used.

lected quantitatively in a volume of 2.5 ml. The enzyme is then allowed to change to its active form by incubation at 20° for 15 min; the activation is very rapid at pH below 7.[5] The enzyme activity can then be measured by the standard assay method. To allow for the dilution on passage through Sephadex, 0.25 ml of enzyme is added to each cell, the volume of water added being decreased appropriately.

Preparation of Rabbit Kidney D-2-Hydroxy Acid Dehydrogenase

Rabbit kidney is the most active source of the enzyme so far discovered. The activity is three times higher in cortex than medulla, but in the large-scale preparation no attempt is made to separate them.

All steps, except step 7, are carried out at 0–5°. The pH is adjusted, where necessary, with either 1 M acetate, pH 4.0 or 1 M Tris chloride, pH 9.0.

Step 1. Preparation of Homogenate. Rabbit kidneys (2.4 kg) removed from freshly killed rabbits and stored in ice, are obtained from a slaughterhouse. They are blended in three batches in a total volume of 6 liters of 0.25 M sucrose–10 mM Tris chloride, pH 8.0; each batch is blended for 2 min at maximum speed in a 1-gallon capacity Waring Blendor and then cooled on ice.

Step 2. Preparation of Lysed Mitochondrial Fraction. The combined homogenates are centrifuged at 350 g for 6 min. The supernatant is carefully poured off through muslin and centrifuged at 22,000 g for 10 min. The cloudy red supernatant is discarded and the buff-colored pellet is resuspended, using a hand-operated homogenizer, in 6 liters of 10 mM potassium phosphate buffer, pH 7.5. The suspension is centrifuged at 22,000 g for 15 min. The pale pink supernatant is discarded, and the precipitate is resuspended in 10 mM phosphate, pH 6.8, to a final volume of 1800 ml.

Step 3. Ultrasonic Treatment. The suspension of lysed mitochondria is divided into two 900-ml fractions, and each is sonicated in a glass beaker surrounded by ice-water, using a Dawe Soniprobe at full power for 60 min.

Step 4. pH 5.5 Precipitation. The suspension is then vigorously stirred, and carefully adjusted to pH 5.5. The thick white precipitate is removed by centrifugation at 22,000 g for 10 min. The yellow, faintly cloudy supernatant is readjusted as soon as possible to pH 7.0.

Step 5. Ammonium Sulfate Precipitation. The solution is stirred as solid ammonium sulfate is slowly added (27% w/v, 44% saturation). After 30 min the suspension is centrifuged at 22,000 g for 10 min. The pellet is resuspended in 10 mM phosphate, pH 6.8, to a final volume of

100 ml. The solution is dialyzed overnight against 10 liters of the same buffer. A precipitate forms during dialysis, and is removed by high speed centrifugation.

Step 6. DEAE-Cellulose Column. The clear supernatant is adjusted to pH 8.0 and applied to a 20 × 2 cm column of Whatman DE 52, equilibrated with 25 mM Tris, pH 8.0. The column is washed with 50 ml of the Tris buffer, followed by 50 ml of 60 mM NaCl in Tris buffer, and then a linear gradient from 0.06 M to 0.2 M NaCl in the same buffer, 400 ml in all. Fractions of 10–20 ml are collected, and their enzyme activity and protein content are measured. The fractions of highest specific activity are usually eluted between 0.10 and 0.15 M NaCl.

Step 7. Polyethylene Glycol Precipitation. The combined active fractions from the DEAE-cellulose column (1.5–4.0 mg protein/ml) are stirred at room temperature, and 0.43 volume of a 50% w/v aqueous solution of polyethylene glycol 6000 (BDH Chemicals, Poole, Dorset, U.K.) adjusted to pH 7.0 is added with stirring, to give a final polyethylene glycol concentration of 15%. The pH is carefully adjusted to 6.8; after 20 min the suspension is centrifuged at 12,000 g for 10 min. The supernatant is collected and adjusted to pH 5.5, and after 20 min centrifuged again. The precipitate from the second centrifugation is redissolved in 25 mM Tris, pH 8.0, to a final volume of about 10 ml.

Step 8. Hydroxyapatite Column. Hydroxyapatite is prepared by the method of Levin,[7] or Bio-Gel-HTP (Bio-Rad Laboratories, Richmond, California) mixed with Celite 545 (1 g/g dry weight of gel) to permit a fairly rapid flow, suspended in water, and packed into a 2 × 20 cm column. The enzyme preparation, adjusted to pH 7.0, is applied to the column, followed by 50 ml of water, then a linear gradient of potassium phosphate, pH 7.4, from 0 to 60 mM, with a total volume of 400 ml. All eluates from the column are checked for enzyme activity. Normally the enzyme is almost totally adsorbed by the column and is eluted by phosphate between 10 mM and 20 mM concentration.

Step 9. Ammonium Sulfate Precipitation. The fractions with highest specific activity are combined and 31.5% w/v ammonium sulfate is added with stirring, and the precipitated enzyme is collected by centrifugation.

The enzyme may be stored as a precipitate in ammonium sulfate at 0° for months, with only small loss of activity.

The purification procedure is summarized in the table. By following this procedure, enzyme preparations with specific activity between 4 and 8 units/mg protein have been obtained. The specific activity of the less pure preparations can be improved by gel filtration on Sephadex G-200.

[7] O. Levin, this series, Vol. 5 [2].

PURIFICATION OF D-2-HYDROXY ACID DEHYDROGENASE FROM
2.4 KG OF RABBIT KIDNEYS[a]

Step	Volume (ml)	Total activity (units)	Specific activity (units/mg)	Purification (fold)	Yield (%)
1. Homogenate	8000	654	0.0011	(1)	(100)
2. Washed mitochondria	1800	333	0.0062	5.7	51
4. Ultrasonic treatment, pH 5.5 precipitation	1490	297	0.040	37	45
5. (NH₄)₂SO₄ precipitation, dialysis	97	251	0.162	148	38
6. DEAE-cellulose column	130	165	0.795	730	25
7. Polyethylene glycol fractionation	9.3	127	1.95	1780	19
8. Hydroxyapatite column	34	56	7.1	6500	8.5
9. (NH₄)₂SO₄ precipitation	2.0	54.5	7.4	6750	8.3

[a] For steps 1 and 2, enzyme was measured by the method described for animal tissues, and protein by the biuret method. At the other stages enzyme was measured in the standard assay and protein by UV absorption.

Properties of the Purified Enzyme

Homogeneity. The purified enzyme runs as one component in the analytical ultracentrifuge. On polyacrylamide gel electrophoresis, it shows one major protein band, and one or two minor bands, all of which have D-2-hydroxy acid dehydrogenase activity. These results indicate that the enzyme is substantially pure; the minor bands are probably polymers.

Stability. The enzyme can be stored in 25 mM Tris, pH 8.0 in the dark at 4° for several weeks with only small loss of activity. The enzyme is stabilized by the competitive inhibitor oxalate.

Molecular Properties. The relative molecular mass, estimated by gel filtration on Sephadex G-200 is 102,000 ± 10,000 daltons. The enzyme contains FAD, since the apoenzyme formed by acid ammonium sulfate treatment can be reactivated by FAD, but not FMN or riboflavin. The flavin is not fluorescent, but gives rise to absorption peaks at 380 and 450 nm, the intensity of which is consistent with 2 molecules of FAD per molecule of enzyme. These peaks are reduced when substrates are added.

Substrate Specificity.[1] The K_m for D-lactate is approximately 2 mM, and the optimum pH is 8.6. D-Glycerate, and longer-chain homologs of D-lactate, at least as far as D-2-hydroxy-n-octanoate are substrates; both

K_m and V_{max} increase with chain length along the series. However, glycolate, the lowest member of the series, is not a substrate, and neither are 3-methylpentanoate, phenylglycolate or phenyllactate, which have bulky side groups.

Certain D-hydroxydicarboxylic acids are substrates. The optimum pH for D-malate oxidation is 6.5 or lower, and the K_m is approximately 1 mM. meso-Tartrate is a good substrate, whereas D-tartrate is only slowly oxidized, indicating that the stereochemistry of carbon-3 is important. Thus the more metabolically important D-2-hydroxy acids, 3-phospho-D-glycerate and 6-phospho-D-gluconate, are not substrates.

Acceptor Specificity.[1,2] The enzyme can use a variety of artificial electron acceptors, including phenazine methosulfate, methylene blue, and ferricyanide. The activity with ferricyanide is stimulated 2.2-fold by 100 mM ferrocyanide in the assay medium. Possible physiological acceptors are ubiquinone (UQ_1), cytochrome c, and oxygen, which are reduced at 72%, 25%, and 15%, respectively, of the activity with 2,6-dichlorophenolindophenol. It is not clear therefore whether the enzyme *in vivo* acts as a dehydrogenase or an oxidase. NAD and NADP will not act as acceptors for the enzyme.

Inhibitors.[8,9] A number of dicarboxylic acids inhibit competitively with substrate; the most active of these is oxalate, with K_i about 5μM. The products of the reaction, pyruvate and oxaloacetate, which are the products of oxidation of D-lactate and D-malate, respectively, inhibit competitively with the acceptor, with K_i about 0.6 mM and 0.17 μM, respectively. L-Lactate is a weak inhibitor, so that the rate with 50 mM DL-lactate as substrate is approximately 80% of that with 25 mM D-lactate.

The inhibition by chelating agents is somewhat complex.[10] Cyanide ion is a reversible inhibitor, competitive with the substrate. Ethylenediamine tetracetate and 1,10-phenanthroline cause a progressive inactivation; this is not due to removal of a metal from the enzyme, but to binding of the chelating agent to the enzyme. The course of inhibition is slower in the presence of D-lactate, or oxalate, but is accelerated by the presence of cyanide. These results suggest that a metal atom may be involved in substrate binding to the enzyme.

Use of the Enzyme for Assay of D-2-Hydroxy Acids. The method used for assay of the enzyme can be modified to measure D-2-hydroxy acids, such as D-lactate or D-malate which are substrates. Ferricyanide is a more suitable acceptor than indophenol, as the reduced form of the latter tends to autoxidize. Enzyme purified as far as step 6 is suitable

[8] P. K. Tubbs, *Biochem. J.* **82**, 36 (1962).

[9] R. Cammack, *Biochem. J.* **118**, 405 (1970).

[10] P. K. Tubbs, *Biochem. Biophys Res Commun* **3**, 513 (1960).

for this purpose. A typical reaction medium would contain potassium ferricyanide, 2 μmoles, and Tris, pH 8.6, 50 μmoles in a 1 cm light path cell at 30°, 0.05–1.0 μmole of D-2-hydroxy acid is added, and A_{420} is followed until it reaches a steady level. Then 0.1 unit of enzyme is added to give a final volume of 2.0 ml. The resulting decrease in A_{420} (allowing for dilution by enzyme solution) is a measure of the quantity of D-2-hydroxy acid; 0.1 μmole should cause ΔA_{420} of 0.104. Large quantities of reducing agents such as thiols will interfere with this assay. The method of Britten[11] avoids this difficulty by using a large excess of ferricyanide, and converting the oxo acids formed to their 2,4-dinitrophenylhydrazones which can be estimated spectrophotometrically. In this way the hydroxy acids can be identified by chromatography of the derivatives.

[11] J. S. Britten, *Anal. Biochem.* **24**, 330 (1968).

[72] Lactate Oxygenase of *Mycobacterium phlei*

By SHIGEKI TAKEMORI and MASAYUKI KATAGIRI

Lactate oxygenase, which also has been termed "lactate oxidative decarboxylase," is a flavoprotein with FMN as the prosthetic group and catalyzes the incorporation of one atom of oxygen into substrate while the other is reduced to water according to the following reaction[1,2]:

$$CH_3CHOHCOOH + O_2 \rightarrow CH_3COOH + CO_2 + H_2O$$

This enzyme was first crystallized from cells of *Mycobacterium phlei* by Sutton in 1957.[1] Recently, the procedure for purifying the enzyme to obtain crystalline preparations in an excellent yield has been devised and the nature of the crystalline enzyme has been studied most extensively.[3,4]

Assay Method

Principle. The activity of the enzyme is assayed at 25° by measuring the consumption of molecular oxygen dissolved in the reaction medium with the use of a Clark oxygen electrode from Yellow Springs Instruments Co., Ohio. A manometric technique may also be used.[3]

[1] W. B. Sutton, *J. Biol. Chem.* **226**, 395 (1957).
[2] O. Hayaishi and W. B. Sutton, *J. Amer. Chem. Soc.* **79**, 4809 (1957).
[3] S. Takemori, K. Nakazawa, Y. Nakai, K. Suzuki, and M. Katagiri, *J. Biol. Chem.* **243**, 313 (1968).
[4] S. Takemori, Y. Nakai, K. Nakazawa, M. Katagiri, and T. Nakamura, *Arch. Biochem. Biophys.* **154**, 137 (1973).

Reagents

L-Lithium lactate, 750 mM
Potassium phosphate buffer, 100 mM, pH 6.0

Procedure. Mixtures are placed in the reaction vessel that contain 72 μmoles of the buffer,[5] 270 μmoles of lactate, as well as an appropriate amount of the enzyme in a total volume of 3.6 ml. The reactions are initiated by adding enzyme.

Definition of Unit and Specific Activity. A unit is defined as the amount of enzyme which consumes 1 μmole of molecular oxygen per minute at 25°. Specific activity is defined as the number of units per milligram of protein determined by measurement of absorbance at 280 or 454 nm.[6] For the crystalline enzyme, the absorbance indexes $\epsilon_{0.1\%}$ at 280 and 454 nm are 2.04 and 0.226, respectively.

Preparation of the Enzyme

Growth of Cells

The cells of *M. phlei* are grown in a medium at pH 7.0 containing 0.5% L-sodium glutamate, 0.5% KH_2PO_4, 2% glycerol, 2% soluble starch,[7] 0.5% yeast extract, and 0.15% magnesium citrate. Eight 500-ml Roux bottles, each containing 120 ml of growth medium, are inoculated with a culture of *M. phlei*. After 6 days of horizontal standing at 37°, the bottles are vigorously shaken to suspend the surface growth, and the contents are then distributed to ninety 500-ml Roux bottles containing 120 ml of the same medium. The culture is further incubated for 6 days at 37°. Cells are harvested in a continuous flow centrifuge, washed twice with distilled water and are kept frozen. The yield of cells is approximately 30 g of cell paste per liter of culture medium.

Purification Procedure

The enzyme purification presented below is a modification of the method previously described.[3] The DEAE-Sephadex step in the procedure

---

[5] In the assay of lactate oxygenase from *M. smegmatis,* the imidazole-HCl buffer is recommended by P. Sullivan ["Favins and Flavoproteins" (H. Kamin, ed.), p. 470, University Park Press, Baltimore, Maryland, 1971]. We can not find any appreciable difference in the *M. phlei* enzyme activity between imidazole-HCl and potassium phosphate buffer at a concentration of 20 mM.

[6] The estimation of the protein content from absorbances at 280 and 260 nm gives a faulty result.

[7] The commercial product is again hydrolyzed with 0.1 N H_2SO_4 for 30 min at 100°.

can be satisfactorily omitted, and the enzyme is directly crystallized from the preparation obtained with ammonium sulfate fractionation.

All operations are carried out at room temperature unless otherwise noted.

Step 1. Cell Extract. The frozen cells (200 g) are thawed and ground with 400 g of aluminium oxide (Wako W-800, Wako Pure Chemical Co, Osaka, Japan) in a mechanical mortar (20.5-cm diameter) for 30 minutes. The paste is extracted with 600 ml of distilled water by additional grinding for a few minutes. The resulting slurry is centrifuged at 22,000 g for 20 min and the supernatant fluid is saved. The debris together with alumina is reground for 10 min and mixed with 400 ml of distilled water. The supernatant fluid obtained after centrifugation is combined with the initial extract.

Step 2. First Ammonium Sulfate Fractionation. To the crude cell-free extract is added 24.3 g of solid ammonium sulfate for each 100 ml of solution and the suspension is stirred for 30 min. The precipitate is removed by centrifugation at 22,000 g for 30 min. The enzyme is then precipitated from the supernatant by additional 16.5 g of ammonium sulfate for each 100 ml of solution, and the suspension is centrifuged at 77,500 g for 30 min.[8] The clear supernatant is discarded, and the precipitate is dissolved in a small volume of 0.1 M potassium phosphate buffer, pH 6.0 (about 100 ml).

Step 3. Reprecipitation with Ammonium Sulfate. To each 100 ml of the enzyme solution which is viscous and turbid, 10 g of ammonium sulfate are slowly added with stirring. The precipitate, which is allowed to accumulate overnight, is collected and dissolved in a small volume of 0.1 M phosphate buffer (about 60 ml).

Step 4. Crystallization. The turbid solution resulting from the above step is subjected immediately to crystallization. Finely powdered ammonium sulfate is slowly added to the enzyme solution until a turbid solution becomes viscous. The enzyme solution is constantly stirred during the addition of ammonium sulfate and then is allowed to stand at 5° overnight. At this time microscopically visible crystals (rectangular transparent plates) begin to appear in such a viscous turbid solution. If the mixture is allowed to stand at 5° for several days, the yellow crystals settle to the bottom leaving a brown turbid solution. The crystals are collected by brief centrifugation (2800 g for 10 min), and the viscous turbid supernatant is discarded. The yellow crystals are dissolved in a minimum volume of 0.1 M phosphate buffer, pH 6.0. Dissolution of the

---

[8] The mixture must be centrifuged hard to cause a clear precipitation. A part of the enzyme remains in the supernatant if the suspension is centrifuged at a low speed.

crystals occurs slowly, and it is sometimes necessary to let the preparation stand for 1 hr at room temperature. The insoluble precipitate is removed by centrifugation and discarded. The supernatant solution is allowed to stand in the ice bath and crystallization is accomplished by the addition of finely powdered ammonium sulfate. The enzyme starts to crystallize almost immediately, and a silky sheen begins to appear. After all the enzyme has crystallized by standing overnight at 5°, the crystals are collected and dissolved in the same buffer. It is sometimes found that, during recrystallization, white precipitates remain associated with the crystals. Amorphous contaminating protein is removed by centrifugation, and the enzyme is recrystallized by the addition of ammonium sulfate. Usually, four recrystallizations are required to obtain electrophoretically and ultracentrifugally homogeneous preparations. The specific activity of the crystalline enzyme at this stage is about 32 units/mg. The overall recovery is approximately 50% and 70 mg of the crystalline enzyme is usually obtained from 100 g of frozen $M.$ $phlei$ cell paste. The crystals are stored at 5° as a suspension in the presence of ammonium sulfate for several months without appreciable loss in activity.

Properties

$Absorption$ $Spectrum.$ The enzyme has a typical flavoprotein absorption spectrum with maxima at 280, 375, and 454 nm with shoulders at 430 and 480 nm. The absorbance indexes ($a_{(mM\ FMN)^{-1}}^{cm^{-1}}$) at 280, 375, and 454 nm are 113.0, 7.06, and 12.5, respectively.

$Coenzymes.$ The enzyme is reversibly resolved into FMN and apoenzyme moieties by the acid ammonium sulfate treatment.[3] The apoenzyme

SUMMARY OF MOLECULAR WEIGHT DETERMINATION OF LACTATE OXYGENASE

Methods	Molecular weight
Sedimentation and diffusion[a]	370,000
Archibald method	352,000
Gel filtration	340,000
Electron microscopic analysis ($N \times V \times \rho$)[b]	58,600 × 6 (352,000)
Flavin and terminal amino acid analyses	56,000 × 6 (336,000)

[a] Estimated on the basis of the sedimentation constant of 12.5 S, the diffusion constant of 3.01×10^{-7} cm²/sec, and the partial specific volume of 0.732 ml/g which is calculated from the amino acid composition.

[b] M. Katagiri, S. Takemori, T. Oda, and T. Matsumoto, $Arch.$ $Biochem.$ $Biophys.$ **160,** 295 (1974). N, Avogadro's number; V, calculated volume of the subunit sphere by using $r = 25.6$ Å, and ρ, density of the enzyme, 1.39.

can be readily converted to a holoenzyme by the addition of either FMN or riboflavin. FAD has no effect. No transition metal is detectable in the crystalline enzyme. A very small amount of sugar (hexose, $0.158 \pm 0.004\%$ and hexosamine, $0.120 \pm 0.005\%$) is found in the enzyme.

Oligomeric Structure. The molecular weight values of the enzyme determined by various methods are summarized in the table. The enzyme is dissociated into subunits with molecular weight of 54,000–57,000. This value corresponds to the minimum molecular weight (56,000) obtained from flavin content. The enzyme contains 1 mole of serine and arginine per 56,000 g of the protein as amino and carboxyl termini, respectively. Thus, lactate oxygenase exists in an oligomeric structure composed of six polypeptide chains beginning at serine and terminating at arginine. There are no interchain disulfide bridges in the molecule.[9]

Specificity. The maximal turnover number for the reaction with L-lactate is 3000 moles of O_2 consumed per minute per mole of the enzyme flavin at optimal pH of 6.0. The apparent K_m value for L-lactate is 34 mM. When D-lactate is used as a substrate, the activity is competitively inhibited by D-lactate, so that DL-lactate is not suitable as a substrate for kinetic studies.

The following aliphatic acids with various carbon chain lengths and aromatic L-α-hydroxy acids serve as oxygenatable substrate for the enzyme: α-hydroxybutyrate, α-hydroxyvalerate, α-hydroxycaproate, mandelate, and β-phenyllactate, whereas the compounds such as α-hydroxyisobutyrate and β-hydroxy-n-butyrate are inactive. When L-malate is used as a substrate, the enzyme acts as a malate oxidase, producing oxaloacetate and hydrogen peroxide.[10]

Anaerobic Reaction. When a substrate amount of the enzyme is incubated anaerobically with an equimolar amount of L-lactate, the enzyme-bound FMN is fully reduced. Upon exposure of the reduced enzyme to air, all the added L-lactate is converted to the dehydrogenated product, pyruvate, corresponding to the amount of the $FMNH_2$ moiety. However, the oxygenated product, acetate, is not produced by adding air to the reduced enzyme. The oxygenation of substrate and the reduction of flavin appear to be coupled tightly in the enzyme.

[9] S. Takemori, H. Tajima, F. Kawahara, Y. Nakai, and M. Katagiri, *Arch. Biochem. Biophys.* **160**, 289 (1974).
[10] S. Takemori, M. Nakanishi, and M. Katagiri, *Seikagaku* **44**, 431 (1972).

[73] Pyruvate-Ferredoxin Oxidoreductase from *Clostridium acidi-urici*[1]

By JESSE C. RABINOWITZ

$$\text{Pyruvate} + \text{CoA} + \text{ferredoxin}_{ox} \rightleftharpoons \text{acetyl-CoA} + CO_2 + \text{ferredoxin}_{red}$$

Assay Method

Principle. The enzyme activity can be determined by measuring the radioactivity of pyruvate formed in the exchange reaction between pyruvate and $^{14}CO_2$,[1,2] or by the decrease in absorption at 450 nm caused by the reduction of FAD in the presence of pyruvate.[1]

Reagents and Procedure of Measurement of Oxidoreductase Activity

Sodium pyruvate (recrystallized twice from ethanol), 55 μmoles
Potassium phosphate buffer, pH 7.0, 225 μmoles
Sodium bicarbonate, pH 7.0, 11.0 μmoles
$[^{14}C]NaHCO_3$, 30,000 cpm
Coenzyme A, 10 nmoles
2-Mercaptoethanol, 200 μmoles
Cobinamide (Factor B),[3] 0.02 nmole
Enzyme solution

Total volume of 1.0 ml; incubation in air at 37°; reaction started by addition of enzyme. After 30 min, add 0.1 ml of 20% perchloric acid and purge the acidified solution with CO_2 for 10 min to remove unreacted $^{14}CO_2$. Neutralize the reaction mixture with 7 N KOH and centrifuge. Transfer a 0.4-ml aliquot of the supernatant solution into 10 ml of Triton-X Aqueous Fluor[4] and determine the radioactivity in a scintillation counter.

Purification Procedure

All steps are carried out at 0–4° unless otherwise noted. Buffer I contains 50 mM potassium phosphate buffer, pH 7.4, 40 mM EDTA, 0.1

---

[1] K. Uyeda and J. C. Rabinowitz, *J. Biol. Chem.* **246**, 3111 (1971).
[2] S. Raeburn and J. C. Rabinowitz, *Arch. Biochem. Biophys.* **146**, 9 (1971).
[3] W. Friedrich and K. Bernhauer, *Chem. Ber.* **89**, 2507 (1956).
[4] D. C. Tiemeier and G. Milman, *J. Biol. Chem.* **247**, 2272 (1972). Prepare by dissolving 9.7 g of New England Nuclear Omnifluor {or 9.1 g of PPO, 2,5-diphenyloxazole, and 0.61 g of POPOP, 1,4-bis[2-(5-phenyloxazoly)]benzene} in 2140 ml of toluene, and then add 1250 ml of Triton X-100 (Rohm and Haas).

mM ferrous ammonium sulfate, and 0.5 mg of reduced glutathione per milliliter. It was freshly prepared for use. Buffer II has the same composition as buffer I except that the concentration of potassium phosphate was reduced to 20 mM and the ferrous ammonium sulfate concentration was increased to 1 mM.

Growth of Bacterial Cells. Clostridium acidi-urici (ATCC No. 7906) was grown in the same manner described for *Clostridium cylindrosporum*,[5] except that the sodium carbonate and sulfuric acid were omitted from the medium, and the wet unwashed cell paste obtained by centrifugation was stored frozen at —90°. The yield of wet cells is 0.4 to 0.5 g per liter of medium.

1. Sonic Extract. Frozen cells (20 g) were suspended in 30 ml of buffer I. The suspension was subjected to sonic oscillation for 2 min with the Branson Sonifier Model W185D. The extract was then centrifuged for 20 min at 144,000 g, and the precipitate was discarded.

2. DEAE-Cellulose I. In order to remove ferredoxin, the extract was placed on a DEAE-cellulose column (0.8 cm² $\times$ 15 cm) previously equilibrated with buffer I. The column was washed with about 15 ml of buffer I, and the pass-through and the wash were combined.

3. Ammonium Sulfate I. To the combined enzyme solution (40 ml) from the preceding step were added 30 ml of Buffer I, 28 g of solid ammonium sulfate, and a sufficient amount of 15 N ammonium hydroxide to bring the pH to 7.0. The mixture was allowed to stand 10 min and was centrifuged for 20 min at 30,000 g. The precipitate was discarded.

4. Ammonium Sulfate II. Solid ammonium sulfate (9.3 g) was added to the supernatant solution from the preceding step, and the pH was adjusted to 7.0 as before. The solution was allowed to stand 10 min and was centrifuged for 10 min at 30,000 g. The precipitate was dissolved in 2 ml of buffer I and the solution was dialyzed for 2 hr against 300 ml of buffer II.

5. Sephadex G-150. The dialyzed enzyme (5.6 ml) was placed on a Sephadex G-150 column (1.13 cm² $\times$ 25 cm) that had been equilibrated with buffer II. It was developed with this same buffer, and 3-ml fractions were collected. A yellow band containing the enzyme activity was eluted immediately after the void volume of the column. The fractions containing the enzyme, usually about eight in number, were pooled.

6. DEAE-Cellulose II. The pooled enzyme fractions were applied to a DEAE-cellulose column (1.13 cm² $\times$ 25 cm) equilibrated with buffer II. The column was washed with 30 ml of buffer II modified to contain 50 mM phosphate buffer. The enzyme was eluted with a linear gradient

---

[5] J. C. Rabinowitz, see this series, Vol. 6 [97].

PURIFICATION OF PYRUVATE-FERREDOXIN OXIDOREDUCTASE FROM
Clostridium acidi-urici

Fraction	Volume (ml)	Total protein (mg)	Total exchange activity ($10^{-6} \times$ units)	Specific activity (units/mg)
1. Sonic extract	33	1155	28.4	25,000
2. DEAE-cellulose I	40	1120	26.8	24,000
3. Ammonium sulfate	5.8	110	8.1	75,000
4. Sephadex G-150	24	43.2	4.3	100,000
5. DEAE-cellulose II	1.8	3.6	1.3	370,000

obtained by allowing 100 ml of buffer II containing 0.25 M potassium phosphate buffer, pH 7.4, to run into a mixing chamber containing 100 ml of buffer II containing 50 mM potassium phosphate buffer. The flow rate was about 30 ml per hour, and 3-ml fractions were collected. The fractions that contained the enzyme, usually fractions 35 to 48, were pooled and concentrated with a Diaflo ultrafiltration apparatus using an XM-50 membrane.

The purification is summarized in the table.

The same procedure has been successfully applied to larger preparations starting with 200 g of cells. The reagents and sizes of the columns were increased proportionally. The specific activity of the purified enzyme varies from 270,000 to 830,000 units/mg depending upon the preparation and on the particular batch of cells used. The chief reason for the wide variation of specific activity is the lability of the enzyme, and the final specific activity obtained depends on the rapidity with which the purification is executed.

Properties

Physical Properties and Composition. The enzyme has a molecular weight of approximately 240,000 based on sedimentation studies. It contains nonheme iron and inorganic sulfide and thiamine. Its absorption spectrum shows, in addition to the normal maximum at 278 nm, a broad absorption band in the region of 400 nm. This latter absorption band, similar to that of the iron–sulfur protein ferredoxin, is bleached when the protein is incubated with pyruvate and CoA. Unlike other iron–sulfur proteins with intrinsic enzymic activity, this enzyme is free of riboflavin.

Reactions Catalyzed. The enzyme catalyzes (a) a pyruvate-CO_2 exchange reaction, (b) oxidation of pyruvate with reduction of a variety of electron acceptors, (c) acetoin formation from pyruvate and acetaldehyde,[6] and (d) the synthesis of pyruvate from acetyl-CoA and bicarbonate in the presence of reduced ferredoxin.[7,8]

Specificity of Electron Acceptor. FAD, FMN, clostridial ferredoxin, rubredoxin, and various viologen and tetrazolium derivatives can function as electron acceptors in the enzymic oxidation of pyruvate catalyzed by the enzyme. The pyridine nucleotides, however, are inactive.

pH Optimum and Buffer Effects. The enzyme shows activity in phosphate buffers from pH 4.5 to 8.7 with an optimum between pH 6.7 and 7.4. The activity in Tris-chloride buffer at pH 7.7 is about one-fourth of that in phosphate buffer at this pH.

Other Requirements. Anaerobic conditions are necessary for enzyme activity. The cobinamide required under the assay conditions described can be replaced by providing anaerobic assay conditions.

A sulfhydryl compound is required for activity of the enzyme under aerobic or anaerobic assay conditions. This requirement is relatively nonspecific and may be satisfied by 2-mercaptoethanol, dithiothreitol, glutathione, cysteine, but not by ascorbate.

Catalyic amounts of CoA are required for the exchange reaction. Kinetic studies indicate that CoA decreases the apparent K_m for bicarbonate about 5-fold without affecting the K_m for pyruvate.[6]

[6] K. Uyeda and J. C. Rabinowitz, *J. Biol. Chem.* **246**, 3120 (1971).
[7] S. Raeburn and J. C. Rabinowitz, *Arch. Biochem. Biophys.* **146**, 21 (1971).
[8] S. Raeburn and J. C. Rabinowitz, *Biochem. Biophys. Res. Commun.* **18**, 303 (1965).

[74] Glycolic Acid Oxidase from Pig Liver

By MARILYN SCHUMAN JORNS

$$R—CHOH—COOH + O_2 \rightarrow R—CO—COOH + H_2O_2$$

The enzyme exhibits maximal activity with glycolic acid (R = H), but has a wide range of specificity toward L-α-hydroxyacids. Other substrates include: glyoxalic acid (the product of glycolic acid oxidation, R = OH in the hydrated form), L-lactic acid, L-α-hydroxyisocaproic acid, and L-α-hydroxyvaleric acid. Other electron acceptors, such as 2,6-dichlorophenolindophenol or potassium ferricyanide, may be substituted for oxygen.[1–3]

[1] M. Schuman and V. Massey, *Biochim. Biophys. Acta* **227**, 521 (1971).
[2] F. M. Dickinson and V. Massey, *Biochem. J.* **89**, 53P (1963).
[3] F. M. Dickinson, Doctoral Dissertation, University of Sheffield, England, 1965.

Assay Methods

Principle. The oxidation of substrate (usually glycolic acid) is followed spectrophotometrically at 420 nm with potassium ferricyanide or at 600 nm with 2,6-dichlorophenolindophenol as electron acceptor.[4]

Procedures. POTASSIUM FERRICYANIDE ASSAY. The assay is performed at 25° in a cuvette with a 1-cm light path using a Gilford recording spectrophotometer. Each assay contains: 200 μmoles of potassium phosphate (pH 7.0), 1 mg of bovine serum albumin, 3 μmoles of EDTA, 4 μmoles of $K_3Fe(CN)_6$, 2 μmoles of sodium glycolate, enzyme, and water to a volume of 3.0 ml. The reaction is initiated by the addition of enzyme, with the exception of crude preparations (see below). One unit of activity is defined as an absorbance change at 420 nm of 1.0 per minute and corresponds to 2.91 μmoles/min. Specific activity is expressed as units of activity per unit of absorbance at A_{280nm}. This assay is used to monitor the yield and purification during each stage of enzyme purification and for comparing the specific activity of samples isolated in successive preparations.

2,6-DICHLOROPHENOLINDOPHENOL (DCIP) ASSAY. The procedure is identical to the $K_3Fe(CN)_6$ assay except that $K_3Fe(CN)_6$ is replaced by 0.1 μmole of DCIP. A decrease in absorbance at 600 nm of 1.0 per minute is defined as one unit of activity. This corresponds to 0.149 μmoles/min. This assay is approximately 6 times more sensitive than the $K_3Fe(CN)_6$ assay, and it is used to assay dilute enzyme solutions obtained in column fractions.

Application of Assay Procedure to Crude Extracts. A large nonlinear blank rate of dye reduction is observed with crude enzyme fractions prior to calcium phosphate gel-cellulose chromatography. This blank rate decreases and becomes linear after a few minutes of reaction and substrate is then added. Enzyme activity is calculated by subtracting the blank rate observed immediately before substrate addition.

Purification Procedure

The method described below is a modification[4] of a procedure developed by Dickinson.[2,3] All steps of the purification are carried out at 2–4° in the presence of 0.3 mM EDTA. Glass-distilled water is used throughout.

Step 1. Homogenization and pH Precipitation. Pig liver, packed in cracked ice at the slaughterhouse, is cut into 1–2-inch cubes and divided into ten 1-kg batches. Each batch is washed for 5 min in 2 liters of 0.15

[4] M. Schuman and V. Massey, *Biochim. Biophys. Acta* **227**, 500 (1971).

M KCl and the washings are discarded. The tissue is then homogenized in a further 2 liters of 0.15 M KCl for 75 sec at medium speed in a 1-gallon capacity Waring Blendor. The foam is removed, and the pH is adjusted to 4.8, with stirring, using 2 M acetic acid. The homogenate is centrifuged for 30 min at 20,000 g. The precipitate is discarded, and the pH of the supernatant is quickly adjusted to 8.0, with stirring, using 1 M NaOH. The precipitate which forms during neutralization is allowed to settle out overnight, and the bulk of the extract can be siphoned off. The residual material is centrifuged for 20 min at 16,300 g, and the supernatant is pooled with the bulk of the extract.

Step 2. Fractionation with $(NH_4)_2SO_4$. Solid $(NH_4)_2SO_4$ is added to the supernatant from step 1 to 44% saturation and the mixture is stirred for at least 15 min after the salt is dissolved. It is then centrifuged for 20 min at 8000 g and the precipitate is discarded. $(NH_4)_2SO_4$ is added to the supernatant to 58% saturation and the mixture is centrifuged as before. The supernatant is described. The precipitate is dissolved in a minimal volume of 0.1 M sodium phosphate (pH 6.3) and dialyzed for 24 hr against four changes of a 10-fold excess of this buffer. Since the enzyme is sensitive to photoinactivation, this dialysis and all subsequent stages of the preparation and handling are done in darkness. A small precipitate is usually present after dialysis and is removed by centrifugation.

Step 3. Calcium Phosphate Gel-Cellulose Column Chromatography. Columns are prepared according to the method described by Massey.[5] A supension containing 216 g of cellulose powder, 18 g of calcium phosphate gel, and 2.4 liters of water is used to obtain a packed column measuring 5.5 cm $\times$ 50 cm. Continuous stirring during column packing ensures a uniform column and reasonable flow rates (100–200 ml/hr). The column is equilibrated with 0.1 M sodium phosphate (pH 6.3).

The dialyzed enzyme solution from step 2 is applied to a calcium phosphate gel-cellulose column. The column is washed with 0.1 M sodium phosphate (pH 6.3) containing 2% (w/v) $(NH_4)_2SO_4$ until the A_{280nm} of the eluate is less than 0.05. The enzyme is then eluted in 0.1 M sodium phosphate (pH 7.0), containing 5% (w/v) $(NH_4)_2SO_4$. Column fractions which have a specific activity of at least 25% of that of the best fractions are pooled. Glycolic acid oxidase is precipitated from the pooled fractions by addition of $(NH_4)_2SO_4$ to 80% saturation. The precipitate is collected by centrifuging for 60 min at 20,000 g and is resuspended in a minimal volume of 0.1 M sodium phosphate (pH 7.0).

Two successive preparations of enzyme, each made from 10 kg of liver, are routinely pooled for the following purification steps. The com-

[5] V. Massey, this series, Vol. 9 p. 272.

bined preparations are dialyzed for 48 hr versus eight changes of a 50-fold excess of 30 mM sodium phosphate (pH 6.0) prior to chromatography on CM-cellulose. A precipitate, which forms during dialysis, is removed by centrifugation and discarded.

Step 4. *CM-Cellulose Chromatography.* The dialyzed enzyme, obtained in Step 3, is passed through a CM-cellulose column (3.5 cm $\times$ 30 cm), previously equilibrated with 30 mM sodium phosphate (pH 6.0). Under these conditions a yellow protein band, associated with glycolic acid oxidase activity, passes through the column whereas a contaminating hemoprotein is adsorbed as a brown band at the top of the column. The yellow eluate from this column is dialyzed for 48 hr versus eight changes of a 25-fold excess of 5 mM sodium phosphate (pH 7.6).

Step 5. *DEAE-Cellulose Chromatography.* The dialyzed enzyme, obtained in Step 4, is applied to a DEAE-cellulose column (3.5 cm $\times$ 33 cm), previously equilibrated with 5 mM sodium phosphate (pH 7.6). The column is washed with this buffer until the A_{280nm} of the elute is less than 0.005. The enzyme is eluted from the column with a linear gradient, formed from 1 liter of 10 mM sodium phosphate and 1 liter of 50 mM sodium phosphate, both at pH 7.6. The bulk of the enzyme elutes from the column at a constant specific activity, and these fractions are pooled. If the A_{280nm} of the pooled eluate is greater than 1.0, solid $(NH_4)_2SO_4$ is added to 80% saturation and the precipitate is collected by centrifuging for 60 min at 20,000 g. However, if the A_{280nm} of the pooled eluate is less than 1.0, better recovery of enzyme activity is achieved if $(NH_4)_2SO_4$ is added to 100% saturation. The protein precipitate is collected by passing the suspension through a column of Hyflo Super-Cel (Johns-Manville Co. (1.0 cm $\times$ 3.5 cm), previously equilibrated with 0.1 M sodium phosphate (pH 7.0) saturated with $(NH_4)_2SO_4$. The precipitated enzyme is retained at the top of the column and is eluted in a small volume of 0.1 M sodium phosphate (pH 7.0). Side fractions (suitable for preliminary experiments) are similarly pooled and concentrated.

Recovery and specific activity data of a typical preparation are given in Table I.

Properties

Stability. The enzyme is stable at $-20°$ for several months if stored as an $(NH_4)_2SO_4$ precipitate or in the presence of 2.5 M glycerol. It is also stable, catalytically, at $0°$ in the dark for several weeks in the absence of protecting reagents, but a 10–15% loss of absorption at 450 nm generally occurs. This change can be reversed or prevented by treatment with 0.35 M $(NH_4)_2SO_4$.[4]

TABLE I
PURIFICATION OF GLYCOLIC ACID OXIDASE FROM PIG LIVER[a]

Fraction		Volume (ml)	Total activity (units)[b]	Total A_{280nm}	Specific activity (units/ A_{280nm})	Recovery of activity[c] (%)
pH 4.8 Precipitation	A[d]	22,650	420	936×10^3	4.5×10^{-4}	—
supernatant	B	22,750	490	940×10^3	5.2×10^{-4}	—
$(NH_4)_2SO_4$ supernatant	A	23,770	523	401×10^3	1.3×10^{-3}	—
	B	23,930	350	450×10^3	7.7×10^{-4}	—
$(NH_4)_2SO_4$ precipitate	A	1,750	525	124×10^3	4.2×10^{-3}	100
	B	1,530	475	120×10^3	4.0×10^{-3}	97
Calcium phosphate gel	A	64.5	328	800	0.41	63
eluate	B	62.0	304	1060	0.28	62
Pool preparations A and B		126.5	632	1860	0.34	63
CM-cellulose eluate		338	497	1110	0.45	49
DEAE-cellulose eluate						
Best fraction		5.0	206	145	1.42	20
Side fraction		8.3	94.5	90.5	1.04	9.3

[a] Data from M. Schuman and V. Massey, *Biochim. Biophys. Acta* **227**, 500 (1971).
[b] Units of enzyme activity are based on the $K_3Fe(CN)_6$ assay.
[c] Enzyme activity, prior to calcium phosphate gel chromatography, can only be roughly estimated due to the presence of a large blank reaction. Estimates of percent activity recovered are based on the highest total activity measured prior to calcium phosphate gel chromatography.
[d] A and B refer to two separate 10-kg preparations which are pooled for final purification.

Prosthetic Groups. Glycolic acid oxidase is a flavoprotein with a molecular weight of 100,000 as determined by molecular sieve chromatography.[3] The enzyme contains two chromophores: FMN and 6-hydroxyFMN.[4,6] There is one mole of FMN per 50,000 g of protein, and thus 2 moles of FMN per mole of enzyme.[2-4] The content of 6-hydroxyFMN varies between 13 and 20% of the FMN content in different preparations. The spectral properties of the native enzyme also vary somewhat with different preparations depending, in part, on the amount of 6-hydroxyFMN present. Enzyme preparations containing the minimum observed amounts of 6-hydroxyFMN show an A_{280nm}/A_{450nm} ratio of 7.3 and a molar extinction coefficient at 450 nm of 11.7×10^3. The visible absorption spectrum, dominated by the FMN component, shows maxima at 375 and 452 nm with marked shoulders at 330, 430, and 480

[6] S. G. Mayhew, C. D. Whitfield, S. Ghisla, and M. S. Jorns, *Eur. J. Biochem.*, **44**, 579 (1974).

TABLE II
STEADY-STATE KINETIC CONSTANTS

Electron acceptor	K_m glycolate (mM)	K_m acceptor (mM)	Turnover number (mole/mole FMN/min)
DCIP[a]	0.42	0.28	1250
Oxygen[b]	0.32	0.27	620

[a] pH = 7.0 [M. Schuman and V. Massey, *Biochim. Biophys. Acta* **227**, 521 (1971)].
[b] pH = 8.3 [F. M. Dickinson and V. Massey, *Biochem. J.* **89**, 53P (1963); F. M. Dickinson, Doctoral Dissertation, University of Sheffield, England, 1965].

nm. It also has a broad and weak absorption band at 600 nm. The spectrum of the enzyme-bound 6-hydroxyFMN component, obtained after selective removal of FMN, shows absorption maxima at 328, 425, and 600 nm. The possible function of the 6-hydroxyFMN component in catalysis remains unclear since substrate will only reduce the FMN component in the native enzyme and selective removal of FMN results in total loss of catalytic activity.[4] Similar results are obtained with electron-transferring flavoprotein, which contains both FAD and 6-hydroxyFAD.[6]

Nature of the Active Site. Considerable evidence indicates that the FMN prosthetic group is located near one or more positively charged amino acid residues (which function in binding α-hydroxyacid substrates) and also a hydrophobic region of the protein. The enzyme binds a variety of anions (e.g., sulfate, chloride, carboxylic acids) which disrupt flavin interaction with the positively charged groups, altering chemical and spectral properties of the bound FMN. For example, the pK for the ionization of the 3-imino nitrogen in free FMN (pK = 10.3) is shifted to about pH 8 in FMN bound to glycolic acid oxidase. However, in the enzyme–oxalate complex the observed pK is increased to pH 9.3.[1,4]

Similar to other flavoprotein oxidases, this enzyme forms a red anionic flavin semiquinone and reacts readily with sulfite to form a covalent FMN-sulfite adduct, $K_D = 0.8$ μM.[4,7]

Kinetic Properties. The oxidation of glycolic acid, coupled with the reduction of DCIP, shows converging-line steady-state kinetics[1] whereas the reaction with oxygen as the electron acceptor shows parallel-line kinetics.[2,3] The kinetic constants are summarized in Table II. At glycolic acid concentrations higher than 1.7 mM, excess substrate inhibition is observed in the DCIP system.[1]

Inhibitors and Activators. Straight-chain monocarboxylic acids

[7] V. Massey, F. Müller, R. Feldberg, M. Schuman, P. A. Sullivan, L. G. Howell, S. G. Mayhew, R. G. Matthews, and G. P. Foust, *J. Biol. Chem.* **244**, 3999 (1969).

$[CH_3-(CH_2)_n-CO_2H]$ are noncompetitive inhibitors with either glycolic or glyoxalic acid as the variable substrate and DCIP as the electron acceptor. Binding affinity increases as n is increased (e.g., $K_{i\,(n=0)} = 37$ mM, $K_{i\,(n=5)} = 92$ μM) and log K_1 is a linear function of n. Dicarboxylic acids $[HO_2C-(CH_2)_n-CO_2H]$ are competitive inhibitors when glycolic acid is the variable substrate and DCIP is the electron acceptor but are noncompetitive inhibitors when glyoxalic acid is the variable substrate. Binding affinity decreases as n is increased (e.g., $K_{i\,(n=0)} = 0.44$ mM, $K_{i\,(n=2)} = 14$ mM). Most inorganic anions, such as chloride and sulfate, are also inhibitors, but phosphate and arsenate cause enzyme activation, maximal at 0.1 M salt concentrations.[1]

[75] Glyoxylate Reductase, Two Forms from *Pseudomonas*[1]

By R. P. HULLIN

Glyoxylate + NAD(P)H + H$^+$ → glycolate + NAD(P)$^+$

Two enzymes which catalyze the reduction of glyoxylate to glycolate may be obtained from *Pseudomonas*. Reduced pyridine nucleotides act as hydrogen donors for the enzymes. The NADH-linked glyoxylate reductase is entirely specific for its coenzyme, but the NADPH-linked reductase shows some affinity toward NADH.

Assay Method

Principle. The enzymes are assayed by measuring the rate of oxidation of either NADH or NADPH in the presence of glyoxylate. The oxidation of the coenzyme is followed spectrophotometrically at 340 nm.

Reagents

Sodium glyoxylate, 1.0 M
Phosphate buffer, 1.0 M, pH 6.8
NADH, 4 mM, in 1.0 M phosphate buffer, pH 6.8
NADPH, 4 mM, in 1.0 M phosphate buffer, pH 6.8

Procedure. The reaction mixture in 3.0 ml silica cells ($l = 1$) contains 0.3 ml of 1.0 M phosphate buffer pH 6.8, 0.1 ml of 1.0 M sodium glyoxylate, and an appropriate amount of enzyme. The reaction is started by the addition of 0.1 ml of 4 mM NADH or 4 mM NADPH, and the extinc-

[1] L. N. Cartwright and R. P. Hullin, *Biochem. J.* **101**, 781 (1966).

tion at 340 nm is observed using a recording spectrophotometer at 25°. At some stages, it is necessary to measure the rates of oxidation of NADH and NADPH in the absence of glyoxylate and use the values to determine the true enzyme activity.

Activity. One unit of enzyme is defined as that quantity which gives $\Delta A_{340nm} = -0.001$ unit/min. Specific activity is defined as units of enzyme per milligram of soluble protein as determined by the optical density at 280 nm after correcting for nucleic acid absorption at 260 nm.[2]

Purification Procedure

The enzymes are obtained from a strain of *Pseudomonas fluorescens* isolated from soil which has the ability to grow on butane-2,3-diol as sole carbon source. Growth medium contains the following per liter: KH_2PO_4, 5.0 g; $(NH_4)_2SO_4$, 2.0 g; $MgSO_4 \cdot 7H_2O$, 0.4 g; butane-2,3-diol, 2.0 g.

The medium made up without the magnesium salt is adjusted to pH 7.0 with 5 M sodium hydroxide. After sterilization, the $MgSO_4$ is added aseptically as a 10% (w/v) solution. Cells from a fresh agar slope culture are used to inoculate two 50-ml samples of medium which after 24 hr of growth at 30°, serve as inocula for 10 liters of medium in a 12-liter flask. After 24 hr of vigorous aeration, the crop of cells obtained is used to inoculate 40 liters of medium in a 55-liter aspirator containing 10 fish-tank aerators, which is vigorously aerated at room temperature with a small amount of silicone MS antifoam to prevent excessive frothing. After 36 hr, the cells are brought into the logarithmic phase of growth by adding 5 liters of fresh medium and harvested at room temperature with a de Laval centrifugal separator. The cells, which have a distinct pink color en masse and wet weight of 1.0–1.5 g per liter of medium, are packed in stainless steel presses,[3] precooled to −14°. After at least 6 hr at −14°, the cells are crushed with the aid of a fly press. The crushed cells can be stored at −14° for up to 3 months without deterioration.

Step 1. Extract. Frozen crushed cells (35 g) are thawed and diluted to 110 ml with 5 mM phosphate buffer, pH 7.0. Deoxyribonuclease (1 mg) and ribonuclease (1 mg) are added during the preparation of the homogeneous suspension. The extract is centrifuged at 70,000 g for 30 min in a Beckman, Model L, preparative ultracentrifuge using a type 30 rotor at 2° and the precipitate is discarded.

All subsequent operations are conducted at 2°.

[2] O. Warburg and W. Christian, *Biochem. Z.* **310**, 384 (1941).
[3] D. E. Hughes, *Brit. J. Exp. Pathol.* **32**, 97 (1951).

Step 2. Ultracentrifugation. The supernatant solution is centrifuged at 140,000 *g* for 3 hr in a Beckman, Model L, preparative ultracentrifuge using a type 50 rotor and the ribosomal pellet is discarded.

Step 3. Ammonium Sulfate. Solid ammonium sulfate is added to 30% saturation (17.6 g per 100 ml of extract), and the solution is continuously agitated. After allowing 20 min for equilibration, the precipitate is removed by centrifugation and discarded. More solid ammonium sulfate is added to the supernatant solution to 60% saturation (19.8 g per 100 ml of extract), and the previous procedure is repeated except that the precipitate is retained and dissolved in 30 ml of 5 mM phosphate buffer, pH 7.0. The solution obtained is dialyzed against 10 liters of 5 mM phosphate buffer, pH 7.0, which is replaced three times with 10 liters of fresh buffer at hourly intervals. Continuous agitation is essential to prevent excessive precipitation of the protein, and this is achieved by placing a 10-liter flask on a magnetic stirrer and allowing a plastic-coated magnet to rotate inside the flask. Any small quantity of material which does precipitate during dialysis is removed by centrifuging.

Step 4. Column Chromatography. DEAE-cellulose treated before use according to the method of Peterson and Sober[4] is suspended in 5 mM phosphate buffer, pH 7.0, and poured into a chromatography column (1.8 × 20 cm), the lower end of which is closed with a coarse sintered-glass disk. The column contents are equilibrated by allowing 1 liter of 5 mM phosphate buffer, pH 7.0, to run through it at 2°. The dialyzed material is applied to the top of the column, at a rate just sufficient to keep the top moist; by this means the material can be adsorbed in a narrow band at the top of the column; unadsorbed material is removed by passing 60 ml of 5 mM phosphate buffer, pH 7.0, through the column.

A linear gradient of NaCl is now applied to the column by allowing 200 ml of a solution containing 5 mM phosphate buffer, pH 7.0, to flow, with constant stirring, into 200 ml of a solution containing 5 mM phosphate buffer, pH 7.0, and 0.5 M NaCl; the resultant mixture is allowed to flow through the column at approximately 30 ml per hour, the effluent being collected continuously at 7.75 ml per fraction. The chloride content of the eluents is determined by the method of West and Coll.[5]

Two glyoxylate reductases are eluted from the column: (a) NADPH-linked enzyme is eluted between 40 and 100 mM NaCl, the peak of activity appearing at approximately 65 mM NaCl. Fractions 12–15 inclusive are used for further purification. (b) NADH-linked enzyme is eluted between 150 and 230 mM NaCl with the peak appearing at 175 mM NaCl; fractions 23–26 are used for further purification. Those fractions contain-

[4] E. A. Peterson and H. A. Sober, this series, Vol. 5, p. 3.
[5] P. W. West and H. Coll, *J. Amer. Water Works Ass.* **49**, 1485 (1957).

ing each of the enzymes at a specific activity greater than in step 3 are combined, and the protein is precipitated with solid $(NH_4)_2SO_4$ to 60% saturation (39 g per 100 ml of solution).

Purification of the NADPH-Linked Glyoxylate Reductase

Step 5. Gel Filtration, Sephadex G-200. The precipitate containing the NADPH-linked enzyme is dissolved in 1 ml of 0.1 M phosphate buffer, pH 7.0, is carefully placed on top of a column of Sephadex G-200, 55 × 2.25 cm, pretreated according to the method of Flodin[6] and equilibrated with 0.1 M phosphate buffer, pH 7.0. The bottom outlet of the column is then opened, and the enzyme sample is allowed to enter the bed surface; at the moment it disappears through the surface, a small amount of buffer is added to wash the surface. When this disappears a larger volume of buffer is added. The column is eluted by passing 300 ml of 0.1 M phosphate buffer, pH 7.0, at a rate of 5.5 ml per hour, the effluent being collected continuously at 2.75 ml per fraction. The enzyme is eluted between 115 and 154 ml of buffer collected and fractions 44–51 inclusive, which possess activity greater than 300 units/ml, are combined.

Step 6. Precipitation with Ammonium Sulfate. The pooled enzyme fractions are brought to 60% of saturation with solid $(NH_4)_2SO_4$ (39 g/100 ml.) After centrifugation, the precipitate is dissolved in 2 ml of water and diluted in the ratio 1:10 with buffer solution when required.

A summary of the purification procedure, as presented in Table I, represents a 220-fold purification.

Purification of the NADH-Linked Glyoxylate Reductase

Step 5. Gel Filtration, Sephadex G-200. The precipitate containing the NADH-linked enzyme is dissolved in 1 ml of 0.1 M phosphate buffer, pH 7.0, applied to the surface of a Sephadex G-200 column, 57 × 2.25 cm, and the column eluted with 300 ml of 0.1 M phosphate buffer, pH 7.0, as described in Step 5 of the purification of the NADPH-linked enzyme. The NADH-linked enzyme is eluted between 100 and 132 ml of buffer collected and fractions 40–44 inclusive possessing activity greater than 3000 units/ml are combined.

Step 6. Precipitation with Ammonium Sulfate. The pooled fractions are brought to 60% of saturation with $(NH_4)_2SO_4$ (39 g/100 ml), and the enzyme precipitate, after centrifuging, is dissolved in 2 ml of water and diluted when required in the ratio 1:20 with buffer solution.

[6] P. Flodin, *in* "Dextran Gels and Their applications in Gel Filtration." Pharmacia, Uppsala, Sweden, 1962.

TABLE I
PURIFICATION OF NADPH-LINKED GLYOXYLATE REDUCTASE

Step	Volume (ml)	Total activity (units)	Total protein (mg)	Specific activity (units/mg)	Recovery (%)
1. Crude extract	94	46,800	1860	25.2	100
2. Ultracentrifugation	84	43,700	1250	35.0	93
3. (NH$_4$)$_2$SO$_4$ ppt. (30–60% sat.)	38	39,600	538	73.6	85
4. DEAE-cellulose	27	22,400	32	692	48
5. Gel filtration	21	13,400	2.7	4920	29
6. (NH$_4$)$_2$SO$_4$ ppt. (60% sat.)	2	12,200	2.2	6660	26

A summary of the purification procedure is presented in Table II; it represents a 79-fold purification.

Properties

Specificity and Kinetic Properties. The NADH-linked enzyme is entirely specific for its coenzyme, but the NADPH-linked reductase shows some affinity toward NADH (relative activities toward NADPH and NADH in the ratio 4.4:1). Both enzymes convert hydroxypyruvate to glycerate. The preparation containing the NADH-linked enzyme is contaminated with malate and lactate dehydrogenases; a gradual but not total removal of the contaminating enzymes is achieved during the purification procedure. The K_m values for glyoxylate are 7 mM and 14 mM

TABLE II
PURIFICATION OF NADH-LINKED GLYOXYLATE REDUCTASE

Step	Volume (ml)	Total activity (units)	Total protein (mg)	Specific activity (units/mg)	Recovery (%)
1. Crude extract	94	231,800	1860	124.7	100
2. Ultracentrifugation	84	219,300	1250	175.5	95
3. (NH$_4$)$_2$SO$_4$ ppt. (30–60% sat.)	38	176,800	538	328	76
4. DEAE-cellulose	28	103,500	90	1150	45
5. Gel filtration	13	61,800	7.8	7930	27
6. (NH$_4$)$_2$SO$_4$ ppt. (60% sat.)	2	58,200	5.9	9870	25

for the NADH- and NADPH-linked enzymes, respectively; for hydroxy pyruvate, the corresponding K_m values are 5 mM and 7 mM. Maximal rate of glyoxylate reduction by both enzymes occurs with 33.3 mM sodium glyoxylate.

Effect of pH. Both reductases show maximal activity between pH 6.0 and 6.8, but the NADPH-linked enzyme retains its activity over a wider range of pH values.

Activators and Inhibitors. Oxo acids inhibit enzyme activity by combining with the active centers. Both enzymes are strongly dependant on free thiol groups for activity, as shown by inhibition with p-chloromercuribenzoate. The reduction of glyoxylate and hydroxypyruvate is not stimulated by anions.

Physical Constants. The equilibrium constants for the reactions

$$\text{Glycolate} + \text{NAD}^+ \rightarrow \text{glyoxylate} + \text{NADH} + \text{H}^+$$

and

$$\text{Glycolate} + \text{NADP}^+ \rightarrow \text{glyoxylate} + \text{NADPH} + \text{H}^+$$

are 6.0×10^{-18} M and 3.0×10^{-18} M, respectively. The molecular weights of the reductases are estimated by thin-layer gel filtration to be in the region of 180,000.

[76] Aldehyde Dehydrogenase from *Pseudomonas aeruginosa*

By W. E. RAZZELL

$$\text{Glycolaldehyde} + \text{NAD}^+ \xrightarrow{\text{K}^+} \text{glycolic acid} + \text{NADH} + \text{H}^+$$

Assay Method[1]

NADH formation is determined in a silica cuvette containing the following components in a final volume of 1 ml: potassium phosphate, pH 7.2, 100 mM; 2-mercaptoethanol, 10 mM; NAD$^+$, 2 mM; enzyme. The reaction is started by addition of glycolaldehyde, 1 mM, after prior incubation for 5 min at 35° of all other components. Reaction is obtained with approximately 20 μl of extract from cells grown on ethanol, or 1 μg of purified enzyme.

[1] R. G. von Tigerstrom and W. E. Razzell, *J. Biol. Chem.* **243**, 2691 (1968).

Purification Procedures

Growth of Cells and Preparation of Extracts. Although details are given for the procedures known to be successful with *P. aeruginosa*, ATCC No. 9027 the response of a number of species and strains of pseudomonads which can utilize ethanol and which possess the potassium-activated enzyme[2] suggests that the method is generally applicable to them.

Cells are grown in Roux flasks[3] containing 100 ml of medium consisting of $NH_4H_2PO_4$, 0.3%; K_2HPO_4, 0.4%, $FeSO_45H_2O$, 5 ppm; yeast extract, 0.1%; and tryptone, 0.1%; all adjusted to pH 7.4 before sterilization. $MgSO_4 \cdot 7H_2O$ solution is sterilized separately and added to a concentration of 0.05% at the time of inoculation, together with 95% ethanol (to 0.3% concentration). A 1% inoculum is grown in the same medium overnight, added, and the culture incubated for 24 hr at 30°. A further addition of ethanol equal to 0.4% of the medium is added after 10 hr of growth. Cells are harvested in the cold by centrifugation at 13,000 *g*.

All subsequent operations (to step 6) are performed at ice-bath temperatures. The cells are washed once with 50 mM potassium phosphate–10 mM mercaptoethanol, pH 7.0, and suspended in 100 mM sodium bisulfite–10 mM mercaptoethanol, pH 7.0. The cell yield is usually 13 g wet weight per liter of medium, and the final suspension should contain about 200 mg/ml. The cells are disrupted by sonic treatment, the temperature of the suspension being kept below 10°. The sonic extract is diluted by the addition of about one-half volume of 100 mM sodium bisulfite–10 mM mercaptoethanol, pH 7.0, and sufficient potassium phosphate, pH 7.0, to yield a final phosphate concentration of 50 mM. After centrifugation at 32,000 *g* for 60 min, the supernatant cell extract contains about 20 mg of protein and 10–12 units of enzyme per milliliter.

The final cell extract can be stored at −20° for several weeks or at 3° for several days.

Cell extracts from the above step are thawed when necessary, without allowing their temperature to rise above 3°.

Step 1. Protamine Sulfate Treatment. As is customary, each batch of cell extract or pool thereof is tested on a small scale for its response to this treatment since excess protamine sulfate will precipitate the enzyme, and it is possible that some lots of protamine sulfate will prove useless. It is important to ensure that the protamine sulfate solution is thoroughly mixed and at 18–22° before adding it to the extract. The protamine sulfate solution (20 mg/ml, pH 5.0) is added with stirring to the

---

[2] W. E. Razzell and R. W. Blackmore, *Can. J. Microbiol.* **15**, 645 (1969).

[3] Shake cultures or aerated vessels do not permit good growth of most *Pseudomonas* species.

cell extract to a concentration of about 20% by volume. Stirring is continued for 20 min, and the precipitate is discarded after centrifugation at 32,000 g for 30 min.

Step 2. Ammonium Sulfate Fractionation. Solid ammonium sulfate is added with stirring in the proportion of 21 g per 100 ml of the solution. Stirring is continued for an additional 30 min, the precipitate is removed by centrifugation at 38,000 g for 30 min and discarded. A further 10.1 g of ammonium sulfate are added for every 100 ml of original protamine sulfate supernatant solution, with stirring as above. After centrifugation, the precipitate is dissolved in 5 mM potassium phosphate–10 mM sodium bisulfite–10 mM mercaptoethanol–0.5% ethylene glycol, pH 7.0, to give a total volume of 16 ml for every 100 ml of protamine sulfate supernatant used. Storage at −20° is possible.

Step 3. Acetone Fractionation. Acetone, 60 ml, chilled to −15°, is added slowly with stirring to every 100 ml of the dissolved ammonium sulfate precipitate fraction. The precipitate obtained is recovered by centrifugation at 23,000 g for 10 min at −5° and set aside (A). This procedure is repeated on the supernatant fraction by the addition of a further 25 ml (B) and then 40 ml (C) of cold acetone. Precipitates A, B, and C are individually suspended in 100 mM potassium phosphate–100 mM sodium bisulfite–10 mM mercaptoethanol–1.0% ethylene glycol, pH 7.0, to a final volume of 100 ml. Fraction C alone contains most of the enzyme activity, but if more than 30% of the enzyme is in fraction B at an equivalent specific activity, it can be pooled with fraction C.

Step 4. Ammonium Sulfate Precipitate. To each 100 ml from the above step, 43 g of solid ammonium sulfate are added, the mixture is stirred for 2 hr, and the precipitate is recovered by centrifugation at 27,000 g for 20 min. The precipitate is suspended in 50 ml of 100 mM potassium phosphate–10 mM mercaptoethanol–5 mM EDTA, pH 7.0, and the suspension is centrifuged as before. More buffer is added to adjust the volume to 78 ml.

Step 5. Fractionation between pH 5.4 and 4.8. For every 100 ml of the dissolved ammonium sulfate precipitate from the acetone fraction, 16.5 g of ammonium sulfate are added. The slight amount of precipitate which forms after 30 min of stirring is removed by centrifugation at 27,000 g for 10 min. Acetic acid (1 M, containing 16.5 g of ammonium sulfate per 100 ml) is used to lower the pH of the preparation. About 6.5 ml is added with stirring to bring the pH from 6.55 to 5.4; and after the inert precipitate is removed by centrifugation at 27,000 g for 10 min, 6.7 ml of the acid mixture is added to adjust the pH to 4.8. The fraction precipitating between pH 5.4 and pH 4.8, after a further 10 min of stirring, is recovered by centrifugation as above. It is dissolved in 100 mM potassium phosphate–10 mM mercaptoethanol–5 mM EDTA–1.0% ethylene glycol, pH 7.0, to yield 19 ml of solution. Storage at —20° is possible.

Step 6. Chromatography. The enzyme is now sufficiently stable to permit rapid handling at room temperature, although storage should be maintained at 3° or —20°, and 50 mM bisulfite or 1% glycol is necessary to protect against activity losses during freezing and thawing.

Cellex-T (chloride form) is converted to the phosphate form in a column (2 × 50 cm) by passing through 0.1 M potassium phosphate, pH 7.0, until the effluent is chloride-free. The column is equilibrated with the starting buffer (5 mM potassium phosphate–10 mM mercaptoethanol–1 mM EDTA–1.0% ethylene glycol, pH 7.0). The enzyme preparation (20 ml or less) obtained in step 5 is dialyzed against 500 ml of the starting buffer for 4 hr, with one change of buffer at 2 hr, applied to the column, and washed into the column with 20 ml of starting buffer. A linear gradient is used to elute the enzyme: 450 ml of starting buffer to which is introduced 450 ml of 300 mM potassium phosphate–10 mM mercaptoethanol–1 mM EDTA–1.0% ethylene glycol, pH 7.0. The column is run at 3 ml/min, and the enzyme peak[4] appears between 440 and 560 ml. To each 100 ml of peak eluate is added, with stirring, 5.0 ml of 1 M potassium phosphate, pH 7.0, and 45 g of ammonium sulfate. The suspension is kept at 4° for 15 hr, the precipitated protein (about 120 mg) collected by centrifugation, and taken up to a volume of 4.0 ml with 50 mM potassium phosphate–10 mM mercaptoethanol–1 mM EDTA, pH 7.0. If frozen storage is planned, however, the precipitate should be dissolved in the buffer described at the end of the next paragraph.

The enzyme can be further purified by gel filtration, although little is achieved compared with the previous steps, other than complete removal of colored material. A column packed with 190 ml of Sephadex G-100 in 50 mM potassium phosphate–10 mM mercaptoethanol–1 mM EDTA, pH 7.0, is washed with 2 volumes of the same buffer. The enzyme (4 ml) is applied to the column and washed through with the above buffer at a flow rate of 0.6 ml/min. Both enzyme and protein emerge between 85 and 135 ml, and to 50 ml of solution, in an ice bath, is added, with stirring, 5.0 ml of 1 M potassium phosphate, pH 7, and 24 g of ammonium sulfate. After being kept at 4° for 15 hr the suspension is centrifuged and the precipitate taken up to 4.0 ml with 100 mM potassium phosphate–10 mM dithiothreitol–1 mM EDTA, pH 7.0, and the solution stored at —22°.

Table I summarizes a representative purification process, which yields a preparation homogeneous except for some dissociated enzyme.[1]

---

[4] The preparation is sufficiently pure so that protein can be estimated from columns by absorbance at 280 nm.

TABLE I

SUMMARY OF PURIFICATION OF ALDEHYDE DEHYDROGENASE

Procedure	Specific activity (units/mg)	Recovery (%)
Cell extract	0.65	100[a]
Protamine sulfate	0.84	98
Ammonium sulfate	1.9	81
Acetone fractionation	5.2	64
Fractionation, pH 5.4 to 4.8	7.3	42
Cellex-T chromatography	14.5	30
Sephadex G-200	14.7	23

[a] Total units per 100 g wet weight of cells = 7000.

Physical Properties[5]

Table II lists the physicochemical properties of the enzyme and its dissociation product.

Table III contains the amino acid analysis.

Enzymic Properties[1,5]

Only KCl and NH_4Cl satisfy the ionic requirements of the enzyme for activity, although RbCl is partially effective. Additional studies[2] have shown that pseudomonads in general possess aldehyde dehydrogenase activated by potassium, whereas phosphate has no such effect. In the

[5] R. G. von Tigerstrom and W. E. Razzell, *J. Biol. Chem.* **243**, 6495 (1968).

TABLE II

PHYSICOCHEMICAL PROPERTIES OF ALDEHYDE DEHYDROGENASE

Property	Observation
Electrophoretic mobility (pH 6.8–8.5)	Migrates to anode (starch)
Isoelectric precipitation point	pH 4.8
Protein contaminants (electrophoresis)	Less than 2.5%
$D_{20,w}$	4.4×10^{-7} cm^2 sec
MW, sedimentation, diffusion	$187,000 \pm 4000$
$s_{20,w}^{\circ}$ enzyme (dissociated form)	9.2 S (5.5 S)
$E_{1cm}^{1\%}$ at 280 nm (pH 7.0)	10.4
$A_{280}:A_{260}$	1.83
Zinc content	Less than 0.6 atom/mole
Molecular activity[a]	3140 mole min^{-1} mole^{-1}

[a] If active sites per mole = 1, then molecular activity = turnover number.

presence of 0.1 M potassium phosphate, pH 7.0, other salts had negligible effect, whereas sodium arsenate inhibition was 50% at 0.1 M.

Apparent Michaelis constants, calculated from results obtained when all other components of the assay were maintained at standard concentrations, were NAD+, 0.37 mM; NADP+, 3.0 mM (4% of rate observed with NAD+); glycolaldehyde, 0.4 mM; glyceraldehyde, 14 mM; potassium 6 mM; ammonium, 12 mM.

The pH optimum in the standard assay, supplemented with 0.1 M Tris chloride, was pH 8.3; but the enzyme is stable under assay conditions for longer periods at pH 7.2 (the preferred pH for storage in the cold).

Substrates showing significant activity at concentrations between 0.1 mM and 1.0 mM include propionaldehyde, butyraldehyde, isobutyraldehyde, and benzaldehyde as well as acetaldehyde, which is as good a substrate as glycolaldehyde but inhibits at the higher concentrations.

The enzyme is sensitive to —SH reagents: 1.0 mM iodoacetate and 0.1 mM iodoacetamide inhibit about 50% in the absence of potassium and NAD+. In the presence of potassium the inhibition is greater, whereas

TABLE III
AMINO ACID COMPOSITION OF ALDEHYDE DEHYDROGENASE

Amino acid	g/100 g	Moles/mole
Lysine	5.80	74
Histidine	2.81	34
Arginine	5.97	64
Aspartic acid + asparagine	9.68	136
Threonine	5.24	82
Serine	3.88	69
Glutamic acid + glutamine	13.94	177
Proline	4.39	71
Glycine	6.14	153
Alanine	9.17	192
Valine	6.44	103
Isoleucine	7.02	100
Leucine	9.33	132
Tyrosine	5.43	56
Phenylalanine	5.54	62
Cysteic acid	1.43	22
Methionine	2.25	28
Tryptophan[a]	3.16	31
Total	107.6	(MW 187,000)

[a] From spectrophotometric data.

in the presence of NAD^+ or NADH the enzyme is protected—provided potassium is present as well.

Inactivation with trypsin is also prevented by NAD^+ or NADH, provided potassium is also present; but potassium alone also protects to some extent.

Reducing agents are required for activity and for stability. In the presence of 50 mM bisulfite, the enzyme retains 80% of its activity after 10 min at 50°, whereas in the absence of reducing agents only 25% of the activity is retained under these conditions.

Dissociation of the enzyme into inactive subunits (half the original molecular weight) occurs if the enzyme is exposed to low salt concentrations. Provided either a reducing agent or EDTA is present, however, the enzyme reassociates and regains enzymic activity upon incubation (at 30°) in the presence of added salt (regardless of the chemical nature of the salt).

[77] Aldehyde Dehydrogenase from Baker's Yeast

By Shelby L. Bradbury, Julia F. Clark, Charles R. Steinman, and William B. Jakoby

$$RCHO + DPN^+(TPN^+) \xrightarrow{K^+} RCOOH + DPNH(TPNH) + H^+$$

Assay Method

Principle. The reaction is conveniently monitored by measuring the production of DPNH spectrophotometrically at 340 nm and 25° with benzaldehyde as the aldehyde substrate. Benzaldehyde was selected because it did not inhibit the reaction under standard assay conditions. However, by limiting the aldehyde concentration, the method described here can also be used as an assay for a wide variety of aldehydes some of which are inhibitory.

Reagents

KCl, 1 M

Tris chloride, 1 M, adjusted to pH 8.0 at 25° with HCl

DPN, 10 mg/ml

β-Mercaptoethanol, approximately 0.1 M, prepared by diluting 0.07 ml of the mercaptan to 10 ml with water.

Benzaldehyde, 12 mM, prepared by adding 5 μl of benzaldehyde to the bottom of a tube containing 10 ml of cold water and mixing.

Procedure.[1,2] The following volumes of the above reagents are added in the indicated order and are diluted to a final volume of 1.0 ml with water: 0.1 ml Tris chloride; 0.2 ml KCl; 0.05 ml β-mercaptoethanol; 0.1 ml DPN; an appropriate amount of enzyme; 0.1 ml benzaldehyde. Product formation is followed in a spectrophotometer coupled to a recorder; a full-scale deflection of 0.1 absorbance may be used. Under the conditions described here, the reaction is linearly proportional to enzyme concentration for absorbance changes from 0.001 to 0.250 per minute. Occasionally, a lag period of 1–2 min is encountered.[1] This is eliminated by storing enzyme at $-20°$ to $-90°$ in the presence of 0.1 M α-thioglycerol; after prolonged storage, it is often necessary to incubate the thawed enzyme with 50 mM β-mercaptoethanol for a few hours at $4°$ before use.

Definition of Units. One unit of activity produces 1.0 μmole of DPNH per minute under the described assay conditions and is equal to an absorbance change of 6.21 at 340 nm. Specific activity is equal to the number of activity units per milligram of protein. Protein is determined colorimetrically;[3] crystalline bovine albumin serves as the standard.

Purification Procedure

General Comments. Any enzyme isolated from yeast is a potential substrate for intrinsic proteases found in abundance in these cells. Fortunately, proteolysis can usually be controlled by avoiding autolysis as a means of cell disruption, by rapid processing of cell extracts, and by using liberal quantities of serine esterase inhibitors. For example, early studies with the aldehyde dehydrogenase from yeast resulted in a crystalline preparation of apparent homogeneity.[1,4] However, analysis of the amino terminus revealed the presence of a spectrum of amino acids in the N-terminal position.[5] That preparation has been obtained by autolysis and, despite the use of a serine esterase type of protease inhibtor, PMSF,[6] had undergone degradation.

The method detailed below[5] also makes use of PMSF and produces three species of aldehyde dehydrogenase designated as A, B, and C, of

[1] C. R. Steinman and W. B. Jakoby, *J. Biol. Chem.* **242**, 5019 (1967).
[2] S. L. Bradbury and W. B. Jakoby, *J. Biol. Chem.* **246**, 1834 (1971).
[3] O. H. Lowry, N. J. Rosebrough, A. L. Farr, and R. J. Randall, *J. Biol. Chem.* **193**, 265 (1951). Since the presence of a high concentration of mercaptan interferes with the method, it is advisable to first precipitate the protein, discarding the supernatant fluid, in order to obtain accurate values.
[4] C. R. Steinman and W. B. Jakoby, *J. Biol. Chem.* **243**, 730 (1967).
[5] J. F. Clark and W. B. Jakoby, *J. Biol. Chem.* **245**, 6065 (1970).
[6] The following abbreviations are used: PMSF, phenylmethylsulfonylfluoride; DFP, diisopropyl fluorophosphate.

which species A is homogeneous; only serine is found as the N-terminal amino acid. It has been shown[1,5,7] that species B and C are proteolytic degradation products of species A,[8] which are defined by their elution sequence from DEAE-Spehadex.[4] Indeed, if PMSF is omitted, very little if any of species A is found.[1]

A second method of preparation[5] is presented in which DFP[6] serves as the major inhibitor of proteolysis. Here the product consists almost entirely of species A. Two methods of preparation are presented here because some investigators will be reluctant to use a reagent as toxic as DFP in the large amounts which are required.

Purification with PMSF

Baker's yeast, a commercially available product of the Anheuser-Busch Brewing Company, is obtained in 1-lb packages. It is crumbled by hand and is frozen by gradual addition to 4-liter beakers containing liquid nitrogen. As the nitrogen evaporates, it is replaced until approximately 2 kg of frozen yeast have been accumulated in each beaker. The yeast may be stored at $-90°$ for several months without loss of either activity or yield of aldehyde dehydrogenase.

The preparation described below is based on the use of 2 kg of yeast.

Step 1. The frozen cells are suspended at room temperature in 4 liters of 0.1 M potassium phosphate at pH 7.5 containing 50 mM α-thioglycerol; the temperature drops to about $4°$ on addition of the yeast. Immediately after the yeast is added, the buffer is supplemented with 60 ml of a 1.6% solution of PMSF in 1-propanol and mixed rapidly. The cold suspension is passed through a Manton-Gaulin homogenizer at a pressure of 10,000 psi. Because the temperature rises to about $20°$ during the procedure, the suspension is cooled to approximately $4°$ before passage through the homogenizer a second time. While the remainder of the suspension is maintained between $0°$ and $5°$, one batch of 1.5 liters at a time is transferred to a 4-liter, Teflon-lined, stainless steel beaker, and adjusted to pH 5.5 by the dropwise addition of approximately 60 ml of 2 N HCl. The beaker and its contents are heated with stirring in a 50-liter water bath, maintained at $60°$, until the protein solution reaches $52°$; approximately 8 min are required. The beaker is immediately transferred to a salt-ice bath, after which it is gently stirred until the temperature reaches $10°$. The suspension is centrifuged at 13,000 g and $4°$ for 20 min.

[7] J. F. Clark and W. B. Jakoby, *J. Biol. Chem.* **245**, 6072 (1970).

[8] Species B and C have been obtained in a state of macrohomogeneity.[5] When these species are dissociated in 2.4 M guanidine hydrochloride and then allowed to reassociate after dilution, only species A is obtained as active enzyme.[1,5]

The resultant clear supernatant fluid, maintained between 0° and 4°, is adjusted to pH 6.9 with about 120 ml of 1 N KOH. The heating step for the entire 2 kg of yeast should be completed within 2 hr.

Step 2. To each liter of protein solution from step 1 is added 260 g of ammonium sulfate. After stirring for 30 min at 4°, the precipitate is removed by centrifugation at 13,000 g for 30 min. The supernatant fluids are pooled and for each liter of original volume an additional 126 g of ammonium sulfate are added. After stirring for 30 min the precipitate is collected by centrifugation and is dissolved in 0.1 N potassium phosphate, pH 7.0, containing 50 mM α-thioglycerol. The resultant solution, about 100 ml, is treated with 1.5 ml of a 2% solution of PMSF in 1-propanol and is supplemented with glycerol to make a final concentration of 25% (v/v). The preparation is dialyzed immediately, over an 18-hr period, against three changes of 4 liters each of starting buffer[9] made 0.01% in inhibitor.

Step 3. The dialyzed solution is used to charge a DEAE-Sephadex A-50 column (4 × 50 cm) that has previously been equilibrated with starting buffer, i.e., the pH of the wash fluid agrees within 0.15 pH units with that of the buffer. After the column is poured to a height of 25 cm, it is washed with 0.5 column volume of starting buffer. Thereafter, DEAE-Sephadex, previously equilibrated with starting buffer, is centrifuged to remove excess liquid; the gel is added to the dialyzed protein solution so as to form a thick slurry. The slurry is poured on top of the column bed and should result in a total column height of about 50 cm. The elution pattern is the same as that for normal application of the sample when this rather convenient method is used. Enzyme is eluted (Fig. 1A) with a linear gradient consisting of 2.5 liters of starting buffer containing 0.045 M potassium chloride in the mixing flask and an equal volume of starting buffer containing 0.25 M potassium chloride in the reservoir. Fractions chosen for further purification are pooled, made 0.01% in PMSF, and concentrated to about 70 ml with an Amicon ultrafiltration cell and PM-10 filter.[10]

Step 4. Hydroxyapatite[11] is washed twice with water in order to remove fine particles and is then equilibrated with 10 mM potassium phos-

---

[9] Starting buffer is prepared by mixing 100 ml of a stock buffer solution with 250 ml of glycerol, 5 ml of an α-thioglycerol, and 3.36 g of KCl, and diluting to 1.0 liter. The stock buffer solution is prepared from 50 ml of 1 M Tris base, 100 ml of 1 M KH$_2$PO$_4$, 1.86 g of ethylenediaminetetraacetic acid, and 775 ml of water; the pH is adjusted to 8.0 at 4° with 1 N KOH, and the solution is diluted to 1.0 liter.

[10] The PM-10 filter is recommended, although UM-10 was used in earlier work.

[11] O. Levin, this series, Vol. 5 [27].

PURIFICATION OF ALDEHYDE DEHYDROGENASE FROM BAKER'S YEAST[a]

Fraction	Volume (ml)	Total activity (units)	Total protein (mg)	Specific activity (units/mg)
PMFS preparation				
Step 1. Extraction and heating[b]	4400	1590	13,552	0.1
Step 2. Ammonium sulfate[b]	168	1610	5,320	0.3
Step 3. DEAE-Sephadex	69	1220	187	6.5
DFP preparation				
Step 4. Hydroxyapatite	14	1038	115	9.0

[a] Results are based on the use of 2 kg of liquid nitrogen-treated baker's yeast.
[b] At this stage of purification, 1 mM TPN is used in place of DPN for assay of activity. The rate of the reaction with TPN is one-half that of the rate with DPN. The data listed here are in terms of DPN after correction for the differences in rate between the two nucleotides.

phate, pH 6.5, containing 50 mM α-thioglycerol and 25% glycerol (v/v). After the column (2.4 × 35 cm) is poured, it is washed with two column volumes of the same buffer. The protein sample is applied after adjustment, at 4°, to pH 6.6 ± 0.1 with 0.1 N HCl, and one column volume of the buffer is used for washing. A linear gradient is established with 1 liter of the same buffer in the mixing flask and an equal volume of 0.5 M potassium phosphate at pH 7.0, made 50 mM in α-thioglycerol and 25% in glycerol (v/v), in the reservoir. The conductivity at which elution occurs varies somewhat with the particular batch of hydroxyapatite used. For the preparation outlined in the table, enzyme was collected between 960 and 1280 ml. The selected fractions are concentrated with an Amicon cell and PM-10 filter to 10 ml.

Purification with DFP

This method differs only by the addition of DFP to the protein preparation at each of three steps. The initial homogenization is carried out in the same manner as described above, wth PMSF as inhibitor at this stage; DFP, while desirable in terms of preventing proteolysis, is felt to be too dangerous for use during homogenization. However, 0.4 ml of DFP is added immediately before the heat step and an additional 0.13 ml is used in the buffer required to dissolve the ammonium sulfate precipitate. A third addition of DFP, 0.05 ml, is made to the effluent from DEAE-Sephadex before concentration, although the necessity for addition of inhibitor at this stage has not been validated. The result of using DFP is illustrated in Fig. 1B.

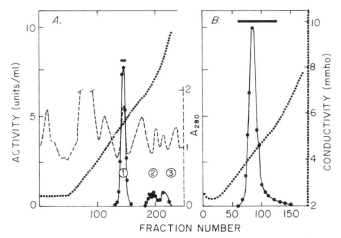

FIG. 1. The elution pattern from DEAE-Sephadex for a preparation treated with phenylmethylsulfonylfluoride (A) or with diisoproyl fluorophosphate (B). Peak 1 refers to the native species A, of aldehyde dehydrogenase; peaks 2 and 3 refer to species B and C (see text), which are the products of proteolysis of species A.

Investigators deciding to use DFP are urged to consult a standard text dealing with the effects of this powerful agent.[12] This laboratory requires the use of expendable rubber gloves, a well-ventilated room, and close proximity to a container of 1 N KOH whenever solutions of DFP are present. Particular care is necessary when work is carried out in a cold room since such enclosures are frequently inadequately ventilated.

Properties of the Enzyme

General. Yeast aldehyde dehydrogenase, a tetramer of about 200,000 daltons, consists of four apparently identical subunits which may be reversibly dissociated in 3 M guanidine hydrochloride.[4,5,7] Hydrodynamic parameters are as follows[5]: $s_{25,w}^{0} = 10.59 \, \text{S}$; $D_{25,w}^{\infty} = 4.70 \times 10^{-7} \, \text{cm}^2 \, \text{sec}^{-1}$; partial specific volume estimated from amino acid composition, 0.74.

In the standard assay, the pH optimum is at 8.7; half of maximal activity is observed at pH 7.3 and 9.3.[4]

The enzyme requires a monovalent cation[4,5]; half of maximal activity is attained at 4 mM KCl. Potassium may be replaced by salts of rubidium, cesium, or ammonium, listed here in order of decreasing activity.[4]

Substrate Specificity. The reaction is that of an ordered Bi Bi mechanism[13] in which an aldehyde is the first substrate to bind to the enzyme and the corresponding acid is the last product to be released.[2] Either

[12] L. S. Goodman and A. Gilman, "The Pharmacological Basis of Therapeutics," 4th ed. Macmillan, New York, 1970.

DPN $(K_m = 29 \ \mu M)$[2] or TPN $(K_m = 50 \ \mu M)$[4] is the pyridine nucleotide. The K_m for DPN does not vary appreciably in the pH range from 6.5 to 9.0.[4] The enzyme has two binding sites for DPNH.[14]

A wide variety of aldehydes serve as substrate, and these are listed[15] in alphabetical order with K_m and, in parentheses, V_{max} expressed as moles per minute per mole of enzyme: aldol, 10 μM (440); benzaldehyde, 53 μM (1700); D-glyceraldehyde, 0.12 M 12,800); DL-glyceraldehyde, 0.12 M (5,100); glycolaldehyde, 48 μM[15] (7500); n-hydroxypentanal, 90 μM (3000); methoxyacetaldehyde, 60 μM (11,400); m-methoxybenzaldehyde, 20 μM (2100); p-methoxybenzaldehyde, 6 μM (280); valeraldehyde, 6 μM (20,000). Additional K_m data have been obtained[4] for the following: acetaldehyde, 9 μM; n-butyraldehyde, 9μM; isobutyraldehyde, 2 μM; formaldehyde, 700 μM; furfural, 5 μM; glutaraldehyde, 60 μM; m-nitrobenzaldehyde, 3 μM; p-nitrobenzaldehyde, 2 μM; propionaldehyde, 3 μM.

Compounds such as chloral, bromal, and glyoxylate which appear inactive as substrates with normal concentrations of enzyme, are oxidized at a detectable rate when the enzyme concentration is increased to about 0.1 mg/ml.

Inhibition. As is the case with all aldehyde dehydrogenases,[16] the enzyme from yeast is inhibited by trivalent arsenicals. In this case, unlike most other aldehyde dehydrogenases, the competitive substrate is DPN.[17] Potassium arsenite and Mapharsen[18] have K_i values of 50 μM and 6 μM, respectively in an otherwise normal assay system.

Stability. The enzyme requires the presence of a polyhydric alcohol for optimum stability; 30% glycerol (v/v) has proved to be satisfactory although an equal concentration of sucrose or ethylene glycol appears to be as good in short-term experiments.[17] As noted under the assay method, a mercaptan is required for storage. With these two supplements, 30% glycerol and 0.1 M α-thioglycerol, the enzyme has been stored at $-80°$ for at least 3 years; whenever activity does decrease after prolonged storage, the enzyme is readily reactivated by incubation with a fresh solution of mercaptan.

[13] W. W. Cleland, *Biochim. Biophys. Acta* **67**, 104 (1963).

[14] S. L. Bradbury and W. B. Jakoby, *J. Biol. Chem.* **246**, 6929 (1971).

[15] This value represents a K_s and was measured on the basis of binding rather than kinetic experiments.

[16] W. B. Jakoby, *in* "The Enzymes" (P. D. Boyer, H. Lardy, and K. Myrbäck, eds.), 2nd ed., Vol. 7, p. 203 (1963).

[17] S. L. Bradbury and W. B. Jakoby *Proc. Nat. Acad. Sci. U.S.* **69**, 2373 (1972).

[18] Mapharsen is the brand name of oxophenarsine hydrochloride which was provided by Dr. J. R. Dice of Parke Davis and Co.

[78] Aldehyde Reductase from *Rhodotorula*[1]

By Gerald H. Sheys, James A. Hayashi, and Clyde C. Doughty

DL-Glyceraldehyde + NADPH + H$^+$ $\rightleftarrows$ glycerol + NADP$^+$

Assay Method

Principle. The method is based upon the decrease in absorbance at 340 nm of reduced nicotinamide adenine dinucleotide phosphate. With excess of each substrate the oxidation of NADPH is directly proportional to the enzyme concentration.

Reagents. The following solutions are used in the assay.

DL-Glyceraldehyde, 100 mM, made fresh daily and kept at 0°
NADPH + H$^+$, 1.2 mM, made fresh daily and kept at 0°
Buffer solution, 50 mM in 0.10 M NaCl, pH 7.5
N-2-Hydroxyethylpiperazine-N'-2-ethanesulfonic acid (HEPES)
Enzyme solution

Procedure.[2] To cuvettes kept in a 37° water bath was added 0.100 ml (10 μmole) of DL-glyceraldehyde solution and 0.100 ml (0.12 μmole) of NADPH solution. Buffer solution, maintained at 37°, was added; in amount from 0.775 to 0.795 ml, depending upon the volume of enzyme to be used, resulting in a final reaction volume of 1.000 ml.

After inserting the cuvette into the cuvette chamber of the spectrophotometer, maintained at 37°, the absorbance base line is observed, 5–25 μl of enzyme are added, and the rate of change in absorbance at 340 nm is recorded. Under these conditions, the initial velocities are linear from 3 to 4 min and highly reproducible.

Definition of Enzyme Units. One unit of enzyme is that amount which catalyzes the oxidation of 1 μmole of NADPH in 1 min at 37°. Specific activity is expressed as units per milligram of protein, the protein being measured spectrophotometrically[3] or by the biuret method.[4]

Purification Procedure

Organism Growth. The pink yeast was isolated from soil on agar plates containing DL-glyceraldehyde as the sole carbon source.

[1] Alditol: NADP$^+$ oxidoreductase, EC 1.1.1.21.
[2] G. H. Sheys, W. J. Arnold, J. A. Watson, J. A. Hayashi, and C. C. Doughty, *J. Biol. Chem.* **246**, 3824 (1971).
[3] O. Warburg and W. Christian, *Biochem. Z.* **310**, 384 (1941).
[4] A. G. Gornall, C. J. Bardawill, and M. M. David, *J. Biol. Chem.* **177**, 751 (1949).

For preparation of the enzyme, the yeast is grown at room temperature with aeration (usually on a rotary shaker) in liquid medium containing 0.7% Yeast nitrogen base (Difco) and 1.0% glucose. The pH is maintained at approximately 7.0 by periodic addition of 10 N NaOH. After 2–3 days, the cells are harvested and washed three times with distilled water. The usual yield of cells is 8–10 g per liter of medium.

Preparation of Cell Extract. Eighty grams wet weight of cells and 80 g of glass beads (0.1 mm diameter) are suspended in 1.5 volumes of 0.05 potassium phthalate buffer, pH 5.8, in the ice-cooled 250-ml homogenizer chamber of a Sorvall Omnimixer. The suspension is ground (30 sec) and cooled (15 sec) alternately with the homogenizer autotransformer set at 75 V for 2.5 hr. The grinding and cooling may be unattended if one uses on-off times such as the Flexopulse times, Eagle Signal Corp., Moline, Illinois.

Cellular debris and small glass particles are removed by centrifuging the suspension at 48,000 g for 10 min. The cell-free extract is stable for many months when stored at −40° and may be frozen and thawed repeatedly without loss of activity.

Step 1. Acid Precipitation. The pH of the cell-free extract is lowered cautiously to 5.0 using 1 N acetic acid and monitoring continuously with a pH meter. Stirring is continued for 2 min followed by centrifugation at 48,000 g for 10 min. The pH of the supernatant solution is readjusted to 6.0 and dialyzed overnight against 10 volumes of 5 mM phosphate buffer, pH 7.4.

Step 2. Protamine Sulfate Treatment. Protamine sulfate, 1.5 mg per unit of enzyme in the dialyzed solution, is added with stirring and with maintenance of the pH at 7.0. The mixture is stirred for 15 min after addition of the protamine sulfate; the precipitate is removed by centrifugation at 48,000 g for 10 min and discarded.

Step 3. Ammonium Sulfate Fractionation. Throughout this fractionation the pH is maintained at 7.0 by addition of 1 N NaOH and monitoring with a pH meter. Solid ammonium sulfate is added to the supernatant solution over a 90-min period to reach 70% of saturation (436 g/100 ml). After centrifugation (48,000 g, 10 min), the precipitate is discarded, and solid salt is added to the supernatant solution, to 90% of saturation (13.4 g/100 ml), to precipitate the enzyme.

The material precipitated between 70 and 90% of saturation is sedimented (6000 g, 30 min) and redissolved in 10 ml of 5 mM phosphate buffer, pH 7.4.

Step 4. Calcium Phosphate Gel Treatment. Calcium phosphate gel is added to the enzyme solution to a gel:protein ratio of 3:1 (w/w).

Stirring is continued for 15 min after the addition, while maintaining the pH of the suspension at 5.5. After centrifugation (5000 g, 2 min), the supernatant portion is decanted, neutralized to pH 7.0, dialyzed 5 hr against distilled water and concentrated to about 4 ml by vacuum ultrafiltration through a cellulose acetate membrane.

Step 5. Electrofocusing. The electrofocusing apparatus was prepared by following the directions given in the LKB Instruments Manual 1-8100-E01 for the Model 8101, 100-ml focusing column. The column was filled with a sucrose gradient and 1% Ampholines from pH 4 to 6. The column is focused for 10–15 hr (cathode at the top, 500 V, < 2 mA after focusing) before the enzyme sample is introduced. A 50-mg sample was made dense by the addition of an appropriate amount of sucrose and carefully introduced into the column at an isodense position near the center of the column through narrow-gauge plastic tubing. Electrofocusing is complete within 24 hr. The column is drained, the contents are collected in 2-ml portions which are assayed for enzyme activity, and the active fractions are pooled (10–14 ml).

Step 6. Sephadex G-100 Column Chromatography. A downward-flow Sephadex column (5 × 100 cm) with a Sephadex G-100 bed volume of 1800 ml is equilibrated with 50 mM Tris·HCl buffer, 0.1 M in NaCl, pH 7.5. The pooled enzyme fractions are applied to the column, eluted with equilibrating buffer at a flow rate of 30 ml per hour, and collected in 7-ml fractions. After assay, the most active fractions are pooled resulting in about 100 ml of enzyme solution, which is reduced by ultrafiltration to about 20 ml, dialyzed 10 hr against water, and lyophilized.

A summary of the purification results is shown in the table.

PURIFICATION OF YEAST ALDOSE REDUCTASE FROM *Rhodotorula*

Step	Protein (mg/ml)	Total protein (mg)	Specific activity	Puri-fication (fold)	Total units	Yield (%)
Cell-free extract	4.8	144	0.14	(1)	42	(100)
pH 5 supernatant	1.9	57	0.35	2.5	40	95
Protamine sulfate super-natant	0.77	45	1.16	8.3	47	112
Ammonium sulfate	3.3	16.5	2.6	19	41	98
Calcium phosphate gel	1.8	12	3.4	24	35	83
Electrofocusing	0.018	0.099	153	1090	28	67
Sephadex G-100 chroma-tography	0.0075	0.070	184	1310	24	57

Properties[5,6]

Purity. One band is obtained on polyacrylamide gel electrophoresis, and one are on immunodiffusion against an antiserum prepared with the purified enzyme as antigen.

Specificity and Substrate Affinity. The enzyme is specific for NADPH. The highest activity is with glyceraldehyde as substrate; slight activity with dihydroxyacetone and p-nitrobenzaldehyde.

The enzyme is not stereospecific since both glyceraldehyde enantiomers serve as substrates. The respective K_m and V_{max} for the glyceraldehydes are: D-glyceraldehyde 0.9 mM and 1.32; L-glyceraldehyde, 4.0 and 1.32; DL-glyceraldehyde, 1.9 and 1.32. The isoelectric pH of the enzyme is 5.05. The enzyme's molecular weight is 61,000. It can be dissociated into two subunits of unequal size. The 37,000 molecular weight subunit has enzymic activity, the 23,000 subunit does not. NADP⁺ appears to stabilize the dimer.

The pH for optimal enzyme activity is 7.5. The reverse reaction was detectable only at about pH 9.0.

NADP⁺ is a noncompetitive inhibitor (K_i = 112 μM) when glyceraldehyde is the variable substrate, and a competitive inhibitor (K_i = 20 μM) of NADPH.

Other inhibitors of the enzyme and their inhibition constants are: 2'-AMP, 0.55 mM; 3'-AMP, 2.5 mM; 5'-AMP, 4.6 mM; sulfate ion, 44 μM; p-hydroxymercuribenzoate, 61 μM; mercuric chloride, 46 μM. There was no apparent cation requirement and no inhibition of EDTA.

The reaction mechanism is ordered Bi Bi.

[5] G. H. Sheys and C. C. Doughty, *Biochim. Biophys. Acta* **242**, 523 (1971).
[6] G. H. Sheys and C. C. Doughty, *Biochim. Biophys. Acta* **235**, 414 (1971).

[79] Alcohol Oxidase from Basidiomycetes

By Frank W. Janssen, Richard M. Kerwin, and Hans W. Ruelius

$$RCH_2OH + O_2 \rightarrow RCHO + H_2O_2$$

Alcohol oxidase[1] catalyzes the oxidation of lower primary alcohols to the corresponding aldehydes and H_2O_2. The isolation of this enzyme from the mycelium of an unidentified basidiomycete (B191039) and the

[1] This name has been designated by the I.U.B. Enzyme Commission and the enzyme has been assigned the number 1.1.3.13.

crystallization by fractional precipitation with polyethylene glycol (PEG) have been reported.[2]

In addition to the above organisms, several other basidiomycetes, including *Polyporus obtusus* and *Radulum casearium*, also form the enzyme.[3] Although the titers produced by the latter organisms are not as high as those obtained with B191039, these organisms are morphologically and metabolically stable and therefore more reliable for production of the enzyme. The fermentation of all three organisms will be briefly described. A simplified PEG fractionation, which was used primarily for isolation of the enzyme from B191039 cultures but is suitable, with appropriate modification, for isolation of the enzyme from the other cultures, will also be described.

Assay Method

Principle. The enzyme is most conveniently assayed by measuring the color produced from H_2O_2 in a coupled reaction utilizing horseradish peroxidase and *o*-dianisidine.

Reagents

Horseradish peroxidase (EC 1.11.1.7) 400 units/mg, 0.1% in distilled water. Store at 4° under toluene.

o-Dianisidine dihydrochloride, 1% solution in 25 mM HCl. Store at 4°.

HCl, 4.0 M

Peroxidase–dianisidine: To 79 ml of distilled water add 10 ml of 0.5 M pH 7.5 sodium phosphate buffer, 10 ml of peroxidase solution, and 1.0 ml of dianisidine solution.

Methanol–peroxidase–dianisidine: Prepare as described for peroxidase–dianisidine except substitute 1.0 ml reagent grade anhydrous methanol for water.

Procedure. Add 1.0 ml of appropriately diluted enzyme preparation to duplicate test tubes in a 25° water bath. At zero time add 4.0 ml methanol–peroxidase–dianisidine reagent. Terminate the reaction at 5 min with 0.2 ml 4.0 M HCl. Read the color produced in a spectrophotometer at 460 nm or a colorimeter with blue filter (Klett-Summerson No. 42). Correct reading for any blank produced by enzyme + peroxidase-dianisidine reagent. The amount of H_2O_2 produced is determined by comparison to standard solutions of H_2O_2 treated in the same way. 420 Klett units (KU) $\cong$ 1 μmole of H_2O_2. The ideal range is 50 to 200 KU.

[2] F. W. Janssen and H. W. Ruelius, *Biochim. Biophys. Acta* 151, 330 (1968).
[3] R. M. Kerwin and H. W. Ruelius, *Appl. Microbiol.* 17, 347 (1969).

Definition of Unit and Specific Activity. One unit of enzyme is defined as the amount of enzyme that produces 1 μmole of H_2O_2 per minute in this system. Specific activity is defined as the number of units per milligram of protein.

Production of Alcohol Oxidase

Organisms and Maintenance. The organisms that were investigated and the sources from which they may be obtained are as follows:

Basidiomycete B191039, Wyeth Labs, West Chester, Pennsylvania

Polyporus obtusus Berk (No. 136), Forest Research Institute, Dehradum, India

Radulum casearium (Morgan) Hoyd, Department of Agriculture, Ottawa, Canada (No. 17532)

The organisms are maintained on slants of Malt or Mycophil Agar (Baltimore Biological Laboratories) at $23° ± 2°$. Inoculum is prepared as described in this volume for carbohydrate oxidase.[4] Five milliliters of inoculum is used to inoculate 500-ml flasks, and 10 ml for 2-liter flasks containing 150 and 1000 ml, respectively, of the media described in Table I. The smaller flasks are incubated on a reciprocating shaker at 90 four-inch strokes per minute. The 2-liter flasks are incubated on a rotary shaker set at 160 rpm. The incubation temperature is $25°$. Production of the enzyme, which is found in the mycelium, is monitored daily beginning with day 5 as follows. The contents of one 500-ml flask are filtered through coarse paper under reduced pressure. The mycelium is added to 150 ml 50 mM sodium phosphate (pH 7.5) and homogenized for 5 min in a water-cooled Micro-Waring Blendor (Central Scientific Co., Chicago, Illinois). The homogenates are filtered and assayed for enzyme content as described above. The 2-liter flasks are monitored similarly except that an aliquot is aseptically removed for the enzyme assay. When optimal enzyme production is achieved (about 10 days) the mycelium is harvested by filtration and stored frozen until extracted.

Purification Procedure

Step 1. The thawed mycelium is homogenized in a Waring Blendor for 3 periods of 4 min each in freshly prepared 5 mM potassium ethyl xanthate in 50 mM pH 7.7 sodium phosphate buffer (7.5 ml/g). The blender cup is cooled in an ice bath for 5 min between each period. The homogenate is centrifuged at 7250 g for 15 min to yield a turbid supernatant

[4] F. W. Janssen and H. W. Ruelius, this volume [39].

TABLE I
PRODUCTION OF ALCOHOL OXIDASE BY BASIDIOMYCETES

Organism	Medium[a]	Fermentation volume (ml)	Maximum titer[b]
B191039	A	1000	7800
B191039	B	150	3000
Polyporus obtusus	C	150	4700
Radulum casearium	D	150	3200

[a] Medium A: cerelose, 10 g (crude glucose, Corn Products, Co., Argo, Illinois); Amber MPH (meat protein hydrolyzate), 10 g (Amber Laboratories, Milwaukee, Wisconsin); $FeSO_4$, 1 g; water, 1 liter. The pH was adjusted to 6.5 before autoclaving.

Medium B: D-glucose, 25 g; L-asparagine, 2.28 g; KH_2PO_4, 2.0 g; $Fe_2(SO_4)_3$, 1 mg; $ZnSO_4 \cdot 7\ H_2O$, 0.88 mg; $MnSO_4 \cdot H_2O$, 0.3 mg; thiamine, 100 μg; biotin, 5 μg; water, 1 liter. The pH was adjusted to 5.0 before autoclaving.

Medium C: Cerelose, 10 g; Amber MPH, 30 g; water, 1 liter. The pH was adjusted to 6.5 before autoclaving.

Medium D: Cerelose, 10 g; Amber MPH, 10 g; KH_2PO_4, 1.0 g; $MgSO_4 \cdot 7\ H_2O$, 1.0 g; NaCl, 0.2 g; $CaCl_2 \cdot 6\ H_2O$, 20 mg; $MnSO_4 \cdot H_2O$, 50 mg; $ZnCl_2$, 1.5 mg; $FeCl_3 \cdot 6\ H_2O$, 5.0 mg; $CuSO_4 \cdot 5\ H_2O$, 0.05 mg; water, 1 liter (salts were added as a solution in water). The pH was adjusted to 6.5 before autoclaving.

[b] The color produced in the alcohol oxidase assay of an appropriately diluted sample of the extract was read on a Klett-Summerson colorimeter. The titer is the colorimeter reading multiplied by the dilution factor.

which contains the enzyme. All subsequent operations are carried out at room temperature.

Step 2. The enzyme in the supernatant (1 volume) is precipitated by the slow addition of polyethylene glycol 6000 (PEG, J. T. Baker Chemical Co.) powder (22.5 g/100 ml supernatant) while mixing on a heavy-duty magnetic stirrer. The precipitate formed during 30 min standing is separated by centrifugation for 15 min at 7250 g. The sediment (fraction 1) is suspended by stirring in about 0.1 volume of the xanthate–phosphate buffer.

Step 3. PEG (3.0 g/100 ml of suspension) is added to the suspension. Centrifugation at 12,900 g yields a clear supernatant (fraction 2) which contains the enzyme and a voluminous precipitate consisting largely of cell fragments.

Step 4. PEG (22.5 g/100 ml) is added as above to the supernatant (fraction 2). The precipitate formed during 30 min standing is recovered by centrifugation at 12,900 g for 15 min. The sediment is dissolved in about 0.1 volume of the xanthate–phosphate buffer and stored at 4°. The

TABLE II
PURIFICATION OF ALCOHOL OXIDASE

Fraction	Volume (ml)	Total activity[a] (units)	Total protein[b] (mg)	Specific activity[c]	Recovery (%)
Crude extract	1920	7960	9300	0.86	—
Fraction 1 (19% PEG ppt.)	—	—	—	—	—
Fraction 2 (3% PEG supernatant)	186	6850	1420	4.82	86
Fraction 3 (second 19% PEG ppt.)	44	5950	625	9.5	75

[a] One unit is the amount of enzyme that produces 1 μmole of H_2O_2 per minute in the system described in the text.

[b] Protein content of fraction is derived from the Kjeldahl nitrogen content by multiplying the latter by 6.25.

[c] Specific activity is the number of units per milligram of protein.

results of a typical fractionation are summarized in Table II. The enzyme is purified 11-fold with recovery of 75%. Such preparations are stable for at least 2 weeks when stored at 4°. The enzyme may be purified to the crystalline state by further fractionation with PEG.[2] If a dry preparation is desired, the solution is dialyzed against 1 mM, pH 8.0 tris buffer, shell frozen, and lypohilized. Such a preparation loses about 20% of its activity when stored over anhydrous silica gel at 4° for 1 year.

Alternate Purification. The enzyme may also be obtained by PEG fractionation of mycelial extracts of *Polyporus obtusus* and presumably other basidiomycetes. It should be noted, however, that higher concentrations of PEG are required if the titer is low. For example a *Polyporus obtusus* mycelial extract containing about 0.025 unit/ml required 35 g of PEG per 100 ml of extract to precipitate the enzyme whereas B191039 mycelial extracts, which contained 0.5–0.9 unit/ml, required only 22.5 g/100 ml to precipitate the enzyme. It is recommended, for this reason, to assay for enzyme activity after each purification step.

Properties

Specificity. In the presence of molecular O_2, alcohol oxidase catalyzes the oxidation of lower primary alcohols to the corresponding aldehydes and hydrogen peroxide. Methanol is the most active substrate. The activity declines with increasing chain length. Unsaturated alcohols are also good substrates, but branched and secondary alcohols are not attacked. Halogenated ethanols are oxidized, but other substituted ethanols are not.

Other Properties. Alcohol oxidase is a flavoprotein with a molecular weight of at least 300,000 and contains several FAD residues per mole of enzyme. The enzyme has a broad pH optimum from 6.5 to 9.0. The activity declines rapidly below pH 6.5. Stability is also markedly reduced in solutions below pH 7.0. Solutions of the enzyme lose activity when frozen.

[80] Alcohol Dehydrogenase[1] from Human Liver[2]

By CHARLES L. WORONICK

$$R\text{—}CH_2OH + NAD^+ \rightleftarrows RCHO + NADH + H^+$$

Assay Method

Principle. The assay is based on the measurement of the reduction of NAD by alcohol in the presence of the enzyme. A modification of the procedure of Dalziel is used.[3]

Reagents

Glycine solution adjusted to pH 10.0 with NaOH before final dilution to 0.1 M

NAD, 1 mg/ml in distilled water

Ethanol solution: 1 ml redistilled 95% alcohol diluted to 100 ml with distilled water.

Procedure. To a cuvette having a 1-cm light path, add 1.85 ml of glycine buffer, 0.15 ml of ethanol solution, and 1.00 ml of NAD solution. The reaction is initiated by the addition of up to 0.010 ml of enzyme solution. The time required to produce an initial change in absorbance of 0.100 at 340 nm, at a temperature of 23°, is measured. The initial reaction for the cruder extracts is not linear with time,[4] but the initial reaction of more highly purified preparations is linear.

Definition of Unit and Specific Activity. One unit of alcohol dehydrogenase is defined as the amount of enzyme that will reduce 1 μmole of NAD per milliliter per minute under the conditions specified in the assay. The number of enzyme units per milligram of protein is defined as the specific

[1] Alcohol:NAD$^+$ oxidoreductase, EC 1.1.1.1.
[2] This work was supported in part by U. S. Public Health Service grant GM 11463.
[3] K. Dalziel, *Acta Chem. Scand.* **11**, 397 (1957).
[4] N. Mourad and C. L. Woronick, *Arch. Biochem. Biophys.* **121**, 431 (1967).

activity. The protein concentration is determined at 280 nm. The absorbance at 280 nm of 1.0 mg of the enzyme per milliliter of 0.1 μ sodium phosphate at pH 7.0 is 0.61 for a 1-cm light path as determined by Mourad and Woronick.[4]

Purification Procedure

The method of purification is a modification of the procedure described by Mourad and Woronick.[4] Adult human livers which appeared to be normal were obtained by autopsy within 16 hr after death. The livers are used immediately or may be stored in the frozen state and thawed overnight at 4° before use. The livers ranged in weight from 900 to 2000 g.

All the following steps are performed at 4° unless otherwise specified, the enzyme being somewhat unstable even at room temperature. Enzyme grade $(NH_4)_2SO_4$ is used for the precipitation steps.

Step 1. Extraction Procedure. The liver is trimmed to remove the gall bladder, larger blood vessels and any other extraneous tissue. It is then cut into strips about 2–3 cm wide, which are washed with squeezing under cold running water to remove blood and blood clots. The pieces of liver are weighed, and ground with a meat grinder. To each kilogram of liver is added 1.5 liters of distilled water; the mixture is allowed to stand at room temperature for 2 hr with occasional stirring. The mixture is then filtered through cheesecloth, and an aliquot of the crude extract is centrifuged and assayed for enzyme and soluble protein.

Step 2. Ammonium Sulfate Fractionation. Dissolve 225 g of solid $(NH_4)_2SO_4$ in each liter of extract. The pH is adjusted to 7.5 (pH paper) by the slow addition of concentrated ammonium hydroxide while stirring, and kept at 4° for 2 hr. The mixture is then centrifuged at 7000 g for 50 min and the precipitate is discarded.

Dissolve 175 g of solid $(NH_4)_2SO_4$ in each liter of supernatant solution, adjust the pH to 7.5 as above, and allow to stand overnight. Centrifuge at 7000 g for 1 hr and suspend the precipitate in a sufficient volume of 0.05 μ sodium phosphate buffer, pH 7.0, to form a slurry. Dialyze against 4 changes of 4 liters of 0.05 μ sodium phosphate buffer pH 7.0 for 2–3 days.

Step 3. Ethanol-Chloroform Treatment. This procedure, which was developed by Tsuchihashi,[5] uses a mixture of 2 volumes of ethanol to 1 volume of chloroform for the denaturation of hemoglobin and probably certain other proteins. To each liter of the enzyme solution is slowly

---

[5] M. Tsuchihashi, *Biochem. Z.* **140**, 62 (1923).

added 200 ml of cold ethanol–chloroform mixture while stirring vigorously. The vigorous stirring is continued for 10 min. The mixture is centrifuged at 15,000 g for 15 min, and the supernatant solution is transferred to a flash evaporator, and a few drops of octanol are added. The ethanol–chloroform remaining in solution is evaporated under vacuum using a bath temperature of less than 30°. The start of excessive foaming is an indication that most of the organic solvents have been removed, at which time the evaporation is stopped. Continued evaporation to concentrate the solution causes a significant loss of enzyme. The solution is dialyzed overnight against 10 liters of 0.05 μ sodium phosphate buffer pH 7.0.

Step 4. Carboxymethyl Cellulose Chromatography. Carboxymethyl cellulose (Sigma Chemical Co.) equilibrated with 0.01 μ sodium phosphate buffer at pH 7.0 is poured into a 5 × 50 cm column. To each 100 ml of enzyme solution is added 400 ml of cold distilled water to reduce the ionic strength to 0.01 μ. The diluted enzyme solution is applied to the column (flow rate up to 300 ml/hr) and all the enzyme is adsorbed. The column is eluted with a sufficient volume of 0.01 μ sodium phosphate buffer pH 7.0 (about 8 liters) to reduce the absorbance of the eluate to less than 0.05 at 280 nm. The enzyme is then eluted in association with a pink band using 0.1 μ sodium phosphate buffer, pH 7.0. The fractions with the highest specific activities are combined, and the enzyme is precipitated by dissolving 45 g of solid $(NH_4)_2SO_4$ per 100 ml of enzyme solution and adjusting to pH 8 with concentrated ammonium hydroxide. The mixture is allowed to stand overnight.

Step 5. Diethylaminoethylcellulose Chromatography. The $(NH_4)_2SO_4$ precipitate is centrifuged at 12,000 g for 40 min, and the precipitate is dissolved in a minimum volume of 0.05 μ sodium phosphate buffer, pH 8.0. The solution is dialyzed against 4 liters of the same buffer for 8 hr and against 4 liters of 0.01 μ glycine–NaOH buffer, pH 9.5 for 8 hr. Prolonged dialysis against the glycine buffer causes excessive destruction of the enzyme. The dialyzed enzyme is applied to a 3 × 30 cm diethylaminoethylcellulose (Calbiochem) column (flow rate 300 ml/hr) equilibrated with 0.01 μ glycine-NaOH buffer, pH 9.5. The enzyme is not adsorbed by the column and is recovered as a colorless solution.[6] The enzyme is precipitated by the addition of 56 g of solid $(NH_4)_2SO_4$ per 100 ml of solution, and adjusted to pH 8.0 with concentrated ammonium hydroxide. The precipitated enzyme is stable at 4° for several months.

---

[6] Some DEAE preparations adsorb the enzyme under these conditions, and give a poor yield and purification after elution with more concentrated buffer. To avoid this, the DEAE-cellulose should be tested in a small column, and used only if it performs as described in the procedure above.

Step 6. Crystallization of the Enzyme. The $(NH_4)_2SO_4$ precipitate is centrifuged at 12,000 g for 40 min and the precipitate is dissolved in a minimum volume of 0.1 μ sodium phosphate buffer pH 8. After 30–40 min, unidentified crystals begin to form. When crystallization is completed (about 1 hr) the suspension is centrifuged and the crystals are discarded. The supernatant solution is dialyzed against 4 liters of 0.05 μ sodium phosphate buffer, pH 8.0, overnight, and against 1 liter of 0.02 μ sodium phosphate buffer, pH 10.6, for 6 hr. The protein concentration is diluted to 10–20 mg/ml and cooled to 0° in a salt-ice bath at −7°. While the enzyme solution remains in the salt-ice, 90% ethanol at −7° is slowly added with stirring until the final alcohol concentration is 15% and the temperature is below 0°. The mixture is maintained at −7° for 24 hr and centrifuged at 12,000 g for 15 min at −7°. Any precipitate obtained at this point is discarded. Again using the salt-ice bath at −7°, 90% alcohol at −7° is added until the final alcohol concentration is 30% and the temperature is no higher than −3°. The clear solution is transferred to a freezer at −18° and crystals of pure alcohol dehydrogenase usually form within 24 hr. The crystals have the appearance of flat triangles and have a specific activity of 3.3 enzyme units per milligram of protein.[4]

Occasionally an amorphous precipitate may form. This precipitate is centrifuged at −18° and dissolved in a minimum volume of 0.1 μ sodium phosphate buffer, pH 10.6, at 0°. Crystal formation usually begins within a few minutes, but the crystals may not be as pure as those produced above because the enzyme has a strong tendency to cocrystallize with other proteins when the total protein concentration is high.[4]

The results of a typical purification are presented in the table.[4]

Properties

Substrate Specificity. The crystalline enzyme reacts with ethanol and NAD as substrates.[4] Studies with partially purified preparations indicate that the enzyme also reacts with a variety of alcohols including methanol,[7–9] certain monohalo-substituted ethanols,[9] ethylene glycol,[9] retinol,[10] and others.[9] NADP is less active than NAD.[10]

Kinetic Properties. The specific activity of the crystalline enzyme is 3.3 units/mg.[4] As determined with a highly purified preparation, the K_m for ethanol is 1.97 mM at 23.5° in pH 10 glycine buffer at an NAD con-

[7] J. P. von Wartburg, J. L. Bethune, and B. L. Vallee, *Biochemistry* 3, 1775 (1964).
[8] J. P. von Wartburg, J. Papenberg, and H. Aebi, *Can. J. Biochem.* 43, 889 (1965).
[9] A. H. Blair and B. L. Vallee, *Biochemistry* 5, 2026 (1966).
[10] E. Mezey, and P. R. Holt, *Exp. Mol. Pathol.* 15, 148 (1971).

CRYSTALLIZATION OF ALCOHOL DEHYDROGENASE FROM HUMAN LIVER[a]

Step	Total yield (%)	Total enzyme		Specific activity (units/mg protein)	Purification (fold)
		Units	Mg		
1. Crude extract	100	2200	670	0.0033	1
2. (NH$_4$)$_2$SO$_4$ Fractionation	70	1500	460	0.017	5
3. Ethanol–chloroform	55	1200	360	0.050	15
4. CM-cellulose chromatography	33	710	220	0.50	150
5. DEAE-cellulose chromatography	30	650	200	1.5	450
6. Crystallization	15	300	90	3.3	1000

[a] These data are taken from N. Mourad and C. L. Woronick, *Arch. Biochem. Biophys.* **121**, 431 (1967).

centration of 0.46 mM, and the turnover number is 1.5 per second per active site.[11] Under similar conditions the crystalline enzyme was reported to have a turnover number for ethanol of 2.5 per second per active site.[4] The K_m for methanol is 10.4 mM, and the turnover number is 0.102 per second per active site.[11] As determined with a partially purified preparation, the K_m for ethanol is 0.45 mM at 23° in 0.15 μ sodium phosphate buffer, pH 7.4, at an NAD concentration of 0.35 mM.[12] At the same temperature and in the same buffer the K_m for NAD is 0.022 mM at an ethanol concentration of 5 mM.[12]

Inhibitors. The enzyme forms a fluorescent ternary complex with NADH and isobutyramide, from which the active site concentration can be determined.[11] The titration of the active sites apparently produces agreement with the report that the enzyme contains 2 active sites per molecule.[7] Pyrazole and NAD also form a ternary complex with the enzyme, from which the active site concentration can be determined by spectrophotometry.[12] The enzyme is inhibited by pyrazole and certain substituted pyrazoles,[12] and by 8-amino-6-methoxyquinoline and a number of its 8-amino-substituted derivatives.[13] Substrate inhibition by alcohol is observed when the alcohol concentration is greater than 0.3 M.[12]

pH Optimum. The pH optimum for ethanol oxidation by partially purified preparations is 10.8 in glycine-NaOH buffer.[8] An atypical form with a pH optimum of 8.5–8.8 has also been reported to occur in 10–20% of certain populations.[8,14]

[11] J. C. Mani, R. Pietruszko, and H. Theorell, *Arch. Biochem. Biophys.* **140**, 52 (1970).

[12] T. K. Li and H. Theorell, *Acta Chem. Scand.* **23**, 892 (1969).

[13] T. K. Li and L. J. Magnes, *Biochem. Pharmacol.* **21**, 17 (1972).

[14] M. Smith, D. A. Hopkinson, and H. Harris, *Ann. Hum. Genet.* **34**, 251 (1971).

Stability. The enzyme is unstable at room temperature and above. It is usually stable between pH 7 and 10.6 providing that the ionic strength is 0.05 or greater, but optimal stability occurs at about pH 8.[4] Enzyme that is more than 100-fold purified is destroyed by freezing.[4] The crystalline enzyme can only be stored a short time in the crystallization medium.[4] Under certain conditions 500-fold purified enzyme preparations are converted into fairly rigid, insoluble gels.[4]

Structure. The amino acid composition of the crystalline enzyme has been reported by Mourad and Woronick.[4] It is a zinc-containing[7] dimeric[15] protein with a molecular weight reported[7] to be 87,000. More recently[16] the partial amino acid sequence has been determined for the 2 subunits; each of their molecular weights has been estimated to be about 40,000, and the dimeric molecular weight to be about 80,000.

Polymorphism and Molecular Heterogeneity. An atypical and a typical form of the enzyme have been reported.[8] Several investigators have reported separating one or both forms of the enzyme into several fractions.[9,17–19] Evidence for at least 7 isoenzymes can be obtained, and 6 of these have been prepared as partially purified but isoenzymically separate forms.[20] It is proposed that the 6 isoenzymes are formed by the random combination of 3 different subunits to form dimers.[20] All 3 subunits appear to be basic polypeptides[20] in accordance of the report[4] that the crystalline enzyme preparations appear to have an isoelectric point above pH 9.5.

[15] T. M. Schenker and J. P. von Wartburg, *Experientia* **26**, 687 (1970).
[16] H. Jörnvall, and R. Pietruszko, *Eur. J. Biochem.* **25**, 283 (1972).
[17] J. P. von Wartburg, *Abstr. Commun. 3rd Meet. Fed. Eur. Biochem. Soc.,* F 180 (1966).
[18] K. Moser, J. Papenberg, and J. P. von Wartburg, *Enzymol. Biol. Clin.* **9**, 447 (1968).
[19] P. Pikkarainen, and N. C. R. Räihä, *Nature* (*London*) **222**, 563 (1969).
[20] T. M. Schenker, L. J. Teeple, and J. P. von Wartburg, *Eur. J. Biochem.* **24**, 271 (1971).

[81] Alcohol Dehydrogenase from *Drosophila melanogaster*

By James I. Elliott and James A. Knopp

$$\text{Sec-alcohol (pri-alcohol)} + NAD^+ \rightleftharpoons \text{ketone(aldehyde)} + NADH + H^+$$

Although the alcohol dehydrogenases (alcohol:NAD^+ oxidoreductase, EC 1.1.1.1) from horse liver and yeast have been extensively studied,

with crystalline preparations obtained in 1948[1] and 1937,[2] respectively
purification schemes for *Drosophila melanogaster* were first reported in
1968 by Sofer and Ursprung[3] and 1969 by Jacobson and co-workers.[4] Initial characterizations of the *Drosophila* enzyme have been reported,[3-8]
and some additional properties are presented here.

Drosophila alcohol dehydrogenase (ADH) exhibits multiple enzymic
forms when electrophoresed and stained for activity using a reduced
tetrazolium staining procedure.[9,10] The bands are numbered sequentially,
band 5 being the band nearest the cathode. Three bands (Nos. 5, 3, and
1) predominate in a homozygous strain, and two minor bands (Nos. 4
and 2) are detectable. Two homozygous strains are known, termed fast
and slow, both displaying the same pattern of bands; however, the bands
detected in the fast strain are shifted more toward the anode. Most of
the ADH studies are done using the fast strain, due to the greater ADH
activity per gram body weight and faster electrophoretic separation.

The bands of a homozygous strain appear identical in molecular
weight by sedimentation velocity[5] and electrophoresis in different concentrations of polyacrylamide gels.[4] The system becomes very interesting
when these results are coupled with the results of genetic analysis that
show a single allele system[11]; thus recombination of dissimilar subunits,
as seen in the lactate dehydrogenase system, is ruled out. A proposal that
relates the migration rate of the individual bands to the number of NAD^+
molecules bound has been shown to be incorrect.[7] It is currently thought
that the different forms are due to conformational changes, which would
prove to be of interest since the forms are slow to equilibrate with each
other and, therefore, could be studied individually.

Raising and Collection of *Drosophila*

The flies are grown in 16-oz jars, with cheesecloth and cotton stoppers,
containing a standard banana-agar media to a depth of 2 cm.

[1] R. K. Bonnichsen and A. M. Wassen, *Arch. Biochem.* **18**, 361 (1948).
[2] E. Negelein and H. J. Wulff, *Biochem. Z.* **293**, 351 (1937).
[3] W. Sofer and H. Ursprung, *J. Biol. Chem.* **243**, 3110 (1968).
[4] K. B. Jacobson, J. B. Murphy, and F. C. Hartman, *J. Biol. Chem.* **245**, 1075 (1970).
[5] K. B. Jacobson and P. Pfuderer, *J. Biol. Chem.* **245**, 3938 (1970).
[6] K. B. Jacobson, J. B. Murphy, J. A. Knopp, and J. R. Ortiz, *Arch. Biochem. Biophys.* **149**, 22 (1972).
[7] J. A. Knopp and K. B. Jacobson, *Arch. Biochem. Biphys.* **149**, 36 (1972).
[8] K. B. Jacobson, *Science* **159**, 324 (1968).
[9] F. M. Johnson and C. Denniston, *Nature (London)* **204**, 906 (1964).
[10] H. Ursprung and J. Leone, *J. Exp. Zool.* **160**, 147 (1965).
[11] E. H. Grell, K. B. Jacobson, and J. B. Murphy, *Ann. N.Y. Acad. Sci.* **151**, 444 (1968).

The flies are tamped to the bottom of the jar, the stopper is removed, and the jar is quickly inverted into a funnel which is inserted in a plastic bottle on ice. Further tamping dislodges most of the flies down the funnel, and the cold bottle stuns them. The flies can be frozen for months at approximately −12° in closed containers with little loss in the ADH enzymic activity.

Assay Procedure

As with other dehydrogenase enzymes, a spectrophotometric assay is used in which the reduction of a coenzyme, in this case NAD^+, is followed as an increase in absorption at 340 nm as a function of time. Since *Drosophilia* ADH shows greater reactivity with secondary alcohols than with primary alcohols, the substrate oxidized is usually 2-butanol or 2-propanol. The assay mixture contains as final concentrations: 0.1 M Tris-glycin (0.05 M of each), pH 9.3 (adjusted with HCl); 1.68 mM NAD^+ (1 mg/ml); and 0.218 M 2-butanol.

A 2-ml aliquot of the above mixture is put in a cuvette and allowed to equilibrate at 25°. To avoid significant volume changes, ADH is added in 1–25 μl amounts; the resulting enzyme concentration should be 0.1–100 nM (6 ng/ml–6 μg/ml) for measurable results. After rapid mixing, and while maintaining the assay mixture at 25°, the absorption increases are recorded, and the activity is determined by measuring the initial slope of the line and converting this result to micromoles of NAD^+ reduced per minute.

Purification Procedure

The original purification procedures of Sofer and Ursprung[3] and Jacobson and co-workers[4] use standard methods and obtain a homogeneous preparation of ADH in good yields. The scheme described below gives similar results, but a homogeneous preparation is obtained in less time. If the steps are run without interruption, the procedure is completed in approximately 24 hr.

A fast strain isolated from a North Carolina wild-type *Drosophila melanogaster* is used for studies in this laboratory. All purification steps are run at 0–4°; unless otherwise specified, the centrifugation is at 20,000 g. The column sizes have been used for as much as 70 g of flies as starting material; larger preparations would necessitate scaling up of these columns.

Step 1. Using a Sorvall Omni-mixer, 10–70 g of flies are homogenized for 20 sec in distilled water, the final volume being 4 ml per gram body weight of flies. After adjusting the pH to 5.0 with 1 M phosphoric acid and stirring for 5 min, the homogenate is centrifuged for 20 min. The deep orange-yellow supernatant fraction is retained.

Step 2. The pH is adjusted to 7.0 with 1 M NaOH, and the solution is passed through a carboxymethyl cellulose column (Whatman CM 52). The bed volume of the column is 21 cm by 5 cm, previously equilibrated with 10 mM phosphate buffer, pH 7.0, and eluted at 120 ml/hr. The majority of the colored contaminants is retarded by the column, while the ADH is not. The ADH enzymic activity elutes after approximately 270 ml of washing with distilled water; the resulting pooled fractions appear pale yellow.

Step 3. Solid ammonium sulfate is added with stirring to a concentration of 500 g/liter, the pH being maintained at pH 7.0 by additions of NaOH or HCl. After stirring an additional 5 min, the solution is centrifuged for 15 min. The supernatant fraction is decanted, with care being taken to remove as much as possible; a sufficient amount of 50 mM Tris· HCl, pH 9.3, is then added to just dissolve the pellet containing the ADH.

Step 4. Solid calcium chloride (dihydrate) is added to a concentration of 1.47 g/ml. Gentle mixing immediately precipitates the majority of the sulfate ions present in the enzyme preparation as calcium sulfate, which is then removed by centrifugation at 1000 g for 2 min. The precipitate is washed with a small amount of buffer and recentrifuged, and the two supernatant fractions are combined.

Step 5. The solution is added to a Sephadex G-200 column, 60 cm by 2.2 cm, previously equilibrated with 10 mM Tris·HCl, pH 9.3. The column is eluted with the same buffer at a flow rate of 35 ml/hr. The void volume by blue dextran peaks at 80 ml, while the peak of ADH enzymic activity occurs at 110 ml. The tubes containing ADH activity are pooled.

Step 6. The ADH is adsorbed to a diethylaminoethylcellulose column (Whatman DE 52), 23 cm by 2.2 cm, previously equilibrated with the same buffer as in Step 5. A linear gradient using 100 ml of 10 and 130 mM Tris·HCl, pH 9.3, elutes the enzyme as a single peak at about 70 mM, a flow rate of 20–40 ml/hr being used. Most of the peak is homogeneous for ADH band No. 5 (fast strain) by disc gel electrophoresis.[12] ADH activity bands are located by a reduced tetrazolium stain[9,10] and protein bands are located by treating gels with 0.01% Coomassie blue in 20% acetic acid and destaining with 7% acetic acid. The trailing 5–10% of the ADH enzymic activity overlaps another protein peak.

Comments on Purification. The precipitation of sulfate ions in step

[12] B J. Davis, *Ann. N.Y. Acad. Sci.* **121,** 404 (1964).

Rapid Purification of ADH from *Drosophila melanogaster*

Step	Total activity (units)	Specific activity (units/mg protein)	Yield (%)
1	678	0.40[a]	
2	660	1.60	97.6
3	598	2.00	87.5
4	487	1.81	71.9
5	412	9.03	61.0
6	309	37.4[a]	45.7

[a] The homogenate was from 32 g of flies; the specific activity prior to lowering the pH in step 1 was approximately 0.24.

4 is necessary because of an apparent affinity which this laboratory's strain of *Drosophilia melanogaster* exhibits for these anions. Failure to remove the sulfate results in the elution of ADH in a broad peak from the Sephadex column in step 5, which may overlap the sulfate peak to such an extent that adsorption to the DEAE column will not occur. Because Ursprung and Sofer did not appear to encounter this problem in their purification procedure,[3] the apparent affinity may not be seen in all strains.

The final homogeneous preparation is most stable in the refrigerator; freezing results in a large amount of denaturization.

Properties

The enzyme is stable over a large pH range, pH 5.0 to almost 10, but is most stable from 8.5 to 9.5. Band 5 denatures rapidly at 40–45°, while bands 3 and 1 retain their activity for short periods of time in this temperature range. Similarly, band 5 loses enzymic activity at a much faster rate at room temperature than do the other two main bands.

Besides showing greater activity with secondary alcohols, *Drosophila* ADH also exhibits "substrate activation" at high concentrations of secondary alcohols. For 2-butanol, two linear portions of a Lineweaver-Burk plot are observed, with the K_m in the region of high substrate concentration (20 mM–1 M) being 5–6 mM and the K_m in the region of lower substrate concentration (less than 1 mM) being 0.33–0.38 mM. Primary alcohols, as well as aldehydes and ketones used in the reverse reaction, do not show this substrate activation. The K_m for ethanol is 2.1–2.2 mM.[13]

Two different molecular weights for *Drosophila* ADH have been re-

[13] J. I. Elliott, Ph.D. Thesis, North Carolina State University, Raleigh, North Carolina, 1974.

ported. One value of 60,000 is based on sedimentation equilibrium and is in good agreement with the amino acid analysis.[4] The other value is 44,000, determined by sedimentation velocity and gel permeation.[3] Subsequent work based on sedimentation equilibrium measurements have yielded a molecular weight of 52,000 with a subunit size of 25,000 using SDS polyacrylamide electrophoresis.[13]

The electrophoretic pattern of *Drosophila* ADH can be altered by incubation of the enzyme with NAD^+, or NAD^+ and acetone; the process has been termed conversion.[8,14] Band 5 will progressively convert to band 1, going through all intermediate bands. When band 5 is eventually converted into band 1, only 5–15% of the original activity is detected by the spectrophotometric assay. The conversion is specific for NAD^+, since no changes are observed with NADH, $NADP^+$, or NADPH, and apparently specific for a ketone because both acetone and methylethyl ketone will cause conversion but aldehydes will not.[13] Protein staining after electrophoresis shows that all protein is observed as band 1,[4] so the apparent losses in activity may be due to band 1 having a lower specific activity than band 5, rather than a denaturation of the enzyme. The ADH band 1 produced by conversion appears to be electrophoretically identical to the natural form; but because the isolation of band 1 from fly homogenates has been unsuccessful, insufficient comparisons are available to conclude that they are identical.

The number of active sites and the binding constant for NADH have been determined.[13] As with other dehydrogenases NADH binds to ADH (band 5) with a shift in fluorescent emission and an increase in quantum yield. By titrating the enzymes' active sites with small aliquots of NADH and recording the fluorescent increase with each addition, the number of sites was determined to be two, and two binding constants of 0.7 μM^{-1} and 0.03 μM^{-1} were observed. These results are in agreement with the data obtained by equilibrium dialysis.

[14] H. Ursprung and L. Carlin, *Ann. N.Y. Acad. Sci.* **151**, 456 (1968).

Section V

Epimerases and Isomerases

[82] Glucose-6-phosphate Isomerase from *Bacillus stearothermophilus*[1]

By Yoshiaki Nosoh

Glucose 6-phosphate $\rightleftharpoons$ fructose 6-phosphate

Assay Method

Principle. The D-fructose 6-phosphate produced from D-glucose 6-phosphate by the enzyme is colorimetrically determined by the method of Roe.[2] In this method, fructose 6-phosphate reacts with resorcinol in acid to form a pink color.

Reagents

Tris-acetate buffer, 80 mM, pH 8.0

Glucose 6-phosphate, sodium salt, 12.5 mM. Barium salt of glucose 6-phosphate is converted to sodium salt according to the procedure of Dounce *et al.*[3]

Roe's reagent.[2] Reagent A: To 300 ml of 12 N HCl is added 60 ml of deionized water. Reagent B: 250 ml of thiourea and 100 mg of resorcinol are dissolved in 100 ml of glacial acetic acid. Before use, 350 ml of Reagent A and 50 ml of Reagent B are mixed and used as Roe's reagent.

Enzyme. Dilute the enzyme solution with 40 mM Tris-acetate buffer (pH 8.0) to obtain a concentration of 9×10^{-2} μg enzyme protein per milliliter.

Procedure. The assay mixture contains 0.5 ml of the Tris-acetate buffer, 0.4 ml of the substrate solution, and 0.1 ml of the enzyme solution. The reaction is started by adding the enzyme solution, and terminated by rapidly cooling the assay mixture in ice bath and adding 4.0 ml of Roe's reagent to the mixture. The color is developed by heating the mixture to 80° for 10 min (total elapsed time). After rapid cooling to room temperature, the optical density of the mixture is measured at 405 nm.

The enzyme reaction is usually carried out for 10 min at 65°. When lower concentrations of substrate (below 2 mM) are used or the enzyme reaction is carried out at lower temperature (below 55°), the determination of fructose 6-phosphate is made, respectively, for 3 or 30–60 min,

[1] N. Muramatsu and Y. Nosoh, *Arch. Biochem. Biophys.* **144**, 245 (1971).

[2] J. H. Roe, *J. Biol. Chem.* **107**, 15 (1934).

[3] A. L. Dounce, S. R. Barnett, and G. T. Beyer, *J. Biol. Chem.* **185**, 769 (1950).

after addition of the enzyme. Under all these experimental conditions, the linearities of the fructose 6-phosphate formation with respect to time are observed.

Definition of Unit and Specific Activity. A unit of enzyme is that amount which catalyzes the formation of 1.0 μmole of fructose 6-phosphate per minute under the assay condition at 65°. Specific activity is expressed as units per milligram of protein. Protein concentration is determined by means of either the biuret method,[4] when the concentration is above 1 mg/ml, or the method of Lowry et al.,[5] using crystalline bovine serum albumin as a standard, when the concentration is below 1 mg/ml.

Purification Procedure

Growth of Bacteria. Bacillus stearothermophilus NCA 2184, kindly donated by Professor C. E. Georgi (University of Nebraska, Lincoln) is grown in a liquid medium of the following composition: 15 g of polypeptone, 1 g of yeast extract, 4 g of glucose, 5 g of NaCl, 2 g of K_2HPO_4, 1 g of $MgSO_4 \cdot 7H_2O$, 1 g of KNO_3, a few drops of silicone oil as an antifoam agent, and 1 liter of deionized water; the pH is adjusted to 7.2 with 1 N NaOH. The cells from a fresh slant culture are preincubated aerobically in a 25-ml medium at 65° for 3 hr, then transferred into a 5-liter medium and cultured for 6–8 hr. The cells harvested by centrifugation are washed twice with 10 mM Tris·HCl buffer (pH 7.0) containing 0.9% NaCl and 5 mM $MgCl_2$, suspended in the same buffer, and stored at —20°.

Preparation of Cell-Free Extract. The cells (200 g, wet weight), suspended in 200 ml of 10 mM Tris·HCl buffer (pH 7.2) containing 0.9% NaCl and 5 mM $MgCl_2$ are incubated with shaking with 100 mg of lysozyme (EC 3.2.1.17) at 37°. After 2 hr of incubation 500 μg of deoxyribonuclease I (EC 3.1.4.5) are added and the mixture is left to stand for 30 min. Cell debris is removed by centrifugation at 5000 g for 10 min to obtain the cell-free extract. All subsequent manipulations are carried out at 4°, except where indicated.

Ammonium Sulfate Fractionation. After centrifugation of the cell-free extract at 25,000 g for 15 min, the supernatant fraction containing the isomerase is diluted with 10 mM Tris·HCl buffer (pH 7.8) so as to obtain a solution of 20 mg of protein per milliliter, and fractionated with solid ammonium sulfate. The fraction precipitated during 60–80% saturation

[4] A. G. Gornall, C. J. Bardawill, and M. M. David, *J. Biol. Chem.* **177**, 759 (1949).
[5] O. H. Lowry, N. J. Rosebrough, A. L. Farr, and R. J. Randall, *J. Biol. Chem.* **193**, 265 (1951).

with ammonium sulfate is collected and suspended in 50 ml of 40 mM Tris-maleate buffer (pH 7.0), dialyzed against the same buffer overnight, and centrifuged to remove precipitate.

Polyethylene Glycol Fractionation. The protein solution thus obtained is diluted with 40 mM Tris-maleate buffer (pH 7.0) so as to obtain a protein solution of 20 mg/ml, and magnesium acetate and polyethylene glycol (No. 6000, Nihon Rikagaku Co.) are added at final concentrations of 10 mM and 10% (10 g per 100 ml of solution), respectively. The mixture is stirred for 30 min at room temperature, then centrifuged; to the resultant supernatant is added magnesium acetate and polyethylene glycol at final concentrations of 20 mM and 20%, respectively. After stirring for 30 min, the mixture is centrifuged; the resulting precipitates are dissolved in 70 ml of 50 mM Tris-acetate buffer (pH 7.8) containing 0.5 mM EDTA at a protein concentration of 30 mg/ml.

Chromatography on DEAE-Cellulose. The protein solution thus obtained is passed through a column of DEAE-cellulose previously equilibrated with 100 mM Tris-acetate buffer (pH 7.8) containing 1 mM EDTA. The column is washed with the buffer used for equilibration of the column, with the buffer containing 100 mM sodium acetate, and then with the buffer containing 180 mM sodium acetate, respectively, with a volume twice that of the column. The enzyme is eluted with the buffer containing 280 mM sodium acetate. The elution pattern is shown in Fig. 1A.

Gel Filtration on Sephadex G-200. The fractions shown with the black bar in Fig. 1A are collected and brought to 80% saturation with respect to ammonium sulfate. The precipitate is dissolved in 50 mM Tris-acetate buffer (pH 7.8) containing 0.5 mM EDTA to a protein concentration of 10 mg/ml, and subjected to a gel filtration on Sephadex G-200. Tris-acetate buffer (50 mM, pH 7.8) containing 0.5 mM EDTA and 130 mM sodium acetae is used for equilibration and development of the column. The elution pattern is shown in Fig. 1B.

Rechromatography on DEAE-Cellulose. The fractions shown by the black bar in Fig. 1B are collected and subjected to a column chromatography on DEAE-cellulose previously equilibrated with 100 mM Tris-acetate buffer (pH 7.8) containing 1 mM EDTA and 180 mM sodium acetate. The column is washed with the same buffer of twice the volume of the column, and the enzyme is eluted by a linear gradient system of 200–250 mM sodium acetate in 100 mM Tris-acetate buffer (pH 7.8) containing 1 mM EDTA. The elution pattern is shown in Fig. 1C. The fractions shown by the black bar in the figure are collected and concentrated in a collodion bag.

The enzyme thus obtained is considered to be homogeneous, as judged

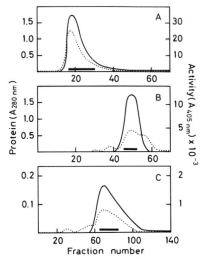

Fig. 1. Elution patterns of glucosephosphate isomerase on DEAE-cellulose and on Sephadex. (A) Chromatography on DEAE-cellulose. (B) Gel filtration on Sephadex G-200. (C) Rechromatography on DEAE-cellulose. Column dimensions (cm × cm): (A) 4.2 × 45, (B) 3.3 × 65, and (C) 2.2 × 30. Volumes collected for each tube (ml): (A) 20, (B) 5, and (C) 10. Amounts of protein charged: (A) 5 g, (B) 50 mg, and (C) 23 mg.······, ($A_{280 \text{ nm}}$); ——, enzyme activity ($A_{405 \text{ nm}}$/min). Reprinted from *Archives of Biochemistry and Biophysics.* **144**, 245–252, © Academic Press, (1971).

from the ultracentrifugal and electrophoretic patterns and the linear sedimentation equilibrium plots of the enzyme. A purification scheme giving the recovery and specific activity at each stage is presented in the table.

PURIFICATION OF GLUCOSEPHOSPHATE ISOMERASE FROM
Bacillus stearothermophilus

Step	Protein (mg)	Total activity (units)	Specific activity (units/mg protein)	Recovery (%)
Crude extract	15,400	146,300	10	100
Ammonium sulfate fractionation, 60–80%	1,840	112,240	61	77
Polyethylene glycol fractionation, 10–20%	1,130	97,180	86	66
DEAE-cellulose chromatography	48	61,400	1,290	42
Sephadex G-200 gel filtration	20	58,110	2,980	40
DEAE-cellulose rechromatography	8	32,600	4,180	22

Properties of the Enzyme

Molecular Weight and Other Physical Properties. The sedimentation coefficient ($s_{20,w}$) is 7.4 S. The molecular weight determined by sedimentation equilibrium method is 172,000. α-Helical conformation is present in the enzyme molecule, and the α-helix content is estimated to be 18 and 24%, respectively, from the optical rotatory dispersion and circular dichroism data.

Optimal pH. The enzyme exhibits a maximum activity at pH 8.0–9.0 in 40 mM Tris-acetate buffer, both at 40° and 65°.

Inhibitors. The enzyme is competitively inhibited by inorganic phosphate and 6-phosphogluconate, both at 40° and 65°.

Thermal Stability. The optimum temperature for the activity is 70°. The enzyme is stable for about 1 month at 4–20° in 20 mM Tris-acetate buffer (pH 8.0), and exposure of the enzyme to 50° for 7 hr caused no inactivation of the enzyme. When the enzyme is exposed to 65°, however, considerable inactivation of the enzyme is observed; 25% loss after 1 hr and 85% loss after 7 hr. The inactivation at 65° is partly prevented by 2 mM glucose 6-phosphate or 50 μM 6-phosphogluconate; 10% loss after 1 hr and 40% loss after 7 hr.

Thermodynamic Quantities.[6] The values of Michaelis constants (K_m) and maximum velocities (V_{max}) estimated at 40 and 65° are 1.98 and 3.68×10^{-4} M and 1.07 and 4.64 mmoles of fructose 6-phosphate formed per minute per milligram of protein, respectively. The values of enthalpy change (ΔH) and entropy change (ΔS) for formation of the ES complex below and above 55° are −2400 and −12,000 cal per mole and 9 and −20 cal per mole per degree, respectively. The values of ΔH and ΔS for activation of the ES complex are 11,000 cal per mole and −7 cal per mole per degree, respectively, over the temperature range, 30–70°. The results may indicate some conformational change in the enzyme protein at a transition temperature, 55°, as suggested for other enzymes,[7-9] especially for ATPase (EC 3.6.1.3),[7-8] from the same bacterium.

---

[6] The unit for the values of V_{max} and the values of ΔS for activation of the ES complex were incorrectly reported in a previous publication (Muramatsu and Nosoh[1]) as micromoles and 9 cal per mole per degree, respectively, instead of millimoles and −7 cal per mole per degree.

[7] A. Hachimori, N. Muramatsu, and Y. Nosoh, *Biochim. Biophys. Acta* **206**, 426 (1970).

[8] A. Hachimori and Y. Nosoh, *Biochim. Biophys. Acta* **315**, 481 (1973).

[9] S. Sugimoto and Y. Nosoh, *Biochim. Biophys. Acta* **235**, 210 (1971).

[83] Glucose-6-phosphate Isomerase from Peas

By Susumu Hizukuri, Yasuhito Takeda, and Ziro Nikuni

Glucose 6-phosphate $\rightleftharpoons$ fructose 6-phosphate

Assay Methods

Principle. The above reaction catalyzed by the enzyme is entirely reversible and the activity is measured with either substrate, glucose 6-phosphate(G-6-P) or fructose 6-phosphate(F-6-P). The G-6-P formed is determined spectrophotometrically with G-6-P dehydrogenase and NADP.[1] F-6-P formed is determined by the colorimetric method of Roe,[2] a fluorescence method,[3,4] or by pH-stat titration coupled with phosphofructokinase and ATP.[5] Among them, the Roe method which measures the pink color developed by heating fructose or F-6-P solution with resorcinol and HCl, is the most convenient, and is accurate enough.

Reagents

Tris·HCl buffer, 0.5 M, pH 8.5
Disodium G-6-P, 0.1 M (stock solution)
HCl, 8 N
Resorcinol, 0.1% in 95% ethanol
F-6-P standard, 1.5 mM, exact molarity should be determined by a proper method.[6]
G-6-P, 20 mM, buffered (test solution). Mix 2 ml 0.1 M stock G-6-P, 1 ml of 0.5 M Tris·HCl buffer, and about 5 ml of distilled water, adjust pH to 8.5 with 0.01 N NaOH, and bring volume to 10 ml with distilled water.

Procedure.[7] A 0.2-ml sample of test solution in a test tube is prewarmed at 37° for a couple of minutes, then the reaction is initiated with an addition of 0.05 ml of enzyme solution diluted properly.[8] The re-

---

[1] E. A. Noltmann, *J. Biol. Chem.* **239**, 1545 (1964).
[2] J. H. Roe, *J. Biol. Chem.* **107**, 15 (1934); J. H. Roe, J. H. Epstein, and N. P. Goldstein, *J. Biol. Chem.* **178**, 839 (1949); M. W. Slein, this series, Vol. 1, p. 304.
[3] G R. Morrison, *Anal. Biochem.* **12**, 150 (1965).
[4] G. G. Vurek and S. E. Pegram, *Anal. Biochem.* **16**, 409 (1966).
[5] J. E. Dyson and E. A. Noltmann, *Anal. Biochem.* **11**, 362 (1965).
[6] Hans-Jurgen Hohorst, *in* "Methods of Enzymatic Analysis" (H.-U. Bergmeyer, ed.), p. 134. Academic Press, New York, 1963.
[7] Y. Takeda, S. Hizukuri, and Z. Nikuni, *Biochim. Biophys. Acta* **146**, 568 (1967).
[8] For the crude extract, about 40-fold dilution is suitable.

action is stopped by an addition of 2 ml of 8 N HCl after 3 or 5 min (not more than 5 min) incubation at 37°. At the same time, 2 ml of 8 N HCl is added to 0.2 ml of test solution before enzyme addition (0 time reaction), and 0.25 ml of G-6-P standard and 50 mM Tris·HCl buffer, pH 8.5 (blank solution of standard). To these solutions, 0.5 ml of 0.1% resorcinol is added and mixed well in test tubes. The test tubes are immersed in a water bath at 80° for 10 min. After rapid cooling of the tubes in a running water bath, the color developed (pink) is measured in a Klett-Summerson colorimeter with filter No. 54 or in a spectrophotometer at 530 nm. F-6-P formed in the reaction is calculated from the difference of the optical density of the incubated and the 0 time tube and the optical density of standard.

Definition of Unit and Specific Activity. One unit of activity is the amount of enzyme producing 1 μmole of F-6-P per minute under the above conditions. Specific activity is expressed as units per milligram of protein.

Purification Procedure

The following procedures include the modification in part of the original paper.[7] All the operations are carried out at 0–4° unless otherwise stated. The enzyme is extractable from dry as well as fresh or deep frozen green peas. Dry peas are soaked in water overnight at about 25° before extraction. The typical results of the procedure starting with 4 kg of soaked pea is shown in Table I.

Step 1. Extraction. Peas are homogenized in a Waring Blendor with 1.5 volume (per wet weight of peas) of cold 0.1 M Tris·HCl–10 mM EDTA, pH 7.5 for 5 min. The homogenate is squeezed through gauze and centrifuged at 14,000 g for 7 min to remove cell debris and some particulate matter (crude extract).

Step 2. MgCl₂ Treatment. To the crude extract, a 0.05 volume of 1 M MgCl₂ is added. The turbid materials are removed by centrifugation at 14,000 g for 7 min.

Step 3. Zinc Acetate Treatment. To the supernatant a 0.05 volume of 1 M zinc acetate is added with stirring, and the pH is adjusted to 7.5 with 1 M Tris. After 30 min of stirring, the precipitate is centrifuged off.

Step 4. First Ammonium Sulfate Fractionation. Immediately after the above centrifugation, 0.67 vol of saturated (4°) ammonium sulfate solution, pH 7.5, is added to the supernatant to give 40% saturation with stirring and the solution is allowed to stand for 30 min. The precipitate

[9] Saturated at 0°.

TABLE I
PURIFICATION OF GLUCOSEPHOSPHATE ISOMERASE FROM PEAS[a]

Purification steps	Volume (ml)	Total protein (mg)	Specific activity (U/mg)	Recovery (%)
1. Crude extract (from 2 kg of dry immature peas[b] or 4 kg of fresh green peas)	5860	194,300	0.81	100
2. MgCl₂ supernatant	5700	139,600	1.05	92.7
3. Zinc acetate supernatant	5660	67,900	1.93	83.3
4. First ammonium sulfate precipitate, 0.4–0.6% salt saturation	900	14,300	7.56	68.5
5. Bentonite supernatant	700	2,150	37.0	52.3
6. Isopropyl alcohol precipitate, 48–63%	123	269	225	38.5
7. Second ammonium sulfate precipitate	60.5	145	333	30.7
8. DEAE-Sephadex column eluate	43	27.3	1260	21.5
9. Crystalline enzyme		15.5	1470	14.5

[a] Y. Takeda, S. Hizukuri, and Z. Nikuni, *Biochim. Biophys. Acta* **146**, 568 (1967).
[b] Peas absorb about their own weight of water while soaking in water at 25° overnight.

formed is centrifuged off at 14,000 g for 10 min. Saturated ammonium sulfate solution, pH 7.5 (0.83 volume), is added to the above supernatant solution to give 60% saturation with stirring and allowed to stand for 30 min. The precipitate formed is collected by centrifugation at 14,000 g for 10 min and is dissolved in 50 mM Tris·HCl–5 mM EDTA, pH 7.5 (about ⅙ volume of the crude extract).

Step 5. Bentonite Treatment. Bentonite, 20 g per gram of protein, is added to the step 4 solution with stirring. After 30 min stirring, the adsorbent is removed by centrifugation. The activity remains in the solution. The enzyme in this state can be stored in a refrigerator about a week without loss of activity.

Step 6. Isopropyl Alcohol Precipitation. Isopropyl alcohol chilled in a freezer (about −10°) is added slowly to the above solution at about −7° to give 48% (v/v), and the precipitate formed is discarded by centrifugation at about −7°. Isopropyl alcohol is added to the supernatant solution to increase the concentration to 63%. The resulting precipitate is collected by centrifugation at about −7° and suspended in 50 mM Tris·HCl–5 mM EDTA, pH 7.5 (1/6 volume of step 5 solution). Insoluble materials are removed by centrifugation.

Step 7. Second Ammonium Sulfate Fractionation. Saturated ammonium sulfate solution, 1.86 volumes, pH 7.5 is added to the step 6 solution to give 65% saturation. The suspension is stored overnight. The precipitate is collected by centrifugation, drained well, and dissolved in 10 mM Tris·HCl —1 mM EDTA, pH 7.5 (a half volume of step 6 solution).

Step 8. Chromatography on DEAE-Sephadex A-50. The step 7 solution is dialyzed against 10 mM Tris·HCl–1 mM EDTA, pH 7.5, at least 5 hr. The dialyzed solution is slowly flowed onto a DEAE-Sephadex A-50 column equilibrated with the buffer. About 40 mg of protein are loaded per milliliter wet volume of the Sephadex. The column is washed thoroughly with the buffer and then with the buffer contained 50 mM NaCl while monitoring the adsorption at 280 nm with an optical device. The enzyme is eluted with the buffer containing 0.1 M NaCl, and the active fractions are combined. The enzyme is precipitated from the combined eluate by the addition of 2.33 volume of saturated ammonium sulfate solution, pH 7.5, to 70% saturation, and the solution stored overnight. The precipitate is recovered by centrifugation and is dissolved in 10 mM Tris·HCl–1 mM EDTA, pH 7.5 (about $\frac{1}{40}$ volume of the combined eluate) to a protein content of about 14 mg/ml. If the specific activity is less than 1100 or so, the enzyme is chromatographed again on DEAE-Sephadex after dialysis as above.

Step 9. Crystallization. Saturated ammonium sulfate solution, pH 7.5, is added slowly by a micropipette to the above solution with stirring. If the solution becomes turbid below 40% saturation (0.67 volume), the turbid materials are removed by centrifugation. The salt concentration is increased further, and the enzyme starts to crystallize in thin needles or rods at about 50% saturation. After standing overnight, the salt concentration is increased to 60% saturation (1.5 volume) to complete crystallization. About 16 mg of crystalline enzyme is obtained from 2 kg of dry, or 4 kg of fresh, green peas.

Properties[7]

Stability. The enzyme is stable at pH 7.0 in Tris·HCl buffer and is unstable at acidic pH even at 6.0. Addition of solid ammonium sulfate in the purification steps causes a considerable loss of the activity even if the pH of the enzyme solution is maintained carefully at 7.5. The crystalline enzyme is moderately stable in ammonium sulfate suspension at 4° and loses about 50% activity in 5 months at about 4°.

Physical Properties. Some physical properties of the enzyme are listed in Table II. A dimeric form of the native enzyme is suggested by centrifugation analysis and by SDS gel electrophoresis (see Table II).[10]

Specificity. The enzyme is capable of isomerizing glucose into fructose

[10] S. Hizukuri and K. Kaimoto, unpublished, 1973.

TABLE II
PHYSICAL PROPERTIES OF PEA GLUCOSEPHOSPHATE ISOMERASE[a]

Properties	Value
Sedimentation constant, $s_{20,w}$	6.8 S
Molecular weight (approach to equilibrium[b])	110,000
Molecular weight (SDS gel electrophoresis[c])	61,000[d]
Extinction coefficient, $E_{1cm}^{1\%}$ (280 nm)	8.75
K_m (G-6-P)	0.27 mM
V_{max} (G-6-P)	1500 U/mg
K_i (6-phosphogluconate)	13 μM
Equilibrium, G-6-P:F-6-P	65:35

[a] Values for the parameters listed have been taken from Y. Takeda, S. Hizukuri, and Z. Nikuni, *Biochim. Biophy. Acta* **146**, 568 (1967).
[b] H. K. Schachman, "Ultracentrifugation in Biochemistry," p. 181. Academic Press, New York, 1959.
[c] K. Weber and M. Osborn, *J. Biol. Chem.* **244**, 4406 (1969).
[d] S. Hizukuri and K. Kaimoto, unpublished.

in the presence of arsenate.[7] The activity of glucose isomerization is about $\frac{1}{50}$ of G-6-P isomerization.

pH Optimum. The pH optimum of the enzyme reaction is at 8.5 in Tris·HCl buffer and at 9.5 in glycine–NaCl–NaOH buffer.[7]

Other Properties. The enzyme is not inactivated with *p*-chloromercuribenzoate and monoiodoacetate and appears to have no —SH groups in the active site.[7] 6-Phosphogluconate is a strong competitive inhibitor of the enzyme.[7]

[84] Phosphoglucose Isomerase[1] of Human Erythrocytes and Cardiac Tissue[2]

By ROBERT W. GRACY and BILL E. TILLEY

D-Glucose 6-phosphate $\rightleftharpoons$ D-fructose 6-phosphate

Assay Method

Principle. The enzyme activity can be measured in either the forward or the reverse direction by using the appropriate coupling enzymes and monitoring the oxidation or reduction of NADH or NADP spectrophoto-

[1] Glucose-6-phosphate ketol-isomerase, EC 5.3.1.9.
[2] The work described in this article was supported in part by grants from the U.S. Public Health Service (AM14638), the Robert A. Welch Foundation (B-502), a Career Development Award from the National Institutes of Health (AM 70198), and North Texas State University Faculty Research.

metrically. The isomerase activity is more conveniently assayed in the reverse direction by coupling with glucose-6-phosphate dehydrogenase and following the rate of reduction of NADP.[3,4] The forward reaction can be measured by coupling with phosphofructokinase, aldolase, triosephosphate isomerase, and α-glycerophosphate dehydrogenase and monitoring the rate of oxidation of NADH. A suitable alternate procedure for the forward reaction is the coupled pH stat assay of Dyson and Noltmann.[5] Either of these methods is more convenient and much preferred to the colorimetric resorcinol procedure of Roe.[6]

Fructose 6-Phosphate as Substrate. The assay mixture contains in a final volume of 1.0 ml: 50 mM triethanolamine buffer, pH 8.3, 1 mM EDTA, 4.0 mM fructose 6-phosphate, 0.5 mM NADP, and 1 unit of glucose-6-phosphate dehydrogenase. The reaction mixture is added to a 1-cm light-path quartz cuvette and preincubated at 30.0° for 5 min in the sample compartment of a thermostatically regulated spectrophotometer. During this time, any glucose-6-phosphate which may be present as an impurity of the substrate is oxidized. The reaction is then initiated by the addition of 25 μl of appropriately prediluted (into 10 mM triethanolamine, 1 mM EDTA, pH 8.3) enzyme sample such that a final rate of change in absorbance at 340 nm of 0.01–0.2 per minute is observed. After correcting for the predilution and dilution of enzyme in the cuvette, the absorbance change per minute is divided by 6.22 (the mM absorbance index for NADPH) to give micromoles of glucose 6-phosphate formed per minute per milliliter of enzyme solution.

Glucose 6-Phosphate as Substrate. The rate of isomerization of glucose 6-phosphate to fructose 6-phosphate is monitored under similar conditions at 340 nm in a recording spectrophotometer at 30.0°. The assay mixture contains in a final volume of 1.0 ml: 50 mM triethanolamine buffer, pH 8.3, 1 mM ATP, 5 mM MgCl$_2$, 50 mM KCl, 0.15 mM NADH, and 1 unit each of phosphofructokinase, fructose-diphosphate aldolase, triosephosphate isomerase, and α-glycerophosphate dehydrogenase. After preincubating as above, 25 μl of the appropriately diluted phosphoglucose isomerase is added, and the oxidation of NADH is recorded as the rate of decrease in absorbance at 340 nm. Since the stoichiometry of this reaction is 2 equivalents of NADH oxidized per equivalent of fructose 6-phosphate formed, the net change in absorbance per minute is divided by 12.44 in order to obtain micromoles of fructose 6-phosphate formed per minute per milliliter.

[3] S. E. Kahana, O. H. Lowry, D. W. Schulz, J. V. Passonneau, and E. J. Crawford, *J. Biol. Chem.* **235**, 2178 (1960).
[4] E. A. Noltmann, *J. Biol. Chem.* **239**, 1545 (1964).
[5] J. E. Dyson and E. A. Noltmann, *Anal. Biochem.* **11**, 362 (1965).
[6] J. H. Roe, *J. Biol. Chem.* **107**, 15 (1934).

In general, the purity of both substrates and all coupling enzymes is adequate to prevent baseline drift or interference during the assay. The small amount of 6-phosphogluconate (a competitive inhibitor of phosphoglucose isomerase) formed in the reverse direction assay prior to addition of the isomerase is usually not sufficient to warrant correction; however, methods are available for further purification of the hexose phosphates.[7-10] Since the forward assay requires more coupling enzymes and coenzymes, it is more susceptible to interfering contaminants in assays of crude extracts and is therefore less desirable than the reverse assay. Also, slight competitive inhibition is present due to the ATP required for the forward assay. Although the coupled pH stat assay[5] requires only a single coupling enzyme, the ATP inhibition is still present. The coupled pH stat assay has been described in detail in Volume 9 of this series,[11] but since that time phosphoglucose isomerase-free preparations of the coupling enzymes required for the spectrophotometric assay have been made available commercially. Therefore, the ability to assay both forward and reverse isomerizations with the same instrument, and the capability to simultaneously monitor several assays with multiple channel spectrophotometers, make the coupled spectrophotometric assay more convenient than the pH stat method for routine use.

In either the forward or reverse assays described above, the substrates are used as their sodium salts, and the concentrations of substrates are determined enzymically by the above methods. One unit of enzyme in either assay method is defined as the amount required to catalyze the isomerization of 1 μmole of substrate per minute, and specific activity is expressed as units per milligram of protein. In crude extracts the protein concentration is determined by a modification[12] of the biuret procedure of Gornall et $al.$[13] The protein concentration of the isolated phosphoglucose isomerase can be conveniently determined from its absorbance at 280 nm using a value for the absorbance index, $\epsilon_{1\,cm}^{1\%}$ of 13.9.

Isolation Procedure

The following method has been developed for the rapid isolation of human phosphoglucose isomerase by substrate elution from phosphocellu-

[7] M. C. Hines and R. G. Wolfe, $Biochemistry$ 2, 770 (1963).
[8] A. Bonsignore, S. Pontremoli, G. Mangiarotti, A. DeFlora, and M. Mangiarotti, $J.$ $Biol.$ $Chem.$ 237, 3597 (1962).
[9] B. Borreback, S. Abraham, and I. L. Chaikoff, $Anal.$ $Biochem.$ 8, 367 (1964).
[10] B. E. Tilley, D. W. Porter, and R. W. Gracy, $Carbohyd.$ $Res.$ 27, 289 (1973).
[11] E. A. Noltmann, this series, Vol. 9, p. 557.
[12] R. W. Gracy and E. A. Noltmann, $J.$ $Biol.$ $Chem.$ 243, 11 (1968).
[13] A. G. Gornall, C. J. Bardawill, and M. M. David, $J.$ $Biol.$ $Chem.$ 177, 751 (1949).

lose. The method can be used to isolate the enzyme from other human tissues with only minor modification. The procedure is superior to other previously published methods with respect to overall recovery, total purification, final specific activity, and ease and speed of isolation. The pure protein can be easily crystallized from ammonium sulfate. The following describes the methods for the isolation of the enzyme from human red blood cells and cardiac muscle. All steps are carried out at 0–4° and 0.1% v/v 2-mercaptoethanol is included in all buffer solutions. Buffer A consists of 10 mM triethanolamine (HCl), pH 7.5, containing 1 mM EDTA and 0.1% 2-mercaptoethanol. Buffer B is 25 mM triethanolamine, pH 8.2, also containing 0.1% 2-mercaptoethanol.

Isolation from Human Erythrocytes

Hemolysate. Whole blood (450 ml) is collected into 65 ml of standard anticoagulant (acid–citrate–dextrose) and sedimented at 15,000 g for 30 min. The plasma and buffy coat are removed, and the erythrocytes are washed by suspending in 500 ml of ice cold 0.145 M NaCl and sedimenting as above. After washing twice more, the erythrocytes are suspended in an equal volume of ice-cold deionized water. The suspension is frozen and thawed repeatedly in a dry ice–acetone bath to ensure complete hemolysis, and the pH of the lysate is adjusted to 7.0.

Removal of Hemoglobin. One hundred-milliliter fractions of the lysate are rapidly mixed with 60 ml of a 1:1 v/v CHCl₃:CH₃OH mixture (prechilled to 0°) and vigorously stirred for 120 sec in an ice bath. This mixture is immediately diluted with 400 ml of ice-cold 10 mM triethanolamine buffer, pH 8.2, containing 5 mM EDTA, and the precipitated hemoglobin is removed by centrifugation for 15 min at 15,000 g. It is important that the pH of the diluting buffer be adjusted such that the pH of the resulting supernatant solution is 7.5. The pink supernatant solution is subjected to ultrafiltration to effect concentration (approximately 100 ml) and buffer exchange (10 mM triethanolamine, pH 8.7). This solution is passed through a 7 × 5 cm Büchner funnel containing DEAE-cellulose (0.8 mEq/g) which has been equilibrated in the above buffer. This filtration removes all traces of hemoglobin.[14]

Substrate Elution from Phosphocellulose. The colorless supernatant solution from the above step is adjusted to pH 7.5 and pumped (1–2 ml/min) onto a 1.5 × 30 cm column of cellulose phosphate (0.91

---

[14] The filtration on DEAE to remove all traces of hemoglobin can be omitted, but this results in the necessity to carry out two successive substrate elution steps from phosphocellulose.

TABLE I

ISOLATION OF PHOSPHOGLUCOSE ISOMERASE FROM HUMAN ERYTHROCYTES

Fraction	Volume (ml)	Total activity (units)	Total protein (mg)	Specific activity (units/mg protein)	Purification (fold)	Recovery (%)
Hemolysate	440	2810	169, 800	0.0165	(1)	(100)
Hemoglobin removal	445	2007	200.7	10.0	606	71.4
Substrate elution	130	1855	2.2	843	51,100	66.0

meq/gram) which has been prewashed by cycling through base and acid[15] and equilibrated in buffer A. After the sample is added, the column is washed with buffer B until the pH of the effluent is 8.2.[16] Subsequently, 200 ml of 7 mM glucose 6-phosphate (in buffer B) is added to the column to effect a specific elution of phosphoglucose isomerase.[17] Under these conditions phosphoglucose isomerase elutes as a sharp peak coincident with the substrate front. The fractions containing phosphoglucose isomerase activity are pooled and dialyzed to remove the substrate.

Table I shows typical results of such an isolation. Phosphoglucose isomerase with a specific activity of approximately 850 units/mg has been routinely obtained, with an overall recovery of approximately 65%. The isolation of the enzyme from whole blood (over 50,000-fold purification) can be carried out in only 2–3 days. The rapidity of this procedure and the inclusion of 2-mercaptoethanol in all solutions prevent the oxidation of SH groups and the resultant multiple forms encountered with previous procedures.[18] The enzyme isolated by the present method has been found to be homogeneous by a variety of criteria including polyacrylamide gel electrophoresis, analytical ultracentrifugation, rechromatography, and isoelectric focusing. The enzyme can be crystallized as described below, but the crystallization does not result in an increase in specific activity.

[15] E. A. Peterson and H. A. Sober, this series, Vol. 5, p. 3.

[16] It should be noted that the pH is critical at this point. Application of the enzyme to the phosphocellulose column at pH values above 7.5 results in some "leakage" of phosphoglucose isomerase with the bulk of protein in the void volume. On the other hand, attempts to carry out the substrate elution at pH values below 8.2 result in a lower recovery of isomerase due to incomplete elution. Note also that pH values given are determined at 4°.

[17] The other substrate, fructose 6-phosphate or the competitive inhibitor 6-phosphogluconate are also effective in achieving elution of phosphoglucose isomerase with essentially identical results.

[18] K. K. Tsuboi, K. Fukanaga, and C. H. Chervenka, J. Biol. Chem. 246, 7586 (1971).

Isolation from Human Cardiac Tissue

A similar procedure has also been used to isolate phosphoglucose isomerase from cardiac muscle. The isolation does not require the solvent fractionation step to remove hemoglobin, and thus the dialyzed homogenate can be directly applied to a phosphocellulose column and eluted with substrate. An overall purification of 600–800-fold with recovery of approximately 70% can be obtained in 3 days.

Homogenate. Cardiac muscle is dissected free of fatty tissue, and the material (376 g) is passed through a meat grinder. The tissue is then suspended in 750 ml of 10 mM triethanolamine, 1 mM EDTA, 10 mM KCl, pH 11.0, and homogenized in a Waring Blendor for 3 min. The mixture is centrifuged at 15,000 g for 30 min; the supernatant solution is adjusted to pH 7.5 and subjected to ultrafiltration to effect concentration and buffer exchange (buffer A).

Substrate Elution. The dialyzed homogenate solution is applied to a 2.4 × 60 cm column of phosphocellulose which has been prewashed and equilibrated in buffer A. The column is then washed with buffer B followed by 7 mM substrate to induce the elution of phosphoglucose isomerase. Figure 1 shows a typical elution profile with the sharp peak of isomerase activity characteristic of substrate elutions.

Although homogeneous phosphoglucose isomerase (specific activity 850 units/mg) can be obtained in a single substrate elution, this procedure requires a larger column and consequently a longer time. Thus, it is usually more convenient and faster to carry out two successive elution steps on smaller columns.

Second Substrate Elution. The fractions obtained from the previous step containing phosphoglucose isomerase are dialyzed to remove the substrate and applied to a 2.4 × 30 cm column of phosphocellulose which has been prewashed and equilibrated in buffer A as above. The column is washed with buffer B until the effluent is pH 8.2 and then 7 mM glucose 6-phosphate is added to the eluting buffer to effect the elution of phosphoglucose isomerase. Table II summarizes a typical isolation of the enzyme from human cardiac muscle.

As in the case of the isolation from erythrocytes, the enzyme is pure at this point as adjudged by various physical criteria. However, if crystallization is deemed necessary for either the erythrocyte or cardiac muscle enzyme, the following procedure can be conducted.

Crystallization. Crystallization can be effected by first concentrating the enzyme solution to approximately 10 mg/ml by dialysis against a saturated solution of ammonium sulfate. The pellet is dissolved in 10 mM triethanolamine buffer, pH 7.7, containing 1 mM EDTA and 0.1%

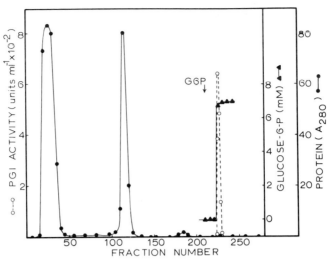

Fig. 1. Substrate elution of human cardiac muscle phosphoglucose isomerase from phosphocellulose. The homogenate, after ultrafiltration, was added to a 2.4 × 60-cm column of cellulose phosphate (equilibrated in buffer A) and washed with buffer B. A flow rate of 80 ml hr^{-1} was maintained, and fractions of 20-ml each were collected. Substrate (G6P; ▲) was added at the indicated point, and fractions were monitored for protein concentration (●) and phosphoglucose isomerase activity (○).

2-mercaptoethanol such that the resulting protein concentration is no less than 10 mg/ml. The ammonium sulfate concentration of the reconstituted pellet is calculated by standard equations,[19] and the proper volume of a saturated solution of ammonium sulfate is added to bring the solution to 0.50 (29.8 g/100 ml) saturation (4°). After a brief centrifugation the

TABLE II

Isolation of Phosphoglucose Isomerase from Human Cardiac Muscle

Fraction	Volume (ml)	Total activity (units)	Total protein (mg)	Specific activity (units/mg protein)	Purification (fold)	Recovery (%)
Homogenate	320	38,180	29,730	1.28	(1)	(100)
First substrate elution	33	28,810	169	170	133	75.5
Second substrate elution	12	28,500	32.6	874	683	74.6
Crystallization	3.4	23,430	27.0	868	678	61.4

[19] E. A. Noltmann, C. J. Gubler, and S. A. Kuby, J. Biol. Chem. 236, 1225 (1961).

ammonium sulfate concentration is slowly increased by further addition of the saturated solution to 0.57 (34.7 g/100 ml) saturation. The solution is allowed to stand overnight at room temperature during which crystallization occurs. Although no difficulty is encountered in crystallization, it should be emphasized that it does not result in a further increase in the specific activity. Furthermore, due to the unavoidable loss of enzyme in handling and transferring solutions, it is not recommended that this step be carried out unless crystals are specifically required.

Properties of Human Phosphoglucose Isomerase

Sources. Previous studies[20] have failed to indicate the existence of tissue-specific variant forms (i.e., isoenzymes) of human phosphoglucose isomerase. Thus, the physical, chemical, and catalytic properties of the enzyme isolated from erythrocytes are identical with the enzyme isolated from muscle tissue. Although erythrocytes are often the only available source of material for the isolation, muscle tissue is preferred, due to its higher concentration of the enzyme.

Physical Properties. Sedimentation velocity ultracentrifugation yields an $s_{20,w}$ value of 72×10^{-13} sec at a protein concentration of 5 mg/ml. Sedimentation equilibrium studies yield a weight average molecular weight of $132,000 \pm 2000$ for the native enzyme and $63,000 \pm 1500$ for the enzyme dissociated in 6 M guanidinium chloride. The enzyme is a dimer consisting of two identical subunits. Human phosphoglucose isomerase is a basic protein with an apparent isoelectric pH of 9.2 as determined by isoelectric focusing.

Catalytic Properties. As with other phosphoglucose isomerases, the enzyme is specific for D-glucose 6-phosphate and shows no activity with mannose 6-phosphate. The K_m for glucose 6-phosphate is 0.12 mM and that for fructose 6-phosphate is 71 μM. No cofactors or activators are required, and the enzyme exhibits a pH optimum at pH 8.3 with half-maximal activities at pH 7 and 10. The enzyme is competitively inhibited by 6-phosphogluconate ($K_i = 63 \mu M$), erythrose 4-phosphate, and to a lesser extent by a variety of other sugar phosphates.[18]

Stability. The enzyme is stable even in dilute solution. However, the inclusion of 1 mM EDTA in the buffer seems to offer additional stabilization to the enzyme. The enzyme is susceptible to oxidation of sulfhydryl groups which gives rise to multiple electrophoretic forms with catalytic activity.[18,20] For this reason it is advisable to store the enzyme in the

[20] D. M. Payne, D. W. Porter, and R. W. Gracy, *Arch. Biochem. Biophys.* **151**, 122 (1972).

presence of 2-mercaptoethanol or dithiothreitol. Under ideal conditions the enzyme should be stored as the crystalline suspension or precipitate in 0.70 saturated ammonium sulfate containing 50 mM triethanolamine buffer, pH 8.3, 1 mM EDTA, and 1 mM dithiothreitol.

[85] Glucosaminephosphate Isomerase from House Flies[1,2]

By S. FRIEDMAN and R. L. BENSON

Fructose 6-phosphate + NH$_3$ $\rightleftharpoons$ glucosamine 6-phosphate + H$_2$O

Assay Methods

Principle. The formation of glucosamine-6-P may be measured by first N-acetylating the amino sugar and then assaying for N-acetyl-aminosugar-6-P by a modified[1] Levvy-McAllan[3] reaction. If N-acetyl-glucosamine-6-P is used to activate the enzyme, the appropriate blank without the acetic anhydride acetylation step (substitute acetic acid) should be run to check for deacetylation of the activator.

The formation of fructose-6-P may be followed by the Davis and

[1] R. L. Benson and S. Friedman, *J. Biol. Chem.* **245**, 2219 (1970).

[2] Rearing: House fly (*Musca domestica* L.) pupae may be obtained from any of a number of entomology departments or agricultural chemical companies. We have successfully used flies of mixed strains as well as the Stauffer susceptible strain (Stauffer Chem. Co., P.O. Box 760, Mountain View, CA 94040; or Department of Entomology, University of Illinois, Urbana, IL 61801).

Pupae are placed in cages (ca. $10 \times 10 \times 10$ inches) at room temperature, and adults permitted to emerge. After emergence, adult flies are fed on sugar cubes and absorbent cotton soaked with water. Two days after eclosion, powdered milk and granulated sugar are mixed (1:1) and placed in plastic dishes in the cages as a source of protein for egg production, which should then take place within 1.5 days. Provision is made for placement of eggs by ovipositing females by putting a large wad of absorbent cotton soaked in sour diluted (1:4) evaporated milk into a 250-ml beaker and covering it with milk-moistened cloth or filter paper. The milk-saturated cotton should fill about two-thirds of the beaker, and be so distributed that a depression in the cloth is provided around the periphery for oviposition. A volume of about 0.5 ml of fly eggs collected from the oviposition site is rinsed with a small amount of water onto the surface of 1.6 liters of a medium composed of ground wheat (mill run), ground barley, finely ground alfalfa, and tap water (10:1:1:6.7, v/v/v/v), contained in a 1-gal plastic drum. The container is incubated at 23–29°, and the mature larvae begin to wander from the medium in about 4.5–5 days. The flies are gathered as either wandering larvae or as pupae and placed in cages for adult emergence. Adult flies (2–5 days old) are harvested and sacrificed by placing the cage in a freezer. Whole flies are stored at −10°.

[3] G. A. Levvy and A. McAllan, *Biochem. J.* **73**, 127 (1959).

Gander[4] modification of the Roe[5] procedure. This method is not accurate with crude enzyme preparations containing glucosephosphate isomerase and/or mannosephosphate isomerase, because fructose-6-P will be converted to glucose-6-P and/or mannose-6-P. Glucose-6-P activation of preparations containing glucosephosphate isomerase cannot be studied for similar reasons. Finally, when glucose-6-P is used to allosterically activate purified enzyme, high blanks are obtained.

Reagents

A. Formation of glucosamine-6-P

Fructose-6-P disodium salt, 0.2 M

NH_4Cl, 0.2 M, titrated to pH 8.5 with KOH

Tris·HCl buffer, 0.25 M, pH 8.5

$K_2H-KH_2PO_4$ buffer, 0.25 M, pH 7.6

EDTA, disodium salt, 25 mM

Enzyme in 1 mM Tris·HCl, pH 7.3

B. Assay for glucosamine-6-P

Glucosamine HCl, 0.25 mM

KOH, 0.56 M in H_3BO_3, 1.12 M

Acetic anhydride solution, 5%, freshly prepared in water at 0°

Ehrlich reagent, freshly prepared:

    1.00 g of p-dimethylaminobenzaldehyde

    1.25 ml of 10 N HCl

    100 ml of glacial acetic acid

C. Formation of fructose-6-P

Glucosamine 6-phosphoric acid, 0.2 M.[6] Commercially available salts of glucosamine-6-P may be converted to the free acid on Dowex 50, H⁺ form.

NaOH, 0.4 M

Tris·HCl buffer, 0.25 M, pH 8.5

EDTA, disodium salt, 25 mM

Enzyme in 1 mM Tris·HCl, pH 7.3

HCl, 6 N

D. Assay for fructose-6-P

Fructose, 0.20 mM

Resorcinol, 0.05% in absolute ethanol

HCl, concentrated

[4] J. S. Davis and J. E. Gander, *Anal. Biochem.* **19**, 72 (1967).

[5] J. H. Roe, *J. Biol. Chem.* **107**, 15 (1934).

[6] Glucosamine-6-phosphoric acid and N-acetylglucosamine 6-phosphoric acid may be prepared according to the methods of J. J. Distler, J. M. Merrick, and S. Roseman, *J. Biol. Chem.* **230**, 497 (1958).

Procedures. a. FORMATION OF GLUCOSAMINE-6-P. The incubation mixture contains the above solutions as follows: 0.05 ml of fructose-6-P, 0.05 ml of $NH_4Cl-KOH$, 0.1 ml of Tris·HCl or PO_4 buffer, 0.05 ml of EDTA, an appropriate amount of enzyme, and water to 0.5 ml. The reaction, usually started by adding the enzyme last, is incubated at 37° for 8 min and terminated by placing the tubes in a boiling water bath for 2 min. Denatured protein is removed by centrifuging the tubes for 10 min at 32,000 *g*. Controls consist of heat-denatured (3 min at 100°) enzyme.

b. ASSAY FOR GLUCOSAMINE-6-P. Four-tenths milliliter of the above supernatant, containing up to 100 nmoles of glucosamine-6-P, is placed in a test tube and treated with 0.25 ml of borate buffer followed by 0.05 ml of cold acetic anhydride solution, mixed within 1 min of the last addition, and incubated a minimum of 4 min at room temperature. Each tube is covered with a marble, and the rack of tubes is placed in a boiling water bath for a minimum of 9 min. During this time, a stream of cold air is directed across the tubes. The tubes are then cooled in ice water and warmed to room temperature; 3 ml of Ehrlich reagent is added. The tubes are mixed, incubated for 20 min at 37°, and the optical densities are read on a spectrophotometer at 585 nm. Standards made up of glucosamine HCl are used for comparison.

c. FORMATION OF FRUCTOSE-6-P. The incubation mixture contains: 0.05 ml of glucosamine-6-P, 0.05 ml of NaOH, 0.1 ml of Tris·HCl, 0.05 ml of EDTA, enzyme, and water to 0.5 ml. The reaction is started by adding the enzyme, incubated at 37° for 8 min, and terminated by the addition of 0.05 ml of 6 *N* HCl. Denatured protein is removed by centrifugation at 20,000 *g* for 10 min. Controls are run with heat-denatured (3 min at 100°) enzyme.

d. ASSAY FOR FRUCTOSE-6-P. Five-tenths milliliter of the supernatant from the above reaction, containing up to 100 nmoles of fructose-6-P, is treated with 0.5 ml of resorcinol followed by 1.5 ml of concentrated HCl. The tubes are shaken, incubated for 8 min at 77°, cooled in ice water, and warmed to room temperature. Optical density is read at 400 nm and compared to fructose standards. Fructose and fructose-6-P yield the same amounts of color.[1]

Difficulties. The apparent K_m's for the substrates of this enzyme are on the order of 20–40 m*M* in the absence of activators. Routine incubation mixtures listed above contain nonsaturating concentrations of substrates (20 m*M*) because (1) higher concentrations of fructose-6-P and NH_4Cl lead to the formation of a nonspecific interfering red color in the hexosamine assay, and (2) routine work with saturating concentrations of sugar phosphates is expensive. Another problem involves the fact that less purified preparations contain glucose-phosphate isomerase, which converts fructose-6-P to glucose-6-P, an activator of the enzyme.

Hexosamines cannot be assayed in the presence of high amounts (50 nmoles) of glycine or glycylglycine with the assay described above. Aminoff et al. (1952)[7] have shown that the time of heating in the boiling water bath must be greatly increased if glycine buffer is substituted for borate buffer of the same pH, and it is probable that a similar effect obtains here. We also suspect that high concentrations of amino acids found in fraction A may inhibit the hexosamine assay and contribute to the greater than 100% yields obtained during purification (see the table).

Assay for Protein. Protein is measured by the biuret method or the Folin-phenol method of Lowry et al.[8]

Definition of Unit. Units of specific activity are given in milli-International Units (mIU, nanomoles/minute/milligram of protein).

Purification Procedure

All operations below are performed at 0–2° unless otherwise stated.

Step 1. Homogenization. In a large chilled mortar, 50 g of flies[2] are homogenized in 100 ml of cold 0.2 M K_2HPO_4 fortified with 1.8 mM mercaptoethanol. The rather viscous homogenate (about pH 6.8) is forced through three layers of cheesecloth to remove large particles and is then centrifuged for 40 min at 32,000 g. The supernatant is poured through 2 layers of cheesecloth to remove floating lipoidal material; the resulting reddish solution is fraction A.

Step 2. Acetone Precipitation. One-hundred milliliters of fraction A is stirred in a beaker immersed in a bath set at −7° to −10°, and 150 ml of acetone cooled to −20° is added to it over a 2-min period (60% acetone v/v). The temperature of the mixture rises to about 7–10° during the addition, and falls to −5° in about 5 min. The resulting suspension is centrifuged at 9500 g for 5 min at −10°, and the supernatant is discarded. The pellet is dissolved in 100 ml of 0.2 M KH_2–K_2HPO_4 buffer, pH 7.0, containing 1.8 mM mercaptoethanol, and recentrifuged for 15 min at 32,000 g to remove insoluble material. The brownish-red supernatant is fraction B.

Step 3. Ammonium Sulfate Precipitation. One-hundred milliliters of fraction B is stirred magnetically while 39.0 g of solid ammonium sulfate

[7] D. Aminoff, W. T. J. Morgan, and W. M. Watkins, *Biochem. J.* **51**, 379 (1952).
[8] See this series, Vol. 3 [73].

is added (60% saturation at 2°). After the salt has dissolved, the mixture is stirred an additional 15 min, the suspension centrifuged for 15 min at 32,000 g, and the supernatant carefully decanted and discarded. The pellet is dissolved in 100 ml of 10 mM potassium phosphate buffer, pH 7.2, containing 1.8 mM mercaptoethanol, and centrifuged 15 min at 32,000 g. The pale red-brown supernatant is fraction C.

Step 4. Ethanol Fractionation. One hundred milliliters of fraction C is placed in a bath at —7° to —10° with constant stirring. Fifty milliliters of cold (—20°) absolute ethanol is added over about a 2-min period (33% ethanol v/v). The temperature of the mixture rises above 0°, and when it falls to —5° it is centrifuged at 9500 g for 5 min at —10°. The precipitate is discarded, the supernatant is returned to the cold bath, and another 50 ml of cold ethanol is added (50% ethanol v/v). The suspension is stirred for 5 min, then centrifuged once more at 9500 g for 5 min at —10°. The precipitate is again discarded, the supernatant is placed in the cold bath, and a third 50-ml portion of ethanol is added (60% ethanol v/v). After being stirred for 5 min, the suspension is centrifuged as before, and the supernatant is discarded. The pellet contains most of the activity; it is suspended in 20 ml of 10 mM potassium phosphate buffer, pH 7.2, containing 1.8 mM mercaptoethanol. The solution is centrifuged for 15 min at 32,000 g to remove particulate matter, and is designated fraction D3.

Step 5. Second Ethanol Fractionation. Fraction D3 is refractionated with ethanol, using the same methods as described in step 4. The third precipitate (50–60% ethanol, v/v) is suspended in 10 ml of 10 mM potassium phosphate buffer, pH 7.3, containing 1.8 mM mercaptoethanol, and centrifuged for 15 min at 32,000 g to remove particulate matter. This slightly colored solution is denoted as fraction E3.

Summary of Purification. Calculations incorporated into the table show that fraction E3 is about 580-fold purified over fraction A. The greater than 100% yields are caused by the removal from fractions A and B of a dialyzable factor which inhibits either the enzyme or the assay for hexosamines.

Fractions A through E3 may be frozen and stored without appreciable loss of activity. Aliquots of fraction E3 may be removed and dialyzed against several changes of 1 mM Tris·HCl buffer, pH 7.3, as needed. The dialyzed extract shows no detectable loss of activity, and may be stored in the freezer. Tris does not seem to affect the enzyme, even at fairly high concentrations.

Difficulties. When fractions A, C, and D3 are treated with organic solvents, appreciable losses of activity occur unless the solvents are kept at —20°. With small amounts of material, the temperature of the solvents

PURIFICATION OF GLUCOSAMINEPHOSPHATE ISOMERASE[a]

Fraction	Volume (ml)	Protein (mg/ml)	Protein (total mg)	Glucosamine 6-phosphate (nmoles/min/mg protein)	Total activity (units)	Purification factor (fold)	Yield (%)
A. Crude extract	498	57	28,400	0.041	1,170	1.0	100
B. Acetone, 0–60% v/v	457	31.5	10,750	0.099	1,070	2.4	92
C. $(NH_4)_2SO_4$, 0–60% v/v	498	8.6	4,280	0.48	2,060	11.6	175
D3. First ethanol, 50–60% v/v	100	2.8	280	5.9	1,650	144.0	142
E3. Second ethanol, 50–60% v/v	50	1.4	70	24.0	1,670	580.0	143

[a] Reaction mixture: 50 μmoles of PO_4 buffer, 2.5 μmoles of EDTA, 20 μmoles of NH_4Cl, 10 μmoles of fructose-6-P, enzyme, in final volume of 1 ml. Incubated at 37° for 16 min.

may be higher ($-7°$), since rapid heat removal minimizes temperature changes upon addition. Losses of activity may also occur during purifica tion if a so far unidentified chelatable metal ion is present in the aqueous medium. This may be prevented by adding 5 mM disodium EDTA to the 0.2 M phosphate buffer, and 1 mM EDTA to the 10 mM buffer (all three media containing 1.8 mM mercaptoethanol).

Properties of the Enzyme

Specificity. The purified enzyme is specific for fructose-6-P and an unidentified form of inorganic ammonia (at pH 8.4, about 78% exists as NH_4^+ and 22% as NH_4OH). Less pure fractions are active with glu cose-6-P because of the presence of glucose-phosphate isomerase which converts this compound to fructose-6-P. The following compounds are not substrates: D-fructose, L-glutamine, L-asparagine, and L-alanine.

Molecular Weight. The molecular weight of the enzyme is about 154,000, as determined by sucrose gradient centrifugation.

Activators and Inhibitors. The reaction in the direction of the forma tion of glucosamine-6-P is activated by glucose-6-P, N-acetylglucos amine-6-P, 2-deoxyglucose-6-P, 6-phosphogluconate, and inorganic phos phate. It is also slightly stimulated by inorganic arsenate, but is unaffected by glucose-1-P. Mannose-6-P is inhibitory.

The rate of formation of fructose-6-P is activated by N-acetylglucos amine-6-P, little affected by glucose-6-P and sodium acetate, and slightly inhibited by inorganic phosphate.

Effects of pH. In the absence of activators, the reaction leading to the formation of glucosamine-6-P has a pH optimum of about 7.1–7.3 in Tris·HCl and Tris·maleate buffers. In the presence of inorganic phos phate and glucose 6-phosphate, the pH optimum is shifted to about 8.2–8.4. The pH optimum for fructose-6-P formation has not been mea sured, but preliminary experiments indicate little change in activity be tween pH 6.8 and 8.3.

Equilibrium Constant. In an aqueous medium, the formation of fructose-6-P and inorganic ammonia are favored, and the equilibrium constant is in the range of 0.1 to 0.2 M for $K_q = $ (fructose-6-P) (NH_4Cl)/ (glucosamine-6-P), water taken as unity.

Kinetic Constants. In the absence of activators, the apparent K_m's are: fructose-6-P, 36 mM; NH_4Cl, 25 mM; glucosamine-6-P, 17 mM. The apparent $K_{activator}$ for glucose-6-P is 9.5 mM. In the presence of 10 mM glucose-6-P, the K_m for fructose-6-P is 4.8 mM and for NH_4Cl is 4.2 mM, the K_m for glucosamine-6-P remaining unchanged.

Occurrence. This enzyme may be found in house fly pupae and in pupae and adults of *Sarcophaga bullata* Parker (flesh fly). It occurs in a variety of other organisms, and three accounts are contained in this series: Vol. 5 [56] and [57],Vol. 9 [100].

[86] N-Acetylglucosamine 2-Epimerase from Hog Kidney

By ASIS DATTA

$$N\text{-Acetylglucosamine} \xleftrightarrow{\text{ATP}} N\text{-acetylmannosamine}$$

An ATP-dependent 2-epimerase which catalyzes the interconversion of N-acetylglucosamine and N-acetylmannosamine has been purified from hog kidney.[1,2] Kinetic studies suggest that ATP is not absolutely essential for the activity of 2-epimerase, but it exerts a stimulatory effect on the enzyme, causing approximately a 20-fold increase in its activity.[2] Furthermore, the 2-epimerase, like a regulatory enzyme, possesses two distinct but mutually interacting sites—a catalytic site that binds the substrate, and an "allosteric" site that binds the effector molecule, ATP.

Reagents

N-Acetyl-D-mannosamine, 0.1 M

ATP, 0.1 M. The crystalline sodium salt is dissolved in water, the pH is adjusted to 7.6 with NaHCO$_3$ and the solution is stored at $-18°$.

MgCl$_2$, 0.1 M

Tris·HCl buffer, 0.5 M, pH 7.4

N-Acetylglucosamine kinase[3]

---

[1] S. Ghosh and S. Roseman, *J. Biol. Chem.* **240**, 1531 (1965).

[2] A. Datta, *Biochemistry* **9**, 3363 (1970).

[3] Preparation of N-acetylglucosamine kinase: N-acetylglucosamine kinase, used in the assay of 2-epimerase is prepared from hog spleen by using the following operations. Hog spleen (100 g) is homogenized with 200 ml of 30 mM potassium phosphate buffer (pH 7.6) containing 1 mM EDTA and 10 mM 2-mercaptoethanol in a Waring Blendor. The supernatant (150 ml), collected by centrifugation at 16,000 g for 30 min is diluted with an equal volume of water, and the enzyme is precipitated by adding 22.5 ml of a 2% protamine sulfate solution. The residue collected by centrifugation is washed with 75 ml of 50 mM potassium phosphate buffer (pH 7.6). The enzyme is extracted from the residue with 75 ml each of 75, 75, and 100 mM potassium phosphate buer (pH 7.6). These three extracts are then combined, brought to 32.5% saturation with solid ammonium sulfate, and any precipitate formed at this stage is rejected by centrifugation. The supernatant

Reagents for the estimation of ADP by pyruvate kinase-lactic dehydrogenase system are described in volume 42 [10].

Assay Method

Principle. The epimerase activity is assayed by measuring the ADP
formed in the reaction, by coupling the 2-epimerase and the specific *N*-
acetylglucosamine kinase in the following manner:

$$N\text{-Acetylmannosamine} \xrightleftharpoons{\text{2-epimerase}} N\text{-acetylglucosamine}$$

$$N\text{-Acetylglucosamine} + \text{ATP} \xrightarrow{\text{kinase}} N\text{-acetylglucosamine-6-P} + \text{ADP}$$

$$N\text{-Acetylmannosamine} + \text{ATP} \rightarrow N\text{-acetylglucosamine-6-P} + \text{ADP}$$

The assay consists of: (1) incubation of the 2-epimerase with *N*-acetyl-
mannosamine (and ATP, if necessary), (2) treatment of the reaction
product with an excess of the specific *N*-acetylglucosamine kinase
(ATPase free) and ATP for the complete conversion of any *N*-acetyl-
glucosamine formed in step 1 into *N*-acetylglucosamine-6-P and the formation of an equivalent amount of ADP, and (3) the estimation of ADP
enzymically by the pyruvate kinase–lactic dehydrogenase assay procedure. Since ATP is not degraded in step 1,[1,2] the amount estimated in
step 3 equals the amount of *N*-acetylglucosamine formed by
2-epimerization.

The sensitivity of the coupling system is such that as little as 0.01
μmole of *N*-acetylglucosamine can be measured by the production of
ADP. The incubation mixture without either 2-epimerase, kinase, or ATP
gives insignificant blank values. Furthermore, *N*-acetylglucosamine
added to the full incubation mixture without 2-epimerase is fully converted into *N*-acetylglucosamine-6-P and ADP.

Procedure. The incubation mixtures for estimating the ATP-independent enzyme activity contain the following components in a final volume

---

is brought to a saturation level of 50% with solid ammonium sulfate and the precipitate formed is collected by centrifugation, and dissolved in 10 ml of 20 m*M* potassium
phosphate buffer (pH 7.6). This concentrated *N*-acetylglucosamine kinase fraction
contains some ATPase activity, which is removed by heating this fraction at 60°
for 2 min. Coagulated proteins are removed by centrifugation and the supernatant
fluid, dispensed in several small test tubes, is kept frozen until used. These enzyme
fractions are free from ATPase and other interfering enzymes such as *N*-acetylman-
nosamine 2-epimerase and *N*-acetylmannosamine kinase and are quite active for
the assay of the 2-epimerase.

[4] J. L. Reissig, J. L. Stromimger, and L. F. Leloir, *J. Biol. Chem.* **217**, 959 (1955).

of 0.25 ml: 0.1 ml of *N*-acetylmannosamine, 0.025 ml of Tris·HCl buffer, pH 7.4; 0.025 ml of $MgCl_2$; enzyme fraction; and distilled water. For the estimation of the ATP-dependent activity, the incubation mixtures contain the following in a final volume of 0.25 ml: 0.02 ml of *N*-acetyl-mannosamine; 0.01 ml of ATP; 0.025 ml of Tris·HCl buffer, pH 7.4; 0.025 ml of $MgCl_2$; and enzyme fraction. After incubation at 37° for 30 min, the reaction is stopped by heating at 100° for 2 min. Each assay mixture is treated with approximately 2 units (1 unit of enzyme activity catalyzes the phosphorylation of 1 μmole of *N*-acetylglucosamine under the assay conditions described above) of specific *N*-acetylglucosamine kinase (freed from ATPase and other interfering enzymes) and 0.01 ml of ATP (when ATP is not added to step 1). The volume of the mixture is made up to 0.35 ml and incubated again at 37° until all *N*-acetylglu-cosamine is converted into its 6-phosphate ester (usually in less than 15 min). The reaction is stopped again by heating at 100° for 2 min; the mixture is cooled, and the volume is made up to 1.0 ml by adding distilled water. An aliquot of this diluted mixture (0.1 ml) is assayed enzymically for the quantitative determination of ADP formed in each assay tube based on the lactic dehydrogenase–pyruvate kinase procedure as described in Vol. 42 [10]. Control tubes contain heat-denatured en-zyme or lack *N*-acetylmannosamine. The missing component is, however, added to the control tube immediately before the reaction is stopped.

Specific activity (both ATP-dependent and ATP-independent) has been defined as micromoles of *N*-acetylglucosamine formed per minute per milligram of protein under the assay conditions described above.

One unit of ATP-dependent enzyme activity catalyzes the epimeriza-tion of 1 μmole of *N*-acetylmannosamine per minute under the assay conditions stated above.

Purification Procedure

Unless otherwise stated, all operations are conducted at temperatures between 0° and 4° and all buffer systems used in the fractionation pro-cedures contain 1 m*M* EDTA and 10 m*M* 2-mercaptoethanol. The first two steps of the purification procedure, the preparation of crude extract and of protamine extract, are similar to those described previously by Ghosh and Roseman.[1]

Crude Extract and Protamine Extract. Kidney cortex (100 g) is homogenized in a Waring Blendor with 200 ml of 3 m*M* potassium phos-phate buffer (pH 7.6), and the supernatant is collected after centrifuga-tion at 16,000 *g* for 30 min (crude extract). The crude extract (150 ml) is diluted with equal volume of cold distilled water and treated slowly with 4.5 ml of 2% protamine sulfate solution with constant stirring; the precipitate formed at this stage is discarded by centrifugation. The supernatant is further treated with 15 ml of a 2% protamine sulfate, and the precipitate formed is collected by centrifugation. After preliminary washing with 100 ml of 10 mM potassium phosphate buffer (pH 7.6), the active enzyme is extracted from the precipitate twice with 45 ml of 25 mM potassium phosphate buffer (pH 7.6).

Bentonite Adsorption. The combined extracts of protamine precipitate (90 ml) are gently stirred for 10 min with 900 mg of bentonite (suspended in 10 ml of 1 mM EDTA), and the supernatant is collected after centrifugation. Concentration of the 2-epimerase in the bentonite supernatant and the protamine extract fractions is too low to permit satisfactory assay of the ATP-independent enzyme activity. Therefore, the enzyme in the supernatant is next concentrated by adsorbing the enzyme on a small DEAE-cellulose column and eluting it with a small volume of 0.2 M potassium phosphate buffer (pH 7.6; see DEAE-cellulose step). The concentrated enzyme eluates are dialyzed for 24 hr against 25 mM potassium phosphate buffer (pH 7.6) and designated as "protamine concentrate" and "bentonite concentrate."

DEAE-Cellulose. Bentonite supernatant fraction (100 ml) is applied to a DEAE-cellulose column (1.5 × 10 cm) which has been previously equilibrated with 20 mM KCl–10 mM potassium phosphate buffer (pH 7.6). The column is washed with 200 ml of 50 mM potassium phosphate buffer (pH 7.6), then the enzyme is eluted using a linear gradient of 50 mM to 0.15 M potassium phosphate buffer (pH 7.6) containing 20 mM KCl, 1 mM EDTA, and 10 mM 2-mercaptoethanol at a flow rate of 2 ml/min (100 ml each of the two respective elution media). Fractions are collected, 5 ml each, and assayed for ATP-dependent epimerase activity. The enzyme is eluted between 90 mM and 0.11 M potassium phosphate buffer (pH 7.6), and these fractions are pooled (DEAE eluate, 45 ml). The pooled fractions are dialyzed against 5 mM potassium phosphate buffer (pH 7.6) for 5 hr, and the dialyzed preparation is applied again to a small DEAE-cellulose column (1 × 2 cm) equilibrated as before; the enzyme is eluted with 3 ml of 0.2 M potassium phosphate buffer (pH 7.6). The concentrated enzyme fraction is again dialyzed against 5 mM potassium phosphate buffer (pH 7.6) for 12 hr, and the dialyzed preparation is designated as DEAE concentrate.

Calcium Phosphate Gel. The DEAE-cellulose concentrate (3 ml) is treated with 35 mg (dry weight) of calcium phosphate gel. After gentle stirring of the mixture for 10 min, the suspension is centrifuged. The supernatant fluid is designated as calcium phosphate gel fraction.

The purification procedure is summarized in the table.

PURIFICATION OF N-ACETYL-D-GLUCOSAMINE 2-EPIMERASE FROM HOG KIDNEY

	Total activity		Specific activity		Ratio of specific activity A:B	Puri-fication factor
Fraction	Units	Yield (%)	ATP-de-pendent (A)	ATP-inde-pendent (B)		
1. Crude extract[a]	20	100	0.004	—	—	1
2. Protamine concentrate	16	80	0.103	0.0053	19.4	26
3. Bentonite concentrate	16	80	1.00	0.05	20	250
4. DEAE-cellulose eluate	9.3	46.6	5.8	0.29	20	1450
5. Calcium phosphate gel	7.5	37.6	6.0	0.3	20	1500

[a] The assay procedure described above is not suitable for measuring the enzyme activity of crude extracts. For the assay of impure preparations, the second incubation mixture is treated with $ZnSO_4$ and $Ba(OH)_2$ solution for the quantitative removal of N-acetylglucosamine-6-P, and then the amount of N-acetylmannosamine disappearing due to epimerization is estimated by the Morgan-Elson color reaction as described in this series, Vol. 42 [10].

Properties

N-Acetyl-D-glucosamine 2-epimerase activity is demonstrated in the absence of ATP by using high enzyme and substrate concentrations. The ATP-dependent and ATP-independent epimerase activities cannot be separated, and the ratio of the specific activities remains nearly constant ($\sim$20) throughout the range of 1500-fold purification of the enzyme. K_m for N-acetylmannosamine was 9 mM in the absence of ATP, and 1.7 mM in the presence of saturating concentration of ATP. pH optima for both ATP-dependent and ATP-independent activities are 7.4. The ATP saturation curve of the 2-epimerase, which is sigmoidal, suggests cooperative binding of ATP molecules. A Hill plot yields an interaction coefficient of $n = 2.6$. Kinetic evidence suggests the presence of homotropic interactions between substrate molecules. The allosteric effect of ATP is confirmed since specific disruption of the ATP binding site is possible without affecting the ATP-independent catalytic activity of the enzyme. The desensitization of 2-epimerase to ATP is brought about by controlled heat and by p-chloromercuribenzoate treatment. The selective destruction of ATP effect on 2-epimerase strongly suggests that the site for binding the stimulator, ATP, is largely independent of the site for binding the substrates.

Acknowledgment

The author acknowledges Dr. S. Ghosh, Bose Institute, Calcutta for his valuable suggestion.

[87] L-Ribulose-5-phosphate 4-Epimerase from *Aerobacter aerogenes*

By Jean Deupree and W. A. Wood

L-Ribulose 5-phosphate $\rightleftarrows$ D-xylulose 5-phosphate

Assay Method

Principle. L-Ribulose 5-phosphate 4-epimerase is assayed spectrophotometrically in a coupled enzyme assay according to the following reaction sequence:

$$\text{L-Ribulose 5-phosphate} \xrightarrow{\text{4-epimerase}} \text{D-xylulose 5-phosphate}$$

$$\text{D-Xylulose 5-phosphate} \xrightarrow[\substack{\text{thiamine pyrophosphate} \\ \text{Mg}^{2+}, \text{ arsenate}}]{\text{phosphoketolase}}$$

$$\text{D-glyceraldehyde 3-phosphate} + \text{acetate}$$

$$\text{D-Glyceraldehyde 3-phosphate} \xrightarrow[\text{isomerase}]{\text{triosephosphate}} \text{dihydroxyacetone phosphate}$$

$$\text{Dihydroxyacetonephosphate} + \text{NADH} \xrightarrow[\text{dehydrogenase}]{\alpha\text{-glycerolphosphate}}$$

$$\alpha\text{-glycerolphosphate} + \text{NAD}$$

With a limiting amount of epimerase and excess coupling enzymes, the rate of D-xylulose 5-phosphate formation is directly proportional to the rate of oxidation of NADH which is determined by measuring the decrease in absorbance at 340 nm.

Reagents

Sodium L-ribulose 5-phosphate, 39 mM [1]
NADH, 8 mM
Imidazole, 0.5 M, adjusted to pH 7.0
MgCl$_2$, 50 mM
Sodium arsenate
Sodium glutathione
Thiamine pyrophosphate

[1] See this series, Vol. 9 [9–11].

α-Glycerolphosphate dehydrogenase–triosephosphate isomerase mixture, 0.33 mg/ml[2]

D-Xylulose-5-phosphate phosphoketolase, 7 units/ml (see below)

All reagents are prepared separately. A stock solution is prepared by mixing 2 ml of imidazole buffer, 1 ml of $MgCl_2$, 30 mg of sodium arsenate, 5 mg of thiamine pyrophosphate, 38 mg of glutathione, and water to a total volume of 10 ml. The solution is stored at 4°. The L-ribulose 5-phosphate, NADH, and phosphoketolase solutions are stored frozen.

The barium salt of L-ribulose 5-phosphate is converted to the sodium salt in the following manner. Approximately 1 ml of Dowex resin 50-X8 (25–50 mesh, H^+ form) is washed three times with triple-distilled water to remove any yellow color which has leached from the resin. The excess water is poured off, and 38 mg of Ba-L-ribulose 5-phosphate (75% pure) is mixed with the resin until the salt dissolves. The L-ribulose 5-phosphate is eluted by washing the resin on a 2-ml funnel containing a filter disk until approximately 1.5 ml of eluate are collected. The pH of the eluate is adjusted to between 5.5 and 6.0 with 1 N NaOH, and the volume is adjusted to 2.0 ml with triple-distilled water.

D-Xylulose-5-phosphate phosphoketolase can be purified from any strain of *Leuconostoc mesenteroides* either by the procedure of Goldberg and Racker[3] or as described below. Cultures are grown at 30° in media containing 10 g of trypticase, 5 g of yeast extract, 6 g of KH_2PO_4, 2 g of ammonium citrate, 20 g of dextrose, 1 g of Tween 80, 20.5 g of anhydrous sodium acetate, 0.6 g of $MgSO_4$, 0.1 g of $MnSO_4$, 0.03 g of $FeSO_4$ per liter and adjusted to pH 5.8 with glacial acetic acid. Ten milliliters of a 12-hr culture is used to inoculate 2 liters of broth in a Fernbach flask. Twenty liters of broth in a carboy are inoculated with 2 liters of a 12-hr growth. After 24 hr, the bacteria are producing gas and are harvested in a small Sharples centrifuge. The cells are stored frozen at −20°. The crude extract is prepared by sonicating 30 g (wet weight) of cells suspended in 60 ml of water in a chilled 200-W Raytheon 10 kHz sonicator. The cell debris is removed by centrifugation at 18,000 g for 15 min. The pH of the crude extract (83 ml) is adjusted to pH 4.55 with 1 N acetic acid. The solution is centrifuged at 18,000 g for 30 min. The pH of the supernatant is readjusted to 6.0 with 1 M phosphate buffer, pH 8.0, and applied to a DEAE-cellulose column (1.5 × 20 cm) which had been equilibrated with 50 mM phosphate buffer, pH 6.0, and 1 mM thioglycerol. The column is washed with 50 mM phosphate buffer, pH 6.0, and 1 mM thioglycerol until the absorption of the eluate

---

[2] CalBiochem, 10933 N. Torrey Pines Road, La Jolla, California 92037.
[3] See this series, Vol. 9 [90].

at 280 nm is less than 0.04. The phosphoketolase is eluted using a linear gradient containing 300 ml of 0.1 M and 300 ml of 0.6 M phosphate buffer, pH 6.0, and 1 mM thioglycerol. Fractions 57 through 76 (7.7 ml fractions) contain the phosphoketolase; these are pooled, and the enzyme is precipitated with 3 M ammonium sulfate. The precipitate is back-extracted with 2.5 M, 2.0 M, and 1.5 M ammonium sulfate in 0.1 M phosphate buffer, pH 6.0, and 10 mM mercaptoethanol. The phosphoketolase activity is recovered in the 2.0 M and the 1.5 M ammonium sulfate fractions. The appropriate fractions are stored frozen in 0.5-ml aliquots and used as such. Enzyme activity is lost on thawing and refreezing. This procedure resulted in a 72-fold purification with a 43% recovery of activity (Table I).

Assay Procedure. The assay is conducted in a microcuvette of 0.5 ml capacity and 1-cm light path containing 0.1 ml of stock solution, 0.01 ml of triosephosphate isomerase–α-glycerolphosphate dehydrogenase, 0.02 ml of L-ribulose 5-phosphate, 0.01 ml of NADH, 0.005 ml of phosphoketolase, suitable amounts of 4-epimerase and water to make 0.2 ml. The reaction is started by adding the 4-epimerase. If the L-ribulose 5-phosphate is contaminated with D-xylulose 5-phosphate, an initial drop in 340 nm absorbance will occur as the D-xylulose 5-phosphate is converted to α-glycerolphosphate. Therefore, it is necessary to monitor the absorbance until the rate of NADH oxidation returns to a negligible or a linear rate prior to the addition of 4-epimerase. The amount of NADH oxidase activity is determined by omitting L-ribulose 5-phosphate from an otherwise complete reaction mixture. The rate due to 4-epimerase is the total rate of NADH oxidation minus the rate of NADH oxidation prior to adding the epimerase and minus that due to NADH oxidase. The activity of phosphoketolase, α-glycerolphosphate dehydrogenase, and triosephosphate isomerase must be ascertained periodically to ensure that an

TABLE I

PURIFICATION OF PHOSPHOKETOLASE FROM *Leuconostoc mesenteroides*

Step	Total volume (ml)	Total activity (units)	Specific activity (units/mg protein)
Crude extract	83	332	0.022
Acid step	90	260	
DEAE-Cellulose	146	216	1.5
Ammonium sulfate, 2.0 M	10.2	55.5	1.8
Ammonium sulfate, 1.5 M	10.2	144	1.6

excess will be present in the assay. The assay is linear up to a rate of 0.08 ΔA per minute (2.6 $\times$ 10⁻⁴ unit).

Alternate Assay. The reaction is carried out in 2 steps: (1) the epimerase is incubated with L-ribulose 5-phosphate and buffer for a given period and the reaction is stopped by heating, and (2) the amount of D-xylulose 5-phosphate formed is determined. The 0.1-ml reaction mixture containing 0.5 μmole of L-ribulose 5-phosphate and 5 μmoles of glycylglycine buffer (pH 8.0) is preincubated at 37° for 10 min before the 4-epimerase is added. The reaction is stopped after 4 min by adding 2 μl of concentrated acetic acid and heating in a boiling water bath for 1 min. The acidity is neutralized with NH₄OH and a 20-μl aliquot is added to a 0.5-ml cuvette containing the constituents of the coupled enzyme assay system except for 4-epimerase and L-ribulose 5-phosphate. Other conditions are: total volume, 0.2 ml; 37°, and pH 7.0. The reaction is started by adding phosphoketolase. The amount of D-xylulose 5-phosphate formed is determined by measuring the total decrease in absorption at 340 nm. Suitable corrections in absorbance values are applied to compensate for dilution by the phosphoketolase. The assay was valid up to 0.4 μmole of D-xylulose 5-phosphate formed per milliliter of initial assay mixture.

Definition of Unit and Specific Activity. A unit of enzyme is defined as 1 μmole of D-xylulose (or NAD) formed per minute under these conditions. Specific activity is expressed as units per milligram of protein. Protein is determined by the ratio of absorbances at 280 and 260 nm.[4]

Purification Procedure[5]

Cell Culture. For preparation of extracts containing L-ribulose-5-phosphate 4-epimerase, mutants of *Aerobacter aerogenes* PRL-R3 may be cultured on media composed of 0.15% of KH₂PO₄, 1.42% Na₂HPO₄, 0.3% (NH₄)₂SO₄, 0.02% MgSO₄, 5 × 10⁻⁴% FeSO₄, 0.005% uracil, and 0.5% L-arabinose. The MgSO₄, FeSO₄, and L-arabinose solutions are sterilized separately in concentrated form and added just before inoculation.

Either wild-type *A. aerogenes* PRL-R3, or the uracil-negative constitutive mutant (u⁻i⁻)[6] can be used. Three to five times more L-ribulose

---

[4] O. Warburg and W. Christian, *Biochem. Z.* 310, 384 (1941).

[5] J. D. Deupree and W. A. Wood, *J. Biol. Chem.* 245, 3988 (1970).

[6] Strain PRL-R3 was originally obtained from The Prairie Regional Research Laboratory, National Research Council, Saskatoon, Saskatchewan, Canada. It has been widely used in the study of carbohydrate and polyol metabolism. The uracil-negative, L-arabinose operon constitutive mutant of *A. aerogenes*, PRL-R3, was supplied by Dr. R. P. Mortlock, Department of Microbiology, University of Massachusetts, Amherst, Massachusetts.

5-phosphate 4-epimerase is produced by the u⁻i⁻ mutant than by the wild-type grown in 0.5% L-arabinose. An initial inoculum is grown with shaking at 37° in four culture tubes each containing 10 ml of medium. After 24 hr, the inoculum is transferred to four 1-liter quantities of growth medium in 3-liter Fernbach flasks and these are shaken at 37° for 12 hr. The four liters of culture are used to inoculate 100 liters of medium in a 130-liter fermenter (New Brunswick Co., Inc.). The culture is grown at 37° at a stirring speed of 200 rpm and an aeration rate of 2.5 ft³/min for 4–5 hr at which time the turbidity at 660 nm with a 1-cm light path is equivalent to 1.8 A. The cells are harvested in a continuous centrifuge (Sharples Corporation) and stored at −20°. The medium is not cooled during harvesting since this organism produces a polysaccharide upon cooling.

Preparation of Cell-Free Extract. Cells are suspended in 2 volumes of cold quartz-distilled water (380 g wet weight + 760 ml H_2O) and ruptured by treatment of 100-ml portions in a chilled 200-W Raytheon 10 kHz sonicator for 20 min. Alternatively, 1 mg of DNase is added, and the entire cell suspension is forced twice through a prechilled laboratory homogenizer (Manton-Gaulin Manufacturing Co., Inc., Everett, Massachusetts) at 7000 psi. The suspension from either process is centrifuged at 20,000 g for 20 min and the pellet is discarded. The pH of the supernatant solution is adjusted to 7.2 with ammonium hydroxide.[7]

The enzyme solution is kept at 4° for this and all subsequent steps.

DEAE-Cellulose Chromatography. The crude extract (1100/ml, 51,500 mg of protein) is stirred for 20 min with approximately 320 g (dry weight) of DEAE-cellulose (Whatman DE-32), which had previously been equilibrated with 75 mM phosphate buffer, pH 7.2. The DEAE-cellulose, with enzymes bound to it, is poured into a column (8.5 × 35 cm) and washed with 75 mM phosphate buffer, pH 7.2, until the absorbance of the eluate at 280 nm is less than 1.0. The protein is eluted from the column with a linear gradient prepared with 1500 ml each of 75 mM and 0.50 M potassium phosphate buffer, pH 7.2. The 4-epimerase is eluted with approximately 0.35 M phosphate.

Ammonium Sulfate Back Extraction. The fractions from the DEAE-cellulose column containing 4-epimerase are pooled and ammonium sulfate and EDTA are added to 2.8 M and 1 mM, respectively. After stirring for 1 hr, the solution is centrifuged at 26,000 g for 15 min. The resulting pellet is extracted with 2.0 M, 1.6 M, 1.1 M, and 0.5 M ammonium sulfate

---

[7] An accurate estimate of specific activity of crude extract is difficult to obtain owing to the NADH oxidase activity present. Occasionally, the specific activity of the fresh crude extract may be three to five times higher than expected. This high activity decreased on storage or on subsequent purification.

in 0.1 M phosphate buffer, pH 7.2, and 1 mM EDTA by stirring with 40 ml of each ammonium sulfate solution for 30 min followed by centrifugation at 26,000 g for 15 min. About 70% of the 4-epimerase is extracted with 1.1 M ammonium sulfate, and this fraction is dialyzed for 6 hr against two 1-liter changes of distilled water.

Calcium Phosphate Gel Fractionation. The protein is diluted to 10 mg/ml (370 ml), and calcium phosphate gel is added slowly at a ratio of 3 mg (dry weight) of gel per milligram of protein. After stirring for 20 min, the gel is pelleted by centrifugation. The 4-epimerase is eluted by washing the gel several times with 60 ml of 4 mM phosphate buffer, pH 7.2. The 4-epimerase is salted out by adjusting the phosphate concentration to 0.1 M, pH 7.2, and adding ammonium sulfate to 2.8 M. The precipitate is collected by centrifugation, dissolved in a minimal volume (5.0 ml) of 50 mM phosphate buffer, pH 7.2, and dialyzed for 6 hr against three 1-liter changes of 50 mM phosphate buffer, pH 7.2.

Chromatography on Sephadex G-200. The enzyme (400 mg) is applied to a column of Sephadex G-200 (2.5 × 100 cm) which had been equilibrated with 50 mM phosphate buffer, pH 7.2. The 4-epimerase elutes after 240 ml of buffer has passed through the column. The fractions containing most of the activity are pooled, and the enzyme is concentrated by the addition of ammonium sulfate to 2.8 M. The precipitate is dissolved in 20 ml of 50 mM phosphate buffer, pH 8.0, and dialyzed against two 1-liter volumes of 50 mM KCl and 50 mM phosphate buffer, pH 8.0.

Chromatography on DEAE-Sephadex G-50. The 4-epimerase (51 mg) is applied to a column of DEAE-Sephadex G-50 (50–150 mesh, 0.9 × 10 cm) which has been equilibrated with 50 mM phosphate buffer, pH 8.0, and 50 mM KCl. The enzyme is eluted with a linear KCl gradient in 50 mM phosphate, pH 8.0, consisting of 40 ml each of 50 mM and 0.40 M KCl. A flow rate of 1.35 ml/hour is maintained, and 2.1 ml fractions are collected. Protein elutes between fractions 68 and 124, and 4-epimerase activity is present between fractions 100 and 124; however, only fractions 106–124 are pooled to eliminate as much protein contamination as possible from the adjacent protein peak.

Crystallization. The 4-epimerase is precipitated by adding ammonium sulfate to 2.5 M as before. The precipitate obtained is back-extracted with 2 ml each of 2.0 M, 1.8 M, 1.4 M, and 1.2 M ammonium sulfate, pH 9.0, in 0.1 M glycine buffer as before. After 24 hr at 0°, fine, colorless, needle-shaped crystals appeared in the 1.6 M ammonium sulfate fraction. The crystals were washed with 1.8 M ammonium sulfate and dissolved in 50 mM glycine buffer, pH 9.0. Recrystallization was carried out by back-extraction in the same manner.

The purification procedure is summarized in Table II. The crystalline

TABLE II
Typical Purification of L-Ribulose-5-P 4-Epimerase from
Aerobacter aerogenes[a,b]

Step	Total volume (ml)	Total activity (units)	Specific activity (units/mg protein)	Recovery (%)	Times purified
Crude extract	1100	1700[c]	0.029	—	—
DEAE-cellulose	820	1700	0.41	100	14
Ammonium sulfate back extraction	200	1200	0.40	70	14
Calcium phosphate gel	5.4	1200	3.0	70	102
Sephadex G-250	23	330	6.5	19	226
DEAE-Sephadex	8.5	320	13.0	19	445
First crystals[d]	5.9	104	12.0	6	445
Second crystals[e]	—	—	14.5	—	—

[a] J. D. Deupree and W. A. Wood, *J. Biol. Chem.* **245**, 3988 (1970).
[b] From 330 g (wet weight) of cells.
[c] Although the assay indicated that the crude extract contained 600 units, it was assumed for the purpose of calculations that 1700 units were present.
[d] Performed on a small portion of the total fraction. The values shown for first crystals are those calculated for the total fraction.
[e] Performed on a portion of first crystals.

protein was 445-fold purified over the crude extract. The 4-epimerase with a specific activity of 13 units/mg protein is pure as determined by recrystallization, polyacrylamide gel electrophoresis, and by high-speed sedimentation equilibrium analysis. However, as noted below, when Mn^{2+} is added, the specific activity increases to 70 units per milligram of protein.

Properties

The crystalline 4-epimerase is devoid of pyridine nucleotide or other coenzymes, and reduced or oxidized pyridine nucleotides had no effect on velocity.[5] No evidence has been found for an oxidation-reduction mechanism using an amino acid as donor-receptor. Following treatment with EDTA, activity is lost, and specific divalent metal ions reactivate in the order: $Mn^{2+} > Co^{2+} > Ni^{2+} > Ca^{2+} > Zn^{2+} > Mg^{5+}$.[8] The crystalline epimerase has a specific activity of 70 when treated with EDTA and then assayed at pH 8.0 in the two-step assay containing optimal Mn^{2+}. This is contrasted with a specific activity of 13 in the continuous assay at pH 7.0 in the presence of Mg^{2+} as required for phosphoketolase.

[8] J. D. Deupree and W. A. Wood, *J. Biol. Chem.* **247**, 3093 (1972)

The K_m for L-ribulose 5-phosphate is 0.1 mM and the ratio of D-xylulose 5-phosphate to L-ribulose 5-phosphate at equilibrium is 1.86.[9] The pH optimum is between 8.5 and 9.5, with about 40% of maximal activity at pH 7.0 and no activity between pH 5.0 and 6.0. The weight average molecular weight obtained by high speed equilibrium analysis was 1.14×10^5.[5]

[9] M. J. Wolin, F. J. Simpson, and W. A. Wood, *J. Biol. Chem.* **232**, 559 (1958).

[88] L-Ribulose-5-phosphate 4-Epimerase from *Escherichia coli*

By WILLIAM O. GIELOW and NANCY LEE

L-Ribulose 5-phosphate $\rightleftarrows$ D-xylulose 5-phosphate

In *Escherichia coli* B/r, L-ribulose-5-phosphate 4-epimerase (EC 5.1.3a), which catalyzes the conversion of L-ribulose 5-phosphate to D-xylulose 5-phosphate, is the third enzyme in the pathway of L-arabinose metabolism.[1]

$$\text{L-Arabinose} \xrightarrow{\text{isomerase (EC 5.3.1.4)}} \text{L-ribulose}$$

$$\text{L-Ribulose} + \text{ATP} \xrightarrow{\text{kinase (EC 2.7.1.16)}} \text{L-ribulose 5-phosphate} + \text{ADP}$$

$$\text{L-Ribulose 5-phosphate} \xrightarrow{\text{4-epimerase}} \text{D-xylulose 5-phosphate}$$

The purification and characterization of these enzymes have been reported.[2-4]

Assay Method

A coupled enzyme assay is used to measure 4-epimerase activity. L-ribulose 5-phosphate is converted to 3-phosphoglycerate with the concomitant reduction of NAD$^+$ by including phosphoketolase (isolated from *Leuconostoc mesenteroides*[5]) and commercial glyceraldehyde phosphate

[1] E. Englesberg, R. L. Anderson, R. Weinberg, N. Lee, P. Hoffee, G. Huttenhauer, and H. Boyer, *J. Bacteriol.* **84**, 137 (1962).

[2] J. W. Patrick and N. Lee, *J. Biol. Chem.* **243**, 4312 (1968).

[3] N. Lee and I. Bendet, *J. Biol. Chem.* **242**, 2043 (1967).

[4] N. Lee, J. W. Patrick, and M. Masson, *J. Biol. Chem.* **243**, 4700 (1968).

[5] E. C. Heath, J. Hurwitz, B. L. Horecker, and A. Ginsberg, *J. Biol. Chem.* **231**, 1009 (1958).

dehydrogenase in the reaction mixture. The 4-epimerase converts L-ribulose 5-phosphate to D-xylulose 5-phosphate, which in turn is converted to glyceraldehyde 3-phosphate and acetate by phosphoketolase (EC 4.1.2.9). Glyceraldehyde-phosphate dehydrogenase (EC 1.2.1.12) and arsenate convert glyceraldehyde 3-phosphate and NAD⁺ to 3-phosphoglycerate and NADH.

The reaction mixture contains (in a total of 0.3 ml) 15 μmoles of sodium glycylglycine (pH 7.5), 3 μmoles of reduced sodium glutathione, 0.4 μmole of NAD⁺, 1 μmole of MgCl$_2$, 1 μmole of ATP, 3 μmoles of Na$_3$AsO$_3$, 0.1 μmole of thiamine pyrophosphate, and excess glyceraldehyde phosphate dehydrogenase (around 20 units) and phosphoketolase (around 4–5 units) in a cuvette at 37°.

The reaction is started by adding 1–10 μl of enzyme solution, mixing well, and following the increase in absorbance at 340 nm in a recording spectrophotometer with the cuvette housing maintained at 37° using a circulating water bath. A control cuvette without added enzyme solution serves to establish endogenous reduction of NAD⁺. An alternative two-step assay useful in studying the properties of 4-epimerase (devoid of the complications inherent in the coupled enzyme reaction mixture) can be used.[4] A unit of 4-epimerase is that amount which transforms 1 μmole of NAD⁺ to NADH per hour (a 1 mM solution of NADH has an absorbance value of 6.22 at 340 nm). Epimerase assays have been successfully performed on small-scale crude extracts of *E. coli* strains after duplicating the MnCl$_2$ precipitation step. Operationally, the specific activity of the extract must exceed 3×10^{-2} unit per milligram of protein so that little enough extract can be added to the reaction mixture to avoid serious interference by contaminating NADH oxidase. Specific activity is expressed as units per milligram of protein. Protein concentration is measured either by the method of Lowry *et al.* with bovine plasma albumin as a standard[6] or, in the case of column eluates, by applying the formula of Layne to optical density readings taken at 260 and 280 nm.[7] Pure 4-epimerase has a specific absorbance value of 1.57 per milligram of protein, dry weight, per milliliter (determined in 10 mM NH$_4$HCO$_3$, pH 7.75).

Purification

Bacterial Strain and Growth Conditions. The L-ribulose-5-phosphate 4-epimerase has been isolated from a mutant strain of *E. coli* B/r (diploid

[6] O. H. Lowry, N. J. Rosebrough, A. L. Farr, and R. J. Randall, *J. Biol. Chem* **193**, 265 (1951).
[7] See this series, Vol. 3 [73].

strain F' *ara*B^{-24}/*ara*B^{-24})[8] which produces no L-ribulokinase activity and increased levels of 4-epimerase as compared with the wild-type strain. Batch cultures are grown at 37° in a Fermacell fermentor, Model F-130 (New Brunswick Scientific Co., Inc.). Approximately 500 g (wet weight) of cells are used for each purification. The growth medium contains 1% Casamino acids (Difco), a mineral base of 1% K$_2$HPO$_4$–KH$_2$PO$_4$ (pH 7), 0.01% MgSO$_4$·7H$_2$O, 0.1% (NH$_4$)$_2$SO$_4$, and 0.1% L-arabinose. The sugar is sterilized separately and added to the fermentor vessel when inoculating. The culture used to inoculate the fermentor is grown to log phase in the same medium without L-arabinose.

The cells are harvested in late log phase in a refrigerated Sharples centrifuge, type AS-14 (Pennsalt Chemicals Corp.). The cell pellet is resuspended in approximately 1 liter of 10 m*M* glycylglycine, pH 7.6, quick-frozen in a dry ice–acetone bath, and stored at −20° until use.

Purification Steps

All steps of the purification are performed at 0–4°. The crude extract is prepared by sonicating 100-ml aliquots of cell suspension for 8 min at level 8 of a Branson Model S-75 Cell Disruptor and centrifuging the sonicated cells for 1 hr at 56,000 *g*. The supernatant is recovered.

Freshly prepared 1.0 *M* MnCl$_2$ is added to the crude extract slowly with stirring to a final concentration of 50 m*M* MnCl$_2$. After 15 min at this concentration, the milky preparation is centrifuged at 66,000 *g* for 1 hr. The supernatant is recovered and immediately adjusted to pH 7.6 with 3% (v/v) NH$_4$OH.

Granular (NH$_4$)$_2$SO$_4$ is slowly added with stirring to 43% of saturation over a 30-min period. After stirring for 30 min more, the preparation is centrifuged for 10 min at 35,000 *g*; the pellet is dissolved in 10 m*M* potassium phosphate, pH 7.6, buffer (hereafter referred to as column buffer) and dialyzed against 4 liters of column buffer. After at least 4 hr (usually overnight), the dialysis buffer is changed and dialysis continued for at least 2 hr.

The dialyzed material is diluted to around 100 ml with column buffer, clarified by centrifugation at 35,000 *g* for 10 min, and applied to a DEAE-cellulose column (3.7 × 92 cm) that has been prepared as described in this volume [96] and equilibrated with column buffer. The sample is washed on with 500 ml of column buffer; then a 0 to 0.5 *M* NaCl linear gradient in column buffer (total volume of 4 liters) is used to fractionate the applied enzyme preparation at a flow rate of 150 ml per hour.

[8] See also this volume [96].

Fractions (15–20 ml each) are pooled on the basis of high specific activity. Usually the absorbance at 280 nm is determined for every third fraction, the absorbance peaks are tested for 4-epimerase activity, and then 4-epimerase assays are performed for every third fraction in the region of the 4-epimerase activity to define the activity peak. Granular $(NH_4)_2SO_4$ is added to the pooled fractions to 80% of saturation. After centrifugation, the pellet is dissolved in 20 ml of column buffer and dialyzed against column buffer as before.

The dialyzed enzyme preparation (clarified by centrifugation if necessary) is applied to a Sephadex G-200 column (2.7×182 cm) that has been equilibrated with column buffer and is eluted with column buffer into 6-ml fractions at a flow rate of 20 ml/hr. Pooling of fractions is again done on the basis of high specific activity (testing every third fraction as before). The pooled enzyme is precipitated by bringing the solution to 80% of $(NH_4)_2SO_4$ saturation as before. After centrifugation the pellet is dissolved in 10 ml of column buffer and dialyzed before the enzyme preparation is reapplied to the Sephadex G-200 column. Pooling of fractions from rechromatography on the Sephadex column can be done on the basis of high specific activity as before or on the basis of homogeneity in the ultracentrifuge and acrylamide gel disc electrophoresis. Pure pooled enzyme sediments as a single symmetrical peak in the analytical ultracentrifuge and migrates as a single band in electrophoresis.

Lyophilization of the 4-epimerase solutions in column buffer results in rapid loss of activity and should not be used instead of 80% $(NH_4)_2SO_4$ precipitation as a means of concentration during the purification procedure. Once purified, the enzyme is best stored frozen.

Crystallization. Crystals of L-ribulose 5-phosphate 4-epimerase have been obtained by bringing 2 ml of purified enzyme (4 mg of protein per milliliter) to 33% of saturation with $(NH_4)_2SO_4$ using a neutral saturated $(NH_4)_2SO_4$ solution at 4°. After 2 days in the refrigerator, crystals were visible, and the crystallization process was complete after 3 weeks. No increase in specific activity was observed.

Results of a typical purification are summarized in the table as shown on page 423.

Properties

No evidence for the requirement of cofactors in 4-epimerase activity has been found. Apparently the 4-epimerase contains no readily dissociable small molecule needed for activity. This is true also for an L-ribulose 5-phosphate 4-epimerase isolated from *Aerobacter aerogenes*.[9] NAD+ (as

[9] J. D. Deupree and W. A. Wood, *J. Biol. Chem.* **247**, 3093 (1972).

PURIFICATION OF L-RIBULOSE-5-PHOSPHATE 4-EPIMERASE FROM
Escherichia coli

Fraction	Volume (ml)	Protein (mg)	Total activity (units × 10^{-5})	Specific activity (units × 10^{-5})	Yield (%)
Crude extract	920	46,200	11.8	25.6	100
Ammonium sulfate fraction	71.5	5,434	5.22	96.1	44.2
DEAE-cellulose column chromatography	13.0	520	2.73	525	23.1
First Sephadex G-200 column	4.5	220	2.16	982	18.3
Second Sephadex G-200 column	22.1	88.4	1.0	1120	8.5

an electron donor and acceptor) is neither associated with the enzyme nor required for activity. Evidence indicating that the mechanism for the 4-epimerase involves no exchange of hydrogen with H_2O has been presented. Suitable double-label experiments indicate that hydrogen transfer in the reaction is intramolecular.[10]

The K_m for L-ribulose 5-phosphate for the 4-epimerase from *E. coli* is 95 μM—identical with that determined for the 4-epimerase purified from *Aerobacter aerogenes*.[11] Activity is maximal at pH 7 and above, and is about 50% of maximal at pH 6, thus resembling the 4-epimerase purified from *Lactobacillus plantarum*.[12] The apparent weight average molecular weight as determined by sedimentation equilibrium at high speed[13] is 105,000 ± 2000. The sedimentation coefficient of the native enzyme (pH 7.6) is 5.8×10^{-13} sec; at pH 2 it is 1.7×10^{-13} sec. Evidence that L-ribulose-5-phosphate 4-epimerase is composed of three identical subunits (molecular weight 35,000 ± 1000) has come from carboxyterminal amino acid analysis, electrophoretic mapping of tryptic peptides, separation of cyanogen bromide cleavage products by polyacrylamide gel disc electrophoresis, and molecular weight determination by sedimentation equilibrium in the analytical ultracentrifuge after dissociation of the enzyme with urea or performic acid oxidation. The amino acid composition of the 4-epimerase has been determined.[4]

[10] L. Davis, N. Lee, and L. Glaser, *J. Biol. Chem.* **247**, 5862 (1972).
[11] M. J. Wolin, F. J. Simpson, and W. A. Wood, *J. Biol. Chem.* **232**, 559 (1958).
[12] D. P. Burma and B. L. Horecker, *J. Biol. Chem.* **231**, 1053 (1958).
[13] D. A. Yphantis, *Biochemistry* **3**, 297 (1964).

[89] D-Ribose-5-phosphate Isomerase from Skeletal Muscle

By G. F. DOMAGK and W. R. ALEXANDER

D-Ribose 5-phosphate ↔ D-ribulose 5-phosphate

Assay Method

Principle. Ribose 5-phosphate is converted by the isomerase into ribulose 5-phosphate, which is measured by the red color formed in the cysteine-carbazole test.[1]

Reagents

Ribose 5-phosphate, sodium salt, 0.03 M. Store in ice bath, freeze overnight.

Tris·HCl buffer, 0.1 M, pH 8.4 containing 1 mM EDTA and 0.5 mM 2-mercaptoethanol

Cysteine, 0.03 M. Dissolve 79 mg of cysteinium hydrochloride in 15 ml of distilled water. This solution should be prepared freshly every day.

H_2SO_4, 75%. Add 75 ml of concentrated sulfuric acid to 25 ml of distilled water, cool, keep in ice bath. This solution must be prepared freshly every day.

Carbazole, 0.1%. Dissolve 50 mg of carbazole, recrystallized from xylene, under warming in 50 ml absolute ethanol.

Procedure. Place 0.3 ml of Tris buffer with exactly 0.10 ml of ribose 5-phosphate and an aliquot of dilute enzyme and incubate at 25°. It is useful to do parallel incubations with two levels of enzyme in a 1:2 ratio; a blank omitting the enzyme must be included in each set of determinations. After 15 min, 0.6 ml of cysteine are added, followed by 5 ml of H_2SO_4. Immediately after thorough mixing, 0.2 ml of the carbazole solution are added. Mix vigorously and read the red color after a 30-min incubation at room temperature in a 1-cm cuvette at 546 nm.

The increase in absorption, corrected for the blank, should not exceed 0.3. With good reagents the absorption of the blank will be below 0.05.

Definition of Unit and Specific Activity. For routine work it has been convenient to use laboratory units defined by an increase of absorption of 1.0 at 546 nm. These values can be converted into international units

---

[1] Z. Dische and E. Borenfreund, *J. Biol. Chem.* **192,** 583 (1951).

by dividing the figures by 15. Specific activity is expressed as units per milligram of protein.

Purification Procedure

Purification of Ribosephosphate Isomerase

Crude Extract. Freshly ground beef muscle, 500 g, is kept frozen for at least 30 min; storage at this stage will be tolerated up to 6 months without loss of activity. Five hundred milliliters of 0.1 M potassium phosphate, pH 7.5–1 mM EDTA–0.5 mM mercaptoethanol are added to the frozen tissue, which is thawed under occasional stirring. The homogenate is centrifuged for 20 min at 20,000 g in a refrigerated centrifuge; the precipitate is reextracted once under identical conditions. The second precipitate is discarded. Recent experiments have shown that the reextraction is not necessary.

Ammonium Sulfate Fractionation by Elution from a Celite Column.[2] Solid ammonium sulfate (29.1 g/100 ml) is added to the crude extract. After 2 hr equilibration in an ice bath, 40 g of Celite 545 + 40 g of cellulose powder (linters) are added per 1000 ml of crude extract. The suspension is homogenized for 15 min on a powerful magnetic stirrer at low speed. Subsequently the suspension is poured into a chromatographic column of 5.5 cm diameter. After the bed has settled, the column is washed with about 500 ml of 20 mM Tris buffer pH 7.5 with a 60% saturation of ammonium sulfate. This washing will remove most of the red pigments. The elution of the enzyme is achieved by further washing of the column with Tris buffer containing a 30% saturation of ammonium sulfate at a speed of 1 ml/min. A fraction collector is connected. The fractions containing isomerase activity are pooled, and the ammonium sulfate saturation of the combined solution is determined. The ammonium sulfate concentration, usually around 30%, is increased to 60%, and the precipitate is collected by centrifugation. The sediment is dissolved in 20 mM potassium phosphate buffer, pH 5.75, and dialyzed against the same buffer. After several changes of the buffer a heavy precipitate is formed, which is inactive and can be discarded after centrifugation.

Chromatography on Phosphocellulose. The enzyme solution is put on a column of phosphocellulose (2.5 cm inner diameter) prepared by equilibrating 20 g of phosphocellulose against the same buffer. Elution of the enzyme is performed by a linear gradient of NaCl increasing from 0 to 0.4 M (300 ml initial volume for each of the 2 solutions). The fractions containing isomerase are pooled, and the enzyme is precipitated by

[2] T. P. King, *Biochemistry* **11**, 367 (1972).

PURIFICATION OF RIBOSEPHOSPHATE ISOMERASE FROM BEEF MUSCLE[a,b]

Step	Volume (ml)	Units per fraction	Protein (mg)	Specific activity (units/mg protein)	Yield (%)
Crude extract	1000	5750	18400	0.31	100
Celite column	108	3600	1340	2.7	62
P-cellulose column	225	3100	114	27	53
Preparative gel electrophoresis	118	2150	6	355	37

[a] Ground beef muscle, 500 g, was used in this experiment.
[b] Reproduced from *Hoppe-Seyler's Z. Physiol. Chem.* **355,** 781 (1974) by permission of the publishers.

adding solid ammonium sulfate to a 60% saturation. The precipitate is dissolved in about 50 ml of 30 mM Tris borate, pH 7.9 and dialyzed against the same buffer.

Concentration and Preparative Electrophoresis. The dialyzate is concentrated to a volume of 5 ml by means of a stirred ultrafiltration cell equipped with an Amicon PM 30 filter, 10% of glycerol (v/v) and a trace of amido black are added to the concentrate, which is then applied to the top of a 5% polyacrylamide gel, sized $16 \times 150 \times 75$ mm. The apparatus used is that of Stegemann[3] (obtained from Labor-Müller, D 351 Hann. Münden, Germany). Collection of fractions is started when the blue stain has reached the bottom of the gel, which is usually after 4 hr at 120 V and 25 mA. The proteins assembling on top of a dialyzing membrane are washed out discontinuously every 20 min by a transverse stream of buffer (10 ml within 3 min). The isomerase activity is found around fraction 25. A summary of the purification procedure is given in the table.

Properties

Stability. Ground tissue, crude extracts, and purified isomerase preparations can be stored frozen at —20° for several weeks without loss of activity.

Physicochemical Constants. The K_m for ribose 5-phosphate is 2 mM. The pH dependency of this isomerase is represented by a broad plateau between pH 7 and 9.

Inhibitors. The enzyme is strongly inhibited by sulfhydryl reagents, particularly organic mercurials.

[3] H. Stegemann, *Z. Anal. Chem.* **261,** 388 (1972).

[90] D-Ribose-5-phosphate Isomerase from *Candida utilis*

By G. F. DOMAGK and K. M. DOERING

D-Ribose 5-phosphate ↔ D-ribulose 5-phosphate

Assay Method

Principle. Ribose 5-phosphate is converted by the isomerase into ribulose 5-phosphate, which is measured by the red color formed in the cysteine-carbazole test.[1]

Reagents

Ribose 5-phosphate, sodium salt, 30 mM. Store in ice bath, freeze overnight.

Tris·HCl buffer, 0.1 M, pH 8.4, containing 1 mM EDTA and 0.5 mM 2-mercaptoethanol

Cysteine, 0.03 M. Dissolve 79 mg of cysteinium hydrochloride in 15 ml of distilled water. This solution should be prepared freshly every day.

H_2SO_4, 75%. Add 75 ml of concentrated sulfuric acid to 25 ml of distilled water, cool, keep in ice bath. This solution must be prepared freshly every day.

Carbazole, 0.1%. Dissolve 50 mg of carbazole recrystallized from xylene under warming in 50 ml of absolute ethanol.

Procedure. Place 0.3 ml of Tris buffer with exactly 0.10 ml of ribose 5-phosphate plus an aliquot of dilute enzyme and incubate at 25°. A blank omitting the enzyme must be included in each set of determinations. After 15 min, 0.6 ml of cysteine is added. After mixing, 5 ml of H_2SO_4 are added by automatic dispenser, followed by 0.2 ml of carbazole solution. Mix vigorously and read the red color after a 30-min incubation at room temperature in a 1-cm cuvette at 546 nm.

The increase in absorption, corrected for the blank, should not exceed 0.3. With good reagents the absorption of the blank will be below 0.05.

Definition of Unit and Specific Activity. For routine work, it has been convenient to use laboratory units defined by an increase of absorption of 1.0 at 546 nm. These values can be converted into international units by dividing the figures by 15. Specific activity is expressed as units per milligram of protein.

[1] Z. Dische and E. Borenfreund, *J. Biol. Chem.* **192,** 583 (1951).

Purification Procedure[2]

Crude Extract. Dried yeast, 80 g, suspended in 400 ml of 1 mM EDTA–0.5 mM mercaptoethanol was incubated for 4 hr at 37° under occasional stirring. After centrifugation for 20 min at 20,000 g, the precipitate was discarded and the supernatant was retained.

Protamine Step. Protamine sulfate, 700 mg dissolved in 50 ml of water, was added dropwise under mechanical stirring to the supernatant. After 30 min in an ice bath, the solution was centrifuged for 20 min at 20,000 g, and the precipitate was discarded.

Fractionation by DEAE Cellulose. The supernatant of the protamine step was treated with a suspension of 10 g of DEAE cellulose in 150 ml of 20 mM Tris buffer pH 7.5. After 30 min in an ice bath, the mixture was centrifuged for 20 minutes at 20,000 g. The supernatant contained about 10% of the ribosephosphate isomerase (RPI) activity and was discarded. The packed DEAE-cellulose was suspended in 20 mM Tris buffer pH 7.5 and applied to a column (25 mm diameter), containing unloaded DEAE-cellulose (3 g of the powder suspended in 50 ml of the same buffer). After the charged cellulose had settled, 100 ml of Tris buffer were used for washing, which did not remove RPI activity. Elution of the enzyme was achieved by a linear NaCl gradient (from 400 ml of Tris buffer to 400 ml of Tris buffer containing 0.3 M NaCl). Fractions of 15 ml were collected, and those containing RPI activity were combined.

Filtration over Phosphocellulose. A suspension of 20 g P-cellulose in 300 ml of 20 mM phosphate buffer pH 6.0 was poured into a column (25 mm diameter). When the bed had settled, the combined ribosephospate isomerase fractions of the previous column were applied to the top of the column and subsequently washed with 150 ml 20 mM phosphate buffer pH 6.0. The combined filtrates contained most of the isomerase activity.

DE-52 Cellulose Column. DE-52-cellulose, 40 g, was equilibrated with 20 mM Tris pH 7.5 and poured into a column of 25 mm diameter. The combined filtrates of the previous column were passed through the DE-52-column, which absorbed all the isomerase activity. The chromatogram was developed with a linear NaCl gradient (400 ml of Tris buffer plus 400 ml of 0.3 M NaCl in Tris); fractions of 15 ml were collected, and those which contained RPI were combined.

Ammonium Sulfate Fractionation. The solution containing ribosephosphate isomerase was treated with solid ammonium sulfate (39.8 g/100 ml) to give a saturation of 65%. After the solution had been kept on

[2] G. F. Domagk, K. M. Doering, and R. Chilla, *Eur. J. Biochem.* **38**, 259 (1973).

PURIFICATION OF RIBOSE-5-PHOSPHATE ISOMERASE FROM *Candida utilis*[a,b]

Step	Volume (ml)	Activity (units)	Protein (mg)	Specific activity (units/mg protein)	Yield (%)
Crude extract	266	27,100	4580	5.9	100
Protamine sulfate	306	26,000	3370	7.7	96
DEAE-cellulose	194	20,400	465	44	74
Phosphocellulose	394	16,300	294	55	60
DE-52 cellulose	200	16,000	163	98	59
Ammonium sulfate, 0.65–0.75	4.5	7,300	29	292	27
Sephadex G-200	44	7,300	20.5	356	27

[a] Eighty grams of dried yeast were used in this preparation.
[b] Reproduced from *Eur. J. Biochem.* **38**, 261 (1973) by permission of the editor.

ice for 2 hr, the precipitate was collected by centrifugation (10 min at 15,000 g) and discarded. The ammonium sulfate concentration of the supernatant was increased to 75% (6.3 g ammonium sulfate for 100 ml), and after 2 hr at 0° the prcipitate was collected by centrifugation and subsequently dissolved in about 4 ml of water.

Sephadex G-200 Column. The ribose-phosphate isomerase solution was cautiously applied to the top of a 25 × 400 mm column of Sephadex G-200 equilibrated with 0.1 *M* Tris pH 7.6. The column was washed with the same buffer. Fractions of 5 ml were collected and monitored for 280 nm absorbance and ribosephosphate isomerase activity. A small fraction of protein following the isomerase peak was discarded.

The data of a typical purification experiment are presented in the table.

Properties

Stability. The purified enzyme was found to be stable over several months when stored at 4° in the presence of 70% saturated ammonium sulfate. Freezing and thawing will destroy the isomerase activity.

Physicochemical Constants. The K_m for ribose 5-phosphate is 2.5 mM. The pH optimum is at 8.4, the isoelectric point of the isomerase is at 4.7.

Inhibitors. The enzyme is strongly inhibited by sulfhydryl reagents, particularly organic mercurials.

[91] Triosephosphate Isomerase from Human and Horse Liver[1]

By ROBERT SNYDER and EUN WOO LEE

Glyceraldehyde 3-phosphate $\rightleftharpoons$ dihydroxyacetone phosphate

Assay Method

Principle. Triosephosphate isomerase converts glyceraldehyde 3-phosphate to dihydroxyacetone phosphate, a substrate for glycerol-1-phosphate dehydrogenase. NADH utilization during glycerol 1-phosphate production is stoichiometric with dihydroxyacetone phosphate production and is measured spectrophotometrically at 340 nm.

Reagents

DL-Glyceraldehyde 3-phosphate, 14.7 mM
NADH, 4.2 mM
Glycerol-1-phosphate dehydrogenase, 0.67 mg/ml (specific activity = 100–200 units/mg protein).
Triethanolamine (2 mM)–EDTA (5.4 mM) buffer, pH 7.9

Procedure. Place 0.1 ml of glyceraldehyde 3-phosphate, 0.1 ml of glycerol-1-phosphate dehydrogenase, 0.15 ml of NADH and 2.55 ml of triethanolamine–EDTA buffer in a 1-cm cuvette. Start the reaction by adding 0.1 ml of isomerase and follow the decrease in extinction at 340 nm.

Definition of Unit and Specific Activity. One unit of triosephosphate isomerase is defined as the amount of enzyme that will isomerize 1 μmole of D-glyceraldehyde 3-phosphate to dihydroxyacetone phosphate in 1 min. Specific activity is expressed as units per milligram of protein. Protein is measured by the method of Lowry et al.[2]

Purification Procedure

Either fresh or deep-frozen livers may be used.

Step 1. Preparation of Water Extract. The liver is freed of connective

[1] E. W. Lee, J. A. Barriso, M. Pepe, and R. Snyder, *Biochim. Biophys. Acta* **242**, 261 (1971).
[2] O. H. Lowry, N. J. Rosebrough, A. L. Farr, and R. J. Randall, *J. Biol. Chem.* **193**, 265 (1951).

tissue and blood vessels, diced, and put through a chilled meat grinder. The ground liver is then weighed and extracted with 2.5 parts of deionized water containing 0.05% EDTA for 4 hr with mechanical stirring at room temperature or overnight in the cold room at 3°. The water extract is obtained after centrifugation for 20 min at 8000 g.

Step 2. Acetone Precipitation. Acetone is added in a fine stream to the mechanically stirred, ice cold, water extract over a 45-min period to a final concentration of 35% (v/v) (horse liver) or over a 60-min period to a final concentration of 50% (human liver). Stirring is continued for an additional 15-min period, and the resulting precipitate is removed by centrifugation for 20 min at 4000 g.

Acetone is again added to the 35% (horse liver) or 50% (human liver) acetone extract over a 45-min period to give a final concentration of 60%, and the resulting precipitate is collected after centrifugation as above. The residue is taken up in 8 volumes of deionized water containing 0.05% EDTA and dialyzed against 0.05% EDTA overnight. Denatured protein formed during dialysis is discarded after centrifugation for 10 min at 27,000 g.

Step 3. Heat Treatment. The dialyzed solution is heated to 45° for 15 min, and the resulting precipitate of denatured protein is removed by centrifugation for 30 min at 16,000 g. The extract is again treated with heat at 50° for 30 min and centrifuged as above to remove denatured protein.

Step 4. Chromatography on QAE-Sephadex. The enzyme is next applied to a 2.3 × 28 cm QAE-Sephadex A-50 column equilibrated with 7 mM phosphate buffer, pH 7.8, at room temperature. Buffer is passed through the column until the extinction at 280 nm is no longer detected. Elution is performed with 0.1 M NaCl in 7 mM phosphate buffer, pH 7.8. Most of the activity is found in the first (human liver) or second (horse liver) peak of the eluate.

Step 5. Ammonium Sulfate Precipitation. The enzyme eluted from the QAE-Sephadex column is successively dialyzed against 0.8 M, 1.3 M, and 2.6 M ammonium sulfate in 20 mM sodium acetate buffer, pH 5.3, for a total period of 2 days. The protein which is precipitated at 2.6 M ammonium sulfate is removed by centrifugation at 13,200 g for 1 hr. The supernatant, which contains most of the enzyme, is again dialyzed against 3.0 M and then 3.5 M ammonium sulfate in 20 mM sodium acetate buffer, pH 5.3 continuously with several changes of the ammonium sulfate solution in the cold room at 3° for 7 days. The resulting crystalline enzyme (crude crystals) is collected after centrifugation as above. The "crude crystals" are dissolved by cautious, dropwise addition of a minimum amount of 0.05% EDTA solution at 3°.

TABLE I

PURIFICATION OF HORSE LIVER AND HUMAN LIVER TRIOSEPHOSPHATE ISOMERASES

Fraction	Specific activity (IU/mg)[a]		Purification factor		Total activity (IU)		Recovery (%)	
	Horse	Human	Horse	Human	Horse	Human	Horse	Human
Water extract	3.5	6.6	1.0	1.0	200,880	735,336	100.0	100.0
Acetone precipitate	15.8	28.3	4.5	4.8	216,992	397,143	58.3	54.0
Horse, 35–60%								
Human, 50–60%								
Supernatant after heat treatment	23.1	181	6.6	27.3	105,000	145,541	52.3	19.8
QAE-Sephadex chromatography	367	2397	105	362	105,350	97,201	52.6	13.2
2.6 M (NH₄)₂SO₄ supernatant	1238	—	354	—	85,656	—	42.6	—
2.6–3.5 M (NH₄)₂SO₄ precipitate	1918	—	548	—	53,071	—	26.4	—
Sephadex G-75 gel filtration	3183	—	909	—	39,420	—	19.6	—

[a] IU = 1 μmole of D-glyceraldehyde 3-phosphate converted per minute.

TABLE II
DISTRIBUTION OF TRIOSEPHOSPHATE ISOMERASE ACTIVITY
AMONG ITS ISOZYMES

Isozymes	Rabbit muscle (% of total activity)	Horse liver (% of total activity)	Human liver (% of total activity)
1	22	50	—
2	44	38	—
3	24	12	—
4	6	—	—
4a	—	—	62
5	4	—	—
5a	—	—	26
6	—	—	11

Step 6. Gel Filtration. The enzyme solution obtained from the preceding step is applied to a 2.5 × 40 cm column of Sephadex G-75 superfine equilibrated with 2 mM phosphate buffer, pH 7.8. The column is then washed with the same buffer until all the protein has passed through the column. Fractions of 8–10 ml are collected, and the enzyme is recovered after approximately 100–120 ml have passed through the column.

A summary of the purification procedure is presented in Table I.

Properties

Isozymes of Horse and Human Liver Triosephosphate Isomerase. Upon polyacrylamide gel electrophoresis, horse and human liver triosephosphate isomerase can each be resolved into 3 isozymes. Table II shows the isozymes listed in order of decreasing electronegativity and the percentage of total activity for each. Rabbit muscle isomerase, which contains five isozymes, is shown for purpose of comparison. Orthogonal gel electrophoresis[3] of the horse liver isozymes results in equal migration ratios for the major isozymes suggesting that they are equal in molecular size and shape, but have different surface charges.

Molecular Weight. Gel filtration and equilibrium sedimentation studies indicate a molecular weight of 46,000–49,000 for human liver triose-phosphate isomerase.

Kinetic Constants. Maximum velocities, Michaelis constants, and equilibrium constants are shown in Table III.

[3] S. Raymond, *Ann. N.Y. Acad. Sci.* **121,** 350 (1964).

TABLE III

MAXIMUM VELOCITIES, MICHAELIS CONSTANTS, AND EQUILIBRIUM CONSTANTS OF
TRIOSEPHOSPHATE ISOMERASE FROM HORSE LIVER AND HUMAN LIVER

| Source | Glyceraldehyde 3-phosphate to dihydroxyacetone phosphate | | Dihydroacetone phosphate to glyceraldehyde 3-phosphate | | K_{eq} |
	$V_{max}{}^a$	K_m (mM)	V_{max}	K_m (mM)	
Horse liver	7935 ± 456^b	0.42 ± 0.06	753 ± 125	0.59 ± 0.01	14.8 ± 2.5
Human liver	2302 ± 119	0.40 ± 0.03	163 ± 5	0.59 ± 0.01	20.9 ± 3.1

[a] V_{max} is expressed as IU/mg protein.

[b] Variability is expressed as the standard deviation and was calculated when the experiment had been repeated in some cases three times but in most cases four times.

[92] Triosephosphate Isomerase[1] from Yeast

By W. K. G. KRIETSCH

D-Glyceraldehyde 3-phosphate $\rightleftharpoons$ dihydroxyacetone phosphate

Assay Method

Principle. The most convenient method of Meyer-Arendt *et al.*[2] is used. The triosephosphate isomerase is assayed by measuring the rate of formation of dihydroxyacetone phosphate (DAP), which is converted to L-α-glycerophosphate (α-GP) by the indicator enzyme L-α-glycerophosphate dehydrogenase. The rate of DAP reduction to α-GP is equivalent to the rate of NADH oxidation, which is measured by the absorbance decrease at 366 nm.

Reagents

Triethanolamine-HCl buffer, 0.1 M, pH 7.5, 0.50 ml
NADH, 14 mM, 0.01 ml
DL-Glyceraldehyde 3-phosphate, 90 mM, 0.02 ml
L-α-Glycerophosphate dehydrogenase, 70 U/ml, 0.01ml

Procédure. Solutions in the indicated quantities were added to a glass

[1] D-Glyceraldehyde-3-phosphate ketol-isomerase, EC 5.3.1.1.
[2] E. Meyer-Arendt, G. Beisenherz, and T. Bücher, *Naturwissenschaften* **40**, 59 (1953).

cuvette with a 1-cm light path. The cuvette compartment was thermostated at 25°. The reaction was initiated by addition of 0.01 ml of triosephosphate isomerase solution with a maximal activity of 2 U/ml.

Tris · HCl buffer cannot be used for the test, because Tris reacts with DL-glyceraldehyde 3-phosphate (DL-GAP).[3]

DL-GAP was prepared from DL-glyceraldehyde 3-phosphate diethylacetal monobarium salt. Barium was removed by Dowex 50-W-X8 (H⁺) exchanger, and the acetal was hydrolyzed for 3 min at 100°.

Definition of the Unit and Specific Activity. One unit of triosephosphate isomerase isomerizes 1 μmole of substrate per minute under the conditions described. Specific activity is expressed in units per milligram of protein. Protein was measured by the biuret reaction[4] using a modified procedure with one-tenth the given volumes. A factor of 37 mg per milliliter of sample per unit absorbance at 546 nm (1 cm light path) was used as an average value for calculating protein concentrations.

Purification Procedure

Step 1. Preparation of Initial Extract. Brewer's yeast[5,6] (260 g net weight) was repeatedly washed with tap water and subsequently dried at room temperature for several days. The cells were disrupted in a ball mill and taken up in 490 ml of 1 M ammonia. The suspension was maintained at room temperature for 20 hr and then diluted with 1.2 liters of distilled water and 34 ml of 0.5 M EDTA and 84 g of ammonium sulfate (concentration in the extract 0.35 M).

Step 2. Heat Treatment. The suspension was incubated at 60° for 10 min and subsequently centrifuged for 20 min at 15,000 g. The supernatant was then further treated with ammonium sulfate.

Step 3. Ammonium Sulfate Fractionation. Finely ground solid ammonium sulfate (180 g) was added slowly over a period of 6 hr to the supernatant with continual stirring (concentration of ammonium sulfate, 1.3 M). After the solution was stirred for an additional 2 hr at room temperature, the precipitate was collected by centrifugation for 10 min at 15,000 g and discarded. An additional 300 g of ammonium sulfate was added to the supernatant slowly with stirring over a period of 4 hr (concentration of ammonium sulfate, 2.9 M). After standing for 16 hr at 4°, the

[3] H. R. Mahler, *Ann. N.Y. Acad. Sci.* **92,** 426 (1961).
[4] G. Beisenherz, H. J. Boltze, T. Bücher, R. Czok, K. H. Garbade, E. Meyer-Arendt, and G. Pfleiderer, *Z. Naturforsch.* **8,** 555 (1953).
[5] With the same procedure the enzyme can be isolated from baker's yeast.
[6] An alternative method for the isolation of the enzyme from baker's yeast is described by I. L. Norton and F. C. Hartman, *Biochemistry* **11,** 4435 (1972).

precipitate was collected by centrifugation (20 min at 40,000 g) and dissolved in 200 ml of distilled water.

Step 4. Methanol Treatment. Ice cold methanol (110 ml) was added dropwise over a 3-hr period with stirring to the redissolved precipitate. The newly formed precipitate was separated by centrifugation (10 min at 40,000 g) and the supernatant fluid reduced to half its volume in a rotating evaporator at 20°. Triosephosphate isomerase was then precipitated by gradual addition of 53 g of ammonium sulfate. After stirring for 2 hr at 20°, the precipitate was collected by centrifugation as above and dialyzed against 5 liters of distilled water overnight at 4°.

Step 5. Protamine Sulfate Precipitation. A protamine sulfate solution (2 g and 3.2 ml of 0.1 N NaOH in 100 ml of water) was added dropwise to the dialyzate until no further sediment was formed. The precipitate was separated by centrifugation as above and discarded.

Step 6. DEAE-Sephadex Chromatography and Crystallization. The protein in the supernatant was precipitated by the addition of ammonium sulfate to 90% saturation. The solution was centrifuged as before, and the precipitate was dissolved in a minimal volume of distilled water and dialyzed against 5 liters of distilled water in preparation for chromatography.

The dialyzed solution was then applied to a DEAE-Sephadex A-50[7] column (65 × 4.5 cm). The column was eluted with a linear gradient between 0 and 0.15 M NaCl in 10 mM triethanolamine–HCl buffer, pH 8.0, at room temperature. The volumes used were 1.5 liters of buffer and 1.5 liters of the same buffer containing 0.15 M NaCl. The enzyme was eluted between 1.5 and 2.0 liters. The peak fractions with 90% of the activity were pooled, and solid ammonium sulfate was added in small portions up to the point of slight turbidity. Within 1 day, stable crystals began to form at 4°.

Step 7. CM-Sephadex Chromatography and Recrystallization. The crystalline protein was collected by centrifugation as above and dialyzed against distilled water. The solution was then added to a CM-Sephadex C-50 column (60 × 4.5 cm) and eluted with a linear NaCl gradient from 0 to 0.15 M in 10 mM sodium phosphate buffer, pH 6.5, at room temperature. The volumes used were 800 ml of each buffer solution without and with NaCl. The protein appeared in the effluent between 300 and 500 ml. The fractions containing the major portion of the enzyme were pooled, and the pH was adjusted to 6 with 0.5 M acetate buffer. Ammonium sulfate was then added in small portions up to the point of slight turbidity. Large stable crystals began to form within 24 hr.

---

[7] The Sephadex exchangers are prepared as described in the Sephadex ion exchangers manual from Pharmacia Fine Chemicals, Uppsala, Sweden.

PURIFICATION OF TRIOSEPHOSPHATE ISOMERASE FROM YEAST

Fraction	Protein (g)	Total units	Specific activity (units/mg protein)	Purification (fold)	Yield (%)
1. Crude homogenate	160	6.55×10^6	41	—	100
2. Heat treatment	10	4.73×10^6	470	11	72
3. Ammonium sulfate fractionation	9	4.73×10^6	600	15	72
4. Methanol treatment	2.7	3.65×10^6	740	17	56
5. Protamine sulfate precipitation	2.6	3.33×10^6	780	19	51
6. DEAE-Sephadex chromatography	0.45	3.10×10^6	6,900	170	47
7. CM-Sephadex chromatography	0.40	$(4.00 \times 10^6)^a$	10,000	250	$(61)^a$

[a] The increase of the enzyme activity after CM-Sephadex chromatography might be due to the separation of an inhibitor fraction.

A summary of the procedure is given in the accompanying table.

Properties

Homogeneity and Constitution. The purified triosephosphate isomerase appears to be homogeneous in polyacrylamide gel electrophoresis and in the ultracentrifuge. In starch gel electrophoresis the yeast enzyme is separated into three forms, the major one of which comprises about 90% of the enzyme protein. The molecular weight of the enzyme, evaluated by equilibrium ultracentrifugation, is 53,000, assuming a partial specific volume of 0.75 ml g^{-1}. The gel filtration method gives a molecular weight of 56,000. Treatment with sodium dodecyl sulfate and maleylation of the enzyme causes dissociation into two subunits. These two subunits are different, as shown by the electrophoretic pattern and the different N-terminal amino acids.[8]

Stability. The pure enzyme was quite stable in 2.5 M ammonium sulfate at pH 6.5 and has been kept at 0–4° for at least two years without loss of activity.

pH Optimum. The isomerase reaction proceeds with maximal velocity in triethanolamine-HCl buffer in the range of pH 7.0–8.5 with L-α-glycerophosphate dehydrogenase as indicator enzyme, and in the range of pH 7.6–9.5 with D-glyceraldehyde-3-phosphate dehydrogenase as indicator enzyme. Accordingly, the real pH optimum of the triosephosphate isomerase reaction is in the same range as that determined for the liver enzyme without indicator enzymes.[8]

Specific and Kinetic Properties. Dihydroxyacetone and D-glyceraldehyde are not isomerized. The K_m values at 25° are 1.27 mM for D-GAP and 1.23 mM for DAP. Recalculated for the free (unhydrated) aldehyde[9] and ketone[10] the K_m values are 42 μM for D-GAP and 0.68 mM for DAP. The molecular activity with D-GAP is 1.0×10^6 moles min^{-1} mole^{-1} at 25°.

Equilibrium Constants. The isomerization of D-GAP is freely reversible with a K_{eq} of 19 at pH 7.5 and 25°. When recalculated for the unhydrated form, the K_{eq} is 316.[10]

Inhibitors. The following compounds act as competitive inhibitors: arsenite, phosphate, phosphoenolpyruvate, D-α-glycerophosphate, 2-phosphoglycolate, and acetylphosphate.[8,11]

The enzyme is not inhibited by p-chloromercuribenzoate, 5,5'-dithiobis(2-nitrobenzoic acid), iodoacetate or 2,4-dinitrofluorbenzene over the pH range of 6–8, and is stable to photooxidation with rose bengal.[8]

Although three sulfhydryl groups could be titrated with 5,5'-dithiobis(2-nitrobenzoic acid), free sulfhydryl groups are probably not necessary for the activity of yeast triosephosphate isomerase.

The enzyme is inactivated by 1-chloro-3-hydroxyacetone phosphate, which selectively esterifies a single glutamyl-γ-carboxylate residue in the activity center.[6]

Ultraviolet Light Absorption. The absorbance of a 1% solution of the purified enzyme at 280 nm is 9.9 (1 cm light path).[12]

[8] W. K. G. Krietsch, P. G. Pentchev, H. Klingenburg, T. Hofstätter, and T. Bücher, *Eur. J. Biochem.* **24,** 289 (1970).

[9] D. R. Trentham, C. H. McMurray, and C. I. Pogson, *Biochem. J.* **114,** 19 (1969).

[10] S. J. Reynolds, D. W. Yates, and C. I. Pogson, *Biochem. J.* **122,** 285 (1971).

[11] H. Krietsch, G. Kuntz, and W. K. G. Krietsch, unpublished results.

[12] W. K. G. Krietsch, Dissertation, Faculty of Science, University of Munich, 1970.

[93] Triosephosphate Isomerase from Rabbit Liver

By W. K. G. KRIETSCH

D-Glyceraldehyde 3-phosphate $\rightleftarrows$ dihydroxyacetone phosphate

Assay Method

The enzyme is assayed spectrophotometrically as described for the yeast enzyme in this volume [92].

Purification Procedure

Step 1. Preparation of Initial Extract. Three kilograms of deep-frozen rabbit liver were partially thawed and passed through a meat grinder. The ground mince was suspended in 9 liters of 10 mM sodium phosphate buffer, pH 7.0 with 10 mM EDTA and 0.5 M ammonium sulfate and stirred overnight at 4°. The cellular debris were removed by centrifugation for 40 min at 15,000 g.

Step 2. Ammonium Sulfate Fractionation. Solid ammonium sulfate (2.4 kg) was slowly added to the supernatant with stirring over a period of 90 min and then subsequently stirred for an additional 12 hr at 4° (ammonium sulfate concentration 2.3 M). The precipitate was centrifuged off (40 min at 15,000 g) and discarded. Additional ammonium sulfate (1.55 kg) was added to the supernatant as before and followed by 9 hr of stirring at 4° (ammonium sulfate concentration 3.2 M). The precipitate was collected by centrifugation as before and dissolved in 500 ml of 10 mM sodium phosphate and 10 mM EDTA buffer pH 7.4.

Step 3. Heat Treatment. The solution was slowly brought to 50° and maintained at this temperatue for 5 min and subsequently cooled in an ice bath. Denatured protein was centrifuged off (20 min at 15,000 g).

Step 4. Chloroform and Ethanol Fractionation. One milliliter of a chloroform–ethanol mixture (2:1, v/v) was added for every 2 ml of the supernatant. After 20 min of stirring at room temperature, the mixture was centrifuged for 10 min at 15,000 g, and the sediment was discarded. The chloroform and ethanol were removed from the solution by a separating funnel and rotating evaporator, respectively, at 20°. Additional precipitate formed during the procedure was centrifuged off (40 min at 15,000 g).

Step 5. DEAE-Sephadex Chromatography. After concentration by dialysis against 20 mM triethanolamine-HCl, 1 mM EDTA buffer pH 8.0 with 30% polyethylenglycol (MW 6000), the solution was added to a DEAE-Sephadex A-50 column (60 × 4.5 cm). A linear 0 to 0.15 M NaCl gradient in 20 mM triethanolamine-HCl and 1 mM EDTA buffer pH 8.0 was passed through the column. The volumes used were 1.5 liters of buffer without NaCl and 1.5 liters of buffer with NaCl. The enzyme eluted between 1.7 and 2.1 liters. The fractions containing enzyme at the highest specific activity were pooled and the protein was precipitated by the addition of solid ammonium sulfate up to saturation. The precipitate was removed by centrifugation for 20 min at 40,000 g and the sediment dissolved in a minimal volume of 20 mM sodium phosphate and 1 mM EDTA buffer pH 6.0. The enzyme solution was dialyzed against 5 liters of the same buffer overnight at 4°.

PURIFICATION OF TRIOSEPHOSPHATE ISOMERASE FROM RABBIT LIVER

Fraction	Specific activity (units/mg protein)	Purification (fold)[a]
1. Crude homogenate	6	—
2. Ammonium sulfate fractionation	24	4
3. Heat treatment	37	6
4. Chloroform–ethanol fractionation	120	20
5. DEAE-Sephadex chromatography	2700	450
6. CM-Sephadex chromatography	3650	608
7. QAE-Sephadex chromatography	6400	1066

[a] The true total activity recovered in the various fractions is not known owing to the progressive increase of total enzyme measured in the initial fractions. It is assumed that the crude homogenates contain an inhibitor activity that was gradually removed. The same observation was made by R. Czok and T. Bücher [*Advan. Protein Chem.* **15**, 315 (1960)] in the preparation of the rabbit muscle triosephosphate isomerase. The total yield of crystalline enzyme which could be isolated from the starting material was 100 mg per kilogram of wet weight rabbit liver.

Step 6. CM-Sephadex Chromatography and Crystallization. The dialyzed solution was added to a CM-Sephadex C-50 column (90 × 4.5 cm). Enzyme activity appeared in the front upon elution with the buffer (20 mM phosphate and 1 mM EDTA, pH 6.0). The fractions containing the bulk of the activity were combined. Solid ammonium sulfate was slowly added up to the point of slight turbidity caused by the appearance of amorphous protein which was centrifuged off (20 min at 40,000 g). Crystallization was induced by further slow addition of ammonium sulfate. Within 3 days at 4°, the enzyme crystallized as fine needles. The crystalline protein was collected by centrifugation and redissolved in 20 mM triethanolamine-HCl buffer pH 8.5.

Step 7. QAE-Sephadex Chromatography and Recrystallization. Following dialysis against 4 liters of 20 mM triethanolamine-HCl buffer pH 8.5, the protein was chromatographed on a QAE-Sephadex A-50 column (60 × 2.0 cm) and eluted with a linear NaCl gradient (0 to 0.15 M) in the same buffer. With volumes of 600 ml of buffer and 600 ml of NaCl buffer solution, the enzyme eluted between 0.9 and 1.1 liters. The fractions which contained 90% of the activity were pooled, and the enzyme was recrystallized as before.

The purification procedure is summarized in the table.

Properties

Homogeneity and Constitution. The purified liver enzyme shows in starch and polyacrylamide gel electrophoresis three major forms (95–98%

of total protein) and two minor forms. The five isoenzymes have the same electrophoretic mobility as those of the rabbit muscle triosephosphate isomerase.[1-3] In the ultracentrifuge the liver enzyme appeared homogeneous. The molecular weight of the enzyme was determined by sucrose gradient centrifugation and by equilibrium ultracentrifugation to be 56,000. For the evaluation of the molecular weight from equilibrium ultracentrifugation, a partial specific volume of 0.75 ml g^{-1} was assumed. In 2 M guanidinium thiocyanate and in a 1% solution of sodium dodecyl sulfate the liver enzyme dissociates into subunits. Like the yeast and muscle enzyme, the liver triosephosphate isomerase is composed of two different subunits.

Stability. The purified enzyme preparation retains all its activity for at least two years if kept at 4° in 2.5 M ammonium sulfate and 1 mM EDTA at pH 7.0.

pH Optimum. Activity as a function of pH is maximal in the rather broad range of pH 6.5–10, when measured directly without indicator enzyme.[1]

Ultraviolet Light Absorption. The absorbance of a 1% solution of the purified enzyme (pH 6.5) at 280 nm is 12.9 (1-cm light path).

Specific and Kinetic Properties. Rabbit liver and muscle triosephosphate isomerase have identical kinetic properties. As the yeast enzyme, they do not isomerize dihydroxyacetone and D-glyceraldehyde. The K_m values for the liver enzyme are 0.42 mM for D-GAP and 0.75 mM for DAP under the standard assay conditions. When recalculated for the unhydrated aldehyde[4] and ketone,[5] the constants are 14 μM for D-GAP and 0.41 mM for DAP. The turnover number with D-GAP is 5.2×10^5 min^{-1} at 25°.

Equilibrium Constants. The equilibrium constant is the same as that of the yeast and muscle enzyme.

Inhibitors. As for the yeast and the muscle triosephosphate isomerase, the following reagents act as inhibitors also for the liver enzyme: arsenite, phosphate, phosphoenolpyruvate, D-α-glycerophosphate, acetylphosphate, phosphoglycollate and D-erythrose-4-phosphate.[1,6]

In contrast to the yeast enzyme, triosephosphate isomerase of liver

[1] W. K. G. Krietsch, P. G. Pentchev, H. Klingenburg, T. Hofstätter, and T. Bücher, *Eur. J. Biochem.* **24**, 289 (1970).
[2] W. K. G. Krietsch, P. G. Pentchev, W. Machleidt, and H. Klingenburg, *FEBS Lett.* **11**, 137 (1970).
[3] W. K. G. Krietsch, P. G. Pentchev, and H. Klingenburg, *Eur. J. Biochem.* **23**, 77 (1971).
[4] D. R. Trentham, C. H. McMurray, and C. I. Pogson, *Biochem. J.* **114**, 19 (1969).
[5] S. J. Reynolds, D. W. Yates, and C. I. Pogson, *Biochem. J.* **122**, 285 (1971).
[6] H. Krietsch, G. Kuntz, and W. K. G. Krietsch, unpublished results.

and muscle is inhibited by sulfhydryl and alkylating agents and by photooxidation. The following group-selective reagents inhibit the liver and muscle enzyme: p-chloromercuribenzoate, 5,5'-dithiobis(2-nitrobenzoic acid), N-ethylmaleimide, o-iodosobenzoate, p-benzoquinone, iodoacetamide, iodoacetate, and 2,4-dinitrofluorobenzene. The liver enzyme contains, as the muscle enzyme, 10 free sulfhydryl groups titratable with 5,5'-dithiobis(2-nitrobenzoic acid). At least one of these groups is essential for the enzyme activity.[1,3]

[94] Triosephosphate Isomerase from Human Erythrocytes[1]

By ROBERT W. GRACY

D-Glyceraldehyde 3-phosphate ⇌ dihydroxyacetone phosphate

Assay Method

Principle. Triosephosphate isomerase activity can be measured in either direction by coupling to the appropriate dehydrogenase and following the rate of oxidation or reduction of NADH or NAD at 340 nm.

Glyceraldehyde 3-Phosphate as Substrate.[2,3] The assay mixture contains in a final volume of 1.0 ml, 50 mM triethanolamine buffer, pH 7.6, 1 mM EDTA, 0.15 mM NADH, 1.5 mM glyceraldehyde 3-phosphate and one unit of crystalline α-glycerophosphate dehydrogenase (EC 1.1.1.8). The reaction mixture is added to a 1-cm light path quartz cuvette and preincubated for 5 min in the sample chamber of a recording spectrophotometer which is thermostatically maintained at 30.0°. The reaction is initiated by the addition of 25 μl of enzyme sample which has been previously prediluted (in 50 mM triethanolamine buffer, pH 7.6) such that the rate of decrease in absorbance at 340 nm is 0.01–0.2 per minute. After correcting for the predilution and the dilution of the enzyme in the assay cuvette, the absorbance change per minute is divided by 6.22

[1] D-Glyceraldehyde-3-phosphate ketol-isomerase, EC 5.3.1.1. The work described here was supported in part by grants from the U.S. Public Health Service (AM14638), the Robert A. Welch Foundation (B-502), a Career Development Award from the National Institutes of Health (AM70198), and North Texas State University Faculty Research.
[2] G. Beisenherz, this series, Vol. 1, p. 387.
[3] E. E. Rozacky, T. H. Sawyer, R. A. Barton, and R. W. Gracy, *Arch. Biochem. Biophys.* **146**, 312 (1971).

(the mM absorbance index for NADH) to give micromoles of dihydroxy-acetone phosphate formed per minute per milliliter of enzyme solution.

Dihydroxyacetone Phosphate as Substrate.[3,4] The reaction can be monitored in the other direction by coupling to D-glyceraldehyde 3-phosphate dehydrogenase (EC 1.2.1.12) and following the reduction of NAD. The assay mixture in a final volume of 1.0 ml contains 50 mM triethanolamine buffer, pH 7.6, 1 mM EDTA, 5 mM dihydroxyacetone phosphate, 0.3 mM NAD, 13 mM sodium arsenate, and 1 unit of crystalline glyceraldehyde 3-phosphate dehydrogenase. The solution is brought to 30°, and the reaction is initiated with prediluted triosephosphate isomerase. The rate of increase in absorbance at 340 nm is recorded, and calculations as above yield the micromoles of D-glyceraldehyde 3-phosphate formed per minute per milliliter of enzyme solution.

Arsenate, a necessary component of the above assay, is a competitive inhibitor of triosephosphate isomerase, and corrections for this inhibition must be taken into consideration. This factor plus the higher cost of dihydroxyacetone phosphate make the former assay using glyceraldehyde 3-phosphate preferable for routine studies. In either of the above assays one unit of enzyme is defined as the amount catalyzing the isomerization of 1 μmole of substrate per minute, and specific activity is defined as units per milligram of protein.

Preparation of Substrates and Reagents. The two coupling enzymes are both commercially available as crystalline suspensions essentially free of triosephosphate isomerase activity. However, since triosephosphate isomerase is inhibited by ammonium sulfate[5] the coupling enzymes are dialyzed prior to use. Thus, a stable baseline and a linear initial velocity are recorded in either assay. The substrates are also available in highly purified form as their acetal or ketal derivatives, which must be hydrolyzed to the free aldehyde or ketone. For DL-glyceraldehyde 3-phosphate (diethylacetal), 100 mg of the barium salt is dissolved in 6.0 ml of water, containing 1.5 g of Dowex 50-X4-200R H-form. The mixture is heated in a boiling water bath for exactly 3 min, followed by cooling in ice and removal of the Dowex by filtration. The solution is adjusted to pH 6.8 with NaOH and decolorized with activated charcoal. The other substrate, dihydroxyacetone phosphate (dimethylketal cyclohexylammonium salt), is hydrolyzed by dissolving 100 mg in 8.0 ml of water and adding 2.0 g of Dowex as above. After stirring for 1 min the Dowex is removed by filtration, and the solution is hydrolyzed at 40° for 6 hr. After neutralization and decolorization, the solutions are stored at 0-2°. It should be

[4] P. K. Chiang, *Life Sci.* **10**, 831 (1971).
[5] D. H. Turner, E. S. Blanch, M. Gibbs, and J. F. Turner, *Plant Physiol.* **40**, 1146 (1965).

pointed out that the lability of the free triosephosphates necessitates that these solutions be prepared fresh weekly and that the concentrations of triosephosphates be determined daily by the coupled assays described above.

Protein Determination. Protein concentrations at various stages of the isolation are determined by the method of Lowry *et al.*[6] The concentration of the pure protein can be estimated from its absorbance at 280 nm using an absorbance index, $\epsilon_{1cm}^{1\%}$, of 12.9.[3]

Isolation Procedure

Fraction I. One unit (450 ml) of human blood is collected into 67.5 ml of standard anticoagulant (acid–citrate–dextrose solution). All subsequent steps are carried out in an ice bath or in a cold room at 0–4°. The plasma and buffy coat are removed after centrifugation at 8000 *g* for 60 min, and the erythrocytes are washed by resuspension in 500 ml of cold 0.145 *M* NaCl followed by centrifugation as above. After three washings, the erythrocytes are suspended in cold deionized water, and frozen and thawed several times; the lysate is centrifuged at 8000 *g* for 90 min to remove the cell debris. The pH of the supernatant solution is adjusted to 7.0 by the addition of cold 1 *M* sodium phosphate buffer.

Fraction II. Phosphocellulose (0.91 meq/g) which has been previously cleaned and equilibrated[7] in 5 m*M* sodium phosphate buffer, pH 7.0, is packed into a 20-cm diameter Büchner funnel to a height of 5 cm. The hemolysate is applied to the ion exchange cellulose and is eluted with 5 m*M* phosphate buffer, pH 7.0, under mild vacuum. Fractions of approximately 200 ml each are collected and analyzed for triosephosphate isomerase activity. Under these conditions triosephosphate isomerase passes through the ion exchanger while most of the other proteins, including hemoglobin, are bound to the cellulose. Fractions containing triosephosphate isomerase activity are pooled and simultaneously concentrated and dialyzed against 5 m*M* triethanolamine buffer, pH 8.0, by ultrafiltration.

Fraction III. The dialyzed fraction II is applied to a 3 × 75-cm column of DEAE-cellulose (0.91 meq/g) which has been previously cleaned and equilibrated in 5 m*M* triethanolamine buffer, pH 8.5. After washing with approximately 600 ml of this same buffer, a linear salt gradient is applied to elute the enzyme. The gradient mixing chamber contains 1 liter of 5 m*M* triethanolamine buffer, pH 8.5, and is connected by a siphon to the reservoir containing 1 liter of the same buffer which

[6] O. H. Lowry, N. J. Rosebrough, A. L. Farr, and R. J. Randall, *J. Biol. Chem.* **193**, 265 (1951).
[7] E. A. Peterson and H. A. Sober, this series, Vol. 5, p. 3

is also 0.4 M in NaCl. Under these conditions triosephosphate isomerase is eluted at approximately 0.15 M NaCl. Fractions containing triosephosphate isomerase activity are collected, concentrated and dialyzed against 5 mM triethanolamine buffer, pH 8.0 by ultrafiltration.

Fraction IV. The dialyzed fraction III is added to a 2.1 × 22-cm column of DEAE-Sephadex which has been equilibrated with 5 mM triethanolamine buffer, pH 8.0. After washing with 300 ml of the same buffer, a shallow salt gradient is initiated (500 ml of 5 mM triethanolamine buffer, pH 8.0, in the mixing chamber, and 500 ml of the same buffer containing 0.1 M NaCl in the mixing chamber). The enzyme is eluted at a flow rate of 30–40 ml per hr and the triosephosphate isomerase activity is found to be coincident with the third protein peak. The appropriate fractions are pooled and concentrated by ultrafiltration.

Fraction V. The enzyme (5–10 mg/ml) is crystallized by dialyzing against a 0.50 saturated solution of ammonium sulfate containing 50 mM triethanolamine buffer, pH 7.8, 1 mM EDTA, and 1 mM 2-mercaptoethanol. A small amount of amorphous precipitate may be formed and should be removed by centrifugation. The supernatant solution is then returned to the dialysis bag, and the ammonium sulfate concentration of the dialysate slowly increased to 0.76 saturation where crystallization begins. After 48 hr crystals are collected by centrifugation and resuspended in 50 mM triethanolamine buffer, pH 7.8 containing 1 mM EDTA and 10 mM 2-mercaptoethanol.

Specific activities of approximately 10,000 μmoles of glyceraldehyde 3-phosphate isomerized per minute per milligram of protein are routinely obtained. The isolation represents an overall purification of 4000–5000-fold with 40–60% recovery. Additional recrystallizations, or rechromatography on DEAE-cellulose, phosphocellulose, or Sephadex do not result in further increases in the specific activity. A typical isolation is presented in the table.

Resolution of the Three Forms. Crystalline human triosephosphate isomerase can be resolved into three forms by electrophoresis or isoelectric focusing.[3,8] For preparative purposes the enzyme is electrofocused in 1% narrow range (pH 5–7) Ampholines for 92 hr at 600 V. Under these conditions a component (I) comprising approximately 5–10% of the total triosephosphate isomerase activity is found with an apparent isoelectric pH of 6.3. A second component (II) comprising 70–75% of the total activity electrofocuses at pH 6.0, and a third component (III) is observed at pH 5.6, which accounts for 20–25% of the isomerase activity.

[8] T. H. Sawyer, B. E. Tilley, and R. W. Gracy, *J. Biol. Chem.* **247**, 6499 (1972).

ISOLATION OF TRIOSEPHOSPHATE ISOMERASE FROM HUMAN ERYTHROCYTES

Fraction	Total activity (units)	Total protein (mg)	Specific activity (units/mg protein)	Purification (fold)	Recovery (%)
Hemolysate	141,700	59,000	2.4	(1)	(100)
Phosphocellulose	99,900	3,000	33.3	13.9	70
First DEAE	94,500	295	320	133	67
Second DEAE	91,800	9.2	9,978	4158	65
Crystals	73,700	7.2	10,236	4265	52

Other Tissues. Triosephosphate isomerase has also been isolated by the above method from human skeletal and cardiac muscle. In contrast to the purification from erythrocytes, the enzyme is not homogeneous after the second chromatographic step on DEAE-cellulose and is contaminated with a protein of large molecular weight. However, this contaminant can be easily removed by gel filtration on a 2.5 × 98 cm column of Sephadex G-100. Since muscle tissue is a much better source of the enzyme it is preferred. However, as is the case with most human enzyme studies, one is often limited to erythrocytes as a tissue source. The three electrophoretic forms of human triosephosphate isomerase found in the erythrocytes are found in essentially the same proportions in muscle tissues.[9]

Physical Properties. Although the basis for the three electrophoretic forms is at this time still uncertain, a variety of studies[8] are consistent with an AA, AB, and BB distribution of dimers. The two types of subunits possess similar but distinguishable amino acid compositions and tryptic fingerprints. The three electrophoretic forms of human triosephosphate isomerase yield essentially identical molecular weights (56,000 ± 3000), sedimentation coefficients ($s_{20,w} = 4.1 \times 10^{-13}$ sec) and subunit molecular weights (28,000 ± 2000) when subjected to analytical ultracentrifugation. The enzyme is completely dissociated in 2.0 M guanidinium chloride, and can be totally renatured into catalytically active enzyme by removal of the dissociating agent.

Catalytic Properties. Normal Michaelis-Menten kinetics are observed in both forward and reverse assays, and K_m values for glyceraldehyde 3-phosphate of 0.34, 0.43, and 0.55 mM are obtained for components I, II, and III, respectively. Apparent K_m values for dihydroxyacetone phosphate of 1.5, 0.82, and 0.69 mM are found for components I, II, and III,

[9] R. M. Snapka, T. H. Sawyer, and R. W. Gracy, unpublished experiments.

respectively. These values are for the total triose phosphates in solution and do not take into consideration the relative distribution of hydrated *gem* diols or enediols which are in equilibrium with the free aldehyde and ketone.[10,11] The enzyme exhibits a broad pH optimum between 7.0 and 9.5 and is competitively inhibited[12] by phosphoglycolate ($K_i = 0.96 \ \mu M$), arsenate ($K_i = 8.15$ mM), α-glycerophosphate ($K_i = 0.42$ mM), and ATP ($K_i = 16$ mM).

Stability. Human triosephosphate isomerase is stable when stored in 0.80 saturated ammonium sulfate at 0–4°. It is recommended that the solution also contain 20 mM triethanolamine buffer, pH 7.5, 1 mM EDTA, and 1 mM dithiothreitol.

The Active Center. Human triosephosphate isomerase is rapidly inactivated by the substrate analog, chloroacetol phosphate, by the selective esterification of a single essential glutamyl γ-carboxylate per catalytic subunit.[13] Peptide maps and autoradiograms suggest the active-site peptide to be identical with the hexapeptide Ala-Tyr-Glu-Pro-Val-Trp isolated from rabbit[14,15] and chicken[16] muscle and yeast.[17]

[10] D. R. Trentham, C. H. McMurray, and C. I. Pogson, *Biochem. J.* **114**, 19 (1969).
[11] S. J. Reynolds, D. W. Yates, and C. I. Pogson, *Biochem. J.* **122**, 285 (1971).
[12] K_i values are given for the unresolved mixture of components I–III.
[13] F. C. Hartman and R. W. Gracy, *Biochem. Biophys. Res. Commun.* **52**, 388 (1973).
[14] F. C. Hartman, *Biochemistry* **10**, 146 (1971).
[15] J. C. Miller and S. G. Waley, *Biochem. J.* **123**, 163 (1971).
[16] S. DeLaMare, A. F. W. Coulson, J. R. Knowles, J. D. Priddle, and R. E. Offord, *Biochem. J.* **129**, 321 (1972).
[17] I. L. Norton, and F. C. Hartman, *Biochemistry* **11**, 4435 (1972).

[95] Triosephosphate Isomerase from Rabbit Muscle[1,2]

By FRED C. HARTMAN and I. LUCILE NORTON

D-Glyceraldehyde 3-phosphate $\rightleftharpoons$ dihydroxyacetone phosphate

Assay Method

Principle. Triosephosphate isomerase activity is most conveniently measured in the direction of formation of dihydroxyacetone phosphate by coupling with α-glycerophosphate dehydrogenase and monitoring the oxidation of NADH at 340 nm.[3] Alternatively, the enzyme can be assayed

[1] Research from the authors' laboratory was sponsored by the U.S. Atomic Energy Commission under contract with the Union Carbide Corporation.
[2] D-Glyceraldehyde-3-phosphate ketol-isomerase, EC 5.3.1.1.
[3] G. Beisenherz, this series, Vol. 1 [57].

in the opposite direction by coupling with glyceraldehyde-3-phosphate dehydrogenase. The latter method is complicated by inhibition of triosephosphate isomerase by arsenate, one of the substrates for the dehydrogenase.[4,5] For routine assays the expense of dihydroxyacetone phosphate relative to that of glyceraldehyde 3-phosphate also favors the former method, which is described here.

Reagents

Triethanolamine hydrochloride, 24 mM, pH 7.9, containing 6 mM EDTA

NADH (Sigma), 2.26 mM. Dissolve 8 mg of NADH in 5 ml of water.

DL-Glyceraldehyde 3-phosphate, 50 mM, pH 5.0. Mix 400 mg of the barium salt of the corresponding diethyl acetal (Sigma) with 10 ml of water and 5 g of Dowex 50 (H$^+$) (200–400 mesh). Shake the mixture vigorously until the barium salt dissolves and then place in a boiling-water bath for 3.5 min. Remove the resin by suction filtration and wash on the filter pad with two 5-ml portions of water. Combine the washings and initial filtrate, bring the resulting solution to pH 5.0 with solid sodium bicarbonate, and store in the freezer. The exact concentration of D-glyceraldehyde 3-phosphate is determined enzymically with the assay described below, with the exception that the aldehyde concentration is reduced to a level below that of NADH.

α-Glycerophosphate dehydrogenase, 10 mg/ml (Sigma). Dilute 0.07 ml to 5 ml with water.

Buffer–substrate mixture. Prepare this solution daily by diluting 0.48 ml of the stock glyceraldehyde 3-phosphate to 20 ml with the triethanolamine buffer.

Procedure. The assay is carried out at 24°. To a 3-ml quartz cuvette with a 1-cm path length are added 2.5 ml of the buffer–substrate solution, 0.2 ml of NADH, and 0.2 ml of α-glycerophosphate dehydrogenase. The contents are thoroughly mixed, and the cuvette is placed in a recording spectrophotometer. The A at 340 nm is set between 0.5 and 1.0, and 0.1 ml of the solution to be assayed (equivalent to 5–15 ng of pure triosephosphate isomerase; 0.033–0.10 unit) is added with rapid mixing. The decrease in A is linear with respect to time for several minutes. A control cuvette, containing all assay ingredients except the isomerase, is also monitored since commercial α-glycerophosphate dehydrogenase sometimes contains slight triosephosphate isomerase activity, thus necessitating a

[4] P. M. Burton and S. G. Waley, *Biochim. Biophys. Acta,* **151,** 714 (1968).
[5] W. K. G. Krietsch, P. G. Pentchev, H. Klingenbürg, T. Hofstätter, and T. Bücher, *Eur. J. Biochem.* **14,** 289 (1970).

correction. The dehydrogenase can be freed of isomerase activity by incubation with 3-chloroacetol phosphate (0.1 mM), an active site-specific reagent for triosephosphate isomerase.[6] The excess reagent is decomposed with β-mercaptoethanol (1 mM) before the dehydrogenase is used in the isomerase assay.

Definition of Enzyme Unit and Specific Activity. One unit of activity is defined as the conversion of 1 μmole of glyceraldehyde 3-phosphate into dihydroxyacetone phosphate per minute and represents a decrease in A at 340 nm of 2.07 absorbancy units per minute. Protein is measured by the optical method of Lowry *et al.*[7] or, in the case of the purified enzyme, by the A at 280 nm using an $\epsilon_{1\,cm}^{1\%}$ of 13.1.[8] Specific activity is defined as units per milligram of protein.

Purification Procedure

The purification procedure given here is essentially that described by Norton *et al.*[8] Another procedure has been described.[5,9,10]

A summary of the purification procedure is given in the table. Unless stated otherwise, all operations are performed at 4°, and aqueous solutions are prepared with glass-distilled water.

Step 1. Extraction. Frozen, deboned muscle tissue from the back and hind legs of young (8–12 weeks old) rabbits is purchased from Pel-Freez Biologicals, Inc. Partially thawed tissue (450 g) is ground twice in a chilled meat grinder. The ground tissue is divided into three equal portions, each of which is homogenized with 200 ml of 1.5 mM EDTA (pH 5.2) for 1 min at high speed in a Waring Blendor. The resulting homogenates are pooled, stirred for 30 min, and centrifuged for 15 min at 10,400 g. The supernatant is strained through glass wool. The sediment is rehomogenized in 400 ml of 1.5 mM EDTA, and the mixture is stirred and centrifuged as above. The combined muscle extracts are adjusted to 930 ml with 1.5 mM EDTA, and β-mercaptoethanol is added to a final concentration of 1 mM.

Step 2. Acetone Fractionation. During a 3-min period, 500 ml of acetone (precooled to −15°) is added with continuous stirring to the muscle extract to give a final acetone concentration of 35% (v/v). During the

---

[6] F. C. Hartman, this series, Vol. 25 [59].

[7] O. H. Lowry, N. J. Rosebrough, A. L. Farr, and R. J. Randall, *J. Biol. Chem.* **193**, 265 (1951).

[8] I. L. Norton, P. Pfuderer, C. D. Stringer, and F. C. Hartman, *Biochemistry* **9**, 4952 (1970).

[9] R. Czok and T. Bücher, *Advan. Protein Chem.* **15**, 315 (1960).

[10] W. K. G. Krietsch, P. G. Pentchev, and H. Kingenbürg, *Eur. J. Biochem.* **23**, 77 (1971).

PURIFICATION OF RABBIT MUSCLE TRIOSEPHOSPHATE ISOMERASE

Fraction	Volume (ml)	Total protein (mg)	Total activity (units)	Specific activity (units/mg)	Yield (%)
1. Extract	930	18,600[a]	3,200,000	170	100
2. 54% Acetone precipitate	100	4,400[a]	2,400,000	545	75
3. Heat treated	96	2,250[b]	2,100,000	950	67
4. Sephadex G-100 pool	206	608[b]	1,810,000	2,980	67
5. Crystals from ammonium sulfate (3.6–3.7 molal)	—	169[b]	1,158,000	6,900	37
6. After SE-Sephadex and repeated crystallization	—	85[b]	663,000	7,800	21

[a] Based on Lowry determinations (see text under *Definition of Enzyme Unit and Speicfic Activity*).

[b] Based on A at 280 nm (see text under *Definition of Enzyme Unit and Specific Activity*).

addition, the temperature of the mixture is maintained below 7°. After centrifugation for 10 min at 10,400 g the precipitate is discarded, and acetone (183 ml/liter of 35% acetone supernatant) is added, with stirring, to the supernatant to a final concentration of 45%. Best results are obtained when this addition is accomplished in about 30 min, and the temperature of the mixture is not allowed to exceed 4°. The 45% acetone mixture is centrifuged as before. The precipitate is discarded, and acetone (195 ml/liter of 45% acetone supernatant) is added to the supernatant, as described in the previous step, to a final concentration of 54%. This mixture is centrifuged, and the precipitate is processed further.

Step 3. Heat Treatment. The 54% acetone precipitate is dissolved in 50 mM sodium phosphate–3 mM EDTA–1 mM β-mercaptoethanol (pH 7.0) to a final volume of 100 ml. The slightly turbid solution is stirred gently in a 100 × 50 mm crystallizing dish with a magnetic stirrer at 25° for 12 hr to remove acetone. The volume is readjusted to 100 ml with the same buffer, and the solution is brought to 51° during 10 min by gentle swirling in a constant-temperature bath and maintained at the elevated temperature for 30 min. Precipitated material is removed by centrifugation at 15,900 g for 15 min.

Step 4. Sephadex G-100. The supernatant (90 ml) from the preceding step is passed through a 6.0 × 185 cm column of Sephadex G-100 equilibrated with 50 mM Tris hydrochloride–3 mM EDTA–1 mM β-mercaptoethanol (pH 7.5). Triosephosphate isomerase elutes at 1550–1800 ml. Fractions containing the enzyme with a specific activity greater than 1300 are pooled and dialyzed against Tris buffer (as used for gel filtration) saturated with ammonium sulfate (ultrapure biological grade from Schwarz Mann). The precipitated protein is collected by centrifugation at 15,900 g for 20 min.

Step 5. Crystallization. The precipitate is redissolved in the same Tris buffer (no ammonium sulfate) to a final protein concentration of 4.25 mg/ml (final volume, 138 ml), and solid ammonium sulfate (62 g) is added to 3.4 molal. Precipitated protein is removed by centrifugation at 15,900 g for 20 min. The supernatant is brought to 3.5 molal ammonium sulfate (1.8 g added), and the precipitated protein is again removed by centrifugation at 15,900 g for 20 min. The supernatant is brought to 3.6 molal ammonium sulfate (1.8 g added) and allowed to stand undisturbed. After 3 days, the needle-shaped crystals of triosephosphate isomerase are collected by centrifugation. A second crop of crystals is obtained by increasing the ammonium sulfate concentration to 3.7 molal (1.8 g added). The crystalline triosephosphate isomerase obtained at this step is approximately 90% pure as judged by disc gel electrophoresis and is therefore suitable for some purposes. If homogeneous material is required, the next step should be carried out.

Step 6. SE-Sephadex Chromatography. Crystalline triosephosphate isomerase from the preceding step is dialyzed against 50 mM sodium phosphate–3 mM EDTA (pH 5.5) and placed on a 2 × 57 cm column of SE-Sephadex C-50 equilibrated with the same phosphate buffer. The column is eluted with a linear gradient of sodium chloride (0–50 mM) in the phosphate buffer. The eluted enzyme is precipitated as in the gel filtration step and crystallized as in the preceding step.

Properties[11]

Purity. The enzyme with a specific activity of 7800 units/mg is homogeneous as judged by analytical ultracentrifugation and disc gel electrophoresis at pH 9.5. This preparation, as well as others that have been characterized, contains isoenzymic forms, which can be detected by various electrophoretic techniques[5,8,12] and separated on a preparative basis by chromatography on DEAE-cellulose.[10] These resolvable isoenzymes apparently have identical primary sequences,[13] and the structural basis for their existence remains uncertain.

[11] A thorough review article on triosephosphate isomerase was published recently [E. A. Noltmann, *in* "The Enzymes," 3rd Ed., Vol. VI (P. D. Boyer, ed.), p. 326. Academic Press, New York, 1972], so only a brief summary of the properties of the enzyme is presented here.

[12] P. M. Burton and S. G. Waley, *Biochem. J.* **107**, 737 (1968).

[13] P. H. Corran and S. G. Waley, *FEBS Lett.* **30**, 97 (1973).

Molecular Properties. The crystalline isomerase has a sedimentation constant of 6.95 S and a molecular weight of 53,000 as determined by sedimentation velocity and sedimentation equilibrium experiments, respectively.[8] A partial specific volume $(\bar{v})$ of 0.737 ml/g, determined from the amino acid composition, was used in the calculation of molecular weight. Based on direct density measurements, $\bar{v}$ is 0.741 ml/g.[14] The exact molecular weight calculated from the recently elucidated primary structure is 53,257, and the chain length is 248 residues.[13]

Triosephosphate isomerase is a dimer,[12,15] and each subunit contains one active site.[16] The two subunits have identical amino acid sequences with the possible exception of partial deamidation of some asparaginyl and glutaminyl residues, which could account for the formation of isoenzymes.[13]

Catalytic Properties. A striking catalytic feature of triose-phosphate isomerase is the high turnover number of the active subunit—approximately 2000 sec^{-1}. The enzyme exhibits maximal activity at pH 6.5–10.[5] At pH 7.8 and 25°, the apparent K_m for D-glyceraldehyde 3-phosphate is 0.32–0.46 mM and for dihydroxyactone phosphate 0.62–0.87 mM.[4,5] The apparent equilibrium constant, (dihydroxyactone phosphate)/ (glyceraldehyde 3-phosphate), is 22.[4,17] To obtain the actual K_m values and equilibrium constant, one must correct for the finding that the true substrates are the free carbonyl forms of glyceraldehyde 3-phosphate and dihydroxyacetone phosphate and that the corresponding hydrated forms (*gem*-diols) are not substrates.[18,19] Thus, the K_m values for glyceraldehyde 3-phosphate and dihydroxyacetone phosphate become 11–15 μM and 0.34–0.48 mM, respectively, and the equilibrium constant becomes 367.[19]

Inhibitors. A wide variety of organic and inorganic anionic substances are competitive inhibitors of the isomerase.[4,5,20] The best ordinary competitive inhibitor found to date is D-α-glycerophosphate with a K_i of 0.12 mM.[4] Two transition-state analogs have been described. These are phosphoglycolic acid[21] and its corresponding hydroxamate,[22] which have

[14] J. D. McVittie, M. P. Esnouf, and A. R. Peacocke, *Eur. J. Biochem.* **29**, 67 (1972).

[15] L. N. Johnson and S. G. Waley, *J. Mol. Biol.* **29**, 321 (1967).

[16] F. C. Hartman, *Biochemistry* **10**, 146 (1971).

[17] R. L. Veech, L. Raijman, K. Dalziel, and H. A. Krebs, *Biochem. J.* **115**, 837 (1969).

[18] D. R. Trentham, C. H. McMurray, and C. I. Pogson, *Biochem. J.* **114**, 19 (1969).

[19] S. J. Reynolds, D. W. Yates, and C. I. Pogson, *Biochem. J.* **122**, 285 (1971).

[20] L. N. Johnson and R. Wolfenden, *J. Mol. Biol.* **47**, 93 (1970).

[21] R. Wolfenden, *Nature (London)* **223**, 704 (1969).

[22] Work of K. D. Collins, quoted by G. E. Lienhard, *Science* **180**, 149 (1973).

K_i's of 6 μM and 3 μM, respectively. It is believed that these compounds more closely resemble the enediol anion intermediate[23,24] than the substrate.

Two kinds of active site-specific reagents for triosephosphate isomerase have been designed—3-haloacetol phosphates[25,26] and glycidol phosphate.[27] Each inactivates the enzyme by a highly selective esterification of a single glutamyl γ-carboxylate at the active site.[6,16,28–31] This carboxyl may be the acid-base group that effects net proton transfer between C1 and C2 of the substrate. Sequence studies show that the active-site glutamyl residue occupies position 165 in the polypeptide chain.[13]

[23] S. V. Rieder and I. A. Rose, *J. Biol. Chem.* **234**, 1007 (1959).
[24] J. R. Knowles, P. F. Leadlay, and S. G. Maister, *Cold Spring Harbor Symp. Quant. Biol.* **36**, 157 (1972).
[25] F. C. Hartman, *Biochem. Biophys. Res. Commun.* **33**, 888 (1968).
[26] A. F. W. Coulson, J. R. Knowles, and R. E. Offord, *Chem. Commun.* **1**, 7 (1970).
[27] I. A. Rose and E. L. O'Connell, *J. Biol. Chem.* **244**, 6548 (1969).
[28] I. L. Norton and F. C. Hartman, *Biochemistry* **11**, 4435 (1972).
[29] F. C. Hartman and R. W. Gracy, *Biochem. Biophys. Res. Commun.* **52**, 388 (1973).
[30] J. C. Miller and S. G. Waley, *Biochem. J.* **123**, 163 (1971).
[31] S. De La Mare, A. F. W. Coulson, J. R. Knowles, J. D. Priddle, and R. E. Offord, *Biochem. J.* **129**, 321 (1972).

[96] L-Arabinose Isomerase

By JIM PATRICK and NANCY LEE

L-Arabinose $\leftrightarrows$ L-ribulose

Assay Method

Principle. L-Arabinose isomerase activity is assayed by measuring the rate of formation of L-ribulose.[1,2] L-ribulose concentration is determined using the cystein-carbazole test.[3]

Reagents

Glycylglycine, 0.25 M, pH 7.6
MnCl$_2$, 50 mM
L-Arabinose, 1 M
HCl, 0.1 M

[1] E. Englesberg, *J. Bacteriol.* **81**, 996 (1961).
[2] N. Lee and E. Englesberg, *Proc. Nat. Acad. Sci. U.S.* **48**, 335 (1962).
[3] Z. Dische and E. Borenfreund, *J. Biol. Chem.* **192**, 583 (1951).

Cysteine
Carbazole
H_2SO_4

Procedure.[4] A reaction mixture containing 5 parts of glycylglycine, 0.5 part of $MnCl_2$, 1.5 parts of L-arabinose, and 3 parts of water is prepared and stored in a freezer. When needed, the prepared reaction mixture is dispensed in 1-ml quantities into Wassermann tubes and warmed to 37°. The reaction is initiated by adding 3–20 units of L-arabinose isomerase activity in a total volume of 10 μl or less. At intervals of 2 or 3 min 0.1-ml aliquots of the reaction mixture are removed to culture tubes (16 × 150 mm) containing 0.9 ml of 0.1 M HCl. The reaction is usually sampled 3 times over 6 or 9 min.

To each tube containing 0.9 ml of HCl and 0.1 ml of reaction mixture is added 0.2 ml of cysteine. Each tube then receives 6 ml of H_2SO_4 followed by 0.2 ml of carbazole and vigorous stirring on a Vortex mixer. After 20 min the optical density at 540 nm is determined and related to standard curves determined from known concentrations of L-ribulose.

Definition. One unit of enzyme activity will convert 1 μmole of L-arabinose to L-ribulose per hour.

Purification Procedures[4]

Source. L-Arabinose isomerase from *Escherichia coli* B/r is perhaps most easily isolated from strain F′ ara B^{-24}/ara B^{-24}. This strain is diploid with respect to the arabinose genes A, B, C, and D and carries the ara B^{-24} mutation on both the chromosome and the episome. As a consequence of this mutation, the strain produces no L-ribulokinase activity. In addition, however, the strain produces amounts of L-arabinose isomerase that account for about 12% of the total soluble protein. This greatly facilitates purification.

Growth and Harvesting of Cells. The bacteria are grown in a New Brunswick Fermacell Fermentor in 40-liter quantities. The growth medium has the following composition: K_2HPO_4–KH_2PO_4 (pH 7.0), 1%; $MgSO_4 \cdot 7H_2O$, 0.01%; $(NH_4)_2SO_4$, 0.1%; Casamino acids (Difco), 1%; $MnCl_2 \cdot 4H_2O$, 0.002%; and L-arabinose, 0.15%. Thirty-six liters of the above medium is sterilized, cooled to 37° and inoculated with 4 liters of a log phase culture grown in the same medium minus L-arabinose. The culture is maintained at 37° with aeration at 2.5 ft³/min for 8 hr and harvested in a refrigerated Sharples centrifuge. The cell pellet is suspended in sufficient cold 10 mM glycylglycine (adjusted to pH 7.6 with

[4] J. Patrick and N. Lee, *J. Biol. Chem.* **243**, 4312 (1968).

1 N KOH) to yield 77 mg dry weight protein per milliliter. The suspension may be used immediately or frozen in a dry-ice acetone bath and stored at $-20°$ until needed. All subsequent steps are performed at about $4°$.

Preparation of Crude Extract. Frozen cells are thawed and sonically disrupted in 100 ml quantities for 6–8 min at level 8 of a Branson 6-kHz sonifier. The sonically treated suspension is centrifuged for 1 hr at 56,000 g in a Spinco Model L ultracentrifuge. The supernatant fluid (crude extract) is recovered.

Precipitation with Manganese. A 1.0 M aqueous solution of $MnCl_2$ is added slowly with stirring to the crude extract to give a final concentration of 50 mM $MnCl_2$. Fifteen minutes after the last addition of $MnCl_2$ the solution is centrifuged at 66,000 g for 1 hr, and the supernatant fluid (Mn^{2+} supernatant) recovered.

Fractionation with Ammonium Sulfate. The Mn^{2+} supernatant fluid is immediately adjusted to pH 7.6 by the addition of an aqueous solution of NH_4OH (3%). Solid $(NH_4)_2SO_4$ is added to 40% saturation over a period of 30 min, and the preparation allowed to equilibrate with stirring for an additional 30 min. The supernatant is recovered after centrifugation for 15 min at 35,000 g and brought to 55% saturation by the addition of solid $(NH_4)_2SO_4$ in the manner just described. The precipitate recovered after centrifugation at 35,000 g is suspended in 10 mM K_2HPO_4–KH_2PO_4 (pH 7.6) (hereafter referred to as column buffer) and dialyzed against 2 liters of the same buffer. The dialysis buffer is changed once after 8 hr, and dialysis is continued for 2 hr. The enzyme solution is removed from the dialysis bag and clarified by centrifugation at 35,000 g for 10 min.

Chromatography on DEAE-Cellulose. About 400 g of DEAE-cellulose are washed with 20 liters of 0.5 N NaOH, rinsed to neutrality with distilled water, and adjusted to pH 3 to 4 with concentrated H_3PO_4. The acidified material is washed to neutrality with distilled water and then suspended in 2 liters of 0.1 M K_2HPO_4–KH_2PO_4 (pH 7.6). After being washed once with 4 liters of distilled water, the DEAE-cellulose is packed in a column (4.9×125 cm) and equilibrated with column buffer. The enzyme recovered from step 3 is applied to the column followed by 500 ml of column buffer. Elution is carried out with a linear gradient of NaCl (0 to 0.3 M in column buffer) in a total volume of 4 liters. Fractions of 17 ml are collected at a flow rate of 150 ml per hour. Protein and enzyme activity are determined in each fraction and fractions of high specific activity are pooled. The pooled fractions are precipitated by the addition of solid $(NH_4)SO_4$ to 70% saturation over a period of 1 hr. The preparation is allowed to equilibrate, with stirring, for 30 min, and

PURIFICATION OF L-ARABINOSE ISOMERASE

Step and fraction	Volume (ml)	Protein (mg)	Total activity (units × 10⁻⁶)	Specific activity (units/mg protein)	A_{280}/A_{260}	Yield (%)
1. Crude extract	335	21,300	22.8	1,070	—	100
2. MnCl₂ supernatant	305	—	21.6	—	—	94.5
3. Ammonium sulfate fraction, 40–55% (dialyzed)	77	6,960	16.2	2,330	—	71.0
4. DEAE-cellulose pooled fractions	234	987	8.15	8,250	1.68	35.8
5. Sephadex G-200 pooled fractions	18	266	3.19	12,000	1.74	14.0

the precipitate recovered after centrifugation at 35,000 g for 15 min. The pellet is suspended in a minimum volume of column buffer and dialyzed against 2 liters of the same buffer overnight. If necessary, the solution is clarified by centrifugation after removal from the dialysis bag.

Gel Filtration on G-200 Sephadex. Sephadex G-200 is allowed to swell in sterile glass-distilled water: after the fines are removed, the Sephadex is packed in a column (2.3 × 45 cm, Pharmacia), and equilibrated with column buffer. Up to 12 ml of the enzyme solution from step 4 are layered on the column and eluted with column buffer. The effluent is collected in fraction of 15–2.0 ml at a flow rate of 7–10 ml/hr and assayed for protein.

Protein-containing fractions from the G-200 column may be assayed for enzyme activity and pooled on the basis of specific activity. Alternatively, each fraction may be examined by sedimentation in an analytical ultracentrifuge and pooled on the basis of homogeneity by this assay. This latter technique is the more sensitive of the two.

Storage of Purified Enzyme. Pooled fractions from the G-200 column were either sterilized by passing through a Millipore filter and stored at 4°, or lyophilized and stored at −20° in sealed evacuated ampules. Enzyme stored in either manner retained activity for at least 1 month.

A summary of a typical purification is found in the accompanying table.

Properties

Purity and Structure.[4,5] Purified L-arabinose isomerase from *E. coli* B/r forms a single band on electrophoresis in acrylamide gels and forms

[5] J. Patrick and N. Lee, *J. Biol. Chem.* **244**, 4277 (1969).

a single symmetrical peak when sedimented in an analytical centrifuge. The $s^o_{20,w}$ is 12.6×10^{-13} sec and shows no dependence on protein concentration. The molecular weight determined by high and low speed sedimentation equilibrium is $3.62 \pm 0.16 \times 10^5$. In the presence of urea (8 M) or pH 2 phosphate buffer, the protein dissociates. Electrophoresis of L-arabinose isomerase on acrylamide gels containing 8 M urea yields a single band. Sedimentation in the ultracentrifuge in the presence of 8 M urea or pH 2 phosphate buffer results in a single symmetrical peak with a $s^o_{20,w}$ of 2.85×10^{-13} sec. The molecular weight of this species determined by both high and low speed sedimentation equilibrium is $6.0 \pm 0.3 \times 10^4$. These observations suggest that L-arabinose isomerase is composed of 6 identical polypeptide chains. This suggestion has received support from peptide mapping of cyanogen bromide-cleaved enzyme.[5]

Substrate Specificity. L-Arabinose isomerase is active on L-arabinose and has a K_m for L-arabinose of 60 mM. Only slight (less than 10% of the activity on L-arabinose) activity is found when D-fucose, L-fucose, or D-xylose are used as substrates. No activity is detected when D-arabinose, D-glucose, L-glucose, D-galactose, D-mannose, L-mannose, or L-rhamnose are used as substrates.

Inhibition. Both L-arabitol and ribitol are effective inhibitors of L-arabinose isomerase activity on L-arabinose. The K_i for arabitol and ribitol, respectively, is 18 mM and 1 mM.

pH optimum. The activity at pH 5 is approximately 5% of the activity at pH 6. No change in activity is found at pH values between 6 and 8.

Effect of Cations on Activity. Although purified L-arabinose isomerase shows no dependency on Mn^{2+} ion, dialysis against buffer containing 1 mM EDTA results in enzyme preparation with decreased activity. The activity can be restored with Mn^{2+} ion, but not with Ca^{2+}, Co^{2+}, Cd^{2+}, Cu^{2+}, or Mg^{2+}.

Effect of Manganous Ion on Synthesis.[6] The specific activity of L-arabinose isomerase from cultures grown in the presence of Mn^{2+} ion is 2-fold greater than that from cultures grown in the absence of Mn^{2+} ion. Added Mn^{2+} ion is without effect on the specific activities of L-ribulokinase or L-ribulose-5-phosphate 4-epimerase enzymes whose structural genes are located on either side of the structural gene for L-arabinose isomerase.[7] The effect of Mn^{2+} ion on L-arabinose isomerase is to cause the synthesis of an enzyme with greater intrinsic activity, but not to change the quantity of L-arabinose isomerase produced. In addition to possessing

[6] J. Patrick, N. Lee, N. B. Barnes, and E. Englesberg, *J. Biol. Chem.* **246,** 5102 (1971).
[7] E. Englesberg, R. L. Anderson, R. Weinberg, N. Lee, P. Hoffee, G. Huttenhauer, and H. Boyer, *J. Bacteriol.* **84,** 137 (1962).

a lower molecular activity, enzyme synthesized without added Mn^{2+} ion is more temperature labile than enzyme synthesized in the presence of added Mn^{2+} ion. Several divalent cations can confer temperature stability, but not full activity, to an enzyme synthesized without added Mn^{2+}. In contrast, of the ions tested, only Mn^{2+} is able to cause the synthesis of a fully active enzyme. Mn^{2+} is, however, unable to convert partially active enzyme to fully active enzyme either *in vitro* or *in vivo*.

[97] L-Arabinose Isomerase from *Lactobacillus gayonii*

By KEI YAMANAKA

L-Arabinose $\rightleftarrows$ L-ribulose

This enzyme from *Lactobacillus gayonii* has been described briefly.[1] Comprehensive data on crystalline enzyme will be discussed in this section.

Assay Method

L-Arabinose isomerase activity can be assayed either by the spectrophotometric method or by the cysteine-carbazole test as described in a previous paper.[1]

Purification Procedure[2]

Culture. The growth medium and culture of the organism are the same as described previously.[1] The composition of medium is as follows: 1% peptone, 1% sodium acetate, 0.2% yeast extract, 0.02% $MgSO_4 \cdot 7H_2O$, 0.01% $MnSO_4 \cdot 4H_2O$, 0.01% $CoCl_2 \cdot 6H_2O$, 1% D-glucose, and 0.1% L-arabinose.

Preparation of Cell-Free Extracts. The washed L-arabinose-grown cells, 60 g in wet weight from 40 liters of medium, are disrupted in small portion by grinding with about 150–200 g of levigated alumina,[3] and the enzyme is extracted with about 650 ml of 20 mM Tris·HCl buffer at pH 7.5. Alumina and cell debris are removed by centrifugation (crude extract, 594 ml). To the crude extract, 30 ml of 1 M $MnCl_2$ is added dropwise and pH is maintained at 7.0–7.5 by adjusting with 1 N NaOH.

---

[1] K. Yamanaka and W. A. Wood, see this series, Vol. 9 [106].

[2] T. Nakamatu and K. Yamanaka, *Biochim. Biophys. Acta* **178**, 156 (1969).

[3] Levigated alumina, about 300 mesh for chromatography was purchased from Wako Pure Chemicals, Osaka, Japan.

After standing for 30 min, the precipitate is centrifuged and discarded (MnCl$_2$-treated fraction, 610 ml).

Ammonium Sulfate Fractionation. The amount of ammonium sulfate is calculated from the table given by Green and Hughes[4] with a temperature correction of 0.92 for conversion to 0°. To the manganese-treated fraction, 176 g of ammonium sulfate is added (50% saturation). The precipitate is discarded. The supernatant is treated with 195 g of ammonium sulfate (95% saturation). The precipitate is dissolved in 20 mM Tris·HCl buffer (pH 7.5) and dialyzed overnight against the same buffer containing 5 mM MnSO$_4$ (102 ml).

Heat Treatment. The enzyme solution in the presence of 5 mM MnCl$_2$ is immersed in a water bath at 80°. When the temperature of the solution reaches 47°, the flask is transferred to another water bath at 50° and maintained for 5 min at the same temperature. The fraction is cooled in an ice bath and the coagulated proteins are removed by centrifugation and discarded (98 ml). To the supernatant is added 70 g of ammonium sulfate to 100% saturation. The precipitate is dissolved in 20 mM Tris·HCl buffer (pH 7.5) and dialyzed overnight with the same procedure as that described above (80 ml).

Column Chromatography on DEAE-Cellulose. The enzyme is chromatographed on DEAE-cellulose (3.0 cm × 70 cm) which had been equilibrated with 20 mM Tris·HCl buffer at pH 7.5. The proteins are eluted with a linear gradient of KCl between 0 and 0.6 M at pH 7.5. Active fractions are pooled and the enzyme precipitated with ammonium sulfate to 90% of saturation. The enzyme is dissolved, passed through a column of Sephadex G-200 (1.5 cm × 90 cm) and eluted with 20 mM Tris·HCl buffer at pH 7.5. Active fractions are combined (17.6 ml), and the enzyme is precipitated with ammonium sulfate to 80% saturation.

Crystallization. Crystallization of the enzyme is carried out with the addition of saturated ammonium sulfate solution. After overnight dialysis, 0.98 ml of a saturated ammonium sulfate is added to the enzyme solution (1.15 ml, protein concentration is about 3.5%). Ammonium sulfate content reached is 46% saturation. The precipitate is removed by centrifugation. The ammonium sulfate content is slowly increased at a rate of not more than 0.05 saturation per day until 56% saturation is reached. A trace of precipitate is removed at each step by centrifugation. The amorphous precipitate between 0.51 and 0.558 saturation contains a small amount of the L-arabinose isomerase activity. Finally, the solution is brought to 63% saturation and the clear solution held at 5° for 4 days. The crystals are collected by centrifugation and dissolved in 2.5

---

[4] A. A. Green and W. L. Hughes, see this series, Vol. 1 [10].

TABLE I
PURIFICATION OF L-ARABINOSE ISOMERASE

Fraction	Total protein (mg)	Total units[a]	Specific activity (μmoles/mg/min)	Yield (%)
Crude extract	2380	1188	0.50	
MnCl$_2$-treated fraction	1340	1390	1.03	117
Ammonium sulfate fraction (50–95% saturation)	1400	1420	1.01	119
Mn-heated fraction	1360	1190	0.88	100
DEAE-cellulose eluate	158	530	3.32	45
Sephadex G-200 eluate	67	364	5.44	33
Crystals	45	239	5.31	20

[a] Colorimetric units.

ml of 20 mM Tris·HCl buffer at pH 7.5 and dialyzed against 100 ml of the same buffer containing 5 mM MnSO$_4$ overnight (first crystals). Recrystallization is performed with the same procedure at 63% saturation. The purification procedure is summarized in Table I.

Properties[2]

Purity. The second crystals show a single symmetrical moving peak on ultracentrifugation.

Molecular Weight. The molecular weight is estimated by centrifugation in a sucrose density gradient by the method of Martin and Ames[5] as 2.71 × 10^5.

Effect of pH. The maximum activity is attained at pH 6.0–7.0 at 35° for 10 min of incubation.

Stability. The enzyme is stable at pH 5.5–9.0 for 10 min of incubation at 50°.

Substrate Specificity. The crystalline enzyme is specific for L-arabinose and L-ribulose. Other pentoses and hexoses are inactive for the enzyme.

Metal Requirement. The enzyme requires Mn^{2+} for activity. The solution of second crystals is dialyzed against 10 mM Tris·HCl buffer at pH 8.0 containing 5 mM EDTA at 2° for 48 hr, then dialyzed against the same buffer without EDTA for 24 hr. The activity of the dialyzed enzyme preparation can be recovered specifically by the addition of Mn^{2+}, and cobaltous ion activates to about half the extent of Mn^{2+}. Potassium

[5] R. G. Martin and B. N. Ames, *J. Biol. Chem.* **236**, 1372 (1961).

TABLE II
INHIBITION CONSTANTS OF PENTITOLS FOR L-ARABINOSE ISOMERASE

Sources	Inhibition constant (K_i) (mM)		
	Ribitol	L-Arabitol	Xylitol
Lactobacillus gayonii[a]	6	7.5	38
Aerobacter aerogenes, PRL-R3[b]	0.35	2.3	ND[c]
A. aerogenes, M-7[d]	1.6	2.2	27
Escherichia coli[e]	6	18	ND
Streptomyces[f]	1.0	1.1	15
Clostridium acetobutylicum[g]	ND	23	ND

[a] T. Nakamatu and K. Yamanaka, *Biochim. Biophys. Acta* **178**, 156 (1969).
[b] K. Yamanaka and W. A. Wood, see this series, Vol. 9 [106].
[c] ND, not determined.
[d] K. Izumori and K. Yamanaka, *J. Ferment. Technol.* **51**, 452 (1973).
[e] J. W. Patrick and N. Lee, *J. Biol. Chem.* **234**, 4312 (1968).
[f] K. Yamanaka and K. Izumori, *Agr. Biol. Chem.* **37**, 521 (1973).
[g] M. Tomoeda, H. Horitsu, and I. Sasaki, *Agr. Biol. Chem.* **33**, 151 (1969).

and strontium ions have a slight activation effect, but Na$^+$, Li$^+$, Mg^{2+}, and Ba^{2+} are ineffective. Zn^{2+}, Cu^{2+}, Fe^{2+}, and Hg^{2+} are inhibitory. The Michaelis constant for Mn^{2+} is 5.25 μM.

Effect of Substrate Concentration. The reaction mechanism of L-arabinose isomerase is compulsory ordered and involves the formation of a ternary complex of enzyme-Mn-substrate. Since dialysis against EDTA appears to yield a Mn-free enzyme preparation, the apparent Michaelis constant for L-arabinose is determined with 5, 10, 20, and 100 μM of Mn^{2+}. These calculated apparent Michaelis constants are then plotted vs the reciprocal of the concentration of Mn^{2+}. From the intercept of this line, the true Michaelis constant for L-arabinose is obtained as 55 mM. The affinity for L-ribulose is calculated as 5.0 mM from the conventional Lineweaver-Burk equation.

Inhibition by Pentitols. Activity of the crystalline L-arabinose isomerase is competitively inhibited by structurally related pentitols. The K_i values of pentitols for the isomerase from several bacterial sources are summarized in Table II. The enzyme is most sensitive to ribitol for all species, and L-arabitol is the most potent inhibitor. D-Arabitol does not inhibit the enzyme activity for all species.

[98] D-Arabinose (L-Fucose) Isomerase from *Aerobacter aerogenes*

By Kei Yamanaka and Ken Izumori

D-Arabinose $\rightleftarrows$ D-ribulose
L-Fucose $\rightleftarrows$ L-fuculose

Free sugar isomerases are characterized by a broader substrate specificity than is found among the isomerases for phosphorylated sugars.[1] This is most evident for D-arabinose (L-fucose) isomerase among four pentose isomerases. This enzyme catalyzes principally two reactions: the interconversion of D-arabinose and D-ribulose and of L-fucose and L-fuculose.[2,3] The enzyme is commonly referred to as D-arabinose isomerase, but Mortlock denoted it as L-fucose isomerase.[2] This enzyme from *Aerobacter aerogenes* PRL-R3 has been briefly described by Mortlock in previous volume.[3] An improved procedure which yields crystalline enzyme is reported in this chapter.

Assay Method

Principle. For routine assay, the colorimetric method described previvously[3] can be employed except that the reaction mixture is incubated for 10 min rather than 20 min.

Reagents

D-Arabinose, 0.1 M
L-Fucose, 0.1 M
MnCl$_2$, 0.01 M
Glycine buffer, 50 mM, pH 9.3

Procedure. The reaction mixture (1.0 ml) contains 0.5 ml of glycine buffer (pH 9.3), 0.05 ml of MnCl$_2$, and 0.01–0.20 ml of appropriately diluted enzyme solution. After equilibrium for 5 min at 35°, the reaction is initiated by addition of 0.05 ml of D-arabinose or L-fucose. The mixture is incubated at 35° for 10 min, and the reaction is terminated by addition of 0.05 ml of 50% trichloroacetic acid. Ketopentose is deter-

---

[1] E. A. Noltmann, *in* "The Enzymes" (P. D. Boyer ed.), Vol. VI, p. 340. Academic Press, New York, 1972.
[2] E. J. Oliver and R. P. Mortlock, *J. Bacteriol.* **108**, 293 (1971).
[3] R. P. Mortlock, see this series, Vol. 9 [102].

mined by the cysteine-carbazole test. Color is developed for 20 min at 20° for ribulose, or at 35° for fuculose, and read at 540 nm or 550 nm, respectively. The reading is corrected for the blank with heated enzyme or with no enzyme.

Definition of Unit and Specific Activity. The unit is defined as the amount of enzyme that produces 1 μmole of ribulose per minute in this assay system.

Purification Procedure[4]

Oliver and Mortlock have purified D-arabinose (L-fucose) isomerase to homogeneity from *A. aerogenes* PRL-R3 and its mutant strain by ammonium sulfate fractionation.[2] Izumori and Yamanaka have purified the enzyme to the crystalline state from *A. aerogenes* M-7, a strain deficient in pentitol dehydrogenases, by a simple procedure of repeated fractional precipitation with polyethylene glycol.[4]

Growth of Cells. Aerobacter aerogenes M-7, which was isolated from sea water in our laboratory, was grown aerobically on the following medium consisting of 0.26% $(NH_4)_2SO_4$, 0.24% KH_2PO_4, 0.56% K_2HPO_4, 0.01% each of $MnCl_2$, $MgCl_2$ and yeast extract, and 0.5% D-arabinose (autoclaved separately). The bacterium is grown at 30° for 20 hr. Cells are harvested by centrifugation, and washed once with 50mM Tris·HCl buffer at pH 7.5 containing each 1 mM $MnCl_2$ and mercaptoethanol.

Extraction. The washed cells (26.5 g from 5 liters of medium) are suspended in 180 ml of 50 mM Tris·HCl buffer at pH 7.5 containing 1 mM of $MnCl_2$ and mercaptoethanol, and disrupted in small portions in a 20-kc sonic oscillator for 20 min. Cellular debris is removed by centrifugation and the clear supernatant is used as the extract.

First PEG Precipitation. All precipitates are dissolved in 50 mM Tris·HCl buffer at pH 7.5 containing 1 mM mercaptoethanol and $MnCl_2$. Prior to the fractionation, one hundredth volume of 1 M $MnCl_2$ (1.6 ml) is added to the crude extract to give 10 mM. To 163 ml of extract, 3.2 g of powdered PEG 6000 (2% w/v) is added over a 30–60 min period while the extract is mixed on a magnetic stirrer; the precipitate is discarded by centrifugation. To the resulting supernatant is added 29.4 g of powdered PEG. The final PEG concentration is 20% (w/v). The suspension is allowed to stand for 30 min and then centrifuged. Precipitate is dissolved in 100 ml of 50 mM Tris·HCl buffer. More than 70% of

[4] K. Izumori and K. Yamanaka, *Agr. Biol. Chem.* **37**, 267 (1974).

[5] PEG = Polyethylene glycol purchased from The Wako Pure Chemical Co., Osaka, Japan.

the original activity is recovered in the precipitate (the first PEG 2–20% precipitate).

Second PEG Precipitation. Powdered PEG (2.2 g, 2% w/v) is added to the first fraction as described above, and precipitate is centrifuged off. To the supernatant (110 ml) is added 14.3 g of powdered PEG (15% w/v). The resulting precipitate is collected by centrifugation and dissolved in 100 ml of buffer. Centrifugation yields a clear supernatant which contains 77% of the original activity (the second PEG 2–15% precipitate).

First Crystallization. Crystallization of the enzyme is accomplished by raising the concentration of PEG slowly by adding a 40% solution of PEG (40 g made up to 100 ml with 50 mM Tris·HCl buffer at pH 7.5 or deionized water). First, powdered PEG (2.3 g, 2% w/v) is added to the second precipitate fraction. A trace of insoluble matter is removed immediately by centrifugation. Powdered PEG is not used for further purification and crystallization to avoid the formation of amorphous material. Ten milliliters of 40% PEG solution is added slowly with gentle stirring. When the PEG content reaches 5.0% (w/v), the solution gradually becomes faintly turbid and is allowed to stand for 1 hr at 0°. After removal of the precipitate by centrifugation, a second 10-ml portion of 40% PEG solution is added drop by drop to the supernatant. The concentration of PEG is raised to 7.6% (w/v). The precipitate is also removed by centrifugation. The third addition of 10-ml of 40% PEG solution is repeated in the same manner and the resulting precipitate is discarded. The final concentration of PEG reaches to 9.8%. The clear supernatant thus obtained is stored overnight at 2°. Examination of the precipitate under a microscope reveals a mixture of fine crystals and a small amount of amorphous material. Centrifugation yields a white sediment which is dissolved in 15 ml of 50 mM Tris·HCl buffer and a trace of insoluble matter which is removed by centrifugation.

PURIFICATION OF D-ARABINOSE(L-FUCOSE) ISOMERASE

Fraction	Total protein (mg)	Total units[a]	Specific activity (μmoles/mg/min)	Yield (%)
Crude extract	4260	935	0.22	—
First PEG 2–20% precipitate	2060	689	0.33	73.7
Second PEG 2–15% precipitate	812	725	0.89	77.5
First crystals	353	711	2.01	76.0
Second crystals	145	659	4.54	70.5
Third crystals	110	660	6.00	70.6

[a] Colorimetric units.

Second Crystallization. Recrystallization is performed by repeating the procedure described above. The final concentration of PEG reaches 5.7% (w/v) by adding 40% PEG solution. The mixture becomes faintly turbid. Insoluble matter is removed immediately by centrifugation. Clear supernatant is stored at 2°. Crystals are collected by centrifugation after standing for several hours or more. Crystals are dissolved in buffer.

Third Crystallization. The same procedure is repeated. PEG solution is added to the point of appearance of turbidity. The solution is stored overnight at 2°, and crystals are collected by centrifugation.

Properties[4]

Purity. The crystalline enzyme is dissolved in a minimum volume of buffer, then adsorbed to a small column of DEAE-Sephadex A-50 at pH 7.5, and eluted by 25 mM Tris·HCl buffer at pH 7.5 containing 0.2 M KCl, 10 mM MnCl$_2$, and 1mM mercaptoethanol. The eluted enzyme shows a single symmetrical peak on ultracentrifugation and also a single band of protein migrating toward cathode in polyacrylamide gel electrophoresis at pH 9.3. This band is clearly identified as D-arabinose isomerase with activity staining.[6]

Stability. The crystalline enzyme is stable at 2° as a sediment in PEG solution for at least 1 month. Enzyme is stable for more than 1 month in 50 mM Tris·HCl buffer at pH 7.5 containing 1 mM of MnCl$_2$ and mercaptoethanol.

Molecular Weight. The sedimentation constant ($s_{20,w}$) is calculated as 15.4 S. The molecular weight is estimated as 250,000 by a molecular exclusion chromatography on Sephadex G-200.

Effect of pH. The pH optimum for the isomerization of both substrates, D-arabinose and L-fucose, lies between 8.0 and 10.0 with a broad peak at about pH 9.3.

Substrate Specificity. Crystalline enzyme is active on two pentoses and the apparent Michaelis constants are 51 mM for L-fucose and 160 mM for D-arabinose. Manganese ion is required specifically for both activities.

Inhibitors. Pentitols inhibit the isomerization of both D-arabinose and L-fucose competitively. The inhibition constants are almost identical for both pentoses: 1.3–1.5 mM for D-arabitol, 2.2–2.7 mM for ribitol, 2.9–3.2 mM for L-arabitol, and 10.0–10.5 mM for xylitol. Tris(hydroxymethyl)-aminomethane and its analog, 2-amino-2-methyl 1,3-propanediol, inhibit equally the enzyme activities on both substrates: The inhibition is noncompetitive; the K_i for Tris is 10 mM for two activities.

---

[6] K. Yamanaka, see this volume [99].

[99] D-Xylose Isomerase from *Lactobacillus brevis*

By Kei Yamanaka

D-Xylose $\rightleftarrows$ D-xylulose
D-Glucose $\rightleftarrows$ D-fructose[1]
D-Ribose $\rightleftarrows$ D-ribulose

The following procedure, a modification of one previously described,[2] provides a method for obtaining the enzyme as crystalline state. This enzyme catalyzed three reactions, as shown above.

Assay Method

Principle. Two assay methoods are employed for D-xylose isomerase activity. The colorimetric assay is most widely used in which D-xylulose is determined by the cysteine-carbazole test. However, color development from xylose may reduce the accuracy of this assay.

Reagents

Maleate buffer, 50 mM, pH 6.0
MnCl$_2$(or MnSO$_4$), 10 mM
D-Xylose, 0.1 M
D-Ribose, 2 M
D-Glucose, 2 M

Procedure. The reaction mixture (1.0 ml) contains 0.5 ml of maleate buffer, 0.05 ml of MnCl$_2$ (or MnSO$_4$), and 0.01–0.1 ml of enzyme preparation. After equilibration for 5 min at 35°, 0.05 ml of D-xylose is added. The mixture is incubated at 35° for 10 min, and the reaction is terminated by the addition of 0.05 ml of 50% trichloroacetic acid. The cysteine–carbazole reaction is then carried out for 20 min at 35° for xylulose.

For the assay of isomerization of D-glucose or D-ribose, a colorimetric procedure is also used. The reaction mixture contains 0.3 ml of maleate buffer, 0.05 ml of MnCl$_2$(or MnSO$_4$), 0.05 ml of CoCl$_2$, and enzyme preparation in total volume of 0.70 ml. After equilibration at 50°, 0.3 ml of 2M D-glucose or D-ribose is added and the mixture is further incubated at 50° for 30 min with D-glucose or for 10 min with D-ribose. Cobalt ion is effective in stabilizing enzyme activity at 50°. Fructose or ribulose

---

[1] K. Yamanaka, *Biochim. Biophys. Acta* **151**, 670 (1968).
[2] K. Yamanaka, see this series, Vol. 9 [104].

are determined by the cysteine–carbazole test. Owing to the disturbance caused by the high concentration of glucose or ribose, an aliquot must be diluted to reduce the color produced by glucose or ribose. For fructose, the color is developed at 50° for 30 min and the absorbance determined at 560 nm. For ribulose, the conditions are color development at 20° for 20 min and measurement at 540 nm. For assay of the isomerization activities on glucose or ribose, 200 or 20 times more enzyme must be used than for D-xylose isomerase.

D-Xylose isomerase activity can be assayed spectrophotometrically. The reduction of D-xylulose with excess D-arabitol dehydrogenase[3] or xylitol dehydrogenase[4] can be followed as the rate of NADH oxidation at 340 nm. Glucose isomerase activity can be assayed with D-mannitol dehydrogenase (crystalline) from *Leuconostoc mesenteroides*[5] or *Lactobacillus brevis*[6] as the coupling enzyme.

Definition of Units and Specific Activity. One unit of D-xylose isomerase produces 1 μmole of xylulose per minute in this assay system. Specific activity is in units per milligram of protein.

Purification Procedure[1]

Growth of Organism. *Lactobacillus brevis* ATCC 8287 or IFO[7] 3960 was grown at 30° in a medium fortified with Mn^{2+} and Co^{2+}.[8] The medium consists of 1% peptone, 1% sodium acetate, 0.2% yeast extract, 0.05% $MnSO_4 \cdot 4H_2O$ (or $MnCl_2 \cdot 4H_2O$), 0.02% $MgSO_4 \cdot 7H_2O$, 0.01% $CoCl_2 \cdot 6H_2O$, 1% D-xylose and 0.1% D-glucose. The pH of the medium is adjusted to 7.0 with NaOH before autoclaving. The sugars and the mixed salt solution (20 to 100 times concentrated solution) are autoclaved separately and added to the medium before inoculation. The organism is inoculated into 5 ml of the above medium in a test tube and incubated for 20 hr. The culture is transferred to 400 ml of the same medium and incubated for 24 hr. The whole culture is transferred to 10 liters of medium, and incubated for 16–18 hr at 30°. Cells are harvested by centrifugation and washed once by suspending the cells in cold 10 mM Tris buffer (pH 7.0).

Extraction. The washed cells, 23.5 g wet weight from 10 liters of medium, are disrupted in small portions by grinding with about 80 g of

[3] K. Yamanaka, *Agr. Biol. Chem.* **33**, 834 (1969).
[4] M. G. Smith. *Biochem. J.* **83**, 135 (1962).
[5] S. Sakai and K. Yamanaka, *Biochim. Biophys. Acta* **151**, 684 (1968).
[6] G. Martinez, H. A. Barker, and B. L. Horecker, *J. Biol. Chem.* **238**, 1598 (1963).
[7] IFO = Institute for Fermentation, Osaka, Japan.
[8] K. Yamanaka, *Agr. Biol. Chem.* **27**, 265 (1963).

levigated alumina.[9] The enzyme is extracted with about 300 ml of 20 mM Tris buffer, pH 7.4, and the alumina and cell debris are removed by centrifugation (crude extract, 210 ml).

Manganese Treatment. To the extract is added dropwise 5% by volume of 1 M $MnCl_2$ (10.5 ml). The pH is maintained between 6.2 and 6.6 with 1 N NaOH or ammonia. After standing for more than 30 min, the precipitate is discarded by centrifugation ($MnCl_2$-treated fraction, 205 ml).

Ammonium Sulfate Fractionation. Amount of ammonium sulfate to be added is calculated from the table given by Green and Hughes[10] after applying a temperature correction of 0.92 for 0°. Solid ammonium sulfate (51 g) is added to Mn-treated fraction to give 45% saturation and the precipitate is discarded. The supernatant is treated with 71 g of ammonium sulfate (95% saturation). The precipitate is collected, dissolved in 15 ml of 20 mM Tris buffer (pH 7.4) and dialyzed overnight against 500 ml of the same buffer containing 5 mM $MnCl_2$.

Heat Treatment. To the dialyzed solution (29 ml) is added 5% by volume of 1 M $MnCl_2$ (1.5 ml) and the enzyme solution is immersed with vigorous stirring in a water bath at 80°. The flask is transferred to another water bath at 50° when the temperature of the solution reaches 47°. The flask is kept at 50° for 5 min. After cooling in an ice-bath, the coagulated proteins are discarded by centrifugation (Mn-heated fraction, 27.5 ml).

Acetone Fractionation. The pH of the heated fraction is adjusted carefully to 5.0 with 0.2 M acetic acid and is chilled to near 0°. Acetone (7.2 ml) previously chilled to −20°, is added dropwise with stirring (20% acetone). The precipitated protein is removed quickly by centrifugation at −10° and discarded. To the supernatant (32.5 ml) is added 11.4 ml of chilled acetone to give 40%. The precipitate is collected and dissolved in 4.0 ml of 20 mM Tris buffer (pH 7.4). Insoluble protein is removed by centrifugation.

Column Chromatography on DEAE-Sephadex. The enzyme solution (4.7 ml) is applied to a column of DEAE-Sephadex A-50 (1.5 cm × 20 cm) which has been equilibrated with 20 mM Tris buffer (pH 7.4). The column is washed with the same buffer, then the protein is eluted stepwise with 20 mM Tris buffer (pH 7.4) containing increasing concentrations of KCl to 0.3 M. A small peak of protein is eluted with 150 ml of buffer containing 0.1 M KCl, but no enzyme activity is recovered. The enzyme is eluted with the next elution by 150 ml of the buffer containing 0.2

[9] Levigated alumina, about 300 mesh for chromatography, was purchased from Wako Pure Chemicals, Osaka, Japan.
[10] A. A. Green and W. L. Hughes, see this series, Vol. 1 [10].

PURIFICATION OF D-XYLOSE ISOMERASE

Fraction	Total protein (mg)	Total units	Specific activity (μmole/mg/min)	Yield (%)
Crude extract	2860	1610	0.56	100
MnCl$_2$-treated	1390	1500	1.08	93
Ammonium sulfate fraction, 45–95% saturation	810	1390	1.71	86.5
Mn-heated fraction	495	1450	2.94	90.3
Acetone, 20–40% precipitate	260	940	4.0	58.4
DEAE-Sephadex eluate	200	1050	5.3	65
Ammonium sulfate fraction, 60–90% saturation	142	800	5.6	50
First crystals	95	580	6.1	36
Second crystals	55	350	6.4	21.7

M KCl. No enzyme activity is observed in the third elution with the buffer containing 0.3 M KCl. Active fractions are collected (55 ml), and the enzyme is precipitated by adding solid ammonium sulfate to 90% saturation. The precipitate is dissolved in a minimum volume of 20 mM Tris buffer (pH 7.4) and dialyzed overnight with the same buffer containing 5 mM MnCl$_2$ (ammonium sulfate fraction, 3.7 ml).

Crystallization. Crystallization is carried out by the addition of saturated solution of ammonium sulfate as follows: 6.4 ml of saturated ammonium sulfate is added dropwise to 3.7 ml of the purified enzyme preparation (protein concentration is about 3.83%). Ammonium sulfate content reaches 64% saturation. A trace of precipitated protein is removed by centrifugation before crystallization. The clear solution is kept at 5° for 1 or 2 days. After the crystals have appeared, the suspension is allowed to stand for an additional 1 or 2 days to complete crystallization. The first crystals are then collected by centrifugation, dissolved in minimum volume of 20 mM Tris buffer (pH 7.4) and dialyzed against the same buffer containing 5 mM MnCl$_2$ as described previously. Recrystallization is performed by repeating the same procedure at 62% ammonium sulfate saturation. The purification procedure is summarized in the table.

Purity. The recrystallized enzyme proves to be pure as evidenced by a single symmetrical moving peak on ultracentrifugation analysis, and a single band of protein migrating toward the cathode on polyacrylamide gel electrophoresis at pH 9.4. This band corresponds to the D-xylose isomerase visualized by activity staining method.

Detection of D-*Xylose Isomerase on Disc Electrophoresis.*[11] D-Xylose isomerase is detected in the disc gels by reduction of triphenyltetrazolium chloride by ketose at alkaline pH to the red formazan. To 0.9 ml of 50 mM maleate buffer (pH 6.0) is added 1.0 ml of 1 M D-xylose and 0.1 ml of 10 mM MnCl₂ solution. The gels are immersed in the above solution at 35° for 10 min, rinsed with water to remove D-xylose, and immersed into 0.1% triphenyltetrazolium chloride in 1 N NaOH at room temperature in the dark. After 1–2 min, the gels are dipped into a 2 N HCl solution to stop the reaction. The isomerase is visualized as a bright reddish band. After washing with water, the gels can be preserved in a 7% acetic acid. Isomerization of other sugars is also detected with 2 M solutions of D-glucose or D-ribose instead of D-xylose.

Properties[1]

Molecular Weight. The sedimentation coefficient (s_0) obtained is 11.46 S. The diffusion coefficient ($D_{20,w}$) is 5.84×10^{-7} cm²·sec⁻¹. Molecular weight is then calculated from s_0 and $D_{20,w}$ to be 191,000 and by centrifugation in a sucrose density gradient by the method of Martin and Ames[12] to be 197,000.

The extinction coefficient ($E_{1\,cm}^{1\%}$) at 280 nm is 15.42, and the ratio of absorbancies at 280 and 260 nm is 2.0.

Effect of pH. The pH optima for the isomerization of three sugars, D-xylose, D-glucose and D-ribose are identical between 6.0 and 7.0.

Substrate Specificity. The enzyme is specific for aldoses of five or six carbons bearing *cis* hydroxyl groups in positions 2 and 4. The single enzyme catalyzes the isomerization of D-xylose, D-glucose, and D-ribose. However, the affinities for the three sugars are quite different. Other pentoses and hexoses, such as D-arabinose, L-arabinose, L-xylose, D-lyxose, D-mannose, and D-galactose are inactive with excess enzyme (100 μg of crystalline enzyme) and with a longer incubation time (60 min) at 35°. No xylulose can be detected from D-lyxose with a longer incubation time (180 min) followed by column chromatography on Dowex 1-borate,[1] and by spectrophotometric assay with NADH and excess amount of D-arabitol dehydrogenase.[3] Xylose is the only product formed from D-xylulose.

Metal Requirement. The enzyme specifically requires Mn²⁺, for activity. Cobalt ions show a slight effect, but Mg²⁺, Zn²⁺, and other metallic ions are ineffective. The dissociation constant for manganese ions is 6.1 μM. This value coincides well with the K_d values of 8 ± 2 μM of Mildvan

---

[11] S. Nakamura, J. Shimizu, and K. Yamanaka, *Bull. Yamaguchi Med. Sch.* **18,** 1 (1971).

and Rose derived from NMR and EPR studies[12,13] in which the binary enzyme–Mn^{2+} complex was detected.[14] This value is also in good agreement with the dissociation constant for Mn^{2+} of crystalline L-arabinose isomerase from *Lactobacillus gayonii* of 5.25 μM.[15,16]

Effect of Substrate Concentration.[17] Among the four pentose isomerases, D-xylose isomerase is the first for which a compulsory order of Mn^{2+} binding in the reaction sequence was shown.[17] The reaction mechanism of D-xylose isomerase involves the formation of a ternary complex of enzyme–Mn substrate or inhibitor. The apparent K_m for D-xylose varies with the Mn^{2+} concentration. By plotting the apparent K_m obtained at 5, 10, and 50 μM Mn^{2+} vs the reciprocal of the Mn^{2+} concentration, the intercept gives the K_m for D-xylose at infinite Mn^{2+} concentration. The K_m decreases to a value of 5 mM. The Michaelis constants for D-glucose and D-ribose are 0.92 M and 0.67 M, respectively.

Inhibitors. Pentitols are typical competitive inhibitors for the isomerase. Of the pentitols, xylitol and D- and L-arabitols inhibit competitively, whereas ribitol is ineffective. The K_i values are 2.7, 130, and 146 mM for xylitol, and D- and L-arabitols, respectively. D-Lyxose, the C-2 analog of D-xylose, is also a competitive inhibitor for which K_i value is 70 mM. These pentitols and D-lyxose combine with the enzyme through a Mn^{2+} bridge and thereby compete for the substrate binding site.[17] This proposed reaction mechanism involving a ternary complex with enzyme, Mn^{2+} and substrate or inhibitor was finally confirmed by ESR analysis.[14]

[12] R. G. Martin and B. N. Ames, *J. Biol. Chem.* **236**, 1372 (1961).
[13] A. S. Mildvan and I. A. Rose, *Fed. Proc., Fed. Amer. Soc. Exp. Biol.* **28**, 534 (1969).
[14] K. J. Schray and A. S. Mildvan, *J. Biol. Chem.* **247**, 2034 (1972).
[15] T. Nakamatu and K. Yamanaka, *Biochim. Biophys. Acta* **178**, 156 (1969).
[16] K. Yamanaka, see this volume [97].
[17] K. Yamanaka, *Arch. Biochem. Biophys.* **131**, 502 (1969).

[100] Mutarotase (Aldose 1-Epimerase) from Kidney Cortex

By J. MARTYN BAILEY, P. H. FISHMAN, J. W. KUSIAK,
S. MULHERN, and P. G. PENTCHEV

$$\alpha\text{-Aldose} \leftrightharpoons \beta\text{-aldose}$$

Mutarotase (aldose 1-epimerase, EC 5.1.3.3) catalyzes the interconversion of the α- and β-anomers of certain sugars including D-glucose and D-galactose. The enzyme was first identified by Bentley and Neuberger in 1948 in preparations of glucose oxidase from *Penicillium notatum*,[1] where it accelerated the oxidation of glucose by converting α-glucose to the β-form, for which glucose oxidase is specific.[2]

In 1954 a similar enzyme was identified in mammalian tissues.[3] It is present in particularly high concentration in kidney cortex comprising about 0.3% of the total soluble protein of this tissue. Enzyme from 1 g of bovine kidney cortex will convert about 2 g of α-glucose to β-glucose per minute.[4]

Purification of the enzyme from mammalian kidney has been reported for hog,[5-7] beef,[8] sheep, lamb, rabbit,[9] and human[10]. The isolation of the enzyme from *P. notatum* and *E. coli* has been reported,[11,12] and the widespread occurrence in higher plants,[13] fish,[14] birds,[9] and amphibians[14] has been noted. The properties of the enzyme from human red blood cells and human liver have also been described.[15,16]

Despite the widespread distribution of mutarotases, no definite function has yet been demonstrated in higher species. It does not appear that the enzyme is involved or required in glucose metabolism. None of the enzymes involved in the phosphorylation or dephosphorylation of glucose (hexokinase, glucokinase, pyrophosphate phosphotransferases, and glucose-6-phosphatase) display any anomeric specificity.[17,18] Glucose 6-phosphate has a very rapid spontaneous mutarotation rate, some 200 times faster than glucose, and at physiological pH is not a substrate for mutarotase. A considerable amount of evidence has been obtained which suggests that mutarotases are involved in sugar transport.[3] The substrate

[1] R. Bentley and A. Neuberger, *Biochem. J.* **45**, 584 (1948).
[2] D. Keilin and E. F. Hartree, *Biochem. J.* **50**, 341 (1952).
[3] A. S. Keston, *Science* **120**, 355 (1954).
[4] J. M. Bailey, and P. G. Pentchev, *Proc. Soc. Exp. Biol. Med.* **115**, 796 (1967).
[5] A. S. Keston, *Fed. Proc., Fed. Amer. Soc. Exp. Biol.* **14**, 234 (1955).
[6] A. M. Chase, S. M. Lapedes, Li Lu-Ku, *Proc. Int. Congr. Biochem. 7th, F*, 141 (1967).
[7] S. M. Lapedes and A. M. Chase, *Biochem. Biophys. Res. Commun.* **31**, 967 (1968).
[8] J. M. Bailey, P. H. Fishman, and P. G. Pentchev, *J. Biol. Chem.* **244**, 781 (1969).
[9] S. Mulhern, P. H. Fishman, J. W. Kusiak, and J. M. Bailey, *J. Biol. Chem.* **248**, 4163 (1973).
[10] J. W. Kusiak, M.Sc. Thesis, George Washington Univ., 1972.
[11] R. Bentley and D. S. Bhate, *J. Biol. Chem.* **235**, 1219 (1959).
[12] K. Wallenfels, and K. Herrman, *Biochem. Z.* **343**, 294 (1965).
[13] J. M. Bailey, P. H. Fishman, and P. G. Pentchev, *J. Biol. Chem.* **242**, 4263 (1967).
[14] J. M. Bailey, P. H. Fishman, and S. Mulhern, *Proc. Soc. Exp. Biol. Med.* **131**, 861 (1969).
[15] W. Sacks, *Arch. Biochem. Biophys.* **123**, 507 (1968).
[16] A. Kahlenberg and G. Miller, *Can. J. Biochem.* **50**, 1028 (1972).
[17] M. Salos, E. Viñuela, and A. Sols, *J. Biol. Chem.* **240**, 561 (1965).
[18] J. M. Bailey, P. H. Fishman, and P. G. Pentchev, *Biochemistry* **9**, 1189 (1970).

specificity of the enzyme is similar to that observed for sugar transport in kidney and intestine.[19,20] All actively transported sugars have been shown either to be substrates for mutarotase or to interact strongly with the active center as competitive inhibitors.

During embryological development it has been found that the time of appearance of the enzyme in kidney and intestine correlates well with the development of a functional capacity to transport sugars. It has also been shown that glucose interacts with the enzyme in some way during the reabsorption process. The inhibitory ability of phlorizin, phloretin and some 30 structurally related compounds correlates closely with their ability to inhibit glucose transport.[21] It has been suggested that the catalysis of mutarotation may be a coincidental consequence of the interaction of the sugar with the enzyme in its function as a transport or binding protein.[22]

Assay Procedures

The assay system for the enzyme is based upon the change in optical rotation of the substrate during the enzyme-catalyzed mutarotation reaction. The specific rotation of α-D-glucose used as substrate is $112.5°$ at 589 nm. The specific rotation of β-D-glucose is $19°$ and that of the equilibrium mixture produced is $52.5°$.

Manual Polarimetric Assay Procedure

The reaction may be followed manually in any conventional laboratory polarimeter equipped with a sodium lamp source and capable of reading with a precision of about 10 millidegrees.

Crystalline α-D-glucose (36 mg, 200 μmoles, ground to pass a 200-mesh screen) is dissolved in 12 ml of EDTA buffer (5 mM, pH 7.4 at 25°) containing the enzyme sample to be assayed. The solution is rapidly introduced into the polarimeter tube, and readings of optical rotation are taken at 2-min intervals for 10 min. Using a 2 dm polarimeter tube, the initial rotation is 675 millidegrees, falling to 315 millidegrees at the completion of the reaction. The first-order rate constant for the mutarotation reaction is obtained from the slope of the straight-line plot obtained by graphing

$$\ln (\alpha_o - \alpha_e)/(\alpha_t - \alpha_e) = Kt$$

[19] A. S. Keston, *J. Biol. Chem.* **239**, 3241 (1964).
[20] J. M. Bailey and P. G. Pentchev, *Amer. J. Physiol.* **208**, 385 (1965).
[21] D. F. Diedrich, and C. H. Stringham, *Arch. Biochem. Biophys.* **138**, 493 (1970).
[22] J. M. Bailey and P. G. Pentchev, *Biochem. Biophys. Res. Commun.* **14**, 161 (1964).

where K is the rate constant and α_o, α_t, and α_e are the observed angular rotations at time zero, t, and equilibrium, respectively.

The spontaneous mutarotation rate of glucose at 25° is about 0.032 min^{-1}, and 1 working unit of enzyme is the amount that gives a 10% increase in the spontaneous mutarotation rate under the above conditions. This unit is close to 1 IU and may be corrected to IU (as outlined below) if the K_m of the particular enzyme is known.

Semiautomated Assay Procedure for Mutarotase

The manual assay procedure involves calculation and plotting of the first-order rate curves for mutarotation of substrate directly from measurements of angular rotation displayed by the polarimeter. The procedure is relatively tedious and is not well suited to multiple assays, nor does it have the precision necessary for kinetic studies.

A semiautomated assay procedure using a recording polarimeter is preferred. The apparatus, illustrated in Fig. 1, utilizes a Bendix type 143A photoelectric polarimeter coupled to a log-linear chart recorder (the Beckman 10-inch or the Sargent SRL are suitable). The recorder span is most conveniently adjusted by means of a decade resistance box coupled to the output terminals of the polarimeter. The polarimeter cell is of flow-through design and water-jacketed, so that the temperature can be maintained to within 0.1° by use of a circulating water bath.

The calibration of the polarimeter-recorder combination is based on the observation that mutarotation follows the first-order kinetic equation given above. It follows, therefore, for any sugar concentration, that if $\alpha_o - \alpha_e = \Delta\alpha$, the total change in rotation occurring during reaction, then, at 90% completion, the fraction remaining is 0.1 $\Delta\alpha$. The fraction in the left-hand side of the equation becomes 10, the decimal logarithm of which is 1.0. By setting the recorder to 1.0 when the polarimeter reading is 0.9 $\Delta\alpha$, the slope of the line recorded is thus equal to the mutarotational rate constant (Fig. 2).

The polarimetric assay procedure thus gives values for the overall first-order rate constant of the enzyme-catalyzed mutarotation of a sugar at a fixed substrate concentration. The unit of mutarotase activity, however, is defined in conventional terms and corresponds to the conversion of 1 μmole of α- to β-glucose per minute at 25°, pH 7.4, and optimum substrate concentration. In order to derive this unit from the polarimetric data, the following calculations are necessary.

1. The rate constant, k_c, due to enzyme catalysis is first obtained by subtraction of the spontaneous rate constant.

2. Since the mutarotation reaction is freely reversible, $k_c = k_1 + k_2$,

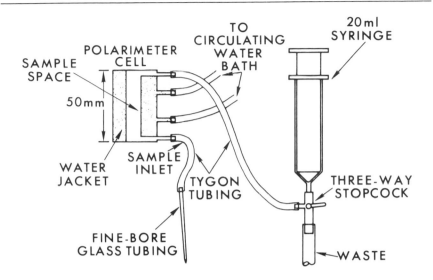

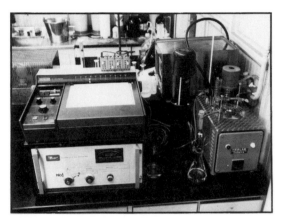

Fig. 1. Recording polarimeter for semiautomatic assay of mutarotase. *Top:* Schematic of polarimeter cell assembly. *Bottom:* Overview of recording polarimeter.

where k_1 and k_2 are the individual rate constants for the forward and back-reactions, respectively. Since β-glucose is also a substrate for the enzyme and the enzyme does not disturb the position of equilibrium, it follows that

$$k_1 = k_2 \times K_{eq}$$

K_{eq} calculated from the specific rotations of α-, β-, and equilibrium glucose (112.5°, 19°, and 52.5°, respectively) is 1.84, from which it may be shown that $k_1 = 0.65 \, k_c$.

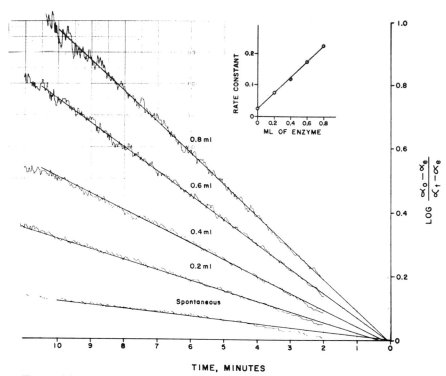

FIG. 2. First-order mutarotation reactions for glucose alone and in the presence of increasing amounts of enzyme are followed for 10 min in the logarithmic mode on a recorder coupled to a Bendix polarimeter. *Inset:* Plot of rate constant versus milliliters of enzyme showing linearity of the semiautomatic assay for mutarotase.

3. Conversion of substrate to product is determined by the velocity of the forward reaction, $v_1 = k_1 S$, where S in the standard assay system is 200 μmoles of α-glucose (16.7 mM). Thus v_1 will have a value of 1 μmole/min when $k_1 = 0.005$ min^{-1}, i.e., when $k_c = 0.0077$ min^{-1}.

4. In order to adjust velocity v_1 to optimum substrate concentration, the K_m for the enzyme acting on glucose is determined (19 mM for most enzymes) and inserted in the equation $K_m = S\,[V_{max}/V - 1]$, giving a value for V_{max}/V of 2.3, from which it follows that an amount of enzyme which gives a k_c of 0.0033 min^{-1} in the standard assay would convert 1 μmole of α-glucose to β-glucose per minute under optimum conditions.

The factor 0.0033 min^{-1} derived above is close to 10% of the spontaneous mutarotational rate of glucose under the assay conditions described. This forms the basis for a useful working unit for measurement of the activity of an uncharacterized enzyme; i.e., 1 unit of enzyme is an amount which gives a 10% increase in the spontaneous mutarotational rate constant of a 12-ml solution of 16.7 mM α-glucose at 25° and pH 7.4. This measurement may be used routinely for the assay and isolation of mutarotases for which the K_m and, hence, the V_{max}/v correction described above are not known.

Colorimetric Procedure

This method lacks the necessary precision for kinetic studies but is suitable for routine assays of the enzyme where a polarimeter is not available. It is based upon the observation that glucose oxidase is specific for the β-anomer of D-glucose. Consequently the rate of oxidation of a sample of pure α-glucose is dependent upon its rate of mutarotation and hence is accelerated by the addition of mutarotase.[23] The oxidation of glucose may be followed using the conventional glucostat reagents by coupling the hydrogen peroxide produced to the formation of colored product via the action of added peroxidase enzyme on the chromogen o-dianisidine. The overall reaction may be represented as follows:

$$\alpha\text{-Glucose} \xrightarrow{\text{mutarotase}} \beta\text{-glucose}$$

$$\beta\text{-Glucose} + O_2 \xrightarrow{\text{glucose oxidase}} \text{gluconic acid} + H_2O_2$$

$$H_2O_2 + \text{reduced chromogen} \xrightarrow{\text{peroxidase}} \text{oxidized chromogen}$$

Reagents. The standard glucose oxidase-peroxidase and chromogen kits sold under the trade name of "Glucostat" (Worthington Biochemical Corp.) may be used. The reagents are made up to 4 times the concentrations normally used in the procedure for glucose analysis.

Dissolve 10 mg of crystalline α-glucose in 25 ml of ice-cold water immediately before use. Into two colorimeter tubes pipette the components tabulated below:

	Blank tube (ml)	Sample tube (ml)
Distilled H$_2$O	2.5	2.0
Chromogen	0.1	0.1
Glucostat	0.3	0.3
Mutarotase sample	0	0.5

Incubate the tubes at 25° in a water bath and at zero time add 0.1

[23] A. S. Keston, *Fed. Proc., Fed. Amer. Soc. Exp. Biol.* 17, 253 (1958).

ml of the α-glucose solution to both the blank and the sample tubes. Read the optical density of the sample tube against the blank at a wavelength of 400 nm after 10 min.

A standard curve should be constructed using varying quantities of mutarotase in the range 0 to 50 units per assay. For accurate work the method should be standardized against the polarimetric procedure. However, it has been found depending upon the individual batch of glucose oxidase reagent that 20 units of mutarotase usually give an optical density of 0.2 to 0.25. If greater sensitivity is desired, the procedure may be run at 0° to decrease the contribution due to the spontaneous mutarotation of glucose.

Purification Procedure

Kidney cortex from a number of mammalian species is a particularly rich source of the enzyme (Table I). Liver tissue is the most convenient source in fish. The following procedure is suitable for purification of the enzyme from most of the sources listed in Table I.

Step 1. Ammonium Sulfate Fractionation. All procedures are carried

TABLE I
Species Distribution and Abundance of Mutarotases[a]

Source	Mutarotase content	
	Units/g fresh wt	Mg enzyme/g tissue protein
Escherichia coli	54	1.2
Mushroom	39	
Green pepper	19	0.46
Catfish kidney	226	
Bullfrog kidney	240	
Sheep embryo kidney	800	2.5
Chicken kidney	700	
Rabbit kidney	950	2.4
Human kidney	1050	1.5
Calf kidney	2100	1.7
Beef kidney	3600	2.9

[a] The various tissues were homogenized in 4 volumes of EDTA buffer (5 mM, pH 7.4), and the mutarotase content was measured in a polarimetric assay system containing 36 mg of freshly dissolved α-D-glucose. One unit of enzyme catalyzes the conversion of 1 μmole of α-glucose to β-glucose per minute at 25°. The tissue content of mutarotase protein was calculated from the specific activities for enzymes which were obtained in pure form by the procedures described.

out at 5°. The cortex is removed from fresh beef kidneys; 1 kg of cortex tissue is homogenized for 1 min in 4000 ml of EDTA buffer (5 mM, pH 7.2). The homogenate is centrifuged at 15,000 g for 30 min, and the precipitate is discarded. The supernatant should contain about 3.5×10^6 units of mutarotase. To each liter of supernatant is added 200 g of ammonium sulfate (35% saturation). The solution is stirred for 30 min and centrifuged at 15,000 g for 30 min. The precipitate is discarded and ammonium sulfate (180 g per liter of supernatant) is added to give 62% saturation. The solution is stirred and centrifuged as before, and the precipitate is dissolved in a minimal volume of EDTA buffer (5 mM, pH 7.2). The solution is dialyzed for 6 hr against each of three changes of 50 volumes of phosphate buffer (1 mM, pH 6.5). The impermeate is centrifuged at 32,000 g to remove any precipitate.

Step 2. DEAE-Cellulose Chromatography, pH 6.5. The supernatant is loaded onto a 200-ml bed volume of DEAE-cellulose (Bio-Rad Laboratories, Cellex D) column (2.5 × 45 cm) preequilibrated with phosphate buffer (1 mM, pH 6.5). The column is eluted with 1200 ml of linear phosphate buffer gradient (1 to 20 mM, pH 6.5). Fractions of 10 or 20 ml are collected using a fraction collector and assayed for mutarotase activity. Those fractions containing the bulk of the enzyme activity are combined.

Step 3. Chromatography on Hydroxyapatite. The combined eluates are applied directly to a 200-ml bed volume column (2.5 × 45 cm) of hydroxyapatite (Bio-Gel HT, Bio-Rad Laboratories) equilibrated with phosphate buffer (1 mM, pH 6.5). The column is eluted with phosphate buffer (36 mM, pH 6.5) at a flow rate of about 50 ml/hr. Fractions 10–20 ml are collected and assayed for enzyme and combined accordingly.

Step 4. Gel Filtration on Bio-Gel P-100. The effluent containing activity is concentrated by ultrafiltration using a Diaflo PM 20 membrane to a final volume of about 15 ml. The ultrafiltrate is applied to a 450-ml bed volume column (2.5 × 100 cm) of Bio-Gel P-100 (100–200 mesh). The column is preequilibrated and eluted with phosphate buffer (1 mM, pH 7.7) at a flow rate of about 20 ml/hr. Fractions of 5–10 ml are collected and assayed for enzyme activity.

Step 5. DEAE-Cellulose Chromatography, pH 7.7. The effluent fractions containing activity are applied to a 45-ml bed volume DEAE-cellulose column (1.5 × 30 cm) equilibrated with phosphate buffer (1 mM, pH 7.7). The column is developed with a linear gradient (1–20 mM phosphate buffer pH 7) in a total elution volume of 300 ml.

Step 6. Buffer Exchange Using Sephadex G-150. Since the enzyme is not stable for prolonged periods in phosphate buffer a final buffer exchange into EDTA buffer for storage is performed using a Sephadex

TABLE II
PURIFICATION OF BOVINE KIDNEY MUTAROTASE

Step	Protein (mg)	Total units $\times 10^6$	Specific activity	Purification (fold)	% Recovery overall
Original homogenate (from 800 g of kidney cortex)	41×10^3	3.11	75.8	—	100
1. Ammonium sulfate	2.7×10^3	2.7	100	1.3	87
2. DEAE, pH 6.5	2750	2.16	785	10.4	70
3. Hydroxyapatite	202	1.88	9300	122.5	61
4. Bio-Gel	80	1.50	18,800	248	48
5. DEAE, pH 7.7	49	1.2	24,200	319	39
6. Sephadex G-150	40	1.06	26,500	350	35

G-150 column (450 ml, 2.5 × 100 cm) equilibrated with EDTA buffer (5 mM, pH 7.2) and developed with the same buffer. The fractions containing enzyme are pooled and stored at 5°. Yield from 1 kg of cortex is 40 mg (about 1 million units) of pure enzyme.

A typical purification is summarized in Table II.

Crystallization. The enzyme obtained from step 6 above should be homogeneous, giving a single band upon disc gel electrophoresis at 3 pH values (pH 9.5, pH 8.0, and pH 4.3). The enzyme may be crystallized by the following procedure, which is based upon the reverse solution technique of Jakoby.[24] A solution of the enzyme (1 mg/ml) in EDTA buffer (5 mM, pH 7.2) is brought to 95% saturation with ammonium sulfate at 2°. After standing for 5 min, the solution is centrifuged and the supernatant is discarded. The precipitate of enzyme is extracted with 1 ml of 60% saturated ammonium sulfate in 5 mM EDTA and 2 mM mercaptoethanol. The solution is centrifuged and the supernatant decanted and saved. The process is repeated with 55% and 50% saturated solutions, the supernatants are combined, and the solution is allowed to stand at 25°. Crystals of enzyme precipitate within 1 hr.

Stability. The enzyme is stable for an indefinite period when stored in EDTA buffer at 4° but slowly inactivates if stored frozen.[8]

Properties

Substrate Specificity. Of 18 sugars tested[8] the mutarotation rate of only five was increased significantly after addition of the enzyme. The five substrates are D-glucose, D-galactose, D-xylose, L-arabinose, and D-

[24] W. B. Jakoby, *Anal. Biochem.* **26**, 295 (1968).

TABLE III
KINETIC PARAMETERS OF CRYSTALLINE BEEF KIDNEY MUTAROTASE

Substrate	K_m (mM)	Turnover number $\times 10^6$ min^{-1}
α-D-Glucose	19	1.0
α-D-Galactose	6.5	1.2
α-D-Xylose	13.2	1.2
α-L-Arabinose	8.3	1.9
α-D-Fucose	2.0	0.35

fucose. The turnover numbers and K_m values are given in Table III. From the survey of sugars a pattern of essential structural features necessary for substrate interaction with the enzyme was derived. The substitution or removal of the equatorial hydroxyl on carbon 2 abolishes all substrate interaction (2-deoxy-D-glucose, D-glucosamine, D-galactosamine, N-acetyl-D-glucosamine, N-acetyl-D-galactosamine, 2-deoxy-D-galactose). A free hydroxyl on carbon 3 is necessary since 3-O-methyl-D-glucose is not a substrate. An axial-equatorial (cis) relationship of the hydroxyl groups on carbons 2 and 3 abolishes substrate interaction (all mannose sugars, D-talose, and D-lyxose), and the steric relationship at these 2 carbons cannot be equatorial-axial (cis) (D-ribose). The required positioning appears to be equatorial-equatorial (trans) in the proper stereoconfiguration, since all the substrates displayed this characteristic. The steric position of hydroxyl 4 is not critical since D-galactose and D-glucose or D-xylose and L-arabinose were substrates. A free hydroxyl on carbon 5 is not necessary since the substrates exist as pyranose ring structures. The hydroxyl on carbon 6 is not critical since 6-dexoy-D-galactose, and the pentoses D-xylose and L-arabinose are substrates. Substitution at carbon 6 with a phosphate group, as in glucose 6-phosphate, however, destroys substrate capacity at physiological pH values.

Inhibitors. The enzyme is inhibited strongly by inhibitors of sugar transport[21] such as phloretin $K_i = 20$ μM phlorizin $K_i = 0.2$ mM, and diethylstilbestrol $K_i = 20$ μM. The enzyme is sensitive to inhibition by Hg^{2+}; 0.1 μM inhibits completely. The addition of a second sugar to the assay often markedly reduced the enzyme-catalyzed mutarotation of α-D-glucose. Of the 50 sugars tested, 30 were inhibitors of the enzyme in varying degree (Table IV). The inhibition in each case follows classical Michaelis kinetics for competitive inhibition, from which the corresponding K_i values were derived. When a sugar which was also a substrate was used as an inhibitor, the K_i values as expected are the same as their respective K_m values. Many sugars that are not substrates were nevertheless ex-

TABLE IV

COMPETITIVE INHIBITORS OF PURIFIED BOVINE KIDNEY MUTAROTASE

Inhibitor sugar	K_i (mM)	Inhibitor sugar	K_i (mM)
D-Fucose	2.0	L-Arabitol	27.7
L-Fucose	2.6	D-Cellobiose	31.5
L-Xylose	3.7	D-Maltose	32.5
2-Deoxy-D-ribose	5.5	3-O-Methyl-D-glucose	35.5
D-Galactose	6.3	L-Xylose	38.0
D-Allose	6.4	L-Glucose	44.5
L-Arabinose	8.0	D-Erythrose	47.7
D-Ribose	9.8	D-Sorbitol	67.5
α-Methyl-D-glucoside	12.6	D-Xylitol	68.5
D-Xylose	14.0	D-Inositol	69.5
D-Galacturonic acid	18.4	D-Arabinose	77
D-Glucuronic acid	21.0	β-Methyl-D-xyloside	82.5
2-Deoxy-D-glucose	24.0	D-Melibiose	110
Galactitol	25.4	2-Deoxy-D-galactose	140
Erythritol	25.5	D-Mannitol	149

[a] The enzyme-catalyzed mutarotation rate of 16.67 mM α-D-glucose was determined in the presence and in the absence of a second sugar added in its anomeric equilibrium form. K_i for the sugars was determined by plotting V/V_I against [I] where V and V_I represent the enzymatic velocities in the presence and in the absence of inhibitor, and [I] represents the concentration of the competitive inhibitor. The intercept of this line with a line representing $V/V_I = K_m/(K_m + [S]) + 1$ corresponds to a point on the [I] axis equal to K_I. The following sugars were tested and were not inhibitors (a noninhibitor being defined for practical purposes as a sugar with a K_i of greater than 150 mM): D-arabitol, D-fructose, N-acetyl-D-galactosamine, D-galactosamine, D-galactose 6-phosphate, D-glucosamine, N-acetyl-D-glucosamine, D-glucose 6-phosphate, D-lactose, D-mannose, L-mannose, α-methyl-D-mannose, D-raffinose, D-rhamnose, D-sedoheptulose, D-sucrose, D-trehalose, and D-turanose.

tremely potent inhibitors. The affinity of two of these inhibitors (L-fucose and L-xylose) is approximately 10 times that of the substrate glucose (Table IV).

Molecular Properties. The enzyme is a monomer, and the molecular weights from many sources are very similar and average 37,000 (Table V). The Stokes radius averages 31.4 Å. Amino acid analysis indicates a total of 336 residues. The frictional ratio in buffers of low osmolarity is about 1.37. Enzymes from all species undergo an identical molecular transition to a more compact form ($f/f_0 = 1.17$) in buffers of high osmolarity. This transformation is specifically reversed by substrate sugars. On the basis of K_m values for glucose, kidney mutarotases fall into two

TABLE V

MOLECULAR WEIGHTS OF MUTAROTASES FROM DIFFERENT SPECIES[a]

Source of kidney enzyme	Molecular weight		
	SDS-gel electrophoresis	Sucrose density gradient	Gel filtration
Beef (adult)	37,800	37,000	37,300
Calf (0–6 months)	36,800	37,000	38,100
Calf (embryo)	—	—	38,000
Human	—	37,500	—
Human[b]	—	37,500	—
Sheep	—	—	38,100
Sheep (embryo)	36,000	—	—
Rabbit	36,000	37,500	36,600
Yellow perch	—	37,400	34,200
Chicken	—	—	35,900
Guinea pig	—	—	38,100
Green pepper	—	—	38,300

[a] The molecular weights of the various mutarotases were measured by their relative mobilities in SDS-gels, by sedimentation in sucrose gradients in the ultracentrifuge, and by gel filtration chromatography on columns of Bio-Gel A. For the gel filtration data in column 4, the values are corrected using the bovine kidney enzyme as standard with an assumed molecular weight of 37,300.
[b] Liver enzyme.

distinct groups, those with a K_m of 19 mM (bovine adult, sheep, sheep embryo, hog, hog embryo, mouse, gerbil, and dog) and those with a K_m of 12 mM (human, rabbit, toadfish, yellow perch, chicken, catfish, and newborn calf). In the ox, three different kinetic forms of the enzyme having otherwise indistinguishable properties are present at different stages of development (embryo calf K_m, 5 mM; calf 0–6 months, 12 mM, and adult 19 mM).

The enzyme has a broad pH optimum in the range pH 4–8. The purified enzyme is rapidly photoinactivated in the presence of methylene blue or rose bengal; however, the pH–photooxidation profile indicates that potent nucleophiles such as histidine are absent from the active center. Photoinactivation is accompanied by stoichiometric loss of tryptophan. N-bromosuccinimide inactivates, reacting with 3 of 5 tryptophan residues in water and with all in 8 M urea. N-Acetylimidazole acetylates 8 of 11 tryosine residues. The acetylated enzyme is catalytically active. Mutarotase reduces the inactivation energy of spontaneous mutarotation of glucose by 11.6 kcal, a value similar to that for distortion from the chair to the half-chair or pseudo-acyclic form of the ring. The interaction

TABLE VI
CATALYTIC COEFFICIENTS OF PURIFIED MAMMALIAN MUTAROTASES

Source	Specific activity (units/mg of protein)		Catalytic coefficient[a]
	Initial	Final	
Beef kidney	76.8	26,500	1.00×10^6
Calf kidney	40.4	23,800	8.8×10^5
Human kidney[b]	20.4	13,600	5.1×10^5
Human liver[b]	3.7	10,440	3.9×10^5
Sheep embryo kidney	13.6	5,430	1.9×10^5
Rabbit kidney	18.4	7,650	2.8×10^5

[a] Moles of α-glucose converted to β-glucose per minute per mole of enzyme at 25° and optimum substrate concentration. Calculated from the molecular weights of the enzymes given in Table V and the specific activities.
[b] Human enzyme was prepared from tissues removed 12–24 hr after death.

with sugars is accompanied by a substantial conformational change in the enzyme. It appears that tryptophan functions in the glucose binding site and that catalysis of mutarotation is accomplished by a ring distortion mechanism.[25] Catalytic coefficients of some purified mammalian mutarotases are listed in Table VI.

[25] P. H. Fishman, J. W. Kusiak, and J. M. Bailey, *Biochemistry* **12**, 2540 (1973).

[101] Mutarotase from Higher Plants

By PETER H. FISHMAN, P. G. PENTCHEV, and J. MARTYN BAILEY

$$\alpha\text{-Aldose} = \beta\text{-aldose}$$

Assay Method

The semiautomatic polarimetric assay described for the mammalian enzyme[1] is used to assay the plant enzyme. Plant tissues are finely diced and homogenized in 3 volumes of EDTA buffer (5 mM, pH 7.4) for 2 min at 0° in a high-speed blender. The homogenates when necessary are adjusted to at least pH 6 with 1 M NaOH and then centrifuged at 27,000 g for 20 min at 4°. An aliquot of the supernatant is then assayed for mutarotase activity by the standard method. A boiled aliquot is also

[1] J. M. Bailey, P. H. Fishman, J. W. Kusiak, S. A. Mulhern, and P. G. Pentchev, this volume [100].

assayed to measure any nonenzymic effects. The calculation of activity and definition of an enzyme unit are the same as for the mammalian enzyme.[1]

Purification Procedure

A survey of a number of plants indicated mutarotase activity in a wide variety of species.[2] Bell (green) pepper (*Capsicum frutescens*) had the highest activity per milligram of protein and was purified on a large scale.[3]

Step 1. Preparation of Initial Extract. Fresh green peppers (10 kg) are diced and homogenized in 5 liters of EDTA buffer (5 mM, pH 7.4) in the cold, using a Waring Blendor. The homogenate is then filtered through glass wool.

Step 2. Ammonium Sulfate Fractionation. Solid ammonium sulfate is slowly added to the ice-cold stirred filtrate until 38% (20 g/100 ml) saturation is reached. The solution is centrifuged at 17,000 g for 15 min at 4° and the precipitate is discarded. The supernatant is brought to 60% saturation (38 g/100 ml) with ammonium sulfate and recentrifuged as before. The supernatant is discarded and the precipitate is dissolved in 450 ml of phosphate buffer (1 mM, pH 6.4) and dialyzed in the cold against the same buffer (three changes of four liters each).

Step 3. Hydroxyapatite Column Chromatography. The dialyzate is centrifuged at 27,000 g for 45 min to remove insoluble material. The supernatant is applied to a 65 × 2.1 cm column of hydroxyapatite (Bio-Gel HT from Bio-Rad Laboratories) previously equilibrated with phosphate buffer (1 mM, pH 6.4). The flow rate is 2 ml per minute under 10 psi of N_2. The column is eluted stepwise with increasing concentrations of phosphate buffer (pH 6.4). The plant mutarotase elutes as a single sharp peak in 0.2 M phosphate buffer.

Step 4. Dialysis. The peak fractions are pooled and dialyzed against three changes of EDTA buffer (5 mM, pH 7.4) in the cold. During storage at 2° further precipitation and removal of sedimented proteins occurred and the specific activity of the plant enzyme preparation increased. The purification is summarized in Table I. The final yield is 20%, at which stage the enzyme has a specific activity of 2720 units per milligram of protein for a 555-fold purification. When analyzed by polyacrylamide gel electrophoresis, this enzyme preparation had 5 major protein bands. Further purification by DEAE-cellulose column chroma-

---

[2] J. M. Bailey, P. H. Fishman, and P. G. Pentchev, *Science* **152**, 1270 (1966).

[3] J. M. Bailey, P. H. Fishman, and P. G. Pentchev, *J. Biol. Chem.* **242**, 4263 (1967).

TABLE I
PURIFICATION OF MUTAROTASE FROM *Capsicum frutescens*

Step	Total activity (units)	Total protein (mg)	Specific activity (units/mg)	Yield (%)	Purification (fold)
Original filtrate	119,000	24,286	4.9	100	1
(NH$_4$)$_2$SO$_4$ fractionation	101,000	4,139	24.4	85	5
Dialysis against phosphate buffer	100,000	1,718	58.2	84	12
Chromatography on hydroxy-apatite	30,600	24.5	1250	26	253
Dialysis against EDTA	27,400	10.1	2720	23	555

tography yielded a preparation with a specific activity of 17,000 units per milligram, but with only 2% recovery.

Properties

Specificity. The substrate specificity of the plant enzyme is different from that of other mutarotases (Table II). Pentoses are poor substrates. Maltose is a poor substrate but an excellent inhibitor. D-fructose is not a substrate.

Inhibitors. A number of sugars inhibit the plant enzyme (Table III). Maltose, cellobiose, L-arabinose, D-galactose, and D-xylose are competitive inhibitors.[3] Inhibition by sucrose and fructose is complex and does not obey Michaelis kinetics. Whereas mammalian mutarotases are inhibited by 1-deoxyglucose,[4] phlorizin,[5,6] and phloretin,[3,7] the plant enzyme is not. The inactivation by mercuric chloride indicates that the plant enzyme is a "sulfhydryl" enzyme, as are the mold,[8] bacterial,[9] and mammalian[10] mutarotases.

Other Properties. The Michaelis constant for the substrate D-glucose is 24 mM.[3] The plant enzyme has an apparent molecular weight of 50,000 from sedimentation data.[3] Based on a specific activity of 17,000, the turnover number is 8×10^5 moles of glucose per mole of enzyme which com-

[4] J. M. Bailey and P. G. Pentchev, *Biochem. Biophys. Res. Commun.* **14**, 161 (1964).
[5] A. S. Keston, *Science* **120**, 355 (1954).
[6] J. M. Bailey and P. G. Pentchev, *Amer. J. Physiol.* **208**, 385 (1965).
[7] D. F. Diedrich and C. H. Stringham, *Arch. Biochem. Biophys.* **138**, 497 (1970).
[8] R. Bentley and D. S. Bhate, *J. Biol. Chem.* **235**, 1219, 1225 (1960).
[9] K. Wallenfels, F. Hucho, and K. Herrman, *Biochem. Z.* **343**, 307 (1965).
[10] J. M. Bailey, P. H. Fishman, and P. G. Pentchev, *J. Biol. Chem.* **244**, 781 (1969); see also P. H. Fishman, P. G. Pentchev, and J. M. Bailey, *Biochemistry* **12**, 2490 (1973).

TABLE II

SUBSTRATE SPECIFICITIES OF MUTAROTASES FROM DIFFERENT SOURCES[a]

Substrate	Spontaneous	Penicillium notatum	Escherichia coli	Capsicum frutescens	Bovine kidney
D-Glucose	100	100	100	100	100
D-Galactose	134	107	108	151	215
L-Arabinose	418	128	25	49	190
D-Xylose	329	43	171	18	250
Maltose	61	3	2	6	0

[a] Data from J. M. Bailey, P. H. Fishman, and P. G. Pentchev [*J. Biol. Chem.* **242**, 4263 (1967)]. Relative velocities were calculated from the rate constant of the forward reaction, k_1, and the substrate concentration. The enzymic rate constants were corrected for the spontaneous contribution.

TABLE III

INHIBITORS OF MUTAROTASE FROM *Capsicum frutescens*

Inhibitor[a]	% Inhibition[b]
Maltose	77
L-Arabinose	61
Cellobiose	45
D-Galactose	40
L-Xylose	35
D-Fructose	25
α-Methylglucoside	24
Lactose	23
Sucrose	19
D-Xylose	11
L-Fucose	2
1-Deoxyglucose	0
Phlorizin	0
Phloretin	6
$HgCl_2$	100

[a] All sugars at 16.7 mM. Phlorizin 2 mM, phloretin 0.1 mM, and $HgCl_2$ 0.02 mM.
[b] Calculated K_i values (mM): maltose, 3.1; L-arabinose, 6.3; cellobiose, 12.3; D-galactose, 14.3.

pares favorably to other mutarotases.[11] The pH optimum is 7.4, but the activity profile is very broad (70% of maximum activity between pH 4.5 and 8.5). The temperature optimum is 40°, and the activation energy for the enzyme catalyzed mutarotation of D-glucose is 11,700 cal per mole.[3]

[11] S. A. Mulhern, P. H. Fishman, J. W. Kusiak, and J. M. Bailey, *J. Biol. Chem.* **248**, 4163 (1973).

[102] Glucose-6-phosphate 1-Epimerase from Baker's Yeast

By BERND WURSTER and BENNO HESS

α-D-glucopyranose 6-phosphate $\rightleftharpoons$ β-D-glucopyranose 6-phosphate

Assay Method

Principle. Because of the specificity of glucose-6-phosphate dehydrogenase for β-D-glucopyranose 6-phosphate,[1-5] the spontaneous and enzyme-catalyzed formation of β-D-glucopyranose 6-phosphate from α-D-glucopyranose 6-phosphate can be tested in the following system with glucose-6-phosphate dehydrogenase as indicator enzyme[2,3,6]:

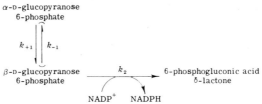

The reactions are started by addition of equilibrated D-glucose 6-phosphate (37% α-D-glucopyranose 6-phosphate and 63% β-D-glucopyranose 6-phosphate[6]) and recorded until the reaction is complete. If the reaction velocity constant of the glucose-6-phosphate dehydrogenase reaction, k_2, is much larger than the reaction velocity constant of the anomerization reaction k_{+1}

$$k_2 = \frac{V_{\text{glucose-6-phosphate dehydrogenase}}}{K_{m(\beta\text{-D-glucopyranose 6-phosphate})}} \gg k_{+1}$$

then, during the fast initial reaction, the concentration of β-D-glucopyranose 6-phosphate approaches zero; the succeeding slow reaction of first order represents the anomerization of α-D-glucopyranose 6-phosphate to β-D-glucopyranose 6-phosphate.

Calculation and Definition of the Activity Constant of Glucose-6-phosphate 1-Epimerase. The reaction velocity constant k_{+1} can be obtained from the slope of the regression line of a first-order plot, plotting log % D-glucose 6-phosphate remaining against the time. The reaction velocity constant k_{+1} is the sum of the velocity constant for the spontaneous anomerization reaction $k_{+1(s)}$ and the velocity constant (activity constant)

[1] M. Salas, E. Viñuela, and A. Sols, *J. Biol. Chem.* **240**, 561 (1965).
[2] J. M. Bailey, P. H. Fishman, and P. G. Pentchev, *J. Biol. Chem.* **243**, 4827 (1968).
[3] J. M. Bailey, P. H. Fishman, and P. G. Pentchev, *Biochemistry* **9**, 1189 (1970).
[4] S. P. Colowick and E. B. Goldberg, *Bull. Res. Counc. Isr. Sect. A.* **11**, 373 (1963).
[5] J. E. Smith and E. Beutler, *Proc. Soc. Exp. Biol. Med.* **122**, 671 (1966).
[6] B. Wurster and B. Hess, *FEBS Lett.* **23**, 341 (1972).

for the anomerization reaction catalyzed by glucose-6-phosphate 1-epi-merase $k_{+1(e)}$, $k_{+1} = k_{+1(s)} + k_{+1(e)}$; $k_{+1(e)}$ is obtained by subtraction of $k_{+1(s)}$ from k_{+1}. Under the experimental conditions described below, a value of $k_{+1(s)} = 3.8 \pm 0.2$ min^{-1} is obtained.

The activity constant of glucose-6-phosphate 1-epimerase is expressed as a pseudo-first order reaction velocity constant. An activity constant of 1 min^{-1} is that amount of enzyme which, in a final volume of 1 ml, causes an increase of the reaction velocity constant by 1 min^{-1}.

Reagents

Buffer: 50 mM imidazole-HCl, 50 mM KCl, 8 mM MgSO$_4$, pH 7.6
The following reagents are dissolved in this buffer:
D-Glucose 6-phosphate (equilibrated), 5 mM NADP$^+$, 100 mM
D-Erythrose 4-phosphate, 2.5 mM (in order to inhibit glucosephos-phate isomerase)
Glucose-6-phosphate dehydrogenase from yeast (grade 1, Boehringer Mannheim GmbH, West Germany), 1000 units/ml, freed of am-monium sulfate by dialysis
Glucose-6-phosphate 1-epimerase, activity constant 100–500 min^{-1} $\times$ ml^{-1}.

Procedure. The reaction mixture (final volume 1 ml), containing 0.91 ml of buffer, 0.02 ml of NADP$^+$ solution, 0.02 ml of D-erythrose 4-phos-phate solution, 0.02 ml of glucose-6-phosphate dehydrogenase, and 0.02 ml of glucose-6-phosphate 1-epimerase, is incubated for 10 min in a cuvette holder (connected to a thermostat) at $25 \pm 0.2°$. The reaction is started by addition of 0.01 ml of equilibrated D-glucose 6-phosphate and recorded until the reaction is complete; velocity of the chart paper of the recorder is 10 cm/min. In analogous experiments in the absence of glucose-6-phosphate 1-epimerase, the velocity constant for the sponta-neous anomerization reaction $k_{+1(s)}$ can be determined.

Applicability of the Test System. In crude extracts the anomerization of α- to β-D-glucopyranose 6-phosphate catalyzed by glucose-6-phosphate 1-epimerase cannot quantitatively be analyzed, because of the presence of gluconate-6-phosphate dehydrogenase, yielding a consecutive reduction of NADP$^+$. Thus, a clear determination of the end point of the primary reaction cannot be obtained. Only after separation of glucose-6-phosphate 1-epimerase from gluconate-6-phosphate dehydrogenase, either by gel filtration on Sephadex G-75 or in the case of the enzyme from baker's yeast by destruction of gluconate-6-phosphate dehydrogenase by acid treatment (pH 4.2, 2 N CH$_3$COOH, see first step of purification proce-dure), a quantitative determination of the anomerization of D-glucose 6-phosphate, catalyzed by glucose-6-phosphate 1-epimerase, is possible.

PURIFICATION OF GLUCOSE-6-PHOSPHATE 1-EPIMERASE FROM BAKER'S YEAST[a]

Fraction	Volume (ml)	Protein (mg × ml⁻¹)	Protein (mg total)	Glucose-6-phosphate 1-epimerase (min⁻¹ × ml⁻¹)	Glucose-6-phosphate 1-epimerase (min⁻¹ total)	Glucose-6-phosphate 1-epimerase (min⁻¹ × mg⁻¹)
Cell-free extract	1230	33	40,500		(412,000)[b]	(10.2)
Fraction pH 4.2	1330	10.2	13,600	310	412,000	30
Fraction pH 5.8	1500	8.8	13,200	270	405,000	31
Ammonium sulfate fractionation	75	74.5	5600	4400	330,000	59
Sephadex G-75 chromatography	188	1.1	207	1300	244,000	1180
DEAE-cellulose chromatography	11.4	0.97	11	13,500	154,000	14,000

[a] Protein was determined by the Biuret reaction [G. Beisenherz, H. J. Boltze, T. Bücher, E. Czok, K. H. Garbade, E. Meyer-Arendt, and G. Pfleiderer, Z. Naturforsch. **8 B**, 555 (1953)], Precinorm S (Boehringer Mannheim GmbH, West Germany) being used as a standard.

[b] In crude extracts the activity constant of the glucose-6-phosphate 1-epimerase catalyzed anomerization of D-glucose 6-phosphate cannot be determined quantitatively (see applicability of the test system). There is no loss of glucose-6-phosphate 1-epimerase activity during the acid treatment, as could be verified by repeated acid treatment.

In the presence of glucosephosphate isomerase the test system is also affected by both its isomerase activity and anomerase activity toward D-glucose 6-phosphate.[1,7-9] However, at low concentrations of D-glucose 6-phosphate, the isomerase activity and anomerase activity of glucose-phosphate isomerase can completely be inhibited by 50 μM D-erythrose 4-phosphate,[1,10,11] whereas glucose-6-phosphate 1-epimerase is not inhibited by this concentration of D-erythrose 4-phosphate.[11] After separation of glucose-6-phosphate 1-epimerase from glucosephosphate isomerase by chromatography on Sephadex G-75 (see purification procedure), D-erythrose 4-phosphate can be omitted from the reaction mixture.

After separation of glucose-6-phosphate 1-epimerase from gluconate-6-phosphate dehydrogenase and inhibition of glucosephosphate isomerase by D-erythrose 4-phosphate, a good proportionality between the concentration of glucose-6-phosphate 1-epimerase and its activity constant is found, as demonstrated earlier for a partially purified preparation of glucose-6-phosphate 1-epimerase from baker's yeast, free from gluconate-6-phosphate dehydrogenase and glucosephosphate isomerase.[6]

Purification Procedure

Cell-Free Extract. One kilogram of fresh baker's yeast (Deutsche Hefewerke) is suspended in 1 liter of 0.2 M sodium acetate. For cell disruption, the suspension is passed through a Gaulin Labor-Hochdruck-Homogenisator 15 M 8 TA (W. G. Schröder Nachfolger GmbH., Lübeck, West Germany) three times at 560 kp/cm². In the first step the temperature increased from 5° to 21°; in the second step from 21° to 36°; then the suspension was cooled to 25°, and in the third step the temperature increased to 40°. As a control of cell destruction the activity of glucosephosphate isomerase was measured[7]: it increased from 47 units/ml in the first step to 97 units/ml in the second step and to 120 units/ml in the third step, and did not increase further with passage of the suspension 4 or 5 times through the homogenizer. The homogenate is centrifuged for 30 min at 15,000 g, yielding 1230 ml of supernatant (cell-free extract).

The following operations are carried out at 4°.

Acid Treatment. With vigorous stirring the pH of the extract is de-

[7] B. Wurster and B. Hess, *Hoppe-Seyler's Z. Physiol. Chem.* **354**, 407 (1973).

[8] B. Wurster and B. Hess, this volume [12].

[9] K. J. Schray, S. J. Benkovic, P. A. Benkovic, and I. A. Rose, *J. Biol. Chem.* **248**, 2219 (1973).

[10] E. Grazi, A. De Flora, and S. Pontremoli, *Biochem. Biophys. Res. Commun.* **2**, 121 (1960).

[11] B. Wurster and B. Hess, *Hoppe-Seyler's Z. Physiol. Chem.* **355**, 255 (1974).

creased to 4.2 by dropwise addition of 350 ml of 2 N CH$_3$COOH (at pH 4.2 glucose-6-phosphate 1-epimerase is stable for hours). The precipitate is removed by centrifugation (30 min at 15,000 g). The pH of the supernatant (1330 ml, fraction pH 4.2) is increased to 5.8 –the isoelectric point of glucose-6-phosphate 1-epimerase[6]—by dropwise addition of 190 ml of 5% NH$_4$OH. The small precipitate is removed by centrifugation (30 min at 15,000 g), yielding 1500 ml of supernatant (fraction pH 5.8).

Ammonium Sulfate Fractionation. To the pH 5.8 fraction, ammonium sulfate (660 g)[12] is slowly added (during 2 hr), and stirring is continued for 1 hr, yielding 1880 ml of suspension of 2.7 M (NH$_4$)$_2$SO$_4$. The suspension is kept in the cold room overnight and centrifuged next morning (30 min at 15,000 g). The precipitate is suspended in 200 ml of 1.6 M (NH$_4$)$_2$SO$_4$ and stirred for 1 hr. The suspension is centrifuged (20 min at 27,000 g) and the ammonium sulfate concentration of the supernatant (230 ml) is increased from 1.6 M to 2.4 M (NH$_4$)$_2$SO$_4$ by addition of 29.4 g of (NH$_4$)$_2$SO$_4$[12] in small portions (during 1 hr). Stirring is continued for 1 hr, then the suspension is centrifuged (20 min at 27,000 g). The precipitate is dissolved in 50 ml of 20 mM sodium phosphate +3 mM sodium azide pH 6.8, yielding a solution of 75 ml volume.

Sephadex G-75 Chromatography. The solution is placed on a Sephadex G-75 column (60 $\times$ 8.8 cm) equilibrated with 20 mM sodium phosphate + 3 mM sodium azide, pH 6.8. Chromatography is conducted overnight; fractions of 17.3 ml volume are collected, flow rate 180 ml/hr. The peak fractions, containing about 90% of the total activity of glucose-6-phosphate 1-epimerase, are combined (188 ml volume).

DEAE-Cellulose Chromatography. The combined fractions of the Sephadex G-75 chromatography are added to a column (24 $\times$ 3.2 cm) of DEAE-cellulose (Whatman DE52), which is equilibrated with the same buffer (20 mM sodium phosphate + 3 mM sodium azide, pH 6.8). Proteins are eluted with 400 ml of this buffer followed by a linear gradient of 600 ml of 20 mM sodium phosphate + 3 mM sodium azide pH 6.8 and 600 ml of 100 mM sodium phosphate +3 mM sodium azide pH 6.8; fractions of 17.5 ml volume are collected, flow rate 100 ml/hr. Peak fractions containing about 90% of the total activity of glucose-6-phosphate 1-epimerase are combined (120 ml volume), adjusted to pH 5.8 by addition of 2 N CH$_3$COOH, and concentrated to 11.5 ml volume by pressure filtration (Amicon PM 10 membrane, 3.5 atm). The concentrate is dialyzed against 1 liter of 2.6 M (NH$_4$)$_2$SO$_4$ and stored at 4°. There is little or no (less than 10%) loss of activity within 1 month.

A summary of the purification procedure is given in the table. In four

---

[12] G. Beisenherz, H. J. Boltze, T. Bücher, F. Czok, K. H. Garbade, E. Meyer-Arendt, and G. Pfleiderer, *Z. Naturforsch.* **8B,** 555 (1953).

preparations or 1100- to 1400-fold purification with a recovery of 35–38% was achieved.

Properties

By gel filtration on Sephadex G-100 the molecular weight of glucose-6-phosphate 1-epimerase from baker's yeast was estimated to be 35,000, and in electrofocusing experiments an isoelectric point of pH 5.8 was obtained.[6] In the test system described above only a pseudo-first order reaction velocity constant (activity constant) of the glucose-6-phosphate 1-epimerase catalyzed anomerization of α-D-glucopyranose 6-phosphate to β-D-glucopyranose 6-phosphate can be determined. An analysis of the reaction progress curves obtained in the test system using a computerized nonlinear optimization technique yielded the Michaelis constants of K_m^α 144 μM and K_m^β 55.5 μM with the turnover numbers of 1950 sec^{-1} and 446 sec^{-1}, respectively.[12a]

The enzyme is inhibited by anions (chloride, sulfate, phosphate, NADP, EDTA)[11]; using a buffer of 50 mM imidazole-HCl, pH 7.6, instead of 50 mM imidazole-HCl + 50 mM KCl + 8 mM MgSO$_4$, pH 7.6, a specific activity constant four times higher was achieved.[11]

Glucose-6-phosphate 1-epimerase from baker's yeast catalyzes to a small extent the anomerization of D-glucose. Using a test system composed of glucose oxidase, peroxidase, and o-dianisidine[2,11,13] analogous to the test system described above, the activity constant of glucose-6-phosphate 1-epimerase toward D-glucose was determined to be 70,000-fold smaller as compared with D-glucose 6-phosphate.

Glucose-6-phosphate 1-Epimerase from Other Biological Sources. In *Escherichia coli* and *Rhodotorula gracilis*, glucose-6-phosphate 1-epimerases with molecular weights of about 30,000 were detected,[14] and in potato tubers two glucose-6-phosphate 1-epimerases with molecular weights of about 30,000 and 45,000, respectively, were discovered.[14] In rat muscle, rat liver, and rat kidney, the occurrence of glucose-6-phosphate 1-epimerase could not be demonstrated.[14]

[12a] E. M. Chance, B. Hess, Th. Plesser, and B. Wurster, *Eur. J. Biochem.* (1974) in press.

[13] H. U. Bergmeyer, "Methoden der enzymatischen Analyse." Verlag Chemie, Weinheim, 1971.

[14] B. Wurster and B. Hess, *FEBS Lett.* 38, 33 (1973).

Section VI

Miscellaneous Enzymes

[103] *N*-Acetylglucosamine-6-phosphate Deacetylase and Glucosamine-6-phosphate Deaminase from *Escherichia coli*

By RICHARD J. WHITE and CHARLES A. PASTERNAK

$$N\text{-Acetylglucosamine-6-P} + H_2O \rightarrow \text{glucosamine-6-P} + \text{acetate} \qquad (1)$$
$$\text{Glucosamine-6-P} \rightarrow \text{fructose-6-P} + \text{ammonia} \qquad (2)$$

N-Acetylglucosamine-6-P deacetylase and glucosamine-6-P deaminase have been shown to play an essential role in the utilization of exogenous amino sugar as a source of carbon and/or nitrogen by *Escherichia coli*.[1,2] Mutants lacking the deaminase are unable to grow on glucosamine or *N*-acetylglucosamine as sole source of carbon or nitrogen but can still incorporate the amino sugar into mucopeptide.[1-3] On the other hand, mutants lacking deacetylase can utilize glucosamine normally, but are unable to grow on *N*-acetylglucosamine or to incorporate it into mucopeptide.[2] Recently the genetic loci for these two enzymes on the *E. coli* K12 genetic map have been determined[4] and found to be very closely linked. Although deaminase catalyzes a potentially reversible reaction, it does not function anabolically in *E. coli*, as mutants lacking glucosamine-6-P synthetase lyse in the absence of exogenous amino sugar, even though they have normal deaminase levels.[5,6] Furthermore the deaminaseless mutants have no requirement for exogenous amino sugar.[2]

N-Acetylglucosamine 6-P deacetylase activity has also been detected in extracts of *Bacillus subtilis*[7] and *Bifidobacterium bifidum* var. *pennsylvanicus*[8] and in bovine parotid gland.[9] Glucosamine 6-P deaminase has been purified from a variety of sources, including *E. coli*,[1,10] *Proteus vulgaris*,[11] and pig kidney.[12]

[1] R. J. White and C. A. Pasternak, *Biochem. J.* **105**, 121 (1967).
[2] R. J. White, *Biochem. J.* **106**, 847 (1968).
[3] J. P. Rol's and C. W. Shuster, *J. Bacteriol.* **112**, 894 (1972).
[4] R. P. Holmes and R. R. B. Russell, *J. Bacteriol.* **111**, 290 (1972).
[5] M. Sarvas, *J. Bacteriol.* **105**, 467 (1971).
[6] H. C. Wu and T. C. Wu, *J. Bacteriol.* **105**, 455 (1971).
[7] C. J. Bates and C. A. Pasternak, *Biochem. J.* **96**, 147 (1965).
[8] J. H. Veerkamp, *Arch. Biochem. Biophys.* **129**, 248 (1969).
[9] Y. Matushita, and Y. Takagi, *Biochim. Biophys. Acta* **124**, 204 (1966).
[10] J. B. Wolfe, B. B. Britton, and H. I. Nakada, *Arch. Biochem. Biophys.* **66**, 333 (1957).
[11] H. I. Nakada, this series, Vol. 5, p. 575.
[12] D. G. Comb and S. Roseman, this series, Vol. 5, p. 422.

Assay Methods

Deacetylase

Principle. Deacetylase can be measured by following the disappearance of substrate, by determining the acetyl amino sugar remaining at the end of the reaction (described below). Alternatively, the appearance of product can be measured by adding an excess of glucosamine-6-P deaminase, glucosephosphate isomerase, and glucose-6-P dehydrogenase and estimating formation of reduced pyridine nucleotide spectrophotometrically.[1]

Reagents and Procedure. Centrifuge tubes contain the following reagents: 0.05 ml of 10 mM N-acetylglucosamine 6-P, 0.10 ml of 0.5 M Tris·HCl buffer, pH 8.5, water, and enzyme to give a total volume of 0.45 ml.

The reaction is started by adding enzyme, allowed to proceed for 10 min at 37°, and then terminated by the addition of 0.05 ml of 50% (w/v) trichloroacetic acid. Precipitated protein is pelleted by centrifuging at 5000 g for 10 min, and samples of the clear supernatant are assayed for acetyl amino sugar using the standard procedure of Levvy and McAllan.[13]

Deaminase

Principle. Deaminase activity is measured by coupling the enzyme to phosphoglucose isomerase and glucose 6-P dehydrogenase, and following the formation of reduced pyridine nucleotide spectrophotometrically.

Reagents and Procedure. The following reagents are added to a quartz cuvette with a 1-cm light path: 0.05 ml of 10 mM glucosamine 6-P, 0.1 ml of 0.2 M sodium phosphate buffer pH 7.5, 0.05 ml of 2 mM NADP, 4 units of phosphoglucose isomerase, 1.5 units of glucose-6-P dehydrogenase, and deaminase to give a total volume of 0.5 ml. The reaction is started by adding deaminase to the cuvette, which is maintained at 37° in a double-beam recording spectrophotometer. The absorbance at 340 nm, relative to a control cuvette containing all components except for enzyme, is recorded.

Units of Activity. One unit of activity is defined as the amount of enzyme that metabolizes 1 μmole of substrate per minute at 37°.

Purification Procedure

Growth of Bacteria. A starter culture is prepared by inoculating 250 ml of minimal medium[14] containing 0.075% (w/v) glucose in a 1 liter

---

[13] G. A. Levvy and A. McAllan, *Biochem. J.* **73**, 127 (1959).
[14] B. D. Davis and E. S. Mingioli, *J. Bacteriol.* **60**, 17 (1950).

conical flask with *E. coli* K12 grown on nutrient agar slants. After over-night incubation with shaking at 37°, the starter culture is used to inoculate 16 liters of the same medium in a 20-liter bottle, which is then incubated for 16 hr at 37° with forced aeration. Growth is limited by 0.075% (w/v) glucose to about 1 mg wet weight per milliliter of culture. Deacetylase and deaminase are then induced by adding 0.05% (w/v) *N*-acetylglucosamine, and incubation is continued for a further 1.5 hr.

Bacteria are harvested with a Sharples Super Centrifuge, washed in 200 ml of 10 m*M* sodium phosphate buffer pH 6.0 containing 10 m*M* mercaptoethanol (hereafter referred to as buffer) and resuspended in 60 ml of the same buffer. This and all subsequent operations are carried out at 4°.

Step 1. Preparation of Crude Extract. The concentrated cell suspension (approximately 25 g wet weight of cells) is disrupted by passage through a French pressure cell (American Instrument Company), and particulate matter is removed by centrifuging at 105,000 *g* for 1 hr and discarding the sediment.

Step 2. Protamine Sulfate Treatment. Forty milliliters of a 2% (w/v) solution of protamine sulfate, pH 6.0, is added to 80 ml of the crude extract and stirred for 30 min. The mixture is then centrifuged for 10 min at 35,000 *g*, and the sediment is discarded.

Step 3. Heat Treatment. The supernatant from the protamine sulfate treatment (105 ml) is heated with constant stirring in a 60° water bath for 6 min and then cooled to 4°. Precipitated protein is removed by centrifuging for 10 min at 35,000 *g* and discarded.

Step 4. Ammonium Sulfate Fractionation. Ammonium sulfate, 36.1 g per 100 ml of solution, is added to the heat-treated supernatant to give a 60% saturated solution, which is allowed to stand for 30 min and then centrifuged for 10 min at 35,000 *g*.

The protein pellet is dissolved in 10 ml of buffer and dialyzed against 3 × 2 liter portions of the same buffer.

Step 5. DEAE-Cellulose Chromatography. The dialyzed 0–60% (NH$_4$)$_2$SO$_4$ insoluble fraction is chromatographed on a 2-cm diameter × 35-cm long column of DEAE-cellulose, equilibrated with buffer. (Protein load was kept to less than 1 mg of protein per milliliter of packed exchanger.) The column is first washed with 100 ml of buffer and then eluted with a 0 to 0.25 *M* gradient of sodium chloride in buffer; 500 ml of 0.25 *M* NaCl in buffer is fed into a mixing chamber containing 500 ml of buffer at a flow rate of 1 ml/min. Fractions (9 ml) are collected until the concentration of chloride[15] in the eluate is 0.125 *M*. The absorbance at 280 nm is used as a measure of protein concentration.[16] Samples

[15] A. L. Tarnoky, "Clinical Biochemical Methods," p. 50. Hilger & Watts, London, 1958.

TABLE I

PURIFICATION OF GLUCOSAMINE-6-PHOSPHATE DEAMINASE AND
ACETYLGLUCOSAMINE-6-PHOSPHATE DEACETYLASE[a]

	Deaminase			Deacetylase		
Fraction	Total activity (units)	Specific activity (units/mg protein)	Re-covery (%)	Total activity (units)	Specific activity (units/mg protein)	Re-covery (%)
Crude cell-free extract	350	0.066	100	805	0.151	100
Protamine sulfate supernatant	98	0.084	28	204	0.174	25
Heat-treatment supernatant	98	0.143	28	170	0.241	21
(NH₄)₂SO₄ precipitate, 0–60% saturated	44	0.113	15	154	0.395	19
DEAE-cellulose eluate	17	27.6	5	34	5.60	

[a] Enzyme units are expressed in micromoles of substrate metabolized per minute. Reproduced by permission of the Editorial Board of the *Biochemical Journal*.

from fractions are assayed for deaminase and deacetylase activity. Fractions corresponding to the separated peaks of deacetylase and deaminase activity are pooled, and enzyme is precipitated by adding solid $(NH_4)_2SO_4$ to give a solution of 90% saturation. After standing overnight, the precipitate is collected by centrifuging at 35,000 g for 10 min. The supernatant is discarded, and the protein pellet is dissolved in 5 ml of buffer and dialyzed against 3 × 2-liter portions of buffer.

Table I summarizes the results of a typical purification experiment.[1] The overall yield of the procedure is rather poor, but a 370-fold purification of deacetylase and a 420-fold purification of deaminase may be obtained. Deacetylase preparations are generally slightly contaminated with deaminase, as a result of tailing of the first peak. This inconvenience can be overcome by using mutants lacking deaminase as source of deacetylase.[2]

Properties

Deacetylase

Specificity. N-Acetylglucosamine-6-P was the only substrate available that is degraded. The purified enzyme does not deacetylate N-acetyl-

[16] O. Warburg and W. Christian, *Biochem. Z.* **310**, 384 (1941).

TABLE II
STIMULATION OF GLUCOSAMINE-6-PHOSPHATE DEAMINASE BY
N-ACETYLGLUCOSAMINE 6-PHOSPHATE[a]

N-Acetylglucosamine-6-P[b] (μM)	Initial rate (arbitrary units)
11	1.00[c]
22	1.00
44	1.28
110	1.63
220	2.33
440	2.79
	2.68

[a] Reproduced by permission of the Editorial Board of the *Biochemical Journal*.
[b] Purified deaminase (Step 5) was added to 1 mM glucosamine-6-P.
[c] Initial rate was 26 units/mg of protein.

glucosamine; crude extracts contain an extremely labile enzyme which does have activity against *N*-acetylglucosamine (see Roseman).[17] Experiments with crude deacetylase preparations from *B. subtilis*[7] and *B. bifidum*[8] indicate a rather low specificity with respect to the acyl group in both cases.

Effect of pH. The enzyme has a well defined optimum at pH 8.55.

Effect of Substrate Concentration. The K_m in sodium phosphate buffer pH 8.5 is 0.4 mM.

Reversibility. No indication of reversibility was obtained under a variety of conditions.

Cofactors. Prolonged dialysis causes insignificant loss of activity, so no essential cofactors appear to be involved.

Inhibitors. Sodium acetate, 15 mM, and ammonium phosphate, 15 mM, have no effect on enzyme activity, but fructose-6-P and glucosamine-6-P are inhibitory (66 mM fructose-6-P and 66 mM glucosamine-6-P cause a 28% and 80% loss in activity, respectively).

Deaminase

Effect of pH. The purified enzyme is more active in 10 mM sodium phosphate buffer than in Tris·HCl at the same pH, the optimum in phosphate buffer being at pH 7.0.

Effect of Substrate Concentration. The K_m in 10 mM sodium phosphate buffer at pH 7.0 is 9 mM.

[17] S. Roseman, *J. Biol. Chem.* **226**, 115 (1957).

Reversibility. In the presence of high concentrations of fructose-6-P and ammonia, some synthesis of glucosamine-6-P can be demonstrated.

Activators. As previously described,[1] the enzyme is stimulated by *N*-acetylglucosamine-6-P. Table II summarizes a typical result.

[104] Methylglyoxal Synthase[1,2]

By R. A. COOPER

Dihydroxyacetone phosphate → methylglyoxal + P_i

Methylglyoxal synthase is formed constitutively by a number of bacteria[3,4] and may be involved in a by-pass sequence[5] for part of the Embden-Meyerhof glycolytic pathway.

Assay Methods

Principle. Methylglyoxal production can be measured continuously by coupling it to the lactoylglutathione lyase reaction and monitoring the formation of *S*-lactoylglutathione at 240 nm (Method 1); or discontinuously, at the end of the reaction period, by measurement of the bis-derivative formed on reaction with the 2,4-dinitrophenylhydrazine reagent (Method 2).

Reagents

Imidazole-HCl buffer, 50 m*M*, pH 7.0
Glutathione, reduced, 33 m*M*, pH 7.0
Dihydroxyacetone phosphate, 15 m*M*, pH 4.5; prepared from dihydroxyacetone phosphate dimethylacetal (dimonocyclohexylamine salt) by acid hydrolysis at 40° for 4 hr
Lactoyl-glutathione lyase (EC 4.4.1.5) from Yeast, Grade III, Sigma Chemical Company, (specific activity 345), 80 units/ml diluted in 50 m*M* imidazole·HCl buffer pH 7.0
2,4-Dinitrophenylhydrazine, 0.1% dissolved in 2 *N* HCl
NaOH, 10%

[1] This work was assisted by grant B/SR/73094 from the Science Research Council.
[2] Dihydroxyacetone-phosphate phospho-lyase, EC 4.2.99.11.
[3] D. J. Hopper and R. A. Cooper, *FEBS Lett.* **13**, 213 (1971).
[4] R. A. Cooper, *Eur. J. Biochem.* **44**, 81 (1974).
[5] R. A. Cooper and A. Anderson, *FEBS Lett.* **11**, 273 (1970).

Method 1

Procedure. The assay mixture in a silica cuvette (10 mm light path, approximately 1.5 ml volume) contains imidazole buffer, 0.8 ml; glutathione, 0.05 ml; lactoyl-glutathione lyase, 0.01 ml; dihydroxyacetone phosphate, 0.05 ml; and distilled water, 0.04 ml. After equilibration at 30°, any blank rate is measured and then the reaction is started by the addition of 0.05 ml of enzyme solution containing sufficient enzyme to catalyze a change in absorbance at 240 nm of approximately 0.025–0.05 unit/min. In this system 1 μmole of lactoyl-glutathione has A_{240} 3.4.[6]

Method 2

Procedure. The assay mixture in a 16 mm $\times$ 125 mm test tube contains imidazole buffer, 0.4 ml; dihydroxyacetone phosphate, 0.025 ml; and distilled water, 0.05 ml. The tubes are equilibrated at 30° and then the reaction is started by the addition of 0.025 ml of enzyme solution containing sufficient enzyme to form 0.01–0.03 μmole of methylglyoxal per minute. An assay mixture with imidazole buffer (0.025 ml) rather than enzyme solution is used as the control. Samples (0.1 ml) of the reaction mixture are removed immediately, after 2.5 min, after 5 min, and after 10 min of incubation and run into test tubes containing 0.9 ml of distilled water plus 0.33 ml of 2,4-dinitrophenylhydrazine reagent. These tubes are incubated at 30° for 15 min then 1.67 ml of 10% NaOH is added to produce the characteristic purple color, and the absorbance at 550 nm is measured after a further 15 min at room temperature. Under these conditions 1 μmole of methylglyoxal has an absorbance of 16.4.

Units. One unit of methylglyoxal synthase is defined as that amount of enzyme catalyzing the formation of 1 μmole of methylglyoxal per minute at 30°. Specific activity is expressed as units per milligram of protein.

Purification Procedure

Preparation from Pseudomonas saccharophila

Methylglyoxal production during glucose catabolism by whole cells of *P. saccharophila* was first reported in 1952.[7] Almost twenty years later an enzyme that was able to form methylglyoxal from the glycolytic intermediate dihydroxyacetone phosphate was identified and partially purified from glucose-grown cells.[4] The buffer used throughout the purification

---

[6] E. Racker, this series, Vol. 3, p. 293.
[7] N. Entner and M. Doudoroff, *J. Biol. Chem.* **196**, 853 (1952).

was 50 mM imidazole·HCl buffer pH 7.0, and protein was determined with the Folin-Ciocalteu reagent.[8]

Growth of Cells. A *P. saccharophila* mutant (obtained from Dr. M. Doudoroff) which could grow well on glucose was grown aerobically at 30° in the minimal salts medium described by Doudoroff[9] with 12.5 mM glucose as carbon source. To ensure vigorous aeration the medium was shaken in tribaffled conical flasks (500 ml medium/2-liter flask). The cells were harvested by centrifugation at 20,000 g for 5 min at 4°.

Step 1. Preparation of Crude Extract. Cells (4.1 g wet weight) were suspended in 20 ml of buffer and disrupted at 0° in an MSE 100 W ultrasonic disintegrator for 5 min. The suspension was centrifuged at 20,000 g for 10 min at 4°; the supernatant was retained.

Step 2. Heat Treatment. The crude extract was heated rapidly in a 30 mm × 180 mm test tube, with continuous stirring, to 64°, kept at that temperature for 1 min, and then cooled in ice-water. The precipitate formed was removed by centrifugation at 20,000 g for 15 min at 2° and the supernatant solution retained.

Step 3. Ammonium Sulfate Precipitation. Solid $(NH_4)_2SO_4$ (209 g/liter) was added slowly to the heat-treated extract at 0° and stirred for 45 min. The precipitate formed was removed by centrifuging at 20,000 g for 15 min and discarded. More $(NH_4)_2SO_4$ (94 g/liter) was added to the supernatant solution at 0°; the solution was stirred for 45 min and centrifuged as above. The supernatant solution was discarded, and the precipitate was dissolved in 2 ml of buffer.

Step 4. Gel Filtration on Bio-Gel P-150. The enzyme solution from step 3 was applied to 2.5 cm × 40 cm column of Bio-Gel P-150 (100–200 mesh) previously equilibrated against buffer containing 1 mM KH_2PO_4. The column was eluted at 4° with buffer containing 1 mM KH_2PO_4 at a flow rate of approximately 25 ml/hr and 2.25-ml fractions were collected. The fractions with the highest specific activities were pooled and stored at 4°.

This procedure gives over 100-fold purification of the enzyme with a reasonable recovery of activity. A summary of such a purification is shown in Table I.

Preparation from *Escherichia coli*

Methylglyoxal synthase is present in *E. coli* at a specific activity higher than that found in *P. saccharophila*, and the enzyme has been

[8] O. H. Lowry, N. J. Rosebrough, A. L. Farr, and R. J. Randall, *J. Biol. Chem.* **193**, 265 (1951).

[9] M. Doudoroff, this series, Vol. 1, p. 225.

TABLE I
PURIFICATION OF *Pseudomonas saccharophila* METHYLGLYOXAL SYNTHASE

Step	Volume (ml)	Total protein (mg)	Total units	Specific activity	Recovery (%)	Purification (fold)
1. Crude extract	17	187.5	9.9	0.053	100	1
2. Heat treatment	15.5	40.6	9.2	0.226	93	4
3. $(NH_4)_2SO_4$ precipitation	2.1	7.7	6.2	0.80	62	15
4. Peak fractions from Bio-Gel P-150 column	4.5	0.33	1.9	5.75	19	108

purified extensively from glycerol-grown cells of *E. coli* K12, strain CA244.[10] These cells are available from Whatman Biochemicals Ltd., Springfield Mill, Maidstone, Kent, U.K. The following method[10] has been used for routine purification of the enzyme. The buffer used throughout the purification was 50 mM imidazole·HCl pH 7.0, and protein was determined spectrophotometrically by the method of Warburg and Christian.[11]

Step 1. Preparation of Crude Extract. A 100-g batch of frozen cells was thawed in 200 ml of buffer, and 60-ml portions were disrupted at 0° in a MSE 100 W ultrasonic disintegrator for 4 min. The suspension was centrifuged at 30,000 g for 30 min at 2°, and the supernatant solution was retained.

Step 2. Heat Treatment. The crude extract (in 50-ml portions) was heated rapidly in a 30 mm × 180 mm test tube, with continuous stirring, to 75°, kept at that temperature for 1 min, and then quickly cooled in ice-water. The extract was then centrifuged at 30,000 g for 15 min at 2° to remove the heavy precipitate, and the supernatant solution was retained.

Step 3. Ammonium Sulfate Precipitation. Solid $(NH_4)_2SO_4$ (209 g/liter) was added to the heat-treated extract at 0° and stirred for 60 min. The precipitate obtained was removed by centrifuging at 30,000 g for 15 min at 2° and discarded. More $(NH_4)_2SO_4$ (111 g/liter) was added to the supernatant solution at 0° and stirred for 60 min; the mixture was centrifuged as above. The precipitate obtained was dissolved in 3 ml of buffer containing 1 mM KH_2PO_4 (to stabilize the enzyme) prior to gel filtration.

Step 4. Gel Filtration on Bio-Gel P-150. The enzyme solution from the preceding step was applied to a 2.5 cm × 92 cm column of Bio-Gel P-150

[10] D. J. Hopper and R. A. Cooper, *Biochem. J.* **128**, 321 (1972).
[11] O. Warburg and W. Christian, *Biochem. Z.* **310**, 384 (1941).

(100–200 mesh) previously equilibrated against buffer containing 1 mM KH_2PO_4. The column was eluted at room temperature with buffer containing 1 mM KH_2PO_4 at a flow-rate of approximately 22 ml/hr, and 2.7-ml fractions were collected. The enzyme was eluted between fractions 52 and 72, and the most active fractions were pooled.

Step 5. Second Ammonium Sulfate Precipitation. Solid $(NH_4)_2SO_4$ (243 g/liter) was added to the pooled fractions at 0° and stirred for 60 min; the solution was centrifuged as described in step 3. The precipitate was discarded, and more $(NH_4)_2SO_4$ (132 g/liter) was added to the supernatant solution at 0° and stirred for 60 min; the solution was centrifuged as above. The precipitate obtained was dissolved in 2 ml of buffer containing 1 mM KH_2PO_4 and retained for the next step.

Step 6. Sephadex G-100 Gel Filtration. The enzyme solution from step 5 was applied to a 2.5 cm $\times$ 89 cm column of Sephadex G-100 (superfine grade) equilibrated against buffer containing 1 mM KH_2PO_4. The column was eluted at room temperature with buffer containing 1 mM KH_2PO_4 at a flow rate of approximately 8 ml/hr, and 2.1-ml fractions were collected. The fractions (83–86) containing the purified enzyme were stored at 4°.

This procedure gives over 1500-fold purification of the enzyme with a reasonable recovery of activity. A summary of such a purification is shown in Table II. The highest specific activities obtained have varied from 350 to 530 in different preparations.

Properties of the *E. coli* Enzyme

Stability. The enzyme in crude extracts loses little activity when stored at 0–4° over several days. However, such preparations show a 90% loss of activity on gel filtration which can be almost wholly prevented by incorporation of 1 mM P_i into the eluting buffer. Removal of P_i from the purified enzyme leads to a rapid loss of activity, which is more pronounced at 0° than at 22°. Dihydroxyacetone phosphate (1 mM) and crystalline bovine serum albumin (10 mg/ml) are able to overcome the inactivation to some extent.[10] Kinetic studies of stabilization by P_i suggest that there may be two P_i-binding sites on the enzyme.[10]

Purity. The most pure enzyme preparation (specific activity 530) shows one major and two minor protein bands[10] on acrylamide gel electrophoresis at pH 8.9. Only the major band had methylglyoxal synthase activity, and it accounted for more than 70% of the total protein. When the electrophoresis was carried out in the presence of 1 mM KH_2PO_4, the mobility of the enzyme band was increased significantly whereas the mobility of the other two bands was not affected.

TABLE II

PURIFICATION OF *Escherichia coli* METHYLGLYOXAL SYNTHASE[a]

Step		Volume (ml)	Total protein (mg)	Total units	Specific activity	Recovery (%)	Purification (fold)
1. Crude extract		230	5405	1695	0.32	100	1
2. Heat treatment		205	496	1440	2.9	85	9
3. First (NH$_4$)$_2$SO$_4$ fractionation		3.5	57	1220	21.4	72	66
4. Pooled fractions from Bio-Gel P-150 column		24.2	5.3	776	141.0	46	440
5. Second (NH$_4$)$_2$SO$_4$ fractionation		2.1	3.5	611	174.0	36	543
6. Peak fractions from Sephadex G-100 column	83	2.1	0.275	86	310.0	5	966
	84	2.1	0.288	127	441.0	7.5	1375
	85	2.1	0.200	106	530.0	6.3	1650
	86	2.1	0.088	44	500.0	2.6	1550

[a] Data from D. J. Hopper and R. A. Cooper, *Biochem. J.* **128**, 321 (1972).

Cofactors. No cofactor requirement has been observed, and the enzyme is not inhibited by 4 mM EDTA.

Reversibility. There is no evidence that the reaction is reversible to any significant extent.

Specificity. Neither dihydroxyacetone nor DL-glyceraldehyde 3-phosphate could substitute for dihydroxyacetone phosphate in the reaction.

Molecular Weight. The *E. coli* and *P. saccharophila* enzymes have molecular weights of approximately 67,000 as estimated by gel filtration.

pH Optimum. The *E. coli* enzyme has optimal activity at pH 7.5 whereas the *P. saccharophila* enzyme is most active at pH 8.2.

Kinetic Properties and Inhibition. At pH 7.0 methylglyoxal synthase shows normal Michaelis kinetics, the K_m for dihydroxyacetone phosphate being 0.47 mM for the *E. coli* enzyme and 0.09 mM for the *P. saccharophila* enzyme. However, in the presence of 0.3–0.5 mM P_i, there is pronounced inhibition of both enzymes[4,10] which is overcome in a cooperative manner by increasing dihydroxyacetone phosphate concentration. In the presence of P_i there appear to be three dihydroxyacetone phosphate-binding sites on the enzyme. PP_i is a potent competitive inhibitor, with K_i values of 95 μM for the *E. coli* enzyme and 48 μM for the *P. saccharophila* enzyme. 3-Phosphoglycerate and phosphoenolpyruvate are also inhibitory, but the kinetics of inhibition are complex for the *E. coli* enzyme[10]; both are apparently simple competitive inhibitors of *P. saccharophila* methylglyoxal synthase[4] with K_i values of 29 μM and 96 μM, respectively.

Distribution. Methylglyoxal synthase has been found in various Enterobacteriaceae[3] and in certain Enterobacteriaceae-like organisms, such as *Aeromonas formicans* and *Obesumbacterium proteus*. It has also been found in the strict anaerobes *Clostridium pasteurianum* and *Clostridium tetanomorphum*. The only strict aerobe so far shown to contain methylglyoxal synthase is *P. saccharophila*.[4]

[105] Pyruvate Formate-lyase from *Escherichia coli* and Its Activation System

By JOACHIM KNAPPE and HANS P. BLASCHKOWSKI

$$\text{Pyruvate} + \text{CoA} \rightleftharpoons \text{acetyl-CoA} + \text{formate} \qquad (1)$$

Acetyl-CoA production according to Reaction (1), a central reaction in the anaerobic metabolism of *Escherichia coli,* is catalyzed by the enzyme pyruvate formate-lyase (CoA-acetylating)(PFL), which is extremely sensitive toward inactivation by oxidation.[1,2] Accordingly, its study requires special precautions. Its purification from the cell extract has only partially been accomplished. On the other hand, the inactive, oxidized form of the enzyme, as it is available either from aerobic or anaerobic cells upon employment of conventional techniques for protein fractionation, is reconvertible to the catalytically active state by an enzymic process.[3] The following components participate in this reaction: A reducing system (reduced flavodoxin[4] or certain metal ion–thiol complexes), a Fe-protein "enzyme II,"[5] S-adenosyl-L-methionine, and pyruvate.[3,6]

$$\text{PFL}_i \text{ (inactive)} \xrightarrow[\substack{\text{enzyme II} \\ \text{adenosyl methionine} \\ \text{pyruvate}}]{\text{reducing system}} \text{PFL}_a \text{ (active)} \tag{2}$$

The process is suggested to be (part of) a regulatory system of protein

[1] This series, Vol. 1 [73].

[2] T. Chase, Jr. and J. C. Rabinowitz, *J. Bacteriol.* **96**, 1065 (1968).

[3] J. Knappe, J. Schacht, W. Möckel, T. Höpner, H. Vetter, Jr., and R. Edenharder, *Eur. J. Biochem.* **11**, 316 (1969).

[4] H. Vetter, Jr. and J. Knappe, *Hoppe-Seyler's Z. Physiol. Chem.* **352**, 433 (1971). The protamine precipitate (see step 2 of the Purification Procedure) from 1 kg of cell paste is thoroughly suspended in 0.1 M potassium phosphate pH 7.4, containing 10 mM mercaptoethanol, to a volume of 3 liters and stirred for 30 min. After centrifugation, the protein solution is filtered within about 10 hr through DEAE-cellulose (2.5 × 6 cm; 0.1 M phosphate pH 7.4). After washing with 100 ml phosphate buffer plus 0.16 M KCl the cellulose bed is removed from the column and the orange-yellow zone of flavodoxin, which has accumulated at the top, is excised. It is eluted by repeatedly suspending the cellulose in an equal volume of 0.28 M potassium acetate, pH 5.0, plus 0.6 M KCl and short centrifugation. The combined extracts are diluted with 3 volumes of water, then applied to DEAE-cellulose (2.5 × 15 cm; 0.28 M acetate pH 5) and chromatographed with a linear gradient from 500 ml of 0.28 M acetate pH 5 and 500 ml of the same buffer plus 0.3 M KCl. The orange-yellow fractions of flavodoxin appearing at 0.13 to 0.16 M KCl are combined and diluted by an equal volume of water. Concentration is achieved by readsorption on DEAE-cellulose (1 × 2 cm) and elution with acetate buffer plus 0.6 M KCl. After buffer exchange for 1 mM potassium phosphate pH 7.2, the preparation is finally chromatographed on hydroxyapatite (2.5 × 8 cm) employing a linear gradient from 1 mM to 100 mM phosphate pH 7.2; total volume 800 ml. The fractions from 35 to 45 mM which contain about 10 mg of pure flavodoxin ($E_{467}/E_{274} = 0.165$; $E_{467}^{0.1\%} = 0.58$), are concentrated by ultrafiltration; they are stored in 10 mM phosphate pH 7 at $-20°$.

[5] The designation enzyme II, as previously used along with enzyme I for the inactive form of pyruvate formate-lyase (see footnote 3), is retained until its nature is better defined.

[6] J. Knappe, H. P. Blaschkowski, P. Gröbner, and T. Schmitt, *Eur. J. Biochem.*, in press.

interconversion which in particular would be responsible for the rapid, protein synthesis-independent appearance of the capacity to catalyze reaction (1) when aerobic cells are deprived of oxygen.

Assay Methods

Principle. Pyruvate formate-lyase activity may be assayed by measuring the production of formate in Reaction (1); CoA is expediently regenerated by the phosphate acetyltransferase reaction.[2,3,7] An optical assay, allowing the continuous recording of pyruvate formate-lyase activity, results from the coupling of the production of acetyl-CoA to the reduction of NAD by the addition of malate, malate dehydrogenase, and citrate synthase[6]:

$$\text{Pyruvate} + \text{malate} + \text{NAD} \xrightarrow[\substack{\text{malate dehydrogenase} \\ \text{citrate synthase}}]{\text{CoA, PFL}} \text{citrate} + \text{formate} + \text{NADH} \quad (3)$$

[Alternatively, the production of pyruvate by the reversal of Reaction (1) can be coupled to oxidation of NADH by addition of lactate dehydrogenase, acetyl-CoA being generated from acetyl phosphate, CoA, and phosphate acetyltransferase.] A further assay, readily applicable also to crude systems and requiring no additional enzymes, employs the isotope exchange reaction between [^{14}C]formate and the carboxyl group of pyruvate (Eq. 4)[1,8]; which comes about by the first, CoA-independent half-reaction of PFL.

$$\text{Pyruvate} + [^{14}\text{C}]\text{formate} \underset{\text{PFL}}{\rightleftharpoons} [^{14}\text{C}]\text{pyruvate} + \text{formate} \quad (4)$$

The activation process (2) is followed by measuring the production of pyruvate formate-lyase activity, and by this means its protein components pyruvate formate-lyase (inactive) and enzyme II are assayed.

Activation Reaction and Pyruvate Formate-lyase Activity Assay by Pyruvate-Formate Exchange

Reagents

2-(*N*-Morpholino)propanesulfonate (M), 0.5 *M*, pH adjusted to 8.0 with KOH
Dithiothreitol, 0.3 *M*

---

[7] R. K. Thauer, F. H. Kirchaniawy, and K. A. Jungermann, *Eur. J. Biochem.* **27**, 282 (1972).
[8] N. P. Wood, this series, Vol. 9 [129].

$Fe(NH_4)_2(SO_4)_2$, 20 mM

S-Adenosyl-L-methionine, 5 mM

Sodium pyruvate, 2 M

[^{14}C]Sodium formate, 0.5 M, 12 μCi/ml

$CoCl_2$, 0.2 M (for procedure A)

Flavodoxin from *E. coli*,[4] 1 mg/ml (for procedure B)

2,6-Dichlorophenolindophenol (DPIP), 0.4 mM (for procedure B)

3-(3,4-Dichlorophenyl)-1,1-dimethylurea (DCMU), 0.1 mM (for procedure B)

Chloroplast fragments (P_{1s1}) from spinach,[9] 0.3 mg of chlorophyll per milliliter; suspended in 35 mM NaCl, 5 mM M buffer (for procedure B)

Perchloric acid, 7.5%

2,4-Dinitrophenylhydrazine, 50 mM in 18 N H_2SO_4

Procedure A, Using Cobalt–Dithiothreitol as Reducing System. Test tubes with two arms, equipped with a ground-glass stopper with an outlet for gas exchange are most convenient for the anaerobic operation and for preincubations prior to mixing the reaction components. One arm (arm E) is filled with the enzymes, the other (arm R) with substrates and reagents. After addition of the bulk of components, the tubes are deaerated by several times evacuating with an oil pump and filling with oxygen-free argon; then they are closed under argon. Any subsequent additions of reagents are made under vigorous flushing with argon. Incubations are at 30°.

To arm E is added: M buffer, 0.05 ml; dithiothreitol, 0.01 ml; pyruvate formate-lyase (inactive), enzyme II, and water 0.18 ml. To arm R is added: M buffer, 0.08 ml; dithiothreitol, 0.01 ml; S-adenosylmethionine, 0.01 ml; pyruvate, 0.01 ml; water, 0.11 ml. After filling with argon, 10 μl of $Fe(NH_4)_2(SO_4)_2$ is added to the mixture in arm E and 10 μl of $CoCl_2$ to the mixture in arm R.[10] After a preincubation of 30 min,[11] the activation reaction is started by mixing the two compartments and is run for 60 min. Subsequently, 20 μl of [^{14}C]formate are introduced to start the pyruvate-formate exchange reaction which is run for an appropriate period of time (see below), usually for 30 min and terminated by adding 0.2 ml of perchloric acid.

Dinitrophenylhydrazine, 0.3 ml, is added to 0.5 ml of the supernatant. The pyruvate dinitrophenylhydrazone is washed in the centrifuge, dis-

[9] F. R. Whatley and D. I. Arnon, this series, Vol. 6 [37].

[10] The precipitate formed with Co^{2+} does not interefere with the assay.

[11] Preincubation with Fe^{2+} and dithiothreitol serves to restore maximal activity of purified enzyme II fractions.

solved in 0.5 ml of 1% sodium carbonate, reprecipitated by addition of 0.05 ml of a mixture containing 10 N HCO_2H and 2 N HCl, and washed again. An aliquot (0.5–2 μmoles) of an aqueous slurry is transferred to a planchet, dried, and counted in a gas-flow counter (no correction for self-absorption is necessary with these amounts applied). The solid is dissolved in 1 ml of 10% aqueous pyridine and its actual amount is determined colorimetrically.[12]

Procedure B, Using Photoreduced Flavodoxin. The composition of the reagent mixture in test tube arm R is changed as follows: Reduce the amount of water to 0.03 ml and add 0.02 ml of chloroplast fragments, 0.03 ml of DCMU, 0.02 ml of DPIP, and 0.02 ml of flavodoxin; omit $CoCl_2$. All other reagents are as used in procedure A. Subsequent to the preincubation period, and the two compartments having been mixed, the tubes are illuminated from a 20-W light source (conditions are not critical) during the 60-min period of the activation reaction.

Definition of Units. One unit of pyruvate formate-lyase activity is defined as that amount of enzyme which catalyzes the conversion of 1 μmole of substrate per minute in the pyruvate-formate exchange reaction.[13] Consequently, 1 unit of the inactive form of pyruvate formate-lyase is that amount which yields 1 unit of pyruvate formate-lyase activity upon complete activation; the inactive lyase is measured by employing enzyme II in excess. If the assay is employed for the measurement of enzyme II, a constant amount of 0.5 unit of pyruvate formate-lyase (inactive) is taken; and one unit of enzyme II is defined as an amount which yields 1 unit of pyruvate formate-lyase activity within the activation period of 60 min. A virtually linear relationship holds up to 0.4 unit.

The progress of the activation reaction and hence the activity of enzyme II is more accurately measured if a shorter time period for the isotope exchange is taken. A means to discontinue the activation, without interfering with the assay of the pyruvate formate-lyase activity which has accumulated, is to add the inhibitor *S*-adenosyl-L-homocysteine (0.2 mM). Provided that the enzyme concentration suffices, one can also quench any further activation by diluting aliquots into separate assay mixtures (from which adenosylmethionine and enzyme II are omitted). Alternatively, the rapid optical assay described below can be employed.

[12] T. E. Friedemann, this series, Vol. 3 [66].
[13] With the amounts of pyruvate and formate employed here, the following equation holds for the calculation of the pyruvate-formate exchange reaction rate from the measured specific radioactivity of pyruvate dinitrophenylhydrazone, p (dpm or cpm per μmole), the starting specific radioactivity of formate, a (dpm or cpm per μmole), and the reaction time period, t (minutes):

$$\mu\text{mole/minute} = -(15.4/t) \log [1 - (3p/a)]$$

Coupled Assay of Pyruvate Formate-lyase Activity Utilizing NAD Reduction

Reagents

Tris·HCl buffer, 50 mM, pH 8.1 (at 30°)
DL-Malate, 0.1 M; neutralized with KOH
NAD, 50 mM
CoA, 5 mM
Citrate synthase, 2 mg/ml, 140 units/ml
Malate dehydrogenase, 5 mg/ml, 5500 units/ml
Bovine serum albumin, 10 mg/ml

Procedure. The assay is applicable for samples containing per milliliter ≥ 0.2 units of pyruvate formate-lyase (active form) and which are from more purified protein fractions. (The DEAE-cellulose steps of the inactive form and enzyme II are suitable; lactate dehydrogenase which would interfere with the optical assay, is absent from these fractions.) The reduced flavodoxin system should be used, if samples are to be analyzed directly from the activation reaction system.

An anaerobic substrate mixture for 10 assays is made as follows: Tris buffer, 8.7 ml; malate, 0.5 ml; NAD, 0.2 ml; CoA, 0.12 ml; dithiothreitol, 0.33 ml; pyruvate, 0.05 ml; serum albumin, 0.1 ml. After deaeration by evacuation and argon flushing, 0.05 ml of $Fe(NH_4)_2(SO_4)_2$ is added; incubate at room temperature until reddish color has disappeared. Kept under argon, this mixture may be used for at least several hours. Reactions are carried out at 30° in 1.4-ml cuvettes (1-cm path length), equipped with a ground-glass joint to which an outlet with a stopcock is connected, the absorption increase at 366 nm being monitored in a recording photometer. Citrate synthase (0.01 ml) and malate dehydrogenase (0.005 ml) are added first, next air is displaced by argon, and 0.98 ml of the substrate mixture is introduced during argon flushing. After the optical density has been checked for constancy, the reaction is started with 5–50 μl of the lyase preparation. A stirring rod is used for mixing.

An increase in absorbancy of 0.05 to 0.4 per minute is a good rate for activity measurements. The absolute rate is calculated by using the extinction coefficient for NADH, $\epsilon_{366} = 3.3$ mM^{-1}cm^{-1}. The rate, measured with this coupled assay according to Eq. (3), is 2.5 times faster than the pyruvate-formate exchange reaction under the conditions given above (where a suboptimal formate concentration is employed).

Purification Procedure

Growth of Bacteria. Escherichia coli K12 is grown in aerobic culture at 37° in a medium which contains per liter: 2.5 g of glucose, 2 g of citric acid·H_2O, 10 g of K_2HPO_4, 3.5 g of $NaNH_4HPO_4$·4 H_2O, 0.2 g of $MgSO_4$·7 H_2O, 0.01 g of EDTA, 0.02 g of $Fe(NH_4)_2SO_4$; pH 7. The cells are harvested near the stationary growth phase. The cell paste can be stored at $-20°$ for at least several months. (Cells grown anaerobically or obtained from commercial sources are also suitable.)

Notes on Purification. The procedure is a slight modification of that described by Knappe *et al.*[3]; pyruvate formate-lyase (inactive form), enzyme II, and flavodoxin are obtained from one batch. All steps are performed at about 4°. The protein fractions of intermediate stages, free of ammonium sulfate, can be stored in the frozen state for at least 1 month. pH measurements are at 20°. Chromatography materials are: DEAE-cellulose, Whatman DE-52; hydroxyapatite prepared according to Levin[14]; dextran gels Sephadex G-25 (fine) and G-150, Pharmacia Fine Chemicals AB.

Protein Determination. The biuret method is employed at the first and second purification step. Thereafter, measurements are made from the absorbances at 280 and 260 nm.

Step 1. Preparation of Cell Extracts. Frozen cells are thawed and suspended in 50 mM potassium phosphate pH 7, using 100 ml of buffer for each 100 g of cells. They are disintegrated, in portions of 75 ml, for 7 min with a Sonifier B-12 (Branson Sonic Power Comp.) with sufficient cooling to hold the temperature below 8°; subsequent centrifugation is at 40,000 g for 2 hr.

Step 2. Protamine and Ammonium Sulfate Precipitations. The supernatant solution is adjusted to a protein concentration of 40 mg/ml with the buffer used in step 1. The pH of the solution, which is rigorously stirred, is adjusted to 6.1 (4°) by adding 1 N HCl; then a 3% solution of protamine sulfate is added, using 7.7 ml for each gram of protein. The final pH should not be below 6.0. After 15 min the mixture is centrifuged. (The precipitate may be used for the isolation of flavodoxin as outlined in footnote 4.) The supernatant solution, which contains 45–50% of the protein of step 1, is adjusted to pH 7 with 1 N KOH; then ammonium sulfate, 37 g for each 100 ml, is added. After 20 min the mixture is centrifuged and the precipitate is dissolved in 0.2 M potassium phosphate pH 6.5, using 25 ml for the fraction from each 100 g of bacteria. Insoluble material is discarded.

Step 3. Sephadex G-150 Gel Filtration for Separating Pyruvate Formate-lyase (Inactive) from Enzyme II. A column with a gel bed of 90 × 10 cm is prepared, and is operated with 0.2 M potassium phosphate pH 6.5 at a hydrostatic pressure of 30 cm. It is loaded with maximally 200 ml of the solution from step 2 (i.e., from about 500 g of cell paste).

[14] Ö. Levin, this series Vol. 5 [2].

The flow rate is 4 ml/min, the effluent being collected in 30-ml fractions. Three major yellow zones develop during chromatography which, by measuring the absorbance at 370 nm, serve as a guideline for monitoring the elution of pyruvate formate-lyase (inactive form) and enzyme II: Those fractions, comprising about 600 ml, which appear between the peak of the first yellow zone and the trough before the second zone, are collected; they contain pyruvate formate-lyase (inactive). Those fractions which appear between 300 ml and 1500 ml after the peak of the second yellow zone are combined; they contain enzyme II.

Pyruvate formate-lyase (inactive form) is concentrated by precipitation with 33 g of ammonium sulfate per 100 ml; the precipitate is dissolved in 30 ml of 40 mM potassium phosphate pH 6.5 and desalted by gel filtration on G-25 with the same buffer. For concentrating enzyme II, 37 g of ammonium sulfate is used, the precipitate being taken up in 40 ml of 10 mM phosphate pH 6.5, and after removing insoluble material by centrifugation desalted by gel filtration with this latter buffer.

Step 4 for Pyruvate Formate-lyase (Inactive Form). DEAE-Cellulose Chromatography. A 20 × 5 cm column bed, equilibrated with 40 mM potassium phosphate pH 6.5, is used for the application of about 2 g of protein (from about 1 kg cells) of the pyruvate formate-lyase (inactive) fraction of step 3. A linear gradient, 6 liters total volume, from 40 mM to 0.2 M potassium phosphate is applied at a flow rate of 3 ml/min. The fractions that contain the enzyme protein are eluted at a buffer concentration between 0.07 and 0.12 M. They are concentrated to a volume of 200 ml by ultrafiltration using a Diaflo PM-10 membrane (Amicon Corp.) and applying a pressure of 30 psi.

Step 5 for Pyruvate Formate-lyase (Inactive Form). Hydroxyapatite Chromatography. The concentrated solution of the preceding step (about 700 mg) is diluted with 2 volumes of water and applied to a 20 × 5 cm column bed, equilibrated with 30 mM potassium phosphate, pH 6.5. A linear gradient, total volume 5 liters, from 10 mM to 0.15 M potassium phosphate, pH 7.4, is then applied at a flow rate of 2 ml per minute. Pyruvate formate-lyase (inactive) is eluted from 90 mM to 0.11 M, the activity coinciding with the 280 nm peak, which appears at this range. The active fractions are combined and by ultrafiltration are concentrated to about 15 ml. The buffer is finally exchanged for 10 mM buffer M, pH 7.3, before storage of the preparation at −20°. It is stable for at least several months.

Step 4 for Enzyme II. DEAE-Cellulose Chromatography. A 20 × 5 cm column bed, equilibrated with 10 mM potassium phosphate pH 6.5, is used for about 3 g of protein (from about 1 kg of cells) of the enzyme II fraction of step 3. A linear gradient, total volume 5 liters, from 10

PURIFICATION OF PYRUVATE FORMATE-LYASE (INACTIVE FORM) AND OF ENZYME II

| | PFL$_i$ | | Enzyme II | |
Fraction	Protein (mg)	Specific activity (units/ mg)	Protein (mg)	Specific activity[b] (units/ mg)
1. Extract[a]	79,000	0.4[c]	79,000	0.5
2. Ammonium sulfate precipitate from protamine supernatant	18,000	1.0	18,000	1.1[e]
3. Sephadex G-150 gel filtration	2,100	4.0	2,900	1.7
4. DEAE-cellulose chromatography	720	8.3	260	6.5[f]
5. Hydroxyapatite chromatography	155	25[d]	32	16[f]

[a] From 900 g of cell paste.
[b] Measured in the Co^{2+}-dithiothreitol system.
[c] Including residual pyruvate formate-lyase, active form.
[d] Purity is about 0.3, as estimated from polyacrylamide gel electrophoresis.
[e,f] Fractional activities without Fe^{2+} are 0.3 and 0.1, respectively.

mM to 0.12 M potassium phosphate pH 6.5 is applied at a flow rate of 2.5 ml/min. Enzyme II activity is eluted from 50 mM to 70 mM buffer, near the elution of a yellow protein. The combined fractions are concentrated 20-fold by ultrafiltration.

Step 5 for Pyruvate Formate-lyase (Inactive Form). *Hydroxyapatite* of step 4 is diluted with 3 volumes of water and applied to a 20 × 2.5 cm column bed, equilibrated with 15 mM phosphate pH 6.5. A linear gradient, total volume 2 liters, from 15 mM to 0.15 M phosphate is applied. The activity appears at a buffer concentration of 80 mM within a volume of 200 ml. After concentrating as before, the buffer is exchanged for 10 mM buffer M, pH 7.3, and the preparation is stored at −20°.

The purification is summarized in the table.

Anaerobic Gel Filtration of Activated Pyruvate Formate-lyase

The activation system with reduced flavodoxin, with quantities scaled up from assay level to preparative quantities is recommended in order to prepare pyruvate formate-lyase (active form) on a larger scale. Its separation from various components of the activation reaction is conveniently achieved by gel filtration. For details for operating anaerobic columns, the article of Repaske[15] should be consulted.

Sephadex G-25 (fine grade; 15 × 2.5 cm) or G-150 (80 × 1.6 cm) is employed in order to separate the enzyme from low molecular weight

[15] R. Repaske, this series, Vol. 22 [28].

constituents (adenosylmethionine, pyruvate) or also from the activating protein(s). Sample volumes of 3 ml can be applied; the heavier chloroplast fragments are previously centrifuged off. The column is operated at 4° with argon as inert gas and the following anaerobic buffer: 50 mM buffer M, pH 7.8, containing 9 mM dithiothreitol and 0.2 mM Fe(NH$_4$)$_2$(SO$_4$)$_2$. The fractions are collected in argon-flushed tubes; activity is determined by the coupled optical assay. With Sephadex G-25, the activity is eluted immediately after the void volume. With G-150, pyruvate formate-lyase (active form) and enzyme II protein migrate according to K_{av} values of 0.15 and 0.48, respectively. The recoveries are at least 95% and 80% with the Sephadex G-25 and G-150 columns, respectively.

Properties

The molecular weight of the inactive species of pyruvate formate-lyase is 140,000; it comprises two polypeptide chains of MW about 70,000. The active form has the same size, as estimated from sedimentation velocity and gel filtration data.[3,16]

Stability. At 0° and pH 8 pyruvate formate-lyase (active form) is virtually stable if kept in media which display a redox potential of ≤ -0.2 V. At 30° the activity declines with a halftime of 50 min. Admittance of oxygen inactivates completely within a few seconds.[6]

Catalytic Parameters. Pyruvate formate-lyase activity is maximal from pH 7.8 to 8.4. A Ping-Pong mechanism, which is suggested from the pyruvate–formate exchange reaction, is corroborated from initial steady-state kinetic data for the forward and reverse reactions [Eq. (1)]. K values are 2 mM, 6.8 μM, 51 μM, and 24 mM for pyruvate, CoA, acetyl-CoA, and formate, respectively; $V_1/V_2 = 2.9$.[6]

Equilibrium Constant. From thermodynamic data, K_{eq} of reaction (1) is 800.[17] Chemical analysis[17] and the above kinetic parameters, using the Haldane equation, yield values of 1300 and 750, respectively.

Activation Reaction. Activation of pyruvate formate-lyase (inactive form) requires reducing systems affording potentials of at least -0.39 V and is maximal only at -0.44 V.[16] The physiological electron-donating system is reduced flavodoxin, which is regenerated *in vivo* by a thiamine-diphosphate-dependent pyruvate:flavodoxin oxidoreductase system.[4] The complexes of Co^{2+} with dithiothreitol and of Fe^{2+} with 2,3-dimercaptopropanol are suitable chemical substitutes. Pyruvate (or its analog

[16] J. Knappe, H. P. Blaschkowski, and R. Edenharder, *in* "Second International Symposium on Metabolic Interconversion of Enzymes" (O. Wieland, E. Helmreich, and H. Holzer, eds.), p. 319. Springer-Verlag, Berlin, Heidelberg, and New York, 1971.

[17] N. Tatanaka and M. J. Johnson, *J. Bacteriol.* **108,** 1107 (1971).

oxamate) is an obligatory component of the activation reaction (apparent $K_m = 0.2$ mM) as well as S-adenosylmethionine (apparent $K_m = 7$ μM); S-adenosylhomocysteine is an inhibitor[7] acting noncompetitively with respect to adenosylmethionine.[18]

The activity of the enzyme II-fraction is due to an Fe–protein of 40,000 molecular weight. Fe^{2+}-dependence is total after the enzyme has been dialyzed against 0.2 M phosphate, pH 6.5, 1 mM o-phenanthroline, and 10 mM mercaptoethanol; Fe^{2+}-recombination has been demonstrated with ^{59}Fe.

Pyruvate Formate-lyase of Other Organisms. Besides in *Escherichia coli*,[2,16,19] the enzyme has also more recently been studied in *Streptococcus faecalis*[20] and in clostridiae.[7,21] In these latter organisms, where its function is probably mainly anabolic, an adenosylmethionine-dependent activation has also been demonstrated.

[18] H. P. Blaschkowski, unpublished results, 1972.
[19] H. Nakayama, G. G. Midwinter, and L. O. Krampitz, *Arch. Biochem. Biophys.* **143**, 526 (1971).
[20] D. G. Lindmark, P. Paolella, and N. P. Wood, *J. Biol. Chem.* **244**, 3605 (1969).
[21] N. P. Wood and K. Jungermann, *FEBS Lett.* **27**, 49 (1972).

[106] 2,3-Butanediol Biosynthetic System in *Aerobacter aerogenes*

By FREDRIK C. STORMER

$$2CH_3-\underset{\substack{\| \\ O}}{C}-COOH \xrightarrow[\text{enzyme}]{\text{pH 6 acetolactate-forming}} CO_2 + CH_3-\underset{\substack{\| \\ O}}{C}-\underset{\substack{| \\ OH}}{\overset{CH_3}{C}}-COOH \quad (1)$$

Pyruvate   α-Acetolactate

$$CH_3-\underset{\substack{\| \\ O}}{C}-\underset{\substack{| \\ OH}}{\overset{CH_3}{C}}-COOH \xrightarrow[\text{decarboxylase}]{\text{acetolactate}} CO_2 + CH_3-\underset{\substack{\| \\ O}}{C}-\underset{\substack{| \\ H}}{\overset{OH}{C}}-CH_3 \quad (2)$$

Acetoin

$$CH_3-\underset{\substack{\| \\ O}}{C}-\underset{\substack{| \\ H}}{\overset{OH}{C}}-CH_3 + NADH + H^+ \underset[\text{reductase}]{\overset{\text{diacetyl-}}{\text{(acetoin)}}}{\rightleftharpoons} CH_3-\underset{\substack{| \\ H}}{\overset{OH}{C}}-\underset{\substack{| \\ H}}{\overset{OH}{C}}-CH_3 + NAD^+ \quad (3)$$

2,3-Butanediol

$$CH_3-\underset{\substack{\| \\ O}}{C}-\underset{\substack{\| \\ O}}{C}-CH_3 + NADH + H^+ \rightarrow CH_3-\underset{\substack{\| \\ O}}{C}-\underset{\substack{| \\ H}}{\overset{OH}{C}}-CH_3 + NAD^+ \quad (4)$$

Diacetyl

2,3-Butanediol synthesis in *A. aerogenes* proceeds in three steps [Reactions (1)–(3)]. Diacetyl(acetoin) reductase catalyzes, in addition to Reaction (3), the irreversible reduction of diacetyl to acetoin [Reaction (4)]. In *A. aerogenes*, the three enzymes are induced by acetate.[1,2] In addition, acetate activates the pH 6 acetolactate-forming enzyme and inhibits diacetyl(acetoin) reductase in the direction from 2,3-butanediol to acetoin at pH 5.8 or lower [Reaction (3)]. This allows maximal activity of this pathway in the presence of acetate and sufficient amounts of NADH in the direction from pyruvate to 2,3-butanediol.

These enzymes constitute approximately 2.5% of the total protein in the cell. Sections I–III describe the purification to homogeneity as well as the properties of the enzymes involved in the formation of 2,3-butanediol from pyruvate and from diacetyl.

I. The pH 6 Acetolactate-Forming Enzyme

In *A. aerogenes* there are two different enzymes producing acetolactate.[3,4] One is the enzyme described herein,[5] the other is acetohydroxy acid synthetase that catalyzes the first and second reaction in the formation of valine and isoleucine respectively. In contrast to the pH 6 acetolactate-forming enzyme,[6] acetohydroxy acid synthetase has been shown to require FAD,[7] and it is inhibited by valine.

Assay Method

The pH 6 acetolactate-forming enzyme activity is determined by measuring the amount of acetolactate formed.

Reagents

Enzyme, diluted in distilled water at 0° prior to assay
Sodium pyruvate, 0.4 M
Cocarboxylase, 0.87 mM
MnCl$_2$, 5 mM
H$_2$SO$_4$, 50% (v/v)

[1] F. C. Størmer, *FEBS Lett.* **2**, 36 (1968).
[2] T. D. K. Brown, C. R. S. Pereira, and F. C. Størmer, *J. Bacteriol.* **112**, 1106 (1972).
[3] Y. S. Halpern and H. E. Umbarger, *J. Biol. Chem.* **234**, 3067 (1959).
[4] Y. S. Halpern and A. Even-Shoshan, *Biochim. Biophys. Acta* **139**, 502 (1967).
[5] E. Juni, this series, Vol. 1, 471.
[6] F. C. Størmer, *J. Biol. Chem.* **243**, 3740 (1968).
[7] F. C. Størmer and H. E. Umbarger, *Biochem. Biophys. Res. Commun.* **17**, 587 (1964).

Creatine, 0.5%
α-Naphthol, 5% in 2.5 N NaOH, freshly prepared
NaOH, 2.5 N
Sodium acetate pH 5.8, 1.0 N

The complete incubation mixture contained in 1.0 ml: 40 μmoles of sodium pyruvate, 50 μmoles of sodium acetate pH 5.8, 0.87 μmole of cocarboxylase, 0.5 μmole of $MnCl_2$, and enzyme.

After incubation for 20 min at 37°, the reaction was terminated by the addition of 0.1 ml of 50% H_2SO_4. After 25 min at 37°, when the acetolactate formed in the assay had been decarboxylated to acetoin, 0.9 ml of 2.5 N NaOH was added, and 0.5 ml of this solution was transferred to a tube and mixed with 2.0 ml of a solution containing creatine and α-naphthol (1:1). During incubation, after the addition of NaOH, a brownish precipitate appeared. This could either be removed by centrifugation or corrected for by using the blank. This mixture was left at room temperature for 20 min. During this period it was shaken every 5 min, and finally 2.5 ml of 2.5 N NaOH was added and immediately read at 540 nm against a blank from which enzyme protein was omitted. Acetoin was used as a standard.

Definition of Specific Activity. Specific activity is defined as micromoles of acetohydroxy acid formed per milligram of protein per hour.

Purification of the Enzyme

Growth and Harvesting of Cells

The stock culture of *A. aerogenes*, strain 1033, was a gift from Professor H. E. Umbarger of Purdue University, Lafayette, Indiana. The cells were routinely maintained on nutrient agar slants.

The bacteria were grown with aeration at 37° in minimal medium with the use of distilled water and supplemented with trace elements[8] and with the addition of yeast extract and tryptose to the media, 5 g per liter of each.[9] Bactodextrose 1%, was used as the source of carbon. The cells, allowed to reach the stationary phase, were harvested when the pH in the culture had dropped to approximately 5.8, usually after 20 hr. The suspension was chilled to 10–15°, and the cells were sedimented in an RC2B Sorvall centrifuge equipped with a continuous flow system. The cells could be stored at −25° for several months without detectable loss of the enzyme activities.

[8] F. C. Størmer, *J. Biol. Chem.* **242**, 1756 (1967).
[9] D. Malthe-Sørenssen and F. C. Størmer, *Eur. J. Biochem.* **14**, 127 (1970).

Procedure

The following procedure for purification of the pH 6 acetolactate-forming enzyme resulted in a 120-fold purification with a 20–30% yield. Unless otherwise stated, all operations were performed at 0–5°.

Buffer. The standard buffer used for preparing solutions for enzyme preparation contained 50 mM potassium phosphate, 1 mM MgCl$_2$, 5 mM sodium pyruvate, and 0.2 mM cocarboxylase, pH = 6.0.[8]

Preparation of Extract. Frozen cells (185 g) were suspended in 5 volumes of the standard buffer. Aliquots of 30 ml were subjected to ultrasound for 10 min with a MSE ultrasonic power unit, Model 60 W. The temperature of the sonic extract increased up to 15° during this operation. The material was centrifuged for 10 min at 16,000 g, and unbroken cells were resuspended in the buffer and treated by sonic disruption. The supernatant was centrifuged for 30 min at 40,000 g, and the clear solution was collected to yield fraction I.

Heat Treatment. The above solution was divided into two portions and heated to 55°, with continuous stirring, by placing the containers in a water bath at 80°. The solutions were kept at 55° for 2 min and cooled to 0° in an ice-water bath with continuous stirring. After 20 min the inactive precipitate was removed by centrifugation and discarded, leaving fraction II.

Streptomycin Precipitation. To the 990 ml of fraction II were added, with stirring 178 ml of a 10% streptomycin sulfate solution adjusted to pH 6 with 1 M KH$_2$PO$_4$. After standing for 15 min, the suspension was centrifuged and the supernatant was collected (fraction III).

First Ammonium Sulfate Fractionation. To fraction III were added, with stirring, 450 g of solid (NH$_4$)$_2$SO$_4$; the pH was adjusted to 6.0 by adding 1 M K$_2$HPO$_4$ dropwise. The suspension was allowed to stand for 20 min after the salt had dissolved. The precipitate was dissolved after centrifugation at 10,000 g for 10 min in sufficient 50 mM phosphate to provide a protein concentration of 20–30 mg/ml. This solution was then dialyzed overnight against 2 liters of the same buffer. A small precipitate, which appeared during dialysis, was removed by centrifugation at 10,000 g and yielded a clear supernatant (fraction IV).

DEAE-Sephadex A-50 Chromatography. Fraction IV was adjusted to pH 7 by adding 1 M K$_2$HPO$_4$ and distilled water to yield a final phosphate concentration of 50 mM. This 350-ml solution was loaded on the column (11 × 5 cm), previously equilibrated with the same buffer, at a flow rate of 3.0 ml/min. The column was eluted with a linear gradient formed from 500 ml of 50 mM and 500 ml of 0.5 M phosphate, pH 7.0. Fractions of 10 ml were collected at a flow rate of 1.5 ml/min, and fractions 45 to 61, which contained the bulk of the activity, were pooled and used for further purification. To this solution was added 75 g of solid $(NH_4)_2SO_4$; the resultant precipitate was collected by centrifugation and dissolved in 100 ml of 50 mM phosphate, pH 6.0 (fraction V).

Hydroxyapatite Chromatography.[10,11] A Büchner funnel with a sintered filter and with a diameter of 5 cm was connected to a suction flask. A test tube was placed in the center at the inside of the flask, where the fractions were collected. A filter paper was placed in the funnel, and a slurry of the hydroxyapatite, suspended in 50 mM phosphate pH 6.0 was added. An aliquot of fraction V containing 125 mg of protein was loaded on the column and washed with a few milliliters of 50 mM phosphate, pH 6.0. The column was eluted three times with 5 ml of 0.35 M phosphate and three times with 5 ml of 0.9 M phosphate, pH 6.0, at the flow rate of 2.5 ml/min. The 0.9 M eluent, fractions 4 to 6, which contained most of the enzyme activity was collected. The gel was discarded, and the procedure was repeated with a new batch of protein. Hydroxyapatite from Bio-Rad Laboratories, 4.3 g, dry weight, was used each time, and during this purification step all buffers contained 2 mM MgCl$_2$, 10 mM sodium pyruvate, and 0.4 mM cocarboxylase. The fractions were combined, volume 250 ml, and solid 97 g of $(NH_4)_2SO_4$ were added, and the pH adjusted as earlier. The active precipitate was dissolved in 50 mM phosphate, pH 6.0, to yield a protein concentration of 17 mg/ml (fraction VI). Although the enzyme is not completely separated from the protein that follows it, it is sufficiently pure for crystallization.

Crystallization

Two different crystalline forms of the enzyme can be obtained, depending on the conditions employed. If the crystallization is performed in the presence of ammonium sulfate, plate-shaped soluble crystals are formed (Fig. 1A). In the absence of the salt, however, needle-shaped insoluble crystals appear (Fig. 1B). The two crystalline forms have been studied by electron microscopy using negative staining technique,[12] showing that the molecular organization was different in the two different crystalline forms of the enzyme.

First Crystallization. The protein concentration from the preceding step should not be lower than 17 mg/ml. In about 8 hr, a "silky shimmer" was observed; after an additional 24 hr, a heavy, white precipitate was revealed, which consisted of thin plates. Occasionally, the time for the

[10] W. F. Anacker and V. Stoy, *Biochem. Z.* **330**, 141 (1958).
[11] W. Velle and L. L. Engel, *Endocrinology* **74**, 429 (1964).
[12] F. C. Størmer, Y. Solberg, and T. Hovig, *Eur. J. Biochem.* **10**, 251 (1969).

appearance of the crystals could be somewhat longer, and the enzyme was crystallized with greater consistency if seed crystals were added to the supernatant solution before storage.

The precipitate was spun down at 12,000 *g* for 10 min and dissolved by gentle shaking at 37°, in a sufficient volume of standard buffer containing 0.2 *M* $(NH_4)_2SO_4$, pH 6.0, to yield a protein concentration of 7.5 mg/ml (fraction VII). More than 85% of the enzyme from the preceding step was crystallized (based on specific activity) in this way.

Second Crystallization. Recrystallization was carried out as follows. Insoluble material in fraction VII was removed by centrifugation, and to the clear supernatant, 1.0 ml of saturated $(NH_4)_2SO_4$ in the standard buffer was added dropwise until the solution became slightly turbid. Crystalline enzyme appeared in a few hours, with the same shape as in the first crystallization, and after standing overnight, the heavy crystalline precipitate was collected by centrifugation (fraction VIII). No further increase in specific activity was observed when fraction VIII was recrystallized by the procedure described above.

Third Crystallization. If the plate-shaped crystals were dissolved in the standard buffer, 50 m*M* acetate, pH 5.8, or distilled water, at a protein concentration of 5 mg/ml or higher, the crystallization usually began within a few minutes and was virtually complete within a few hours. The crystals appeared as long, thin, white needles. The needle-shaped crystals could be formed from fractions VII and VIII, and from fraction VI if the latter solution was dialyzed against the standard buffer. After the needle-shaped crystals had been formed, they were removed from the supernatant by centrifugation and suspended in the standard buffer.

The needle-shaped crystalline material was slightly soluble in H_2O and buffers in the pH range of 5.0 to 11.0. The following compounds had no effect on the solubility of the crystals: sodium lauryl sulfate, sodium deoxycholate, cetyltrimethyl ammonium-bromide, dimethyl sulfoxide, dithiothreitol, EDTA, NaCl, sodium pyruvate, cocarboxylase, and 1 *M* phosphate, pH 6–8.

Since the crystals were not put into soluble form and assayed successfully, the specific activity could not be determined. A washed crystal suspension (even stored in the dry state at room temperature for several months) showed "high" specific activity. When the crystals were brought into solution by decreasing or increasing pH, low specific activity was determined. The observation that the needle-shaped crystals were formed from fraction VI if the solution had previously been dialyzed against the standard buffer led to the speculation that $(NH_4)_2SO_4$ prevented the formation of the insoluble crystals and favored the formation of the soluble plate-shaped crystals. When the plate-shaped crystals were

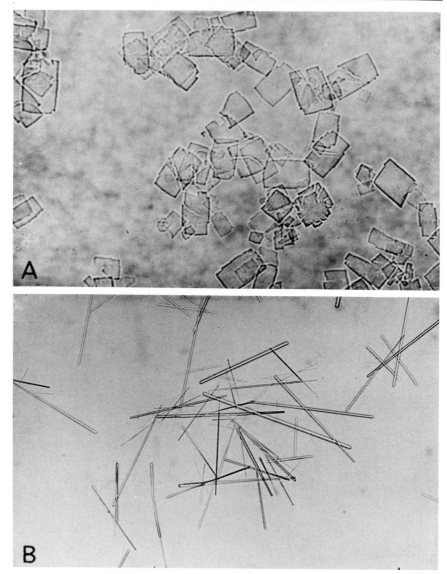

Fig. 1. Photographs of the crystalline pH 6 acetolactate–forming enzyme. (A) soluble crystals, ×204; (B) insoluble crystals, ×450.

dissolved in the standard buffer containing 0.2 M $(NH_4)_2SO_4$ at a protein concentration of 5 mg/ml or higher, the needle-shaped crystals were not immediately formed. After a week, microcrystals of the insoluble material could be observed.

TABLE I
PURIFICATION OF pH 6 ACETOLACTATE-FORMING ENZYME FROM
Aerobacter aerogenes

Fraction and step	Volume (ml)	Protein (mg)	Specific activity (μmoles/mg protein)	Yield (%)	Purification (fold)
I. Sonic extract from 185 g of cells	1150	21,505	264	100	1
II. Heat treatment	990	8,910	550	86	2
III. Streptomycin sulfate treatment	1160				
IV. First (NH$_4$)$_2$SO$_4$	140	3,850	1,250	84	5
V. DEAE-Sephadex chromatography	100	1,700	1,850	56	7
VI. Hydroxyapatite chromatography	14	238	8,823	37	31
VII. First crystallization	4	60	30,600	32	112
VIII. Second crystallization	4	48	31,800	27	120

Table I summarizes the data for a typical preparation.

Properties

Stability. Partially purified enzyme has been stored in the standard buffer containing 50% glycerol at 0° for 18 months without significant change in activity. Suspensions of the plate-shaped crystals have been stored successfully for several years.

Purity and Physical Properties. The purified enzyme appeared to be homogeneous in the ultracentrifuge and in gel electrophoresis. From the sedimentation constant ($s_{20,w}^{\circ} = 10.5 \times 10^{-13}$ sec) and the diffusion coefficient ($D_{20,w}^{\circ} = 5.0 \times 10^{-7}$ cm^2 sec^{-1}) the molecular weight was calculated to be 200,000, assuming a partial specific volume of 0.742 ml/g, calculated from the amino acid composition.[12] The molecular weight based on equilibrium centrifugation, is approximately 220,000.[13]

In 4.8 M guanidine hydrochloride, 8 M urea, and 0.1 M 2-mercaptoethanol, or sodium dodecyl sulfate, the enzyme dissociates into subunits with molecular weight of approximately 58,000.

Electron microscopic studies show that the enzyme is visualized as a dimer, each with a diameter of approximately 5.7 nm. Provided the particles represent spherical bodies of this diameter, the calculated molecular weight is about 80,000, assuming a partial specific volume of 0.742

ml/g (see above). These results indicate that the enzyme is a dimer, each composed of two equal-sized subunits.[13]

Four thiol groups per enzyme molecule has been determined.[14]

Enzymic Properties. Three molecules of cocarboxylase were found to be tightly bound to one molecule of the enzyme, and the enzyme has a requirement for manganese. An absolute requirement for these compounds were not established, but has been demonstrated for the similar enzyme purified from *Serratia marcescens*, where half-maximal velocity was obtained with 0.1 mM MnCl$_2$, and 0.17 mM cocarboxylase.[9]

The enzyme has a sharp pH optimum at 5.8. Apparent K_m is only obtained in acetate, and was estimated to be 5.9–6.6 mM at pH 5.8. Buffers like phosphate, succinate, maleate, and citrate cause sigmoidal kinetics, and lower V_{max}.

Phenylpyruvate and glyoxylate are competitive inhibitors, with K_i values 0.11 mM and 51 μM, respectively. Competitive inhibition was also observed in the presence of ammonium sulfate, but the curves were sigmoidal.[15]

When 2-oxobutyrate was added to the assay, a mixture of acetolactate and acetohydroxybutyrate were formed. 2-Oxobutyrate acted also as a competitive inhibitor with a K_i of 5 mM.[16]

II. Acetolactate Decarboxylase

Assay Method

Enzyme activity was assayed by measuring the production of acetoin (acetylmethyl carbinol).[5]

Reagents

DL-α-Acetolactate, 0.1 M
Potassium phosphate pH 6.2, 1 M
Creatine, 0.5%
α-Naphthol, 5% in 2.5 N NaOH
NaOH, 2.5 N

The incubation mixture contained in 1.0 ml: 200 μmoles of potassium phosphate pH 6.2, 10 μmoles of DL-α-acetolactate,[17] and enzyme. The

[13] N. E. Huseby, T. B. Christensen, B. R. Olsen, and F. C. Størmer, *Eur. J. Biochem.* **14**, 133 (1970).

[14] B. Tveit and F. C. Størmer, *Eur. J. Biochem.* **10**, 249 (1969).

[15] F. C. Størmer, *J. Biol. Chem.* **243**, 3735 (1968).

[16] N. E. Huseby and F. C. Størmer, *Eur. J. Biochem.* **20**, 215 (1971).

[17] K. Kiritani and R. P. Wagner, this series, Vol. 17A, p. 745.

reaction was terminated, after incubation at 37° for 20 min, by the addition of 1.0 ml of 2.5 N NaOH. Acetoin was determined with the creatine-naphthol reagent as described in the section on the pH 6 acetolactate-forming enzyme.

Definition of Specific Activity. Specific activity is defined as micromoles of acetoin formed per milligram of protein per hour at 37°.

Purification of Acetolactate Decarboxylase from *A. aerogenes*[18]

The following procedure for purification of acetolactate decarboxylase resulted in a 145 times purification with a 14% yield. Unless otherwise stated, all operations were performed at 0–5°.

Preparation of Extract. Frozen cells (20 g) were suspended in 5 volumes of 5.4 M NaCl in 30 mM phosphate pH 6.0. The solution was subjected to ultrasound for 10 min with a Branson Sonic Power SW 75 oscillator (fraction I).

Polyethyleneglycol-Dextran Fractionation. To the above suspension (120 ml) was added 120 ml of 5.4 M NaCl in 30 mM phosphate, pH 6.0, 48 ml dextran T 500 (Pharmacia) (20% w/w) in the same buffer, and 96 ml of 3.2 M NaCl in 25% w/w polyethyleneglycol 6000 (Koch-Light Laboratories). The material was centrifuged for 10 min at 500 g, and the top layer was collected (fraction II).

Acid Precipitation. To fraction II, 1 M acetic acid was added until the pH was 4.1. This solution was allowed to stand for 20 min. After centrifugation at 20,000 g for 20 min, the precipitate was suspended in 110 ml of 50 mM phosphate pH 6.0 (fraction III).

Heat Treatment. Fraction III was heated to 55°, with continuous stirring, by placing the container in a water bath at 80°. The solution was kept at 55° for 2 min and cooled to 0° in an ice-water bath with continuous stirring. The inactive precipitate was removed by centrifugation and discarded (fraction IV).

Alcohol Fractionation. To the 107 ml of fraction IV was added, with stirring, 34.5 ml of 96% ethanol at −10°. The resultant precipitate was collected after 20 min, by centrifugation for 20 min at 20,000 g and −10°. The precipitate was dissolved in 27 ml of 0.1 M Tris chloride pH 7.2 (fraction V).

DEAE-Sephadex A-50 Chromatography. Fraction V was mixed with one-third of the material removed from a column (2.3 × 30 cm) previously equilibrated with 0.1 M Tris chloride pH 7.2. The slurry was stirred for 30 min and added on the top of the column, after most of the supernatant had been removed by decantation.

[18] J. P. Løken, and F C. Størmer, *Eur. J. Biochem.* **14**, 133 (1970).

The column was eluted with a linear gradient formed from 250 ml of 50 mM NaCl in 0.1 M Tris chloride, pH 7.2, and of 0.5 M NaCl in 0.1 M Tris chloride, pH 7.2. Fractions of 8 ml were collected at a flow rate of 0.25 ml/min, and fractions 30–34, which contained the bulk of the activity, were collected and used for further purification. To this solution was added 23.5 g of solid $(NH_4)_2SO_4$, and the precipitate was collected by centrifugation and dissolved in 2.0 ml of 20 mM phosphate pH 6.0. This solution was dialyzed overnight against the same buffer (fraction IV).

Isoelectric Focusing. An electrophoresis column of 110 ml capacity and a synthetic carrier ampholyte with isoelectric points distributed between pH 3 and 6 were purchased from LKB instruments. The column was cooled with circulating water at 2° during layering. The bottom of the column was layered with 10 ml of anode—bottom solution (12 g of sucrose dissolved in 14 ml of distilled water and 0.2 ml concentrated phosphoric acid)—with the central tube opened.

A density gradient of sucrose with carrier ampholyte 4% was arranged in the column as recommended by the manufacturer, by stepwise additions of 5-ml fractions of the sucrose solution. The fraction containing 8 mg of the protein to be separated, was mixed with fractions 10–12 before they were introduced into the column. On the top of the sucrose gradient was layered the cathode solution (0.2 ml of ethanolamine + 10 ml of distilled water). The protein to be separated was dissolved in 20 mM phosphate pH 6.0. After the liquid had been introduced and the central tube opened, the electrolysis was started by applying 180 V and continued for 60 hr at 4°. The central tube was then closed, and the contents of the column were drained slowly through a bottom tubing; 2-ml fractions were collected, and the pH was measured at 20° with a Methrom pH meter equipped with semimicro electrode.

Table II summarizes the data for a typical preparation.

Properties

Purity and Physical Properties. The purified acetolactate decarboxylase moves as a major component in disc electrophoresis experiments carried out on polyacrylamide gel, and a molecular weight of approximately 73,000 was determined by high speed equilibrium centrifugation and disc electrophoresis.[19]

Enzymic Properties. The activity of the enzyme was tested over the pH range 5.4–8.2 in phosphate. The curve shows a pH optimum at 6.2–6.4. At lower pH there was an increase in acetoin formation due to

[19] J. P. Løken, T. B. Christensen, and F. C. Størmer, unpublished results.

TABLE II

PURIFICATION OF ACETOLACTATE DECARBOXYLASE FROM *Aerobacter aerogenes*

Fraction and step	Volume (ml)	Protein (mg)	Specific activity (μmoles/mg protein)	Yield (%)	Purification (fold)
I. Sonic extract from 20 g of cells	105	2415	30	100	1
II. Polyethylene-glycol-dextran fractionation	230			85	
III. Acid precipitation	112	784	60	70	2
IV. Heat treatment	107	68	160	60	5
V. Alcohol fractionation	26	90	310	38	10
VI. DEAE-Sephadex chromatography	6	8	1800	20	60
VII. Isoelectric focusing	4	2.4	4300	14	145

the nonenzymic decarboxylation of acetolactate. No substrate is decarboxylated in the assay between pH 5.6 and 8.2 in the absence of enzyme protein. No stimulation of divalent cations was observed with the purified enzyme.

K_m values of 3.4 and 10 mM were calculated for D-α-acetolactate and D-α-acetohydroxybutyrate, respectively. The latter compound had no inhibitory effect upon enzyme activity in the presence of acetolactate.

III. Diacetyl (Acetoin) Reductase

Assay Method

Principle. Diacetyl (acetoin) reductase activity can conveniently be determined, in either the forward or reverse reaction, or in the irreversible reduction of diacetyl, by measuring the initial rates of reduction or oxidation of NAD and NADH, respectively, at 340 nm.

Reagents

For reduction of acetoin:
  Potassium phosphate pH 7.0, 0.5 M
  NADH 1 mM
  Acetoin 50 mM
For oxidation of 2,3-butanediol:
  Potassium phosphate pH 7.0, 0.5 M
  2,3-butanediol, 1 M
  NAD, 20 mM

For reduction of diacetyl:
Potassium phosphate pH 5.8, 1 M
Diacetyl 0.1 M
NADH 0.5 mM
Enzyme. Prior to assay, the enzyme was diluted in 25 mM phosphate
pH 7.5, in the presence of 0.1 mM NAD (or NADH), 25 mM
2-mercaptoethanol, 0.25% albumin, and 20% glycerol.[20]

Procedure. For the reduction of acetoin[21] the reaction mixture contains
0.1 ml of phosphate, 0.1 ml of acetoin, 0.1 ml of NADH, and enzyme
to a final volume of 1.0 ml.

For the oxidation of 2,3-butanediol,[21] NADH was replaced by NAD
(0.1 ml) and acetoin by 2,3-butanediol (0.1 ml).

In the irreversible reduction of diacetyl[22] the reaction mixture contained in 1.0 ml, 0.1 ml diacetyl, 0.1 ml NADH, 0.1 ml phosphate, pH
5.8, and enzyme.

Units. One unit of activity causes NAD to be reduced or NADH to
be oxidized at an initial rate of 1 μmole/min at 25°. Specific activity
is expressed as the number of enzyme units per milligram of protein.

Purification of the Enzyme

Buffers. The standard buffers, prepared at 20° and used throughout
this work were the following: 50 mM phosphate containing 0.1 mM NAD
and 25 mM 2-mercaptoethanol, pH 7.0, and 50 mM Tris chloride containing 0.1 mM NAD and 25 mM 2-mercaptoethanol, pH 7.5. In the last
two fractionation steps (DEAE and Sephadex chromatography), all
buffers contained 20% glycerol. The following procedure for purification
of the enzyme resulted in a 124-fold purification with a 20% yield. Unless
otherwise stated, all operations were performed at 0–5°.

Preparation of Extract. Frozen cells (36 g) were suspended in 5 volumes of the standard phosphate (see above). The solution was subjected
to ultrasound for 15 min with a Branson Sonic Power SW 75 oscillator.
The material was centrifuged for 30 min at 40,000 g, and the clear solution was collected to yield fraction I.

Streptomycin Precipitation. To the 186 ml of fraction I were added,
with stirring, 93 ml of a 10% streptomycin sulfate solution adjusted to
pH 7.0 with 1 M K$_2$HPO$_4$. After standing for 15 min, the suspension was
centrifuged and the supernatant collected (fraction II).

[20] K. Bryn, O. Hetland, and F. C. Størmer, *Eur. J. Biochem.* **18**, 116 (1971).
[21] S. H. Larsen and F. C. Størmer, *Eur. J. Biochem.* **34**, 100 (1973).
[22] L. Johansen, S. H. Larsen, and F. C. Størmer, *Eur. J. Biochem.* **34**, 97 (1973).

Ammonium Sulfate Fractionation. To fraction II were added, with stirring, 243 g per liter of solid $(NH_4)_2SO_4$, and the pH was adjusted to 7.0 by adding 1 M K_2HPO_4 dropwise. The suspension was allowed to stand for 20 min after the salt had dissolved. The precipitate was discarded after centrifugation at 10,000 g for 10 min, and to the solution were added 132 g per liter of solid $(NH_4)_2SO_4$. The precipitate was collected as described above, and dissolved in sufficient standard Tris chloride (see above) to provide a protein concentration of 20 mg/ml. This solution was then dialyzed against 1 liter of the same buffer overnight. A small precipitate, which appeared during dialysis, was removed by centrifugation at 10,000 g and yielded a clear supernatant (fraction III).

DEAE-Sephadex A-50 Chromatography. Fraction III was loaded on the column (40 × 3 cm) previously equilibrated with standard Tris chloride (−NAD). The column was eluted with a linear NaCl gradient formed from 500 ml of standard Tris chloride (−NAD), and of 0.5 M NaCl in standard Tris chloride (−NAD). Fractions of 6.4 ml were collected at a flow rate of 60 ml/hr, and fractions 116–131, which contained the bulk of the activity, were collected (fraction IV).

Hydroxyapatite Chromatography. Fraction IV was made 0.1 mM with NAD and was loaded on the column (8 × 2.5 cm) previously equilibrated with 20 mM phosphate containing 25 mM 2-mercaptoethanol, pH 7.0, at a flow rate of 30 ml/hr. The column was eluted with a 200-ml linear gradient between 20 and 300 mM phosphate, pH 7.0, containing 25 mM 2-mercaptoethanol. Fractions of 6 ml were collected at a flow rate of 30 ml/hr, and fractions 39–53 were collected and made 0.1 mM with NAD (fraction V); 560 g of solid $(NH_4)_2SO_4$ was added per liter to the pooled fractions, and the suspension was centrifuged. The precipitate was dissolved in standard Tris chloride and dialyzed against 1 liter of the same buffer.

Second DEAE-Sephadex Fractionation. The dialyzed solution was loaded on the column and eluted as described above. The fractions were combined (18–22), and solid $(NH_4)_2SO_4$, 506 g/liter was added. The active precipitate was dissolved in standard phosphate (fraction VI).

Sephadex G-100 Fractionation. Fraction VI was made 0.1 mM with NAD and 560 g per liter of solid $(NH_4)_2SO_4$ was added. The precipitate was dissolved in standard phosphate, volume 4 ml, and loaded on a column (20 × 2.3 cm) previously equilibrated with standard phosphate (−NAD). The column was eluted with the same buffer, and fractions of 4 ml were collected at a flow rate of 13 ml/hr. Fractions 9 to 12, which contained the purified enzyme, were collected, and total yield was 8 mg (fraction VII).

Table III summarizes the data for a typical preparation.

TABLE III
PURIFICATION OF DIACETYL (ACETOIN) REDUCTASE FROM *Aerobacter aerogenes*

Fraction and step	Volume (ml)	Protein (mg)	$10^{-2} \times$ Specific activity (units/mg protein)	Yield (%)	Purification (fold)
I. Sonic extract from 36 g of cells	186	5022	44.5	100	1
II. Streptomycin sulfate	268				
III. $(NH_4)_2SO_4$	114	2964	64.6	86	1.5
IV. DEAE-Sephadex chromatography	100	180	862	70	19
V. Hydroxyapatite chromatography	89	39	2640	52	60
VI. DEAE-Sephadex chromatography	28	19	4800	32	107
VII. Sephadex G-100 chromatography	17	8	5530	20	124

Properties

Stability. The enzyme has been stored in a solution containing 0.1 mM NAD, 25 mM 2-mercaptoethanol, 0.20% albumin, 20% glycerol, and 50 mM phosphate, pH 7.0 for several years without significant loss in enzyme activity.[20]

Purity and Physical Properties. The purified diacetyl (acetoin) reductase moves as one component in disc electrophoresis experiments carried out on polyacrylamide gel. The molecular weight of the enzyme is 100,000 (by sedimentation equilibrium and disc electrophoresis methods). The enzyme is composed of four equalized subunits of molecular weight 25,000, and is visualized in the electron microscope as a tetramer with each subunit arranged at the corners of squares. The subunits can reassociate in the presence of bovine serum albumin to yield the native enzyme.[23] When the enzyme is subjected to isoelectric focusing in polyacrylamide gel in the pH range 5–7, at least 12 species, all possessing enzyme activity, are observed.[24]

Substrate Specificity. Diacetyl could be replaced by 2,3-pentanedione, acetoin by acetylethylcarbinol, and 2,3-butanediol by 2,3-pentanediol. The following compounds could not be used as substrates in the pres-

[23] Ø. Hetland, B. R. Olsen, T. B. Christensen, and F. C. Størmer, *Eur. J. Biochem.* **20**, 200 (1971).
[24] Ø. Hetland, K. Bryn, and F. C. Størmer, *Eur. J. Biochem.* **20**, 206 (1971).

ence of NADH: glycolaldehyde, glyoxylic acid, 2,4-pentanedione, methyl and ethyl pyruvate, and acetoacetic ethyl ester. Similarly with NAD as the second substrate, ethylene glycol, 1,2-propanediol, 1,3-, and 1,4-butanediol were ineffective as substrates.[25]

pH Optimum. The enzyme has a pH optimum in the range 4–7 with diacetyl or acetoin as substrates.

The pH-optimum in the presence of 2,3-butanediol, is about 9.5, and very little enzyme activity is detected at pH 5.0.

The K_m value for 2,3-butanediol increases 10-fold in the presence of acetate at pH 5.8. The effect of acetate increases with decreasing pH.[21] No effect of acetate is observed on the other K_m values of the substrates for the enzyme.

[25] S. H. Larsen, L. Johansen, H. J. Storesund, and F. C. Størmer, *FEBS Lett.* **31**, 39 (1973).

Author Index

Numbers in parentheses are reference numbers and indicate that an author's work is referred to, although his name is not cited in the text.

A

Abeles, R. H., 101
Abo-Elnaga, I., 56
Abraham, S., 394
Abramsky, T., 228, 230, 231
Adams, A. D., 320
Adams, E., 115, 116, 118
Adams, J. N., 91
Adams, M. J., 217, 218, 320
Adelman, R. C., 63
Admiraal, J., 49, 50(9)
Aebi, H., 72, 372, 373(8), 374(8)
Aida, K., 84, 85(6), 86(2), 128
Allison, W. S., 265, 267(6), 277
Alroy, Y., 145
Amaral, D., 3(12), 4
Amelunxen, R. E., 264, 265, 266(7), 267, 268, 271, 272, 273
Ames, B. N., 80, 460, 470, 471
Aminoff, D., 403
Anacker, W. F., 167, 197, 522
Anagnostopoulos, C., 305
Anderson, A., 502
Anderson, R. L., 147, 150, 151, 153(1), 419, 457
Anderson, S. R., 289(4), 290
Andreesen, J. R., 100
Ankel, H., 249
Antkowiak, D. H., 293
Arkhangel'skii, I., 56
Armstrong, J. M., 324
Arnold, W. J., 361
Arnon, D. I., 511
Asano, K., 87
Asboe-Hansen, G., 31
Ashwell, G., 3
Avigard, G., 27, 29, 32, 84, 85, 86(11, 14), 87, 88, 89
Axelrod, B., 39

B

Babul, J., 212, 213, 226, 263
Bacon, J. S. D., 11
Baer, E., 290
Bailey, J. M., 57, 472, 473, 480(8), 484, 485, 486, 487, 488, 493(2)
Bain, J. A., 256
Baker, H. A., 138, 141(1), 142(1)
Baker, S. B., 42
Ballou, C. E., 246, 250, 279
Baranowski, T., 259
Bardawill, C. J., 148, 151, 197, 300, 361, 384, 394
Barker, H. A., 300, 467
Barker, R., 65
Barnes, N. B., 457
Barnett, S. R., 383
Barriso, J. A., 430
Barry, S., 322
Bartlett, G. R., 112
Barton, R. A., 442, 443(3), 444(3), 445(3)
Bates, C. J., 497, 501(7)
Bearn, A. G., 47
Beaven, G. H., 226
Bednarz, A. J., 255
Beenakkers, A. M. T., 240, 274
Beisenherz, G., 245, 247(1), 434, 435, 442, 447, 490, 492
BeMiller, J. N., 11, 21(10), 86, 95
Bencze, W. L., 214
Bender, R., 100
Bendet, I., 419
Benkovic, P. A., 57, 491
Benkovic, S. J., 57, 491
Benoiton, L., 117
Benson, J. V., Jr., 11
Benson, R. L., 400, 402(1)
Bentley, R., 3, 472, 486
Benziman, M., 111, 112(7)
Bergmeyer, H. U., 78, 493

Berkower, I., 88, 89(30), 131, 137
Bernaerts, M. J., 155, 158
Bernhard, S. A., 267
Bernhauer, K., 334
Bernsee, G., 85, 86(12)
Bethune, J. L., 372, 373(7), 374(7)
Beutler, E., 57(11), 58, 488
Beyer, G. T., 383
Bhate, D. S., 472, 486
Biaglow, J. E., 37
Bisson, T. M., 104
Blackmore, R. W., 349, 352(2)
Blackwood, J. E., 112
Blair, A. H., 372, 374(9)
Blanch, E. S., 443
Blanco, A., 322
Blaschkowski, H. P., 509, 510(6), 517, 518
Bloch, W., 267
Blumenkrantz, N., 31
Bockelmann, W., 73
Bodansky, O., 47
Bohme, H., 103
Boltze, H. J., 435, 490, 492
Bonnichsen, R. K., 375
Bonsignore, A., 394
Borenfreund, E., 39, 424, 427, 453
Borreback, B., 394
Bouthillier, L. P., 117
Boyer, H., 419, 457
Boyer, P. D., 124
Bradbury, S. L., 355, 359(2), 360
Brandts, J. F., 277, 278(9)
Breitman, T. R., 62
Bridgen, J., 273
Bridges, R. B., 235, 236, 237(6, 9)
Britten, J. S., 329
Britton, B. B., 497
Broad, T. E., 202
Brockman, H. L., 310, 312(3)
Brody, S., 182
Brody, T. M., 256
Brosemer, R. W., 241, 244, 245, 249, 275, 276, 277, 278
Brown, A. T., 235, 236
Brown, D. M., 318
Brown, J. G., 39
Brown, T. D. K., 519
Bruin, W. J., 285
Bruns, F. H., 39
Bryant, M. P., 310

Bryn, K., 530, 532
Bücher, T., 4, 69, 72, 238, 245, 247(1), 249, 259, 260(3), 263(3), 434, 438, 441, 442(1), 448, 449, 451(5), 452(5), 490, 492
Burma, D. P., 423
Burton, P. M., 448, 451, 452(4, 12)
Byrne, W. L., 285

C

Cahn, R. D., 314
Caldwell, P., 94, 97
Cammack, R., 323, 324, 325(5), 328
Cantor, S. M., 25
Carlin, L., 379
Carlson, C. W., 57, 241, 244, 245(10), 275, 276, 277, 278(8)
Carr, T. G., 84
Carter, P., 29
Cartwright, L. N., 343
Castillo, F., 322
Cathcart, E. S., 161, 163
Catravas, G. N., 11
Chaikoff, I. L., 394
Chang, M., 55
Charon, D., 32, 33, 94, 95(2)
Chase, A. M., 472
Chase, T., Jr., 509, 510(2), 518(2)
Chernov, N. N., 37
Chervenka, C. H., 396, 399(18)
Cheshire, M. V., 11
Cheung, G. P., 285, 289
Chiang, P. K., 443
Chiba, H., 293
Chilla, R., 205, 206, 207(1), 428
Chiu, T. H., 91
Chrambach, A., 234, 235
Christen, I., 89
Christensen, D. D., 37
Christensen, T. B., 525(13), 526, 528, 532
Christensson, L., 12, 21
Christian, W., 148, 151, 178, 197, 202, 215, 279, 286, 290, 295, 310, 344, 361, 415, 499(16), 500, 505
Cifonelli, J. A., 96, 97(9), 103
Clark, J., 272
Clark, J. F., 355, 356, 359(5, 7)
Cleland, W. W., 177, 359(13), 360

Subject Index

E

Electrophoresis
of aldolase, 67–73
determination of isoenzyme content in
lactate dehydrogenase, 47
enzyme activity after, 66
of pyruvate kinase, 68–70
Enolase, transition-state analogs, and
active site-specific reagents for,
120–124
Enolase-inhibitor complex, dissociation
constants and stoichiometry of, 122
Enzyme
assay of lactic acid, 41–44
of resazurin and resorufin in, 55
assay procedures, 47–73
microassays, *see* Microassays
oxidation-reduction, 127–379
Epimerase, calculation of activity of, 64
Erythrocyte
glucose-6-phosphate dehydrogenase
from human, assay, purification,
and properties of, 208–214
6-phosphogluconate dehydrogenase
from human, assay, purification,
and properties of, 220–226
phosphoglucose isomerase from
human, assay, isolation, and properties of, 392–396, 399
triosephosphate isomerase from
human, assay, isolation and properties of, 442–447
Escherichia coli
N-acetylglucosamine-6-phosphate
deacetylase from, assay, purification, and properties of, 497–502
L-arabinose isomerase from, assay,
purification, and properties of,
454–458
glucosamine-6-phosphate deaminase
from, assay, purification, and
properties of, 497–502
glucose-6-phosphate 1-epimerase from,
molecular weight of, 493
glucosephosphate isomerase, anomerase activity of, 57
L-glycerol-3-phosphate dehydrogenase
from, assay, purification, and
properties of, 249–254

methylglyoxal synthetase from, preparation of, 505, 506
mutarotase from, substrate specificities
of, 487
pyruvate formate-lyase from, assay,
purification, and properties of,
508–518
L-ribulose-5-phosphate 4-epimerase
from, assay, purification, and
properties of, 419–423

F

Ferrocyanide, determination by UV-
spectrophotometry, 29
Fluorescence coefficient, definition of, 55
Fluorometric determination, of dehydrogenase activity using resorufin, 53–56
Fluorometry, in D-galactose assay, 8, 10
Fructokinase, estimation of, in crude tissue preparations, 61–63
Fructose
chromatographic constants for, 17
degradation on heating in boric
acid/2,3-butanediol or in borax, 21
Fructose 1,6-biphosphate, enzymic preparation of sedoheptulose 1,7-biphos-
phate from, 77–79
Fructose-diphosphate aldolase, detection
of, by tetrazolium dye reduction, 66,
67
Fructose-6-phosphate kinase, phosphorylation of sedoheptulose 7-phos-
phate with, 34–36
Fucose, chromatographic constants for,
16, 17, 20
L-Fucose, enzymic microassay of, 5, 7–10
L-Fucose dehydrogenase
in L-fucose microassay, 5
from sheep liver, 173–177
assay of, 174
chromatography of, 176
molecular weight of, 176
properties of, 176
purification of, 174–176
L-Fucose isomerase, *see* D-Arabinose
isomerase
Fungi, D(−)-lactate dehydrogenase from,
assay, purification, and properties of,
293–298

G

Galactoglycerolipid, assay of, 8

Galactometasaccharinic acid lactone, preparation of, 95, 96

D Galactose
chromatographic constants for, 16, 17, 20
enzymic microassay of, 4, 7–10
fluorometric assay, 8, 10

D-Galactose dehydrogenase, in D-galactose microassay, 4

α-D-Galactose 1,6-diphosphate, preparation of, 83

D-Galacturonic acid, colorimetric assay for, 30, 31

Gluconobacter, 5-keto-D-fructose produced by, 84, 85

Gluconobacter albidus, 5-keto-D-fructose reductase from, 128

Gluconobacter cerinus
aldohexose dehydrogenase from, assay, purification, and properties of, 142–147
5-keto-D-fructose reductase from, assay, purification, and properties of, 127–131

D-Glucosamine, enzymic microassay of, 5–10

Glucosamine-6-phosphate deaminase
from *Escherichia coli,* 497–502
activators for, 502
assay of, 498
chromatography of, 499
properties of, 501, 502
purification of, 498–500
in pig kidney, 497
in *Proteus vulgaris,* 497

Glucosaminephosphate isomerase
from house flies, 400–407
activators and inhibitors for, 406
assay of, 400–403
properties of, 406
purification of, 403–406

D-Glucose
chromatographic constants for, 16, 17, 20
enzymic microassay of, 4, 7–10

α-D-Glucose 1,6-diphosphate
characterization of, 82, 83

preparation of, 81, 82

α-D-[1-^{32}P]Glucose 1,6-diphosphate, preparation of, 83

α-D-[6-^{32}P]Glucose 1,6-diphosphate, preparation of, 83

D-Glucose 6-phosphate, isomerization by glucosephosphate isomerase from baker's yeast, 57

Glucose-6-phosphate dehydrogenase
from bovine mammary gland, 183–188
assay of, 183, 184
chromatography of, 186, 187
purification of, 184–188
from *Candida utilis,* 205–208
activators and inhibitors of, 208
assay of, 205, 206
chromatography of, 207
properties of, 207
purification of, 206–208
from cow adrenal cortex, 188–196
assay of, 188–190
chromatography of, 191, 192
inhibitors for, 195
properties of, 195
purification of, 190–194
in D-glucose microassay, 4
from human erythrocytes, 208–214
amino acids in, 214
assay of, 209
properties of, 212, 213
purification of, 210–212
from *Leuconostoc mesenteroides,* 196–201
assay of, 196, 197
chromatography of, 198
inhibitors for, 201
properties of, 199–201
purification of, 197–200
from *Neurospora crassa,* 177–182
assay of, 177–179
chromatography of, 180
genetics of, 182
properties of, 181, 182
purification of, 179–181
from *Penicillium duponti,* 201–205
assay of, 201, 202
chromatography of, 203
properties of, 204, 205
purification of, 202–204

Glucose-6-phosphate 1-epimerase

A 5
B 6
C 7
D 8
E 9
F 0
G 1
H 2
I 3
J 4